Legume Bioengineering

Legume Bioengineering: Balancing Food Demand and Global Climate Change offers a comprehensive guide to enhancing legume productivity and resilience in the face of global food insecurity and climate change. This book explores the impact of environmental stresses on the legume genome, transcriptome, proteome, and metabolome, and presents cutting-edge biotechnological solutions to mitigate these effects. With contributions from 15 international experts across nine countries, it uniquely combines molecular insights with socioeconomic perspectives, making it a timely resource for advancing sustainable agriculture.

Key Features:

- Global strategies for leveraging legumes to address hunger and food security
- Socioeconomic challenges and adoption of legume cultivation in developing regions
- Genetic resources and breeding strategies for stress tolerance and yield improvement
- Climate-resilient legume production techniques
- Role of legume secondary metabolites in stress adaptation
- Advances in genome editing, epigenetics, microbiome interactions, and miRNA applications
- Current trends and future scope in legume engineering
- Consumer acceptance, environmental impact assessment, and policy frameworks

This book is an essential resource for researchers, graduate students, agricultural scientists, biotechnologists, and policymakers engaged in crop improvement, climate resilience, and food security research. It offers practical insights and scientific innovations critical for shaping the future of legume-based sustainable agriculture.

Legume Bioengineering

Balancing Food Demand and Global Climate Change

Edited by
Praveen Guleria, Vineet Kumar, and Beixin Mo

CRC Press
Taylor & Francis Group
Boca Raton London New York

CRC Press is an imprint of the
Taylor & Francis Group, an **informa** business

Designed cover image: Getty Images

First edition published 2026
by CRC Press
2385 NW Executive Center Drive, Suite 320, Boca Raton FL 33431

and by CRC Press
4 Park Square, Milton Park, Abingdon, Oxon, OX14 4RN

CRC Press is an imprint of Taylor & Francis Group, LLC

ISBN: 9781032757445 (hbk)
ISBN: 9781032761275 (pbk)
ISBN: 9781003477235 (ebk)

DOI: 10.1201/9781003477235

Typeset in Times
by Newgen Publishing UK

Contents

Preface

The first essential component of social justice is adequate food for all mankind.

– Norman Borlaug

Agricultural productivity and sustainability are of prime concern in the present times because of the continuously increasing human population and the prevalent global climate change. Increasing chances of adversities to plants, including legumes, lead to considerable yield losses that risk the sustainability of feeding the growing population. These changing environmental parameters are known to affect the molecular structure of the plant genome, transcriptome, proteome, and metabolome. Legumes are being looked upon as a significant food crop to address food security because of their protein-rich nature and tolerance to certain climatic conditions. Further, it is necessary to understand the effects of changing environmental conditions on the genome, microbiome, and epigenome of legumes. Identification of these changes can help look into the possible traits/genes/target small RNA that possibly regulate the growth, development, and survival of legumes. Hence, this understanding can help in developing a molecular framework to support agricultural sustainability in order to feed the growing human population.

This book, entitled *Legume Bioengineering: Balancing Food Demand and Global Climate Change*, is written by 15 international contributors from nine countries. The first two chapters by Kabir et al. summarize the requirement and importance of legumes to address worldwide hunger and food security. Ngobeni and Molatudi in Chapter 3 illustrate the socioeconomic benefits and challenges of the adoption of legume cultivation in developing countries, with a focus on West Africa. Chapter 4 by Kukreja et al. describes the importance of genetic legume resources to target crop improvement programs focusing on stress resilience and enhanced productivity. Nyaga et al. and Kouighat et al. in chapters 5 and 6 emphasize the importance and strategies of improved legume production in the prevailing conditions of stress and global climate change. Chapter 7 by Mburu et al. explores the potential of legume secondary metabolites to regulate their climate change responses. Chapters 8–11 emphasize the application of genome editing, epigenetic modifications, microbiome interactions, and miRNAs to mitigate climatic variability in legumes. Dey and Ahmed in Chapter 12 offer an overview of the present status, strategies, and applications of legume engineering. Chapters 13–15 summarize the scope of legume engineering for food security, consumer acceptance of engineered legume products, and their environmental impact analysis.

Hence, given the increasing global climate change and stress concerns, biotechnology can offer a sustainable solution to crop improvement, including legumes. The rising human population has further increased pressure on agriculture, thus legumes offer an alternative to address worldwide hunger. This book thus serves as a comprehensive resource for readers to foster a deep understanding of legume engineering, related innovations, and policy initiatives, therefore targeting agricultural sustainability and global food security.

Editors
Praveen Guleria
Vineet Kumar
Beixin Mo

About the Editors

Praveen Guleria is presently working as Assistant Professor in the Department of Biotechnology at DAV University, Jalandhar, Punjab, India. She has worked in the areas of Plant Biotechnology, Plant Metabolic Engineering and Plant Stress Biology at CSIR – Institute of Himalayan Bioresource Technology, Palampur, H.P., India.

Her research interests include plant stress biology, plant small RNA biology, plant epigenomics, and nanotoxicity. She has published several research articles in various peer-reviewed journals. She is also serving as an editorial board member and reviewer for certain international peer-reviewed journals. She has been awarded the SERB – Start Up Grant by DST, GOI. She has also been awarded the prestigious "Bharat Gaurav Award" by the India International Friendship Society, New Delhi. She has also received various awards such as CSIR/ICMR – Junior Research Fellowship, CSIR – Senior Research Fellowship, and state-level merit scholarship awards. She has published nine books with international publishers.

Vineet Kumar is currently working as Professor in the Department of Biotechnology, Lovely Professional University, Jalandhar, Punjab, India. He has worked in different areas of biotechnology and nanotechnology in various institutes and universities in India, namely, Panjab University Chandigarh, CSIR – Institute of Microbial Technology, Chandigarh, India, CSIR – Institute of Himalayan Bioresource Technology, and Himachal Pradesh University. He has published many articles in these areas featuring in peer-reviewed journals. He is also serving as an editorial board member and reviewer for international peer-reviewed journals. He has published more than 12 books with various international publishers.

Beixin Mo is a "Distinguished Professor of Guangdong Province Talent Special Support Plan." She received her Bachelor's degree in Plant Developmental and Molecular Biology from Peking University, China, and received her PhD degree from Guelph University, Canada. Her research areas include elucidation of molecular mechanisms involved in the biogenesis, degradation, and movement of plant small RNAs. She has received extensive research funding and published over 50 papers in distinguished journals, such as *Nature Communications*, *Nature Plants*, *The Plant Cell*, and *Genome Biology*, four books, and three book chapters.

Contributors

Indra Adhikari
Department of Biological Sciences, Texas Tech University, Lubbock, United States

Bulbul Ahmed
Soil Science Division, Bangladesh Wheat and Maize Research Institute, Dinajpur, Bangladesh

Romel Ahmed
Center for Research in Environment, iGen and Livelihood (CREGL), Sylhet, Bangladesh; Department of Forestry & Environmental Science, Shahjalal University of Science & Technology, Sylhet, Bangladesh

Md. Taharat Al Tauhid
Plant Pathology Division, Bangladesh Wheat and Maize Research Institute, Dinajpur, Bangladesh

Fardousi Begum
Agriculture Information Service, Khamarbari, Dhaka, Bangladesh

Khadija Benamar
Department of Plant Protection, Ecole Nationale d'Agriculture de Meknès, Meknes, Morocco
Faculty of Sciences and Technologies, Sidi Mohamed Ben Abdellah University, Fez, Morocco

Rajkumar B. K.
Main Cotton Research Station, Navsari Agricultural University, Surat, Gujarat, India

Savita Budhwar
Department of Nutrition Biology, Central University of Haryana, Mahendergarh, India

Manali Chakraborty
Department of Food Science, University of Copenhagen, Copenhagen, Denmark

Saurabh Chaudhary
Plant, Photosynthesis and Soil, School of Biosciences, University of Sheffield, Sheffield, United Kingdom

Pranav Dawar
Environmental Molecular Sciences Laboratory, Pacific Northwest National Laboratory, Richland, United States

Ilham Dehbi
Department of Plant Protection, Ecole Nationale d'Agriculture de Meknès, Meknes, Morocco
Laboratory of Plant Biotechnology and Molecular Biology, Faculty of Sciences, Moulay Ismail University, Meknes, Morocco

Biplob Dey
Center for Research in Environment, iGen and Livelihood (CREGL), Sylhet, Bangladesh; Department of Tropical and International Forestry, Faculty of Forest Sciences and Forest Ecology, Georg-August-University Göttingen, Göttingen, Germany

Aditi Dwivedi
Department of Plant Molecular Biology and Biotechnology, N. M. College of Agriculture, Navsari Agricultural University, Navsari, Gujarat, India
Department of Biotechnology, College of Agriculture, Junagadh Agricultural University, Junagadh, Gujarat, India

Hajar El Hamss
Department of Plant Protection, Ecole Nationale d'Agriculture de Meknès, Meknes, Morocco

Rachid Ezzouggari
Department of Plant Protection, Ecole Nationale d'Agriculture de Meknès, Meknes, Morocco

Abdelaaziz Farhaoui
Department of Plant Protection, Ecole Nationale d'Agriculture de Meknès, Meknes, Morocco
Equipe de Recherche en Didactique des Sciences et Sciences de l'Education et TICE, Centre Régional des Métiers de l'Education et de la Formation de Fès Meknès, Meknes, Morocco

Umesh Goutam
Department of Biotechnology, Lovely Professional University, Jalandhar, India

Arumoy Halder
Department of Soil Science, Gazipur Agricultural University, Gazipur, Bangladesh

Rasmeih Hamid
Department of Plant Breeding, Cotton Research Institute of Iran (CRII), Agricultural Research, Education and Extension Organization (AREEO), Gorgan, Iran

Akbar Hossain
Soil Science Division, Bangladesh Wheat and Maize Research Institute, Dinajpur, Bangladesh

Md. Monoar Hossain
Wheat Breeding Division, Bangladesh Wheat and Maize Research Institute, Dinajpur, Bangladesh

Tahmina Iasmin
Adaptive Research and Extension Division, Bangladesh Institute of Nuclear Agriculture, Mymensingh, Bangladesh

Chayma Ikan
Laboratory of Agro-Food, Biotechnologies and Valorization of Plant Bioresources (AGROBIOVAL), Department of Biology, Faculty of Science Semlalia, Cadi Ayyad University, Marrakech, Morocco
Centre d'Agrobiotechnologie et Bioingénierie, Université Cadi Ayyad, Marrakech, Morocco

Sopnil Jahin
Department of Biological Sciences, Texas Tech University, Lubbock, United States

Yunfei Jiang
Department of Plant, Food, and Environmental Sciences, Agricultural Campus, Truro, Canada

Md. Parvez Kabir
Soil Science Division, Bangladesh Wheat and Maize Research Institute, Dinajpur, Bangladesh

Grace Ng'endo Kanyita
Department of Biochemistry, Microbiology and Biotechnology, Kenyatta University, Nairobi, Kenya

Ashmita Khanal
Department of Biological Sciences, Texas Tech University, Lubbock, United States

Rahul Khanna
ICAR – Central Institute of Post-Harvest Engineering and Technology, Ludhiana, India

Ekta Khare
Department of Pharmaceutical Sciences, GCRG College of Pharmacy, AKTU, Lucknow, India

Charles Kibet Kirui
Department of Biochemistry, Microbiology and Biotechnology, Kenyatta University, Nairobi, Kenya

Kipkorir Koech
Department of Biochemistry, Microbiology and Biotechnology, Kenyatta University, Nairobi, Kenya

Gilbert Koskey
Training and Research Division, Agroecofarmers Network

Mohamed Kouighat
Research Unit of Plant Breeding and Plant Genetic Resources Conservation, Institut National de la Recherche Agronomique, Rabat, Morocco

Sarvjeet Kukreja
Department of Agronomy, Lovely Professional University, Jalandhar, India

Suneel Kumar
Department of Physics and Astrophysics, Central University of Haryana, Mahendergarh, India

Eric Kimani Kuria
Department of Biological Sciences, Chuka University, Chuka, Kenya

Rachid Lahlali
Phytopathology Unit, Department of Plant Protection, Ecole Nationale d'Agriculture de Meknès, Meknes, Morocco

Komal G. Lakhani
Department of Biotechnology, College of Agriculture, Junagadh Agricultural University, Junagadh, Gujarat, India
Hawkesbury Institute for the Environment, University of Western Sydney, Hawkesbury Campus, Richmond, NSW, Australia

Phetole Mangena
Agricultural Research Council (ARC), Vegetable, Industrial and Medicinal Plants, Pretoria, South Africa

Simon Wambui Mburu
Department of Biological Sciences, Chuka University, Chuka, Kenya

Abelilah Meddich
Centre d'Agrobiotechnologie et Bioingénierie, Université Cadi Ayyad, Marrakech, Morocco

Lucy Molatudi
Department of Plant Production, Soil Science and Agricultural Engineering, University of Limpopo, Sovenga, South Africa

Mourine Mutai
Department of Biochemistry, Microbiology and Biotechnology, Kenyatta University, Nairobi, Kenya

Morris Muthini
Department of Biochemistry, Microbiology and Biotechnology, Kenyatta University, Nairobi, Kenya

Pirtunia Nyadzani Mushadu
Department of Biodiversity, University of Limpopo, Sovenga, South Africa

Tretty Ngobeni
Department of Plant Production, Soil Science and Agricultural Engineering, University of Limpopo, Sovenga, South Africa

Ezekiel Mugendi Njeru
Department of Biochemistry, Microbiology and Biotechnology, Kenyatta University, Nairobi, Kenya

Julian Wanja Nyaga
Department of Biochemistry, Microbiology and Biotechnology, Kenyatta University, Nairobi, Kenya

Bandana Osti
Department of Plant and Soil Science, Davis College of Agricultural Sciences and Natural Resources, Texas Tech University, Lubbock, United States

Bahman Panahi
Department of Genomics, Branch for Northwest & West region, Agricultural Biotechnology Research Institute of Iran (ABRII), Agricultural Research, Education and Extension Organization (AREEO), Tabriz, Iran

Preeti R. Parmar
Main Cotton Research Station, Navsari Agricultural University, Surat, Gujarat, India

Smrutirekha Sahu
Department of Biotechnology, College of Agriculture, Junagadh Agricultural University, Junagadh, Gujarat, India

Samiksha
Department of Biotechnology, Lovely Professional University, Jalandhar, India

Nishu Sharma
Department of Microbiology, Punjab Agricultural University, Ludhiana, India

Shiwani Guleria Sharma
Department of Microbiology, Punjab Agricultural University, Ludhiana, India

Avinash Shrestha
Department of Plant and Soil Science, Davis College of Agricultural Sciences and Natural Resources, Texas Tech University, Lubbock, United States

Fatima-Ezzahra Soussani
Laboratory of Agro-Food, Biotechnologies and Valorization of Plant Bioresources (AGROBIOVAL), Department of Biology, Faculty of Science Semlalia, Cadi Ayyad University, Marrakech, Morocco
Centre d'Agrobiotechnologie et Bioingénierie, Université Cadi Ayyad, Marrakech, Morocco

Kirankumar P. Suthar
Department of Plant Molecular Biology and Biotechnology, N. M. College of Agriculture, Navsari Agricultural University, Navsari, Gujarat, India

Nowrin Islam Toma
Planning and Evaluation Division, Bangladesh Wheat and Maize Research Institute, Dinajpur 5200, Bangladesh

Md. Nasir Uddin
Sustainable Agrifood Systems, International Maize and Wheat Improvement Center, Dhaka, Bangladesh

Priscillah Wakarera
Department of Biochemistry, Microbiology and Biotechnology, Kenyatta University, Nairobi, Kenya

Section I

Importance of Legumes

1 Legumes
Introduction and Utilization

Md. Parvez Kabir, Nowrin Islam Toma, Md. Nasir Uddin, Fardousi Begum, Tahmina Iasmin, Bulbul Ahmed, and Akbar Hossain

1.1 INTRODUCTION

Legumes, members of the Fabaceae family, consist of many plant species important in worldwide food systems, agriculture, and environmental sustainability. These plants produce dry seeds in pods or pulses when harvested, similar to the way food crops are harvested. Beans, peas, lentils, chickpeas, and soybeans are some of the most common types of legumes. Historically, legumes have been cultivated for thousands of years, playing a significant role in meeting the dietary needs of populations globally (Didinger & Thompson, 2022). They are among the most important sources of plant proteins, dietary fibers, vitamins, and minerals, which are nutrients crucial to human health, especially in geographic areas with limited or expensive access to animal protein sources (Affrifah et al., 2023).

In addition to their nutritional value, legumes are known for their environmental advantages, including their ability to fix atmospheric nitrogen to the soil. Legumes increase soil fertility through symbiotic associations with nitrogen-fixing bacteria that convert atmospheric nitrogen, which is unavailable to plants, into an available form (Kebede, 2021). This reduces the dependence on synthetic fertilizers, a fundamental aspect of sustainable agriculture and improvement in soil health. Legumes contribute to biodiversity, erosion control, and water retention when incorporated into crop rotations (Nadeem et al., 2019). Thus, legumes are a key tool for sustainable agricultural practices.

Legumes are also increasingly recognized as important foods in the food industry for non-traditional culinary uses, such as the growing market for plant-based products. The consumer demand for plant-based products (i.e., tofu, tempeh, and plant-based meat analogs) has increased over the past few decades as a result of increased adoption of plant-based diets arising from health, environmental, and socioethical factors (Bakhsh et al., 2021). Legumes are also multifunctional as animal feed, biofuel, and industry products, and their versatility and economic value have motivated the focus on legumes (Guo et al., 2024).

With the reality of increasing climatic change, food insecurity, and a rapidly growing global population, legumes are expected to play a crucial role in meeting future food needs. They can grow in various climates and present nutritional and environmental advantages, which have led to increasing interest in this ideal crop for addressing global food security and environmental sustainability issues (Dutta et al., 2022). The yields and main producing countries for the major legumes cultivated in the world are illustrated in Figure 1.1. In the following section, we provide a detailed overview of legumes, emphasizing their nutritional parameters, sustainable agriculture aspects, wide culinary uses, animal feed, and industrial applicability. This will further emphasize their importance in meeting food security needs and protecting the environment.

DOI: 10.1201/9781003477235-2

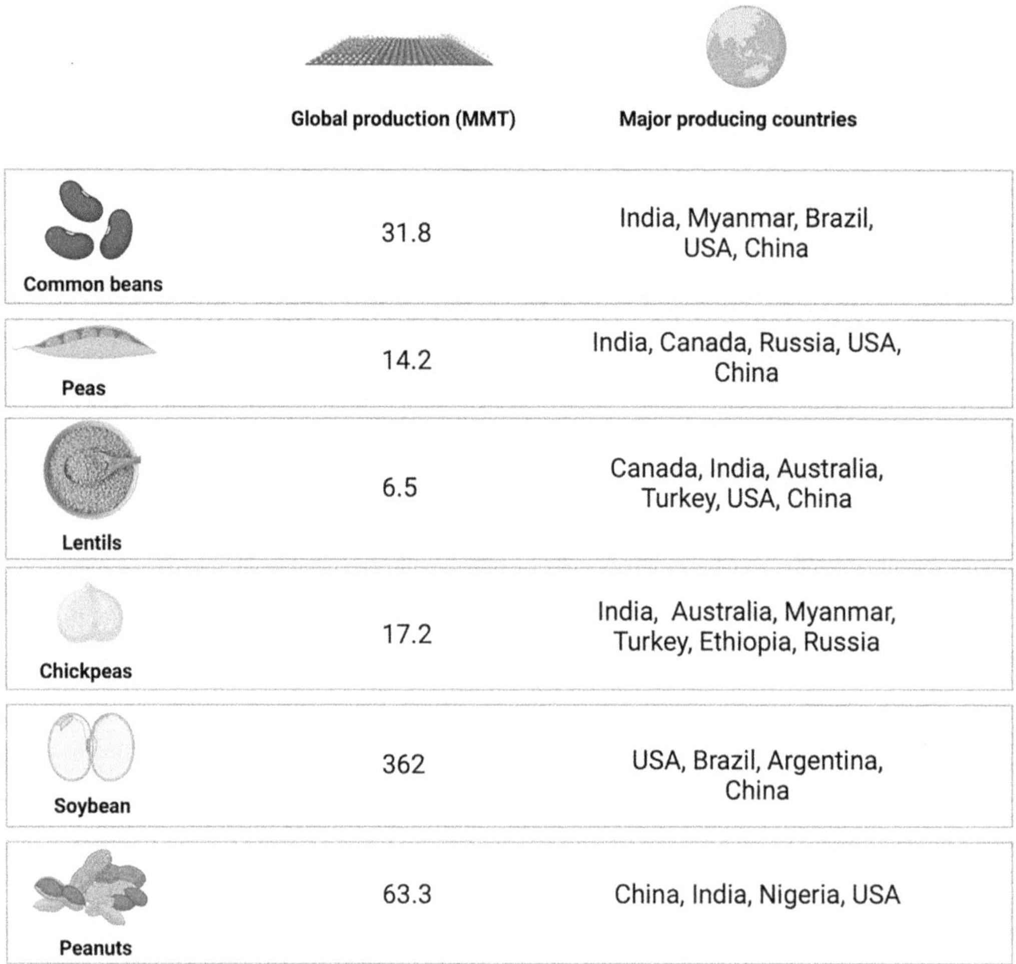

FIGURE 1.1 Yield and main producing countries for the major legumes cultivated in the world. (Adapted from Guo et al., 2024.)

1.2 NUTRITION AND HEALTH BENEFITS

Legumes are considered superfoods because of their nutrient-rich blend of healthy fats, fiber, protein, and 15 different vitamins and minerals (Kumar & Pandey, 2020). These plants are important for many human diets worldwide, especially in regions with limited animal protein availability. The nutritional profile of legumes is excellent for everything from gut health to reducing one's risk of developing diet-related cardiovascular diseases and cancers (Kumar et al., 2022; Didinger et al., 2022).

1.3 KEY NUTRIENTS IN LEGUMES

1.3.1 Protein

Legumes are among the best sources of plant protein. They are also rich in protein (20–25% dry weight) and replace meat in many vegetarian cultures (Shevkani, 2023). For example, a cup

(198 grams) of cooked lentils supplies roughly 18 grams of protein, which is close to 36% of the recommended daily value for adults (Petre & Ajmera, 2023). Beans, peas, and chickpeas also have high protein contents that aid muscle development, tissue repair, and good body functioning (Poteraş et al., 2024).

1.3.2 Carbohydrates

Legumes contain complex carbohydrates, such as fiber and starch, that are suitable for blood sugar regulation. Because of this, the protein and carbohydrates in legumes are absorbed slowly, causing a gradual increase in energy (Banti & Bajo, 2020). For example, a cup of cooked chickpea contains approximately 45 grams of carbohydrates, with 10 grams of fiber (Kakaei et al., 2024). This is again due to the high fiber content supporting stabilized blood sugar and digestive health.

1.3.3 Fiber

Fiber is one of the key nutrients found in legumes. In addition, legumes have soluble and insoluble fibers that are excellent for digestion and heart health. One cup of cooked lentils contains approximately 15.6 grams of fiber, over half (62%) the recommended daily amount for adults (Chelladurai & Erkinbaev, 2020). Fiber is involved in maintaining proper gut health, reducing the risk of colon cancer, and reducing blood cholesterol, which helps reduce the risk of cardiovascular diseases (Grdeń & Jakubczyk, 2023).

Table 1.1 highlights the proximate composition of dry beans and other pulses, demonstrating the nutritional value of legumes, such as protein, carbohydrates, and dietary fiber, which are essential for human health. These key nutrients contribute to a variety of health benefits, such as muscle development, digestive health, and cardiovascular disease prevention, which were discussed earlier in the text. Below the detailed composition of various legumes is outlined.

1.3.4 Vitamins and Minerals

Legumes are full of vitamins and minerals. They provide good sources of folate, iron, magnesium, potassium, and B vitamins, all of which are important for sound health:

- **Iron:** Legumes are also excellent sources of nonheme iron, which is essential for oxygen transport to the blood. For example, one cup of cooked lentils provides approximately 6.6 mg of iron, approximately 37% of the recommended dietary allowance for women (TOI Lifestyle Desk, 2025).
- **Folate:** Legumes are among the richest plant sources of folate (vitamin B9), which is important for cell division and DNA synthesis. A one-cup serving of cooked lentils contains approximately 358 micrograms of folate (90% of the daily recommended daily intake for adults) (EvergreenHealth, 2020).
- **Magnesium and potassium:** Both of these minerals are essential for a healthy heart and muscle function. For example, one cup of cooked black beans contains approximately 120 mg of magnesium and 611 mg of potassium (Medical News Today, 2024).

Table 1.2 lists the mineral contents of various dry beans and pulses, highlighting essential nutrients such as calcium, iron, magnesium, and potassium, which are crucial for healthy heart function, oxygen transport, and cell division, as discussed in the previous section. These minerals further reinforce the nutritional benefits of legumes, making them excellent dietary choices for supporting overall health.

TABLE 1.1
Approximate Composition of Dry Beans and Other Pulses (per 100 g)

Bean/Pulse	Water (g)	Energy (kcal/kJ)	Protein (g)	Total Lipid/ Fat (g)	Ash (g)	Carbohydrate (g)	Dietary Fiber (g)
Adzuki bean	13.40	329/1377	19.90	0.53	3.26	62.90	12.70
Black bean	11.02	341/1427	21.60	1.42	3.60	62.36	15.50
Chickpeas	7.68	378/1582	20.47	6.04	2.85	62.95	12.20
Cowpea	11.05	343/1435	23.85	2.07	3.39	59.64	10.70
Cranberry bean	12.40	335/1402	23.00	1.23	3.31	60.00	24.70
Faba/broad bean	10.98	341/1427	26.12	1.53	3.08	58.29	25.00
Great northern bean	10.70	339/1418	21.90	1.14	3.93	62.40	20.20
Lentil	8.26	352/1473	24.63	1.06	2.71	63.35	10.70
Lima bean, large	10.20	338/1414	21.50	0.69	4.30	63.40	19.00
Lima bean, baby	12.10	335/1402	20.60	0.93	3.55	62.80	20.60
Lupin	10.44	371/1552	36.17	9.74	3.28	40.37	18.90
Moth bean	9.68	343/1435	22.94	1.61	4.26	61.52	--
Mung bean	9.05	347/1452	23.86	1.15	3.32	62.62	16.30
Navy bean	12.10	337/1410	22.33	1.50	3.32	60.75	15.30
Pigeonpea	10.59	343/1435	21.70	0.38	3.45	62.78	15.00
Pink bean	10.10	343/1435	21.00	1.13	3.66	64.20	12.70
Pinto bean	11.33	347/1452	21.42	1.23	3.46	62.55	15.50
Red kidney bean	11.75	337/1410	22.53	1.06	3.37	61.29	15.20
White bean	11.30	333/1393	23.40	0.85	4.20	60.30	15.20
Average	10.74	344/1438	23.10	1.86	3.50	60.76	15.55

Source: USDA (2022).

1.3.5 Antioxidants

Legumes are packed with antioxidants, including flavonoids and polyphenols, which can combat oxidative stress and chronic inflammation in the body (Hernández-Ruiz et al., 2025). These antioxidants also protect against diseases, including cardiovascular disease and cancer, by neutralizing harmful free radicals in the body.

1.4 HEALTH BENEFITS

1.4.1 Cardiovascular Health

Fiber, protein, and antioxidant contents of legumes help reduce the risk of cardiovascular disease (CVD). Research has shown that regular consumption of legumes can lower blood pressure, increase cholesterol levels, and decrease triglyceride levels (Grdeń & Jakubczyk, 2023). A study by Islam et al. (2021) reported that legume consumption has a good effect on reducing cholesterol, including total cholesterol and low-density lipoprotein (LDL), which contribute to cardiovascular risk.

1.4.2 Weight Management

Legumes are rich in fiber and protein and can aid in weight control by preventing hunger and increasing feelings of fullness. Research has shown that individuals who regularly consume legumes

TABLE 1.2
Mineral Contents of Dry Beans and Other Pulses (mg/100 g except μg/100 × g for Selenium, Se)

Bean/Pulse	Ca (mg)	Fe (mg)	Mg (mg)	P (mg)	K (mg)	Na (mg)	Zn (mg)	Cu (mg)	Mn (mg)	Se (μg)
Adzuki bean	66	4.98	127	381	1250	5	5.04	1.09	1.73	3.1
Black bean	123	5.02	171	352	1483	5	3.65	0.84	1.06	3.2
Chickpeas	57	4.31	79	252	718	24	2.76	0.66	21.31	0.0
Cowpea	85	9.95	333	438	1375	58	6.11	1.06	1.54	9.1
Cranberry bean	127	5.00	156	372	1330	6	3.63	0.79	0.92	12.7
Faba/broad bean	22	1.90	38	95	250	50	0.58	0.07	0.32	1.2
Great northern bean	175	5.47	189	447	1390	14	2.31	0.84	1.42	12.9
Lentils	35	6.51	47	281	677	6	3.27	0.75	1.39	0.1
Lima bean (large)	81	7.51	224	385	1720	18	2.83	0.74	1.67	7.2
Lima bean (baby)	81	6.19	188	370	1400	13	2.60	0.67	1.69	7.0
Lupin	176	4.36	198	440	1013	15	4.75	1.02	2.38	8.2
Moth bean	150	10.85	381	489	1191	30	1.92	0.69	1.82	8.1
Mung bean	132	6.74	189	367	1246	15	2.68	0.94	1.04	8.2
Navy bean	147	5.49	175	407	1185	5	3.65	0.83	1.15	11.0
Pigeonpea	130	5.23	183	367	1392	17	2.76	0.13	0.57	1.5
Pink bean	130	6.77	182	415	1460	8	2.55	0.81	1.38	13.0
Pinto bean	113	5.07	176	411	1393	12	2.28	0.89	1.15	27.9
Red kidney bean	83	6.69	138	406	1359	12	2.79	0.70	1.11	3.2
White bean	240	10.40	190	301	1800	16	3.67	0.98	1.80	12.8
Average	113	6.23	177	367	1244	17	3.15	0.76	2.39	7.9

Source: USDA (2022).

as part of their diet generally weigh less and have a lower body fat content than those who do not regularly eat legumes (Tucker, 2023).

1.4.3 Diabetes

Legumes have a low glycemic index (GI) and are packed with high fiber and protein, so they are excellent for keeping blood glucose levels in check. Therefore, there is convincing evidence of an association between the consumption of legumes and a lower risk of type II diabetes (Tang et al., 2020).

1.4.4 Cancer Prevention

Some bioactive compounds in legumes, such as lectins, saponins, and isoflavones, have anticancer properties. Research suggests that legumes, especially soybeans, play a role in decreasing the risk of hormone-dependent cancers, including breast and prostate cancer (Pejčić et al., 2023).

1.5 LEGUMES IN SUSTAINABLE AGRICULTURE

Legumes are vitally important for sustainable agriculture because they can increase soil fertility, reduce the pressure on synthetic fertilizers, and increase crop productivity. Their significance in sustainable agriculture is immense, as they provide ecological and economic advantages.

TABLE 1.3
BNF Capacity (kg ha^{-1}) of Commonly Cultivated Legumes

Common name	Scientific name	BNF (kg ha^{-1})	Cultivation area
Fava bean	*Vicia faba* L.	118.6–311	Greece, Italy
Pea	*Pisum sativum* L.	36.6–125.3	Canada, Greece
Common vetch	*Vicia sativa* L.	107–131	Switzerland
Grass pea	*Lathyrus sativus* L.	101–149	Switzerland
White lupin	*Lupinus albus* L.	53.1–64.1	Italy
Chickpea	*Cicer arietinum* L.	21.0–103.6	Canada
Lentil	*Lens culinaris* Med.	23.0–86.8	Switzerland, Canada
Common bean	*Phaseolus vulgaris* L.	16.3–71.9	Canada
Cowpea	*Vigna unguiculata* (L.) Walp.	36–75	Brazil
Soybean	*Glycine max* (L.) Merr.	90–95	United States
Alfalfa	*Medicago sativa* L.	103–209	Canada, China
Egyptian clover	*Trifolium alexandrinum* L.	35–59	Switzerland
Red clover	*Trifolium pretense* L.	35.4–389	Denmark, United States

Source: Adapted from Kebede (2021).

1.5.1 Fertility and Nitrogen Fixation

There are several reasons to grow legumes in the field, including their capacity to "fix" atmospheric nitrogen in the soil. Nitrogen is a vital nutrient for plant growth and development, but it is often low in agricultural soils. Legumes have a symbiotic relationship with nitrogen-fixing bacteria, including *Rhizobium* species, that live in the root nodules of legumes. These bacteria convert atmospheric nitrogen (N_2) into a form of nitrogen that plants can use, making the soil more fertile naturally. Through this process, legumes contribute up to 200 kg/ha/year of nitrogen to the soil (Kebede, 2021). This decreases the dependence on expensive and ecologically destructive synthetic fertilizers that are both energy intensive to produce and cause soil and water pollution.

N fixation in legumes makes this group a good choice as a rotation crop. When grown on a rotation basis, legumes can increase nutrient levels in the soil and make them more suitable for future crops, thus decreasing the use of chemical fertilizers over time. It has been reported that rotations of cereals with legumes can increase soil organic matter (SOM) buildup, improving soil structure, moisture retention, and biodiversity (Liu et al., 2022). Table 1.3 provides information on the biological nitrogen fixation (BNF) capacity of commonly cultivated legumes, which play a crucial role in improving soil fertility by converting atmospheric nitrogen into a usable form for plants. As discussed earlier, this process helps reduce reliance on synthetic fertilizers, contributing to more sustainable and ecologically friendly farming practices.

1.5.2 Improvement in Soil Organic Matter (SOM)

Legumes do not only increase soil nitrogen but also help to increase organic matter. They contribute to soil carbon sequestration by incorporating organic matter from the decomposed roots, leaves, and shoots of plants, which is an important tool for mitigating climatic change. Soil organic matter is crucial for improving soil health, promoting microbial activity, and decreasing erosion (Tahat et al., 2020). Some legumes, such as alfalfa or clover, are perfect for their long tap root systems, which can penetrate compacted soils to improve the soil structure (Bodner et al., 2021; Jennings & Foster, 2020).

1.5.3 Role in Agroecosystem Sustainability

Legumes increase the biodiversity of agroecosystems by stimulating symbiosis with other soil inhabitants. Legumes increase the N content, facilitating nonleguminous crop growth and decreasing competition for primary nutrients. In addition, legumes can attract beneficial insects, including pollinators, which greatly increase biodiversity (Dingha et al., 2021). The large and extensive root systems of many legumes, mainly cover crops such as clover or vetch, help anchor the soil and reduce soil erosion (Kocira et al., 2020). This is especially important in highly erodible (high rainfall and wind) regions where the loss of topsoil due to erosion leads to reduced agricultural productivity. Some leguminous species also effectively help pest and disease management. For example, it is well known that some plants (e.g., the common bean) have allelopathic effects and the ability to suppress pathogens or pests that affect other crops (Chefonova et al., 2023).

1.5.4 Advantages of Water Use Efficiency

Legumes have also been reported to increase water-use efficiency in agroecosystems (Amiri et al., 2021). Deep roots make much more water available than surface roots do, so they do not need to be watered often. This approach is most useful in areas suffering from water shortages. Leguminous crops, such as chickpeas and lentils, are well adapted to arid conditions (Spalevic et al., 2025). Their ability to grow with little water is suitable for maintaining agricultural productivity in drought-prone regions, and growing plants in places with little rain is becoming more important in the era of climate change.

1.5.5 Economic and Environmental Benefits of Legume-based Farming Systems

The growth of legumes in cropping systems can lead to economic and environmental gains. By reducing the input of chemical fertilizers and improving soil health, legumes contribute to a more resilient agricultural system that can cope with climate variability. Sometimes, legumes also serve as mulch crops. Thus, less dependence on synthetic fertilizers and irrigation means more money in the farmer's bank account. Research reveals that a legume-based system incorporated appropriately into a crop rotation system can make farming operations more profitable through better soil health, lower input costs, and higher yields from the next crop (Yigezu et al., 2019).

Legumes are components of sustainable agriculture systems, fostering carbon sequestration. By increasing soil organic matter and reducing the need for external inputs, legumes reduce greenhouse gas emissions (Oliveira et al., 2021). This is a significant benefit in climate change mitigation, as soil-friendly agricultural practices are increasingly pivotal to global environmental well-being.

1.6 UTILIZATION OF LEGUMES IN FOOD SYSTEMS

Legumes are among the earliest cultivated crops used in human diets globally because of their versatility, nutritional properties, and utilization in different food products. From classic dishes to contemporary plant-based meat substitutes, legumes are crucial components of global food systems. They are used in diverse food production systems, offering a nutritious food source and a sustainable option for future food security. Legumes play a vital role in the development of various food products, as highlighted in the utilization of traditional plant-based meat analogs, as shown in Figure 1.2, which illustrates different forms of plant-based alternatives derived from legumes and other plant sources.

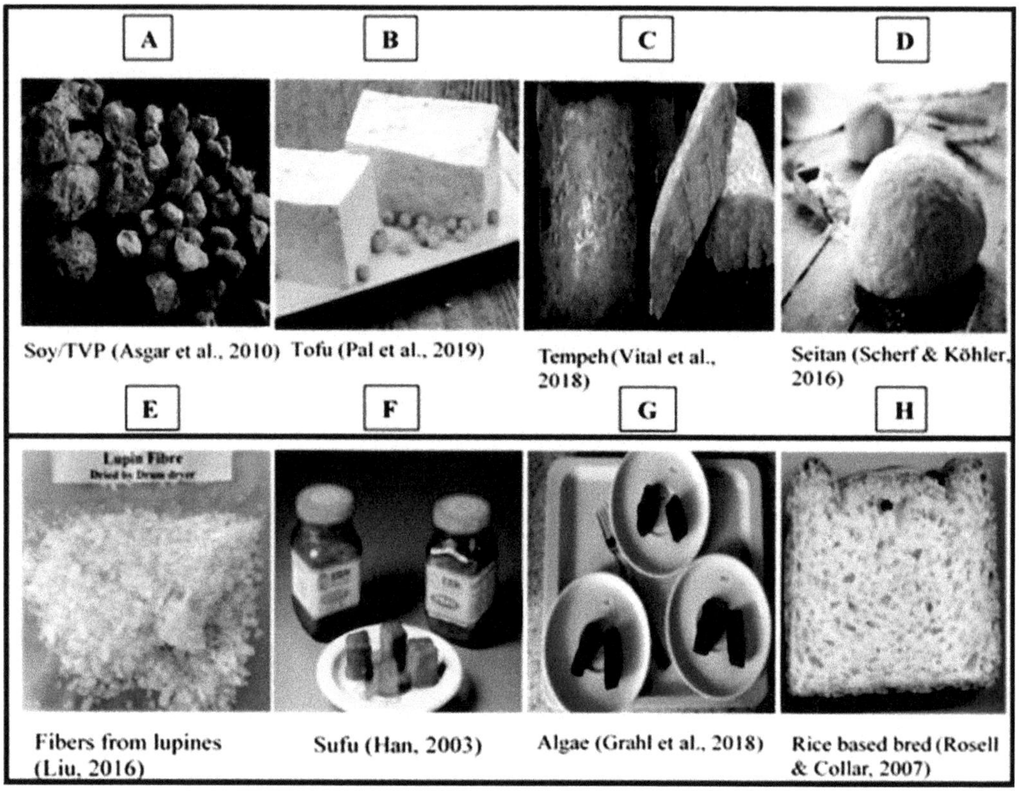

FIGURE 1.2 Traditional plant-based meat analogs. (Adapted from Bakhsh et al., 2021.)

1.6.1 Culinary Uses of Legumes

1.6.1.1 Traditional and Contemporary Foods

In many traditional cuisines around the world, legumes of all shapes and colors play a central role. They are used in a variety of ways, from whole seeds to milled powders and pastes.

Beans such as black beans, kidney beans, and pinto beans form the backbone of cuisines in many different parts of the world, from Latin America to Africa and Asia. They are commonly used in soups, stews, curries, and salads and are often boiled or mashed (Okafor et al., 2024).

Lentils, particularly red, green, and brown lentils, are often used in soups and stews, salads, and Mediterranean and Middle Eastern dishes, such as lentil soups and lentil-based falafels (Ali et al., 2023). They cook up more easily than other legumes, so they are a convenient option.

Chickpeas come in delicious and versatile formats, from hummus and falafel to curries and stews (Rahman, 2021). Chickpea flour is also used in gluten-free baking and cooking and as a replacement for wheat flour in various recipes (Necheporuk et al., 2021). It is common to use chickpeas as snacks or ingredients in salads.

Soybeans can be used to make a variety of products, including tofu, tempeh, soy milk, and edamame (young soybeans) (Modgil et al., 2020). They are also processed into soy protein isolates, which are added to plant-based meat alternatives, such as burgers and sausages (Kyriakopoulou et al., 2021).

1.6.1.2 Legume-based Flours

The use of legume flours such as chickpeas, lentils, peas, and soybeans is becoming increasingly common, especially in gluten-free and high-protein diets (Schmidt & Oliveira, 2023). These flours are found in various baked products, including bread, pancakes, and pasta. They also have more protein and fiber than wheat flour does, making this an excellent option for those looking for more nutrient-dense alternatives. For example, chickpea flour (or besan) is an Indian and Middle Eastern staple used to produce savory pancakes (savory), fritters, and a variety of flatbreads (Sidhu et al., 2022; Basak et al., 2023). It is even used to thicken soups or stews. Lentil flour is becoming a popular ingredient in gluten-free baking. It is incorporated into pasta, bread, and snacks (Emir et al., 2023). The ingredient is rich in protein, fiber, and several essential nutrients, including iron and folate (Kaale et al., 2023; Samaranayaka, & Khazaei, 2024).

1.6.1.3 Legumes in Vegan and Vegetarian Diets

Owing to the increasing popularity of plant-based diets, many people are aware of the high protein content of legumes. They are also an integral part of a vegan or vegetarian diet, supplying delicious, high-quality vegan protein that is easy to digest and rich in branched-chain amino acids (BCAAs), which are crucial for muscle building and overall health (Ahmed et al., 2023; Yanti et al., 2023). Legumes make mock meats, notably soybeans and peas. Brands such as Beyond Meat and Impossible Foods have used pea protein and soy protein to replicate the texture and flavor of meat in burgers, sausages, and other meat alternatives (Maningat et al., 2022; Ahmad et al., 2022). These products are tapping into the increasing desire for plant-based protein due to concerns over health, animal welfare, and sustainability. As mentioned above, burgers and sausages are joined by legume-based varieties of "chicken" nuggets, "fish" fillets, and all other kinds of meat substitutes that offer alternative sources of plant-based protein in place of animal proteins (Kurek et al., 2022).

1.6.2 Value-added Products from Legumes

Legumes are currently processed into value-added products because of their convenience, diversity, and extended shelf-life for consumers.

1.6.2.1 Legume-based Snacks

Legume-based snacks have become highly popular in recent years, and for a good reason, they provide a healthy and filling snack option that is an excellent alternative to traditional sugar- and fat-laden snack foods. For example, roasted chickpeas, lentils, and beans are becoming more popular as healthy, crunchy snacks (Sidhu et al., 2022). They contain protein, fiber, and many other micronutrients, making them a nutritious choice. Another type of ground legume chip is usually promoted as a healthy alternative to regular potato chips (Heid, 2019). They are made with different flavors, and many are baked rather than fried, making them slightly lower-fat, higher-protein alternatives.

1.6.2.2 Legume-derived Beverages

In addition to soymilk, other legume-derived beverages have become available as a dairy alternative (Chaturvedi & Chakraborty, 2021). These include soy milk, pea milk, etc. Soy milk is a popular plant-based drink. It is produced by soaking and grinding soybeans and is often used as a dairy milk alternative in baking and cooking and as a beverage (Davy & Vuong, 2022). Another plant-based alternative for health food store shelves is milk made from yellow peas. It is usually fortified with vitamins and minerals, giving it equivalent nutritional value to dairy milk (Kumar & Pandey, 2020). The diverse legume-derived beverages discussed in the text, such as soy milk and pea milk, have gained popularity as dairy alternatives, as highlighted in Table 1.4, which summarizes the development, objectives, and key findings from various studies on legume-based beverages.

TABLE 1.4
Summary of the Work Done on the Development of Legume-Based Beverages

Legume used	Product Developed	Objective of the Study	Highlights of the Study	References
Soya bean	Soya milk	Enhancing bioactivity of soy milk by enriching it with isoflavone aglycones, using *Lactobacillus rhamnosus* CRL981, which produces β-glucosidase	• GABA-enriched chickpea milk protected PC12 cells from $MnCl_2$-induced injury • Improved cell viability and reduced lactate dehydrogenase release • Fermented soy milk showed 71% higher antioxidant activity • β-Glucosidase activity increased isoflavone aglycone content	Marazza et al. (2012)
Soya bean meal	Plant-based beverage	Investigating the physicochemical, functional, and behavioral properties of the developed product	• The beverages exhibited shear-thinning behavior and stability within pH 3–10 • Soya meal extract-based beverages had high foaming and water absorption capacities	Gerliani et al. (2019)
Red kidney bean	Red kidney bean powder substituted milk	Studying the effect of red kidney bean powder-substituted milk in cinnamon herbal coffee	• Red kidney bean powder-substituted milk (up to 20%) was the most acceptable • Protein content increased by 89.3%, and soluble fiber content rose by 133.0%	Febrianto et al. (2016)
Kidney bean/ common bean	Kidney bean milk and yogurt	Developing common bean milk and probiotic yogurts and characterizing their bioactivities	• Yogurts exhibited higher antioxidant and antiinflammatory activities due to increased phenolic and peptide contents • Yogurts with higher γ-glutamyl-peptide concentrations showed stronger antiinflammatory activity in Caco-2 cells	Chen et al. (2019)
Pea	Probiotic pea milk	Evaluating the effect of skim milk probiotic *L. rhamnosus* AD200 on a fermented beverage supplemented with pea flour (PF) or skim milk	• 1–3% PF-based products had higher viable counts of *L. rhamnosus* compared to SM-supplemented products • PF supplementation reduced syneresis and improved the viscoelastic properties and stability of the gel structure	Zare et al. (2013)
Lentil	Yogurt supplemented with lentil flour	Evaluating the microbial, physical, and sensory properties of yogurt supplemented with lentil flour	• PF-supplemented samples had slightly altered color compared to SM-supplemented and control samples • 1–3% lentil flour milk showed enhanced acid production, lower pH (4.5–4.1), and similar microbial populations of *S. thermophilus* and *L. delbrueckii* subsp. *bulgaricus* • 1–2% lentil flour-supplemented yogurts showed the highest syneresis and sensory properties similar to SM-supplemented yogurt	Zare et al. (2011)

1.6.3 Global Utilization and Cultural Significance

Legumes are a staple in dishes worldwide. For example, in South Asia, lentils are used extensively in dishes such as dhal (lentil soup), a fundamental component of the diet in that region (Ali et al., 2023). In the Middle East, chickpeas power dishes such as hummus and falafel (Al-Khusaibi, 2019). In Latin America, beans, especially black beans, kidney beans, and pinto beans, are the basis of many dishes, such as burritos, tacos, and stews (Amin & Borchgrevink, 2022; Ralat, 2020; Hernandez-Rodriguez, 2021). Legumes are healthy and good for the environment. Because they can fix nitrogen, they can help reduce synthetic fertilizers that can be environmentally harmful. Second, legumes use less water than other crops do, so they are a more sustainable choice, especially in areas where water sources are limited or scarce (Damerau et al., 2019). Finally, legumes, which are also suitable for the planet, are critical to the future of sustainable food systems. They are used by consumers and producers alike. Whether as an ingredient in traditional foods or used to create new and exciting plant-based versions, legumes provide a healthy, sustainable, and affordable solution to the dual challenges of global food security and sustainable food systems.

1.7 LEGUMES IN ANIMAL FEED AND INDUSTRIAL USES

In addition to their role in the human diet, legumes are important in agriculture since they are grown for livestock feed and several industrial purposes. The versatility of legumes extends beyond the kitchen, as legumes are used in animal feed and a variety of industrial applications.

1.7.1 Legumes in Animal Feed

1.7.1.1 Livestock Fodder

Owing to their excellent nutrient contents, legumes are extensively used as high-protein fodder for livestock. Legumes such as alfalfa, clover, and vetch are grown for animal feed; legumes can be much cheaper and healthier than many synthetic feeds (Sowiński & Adamczewska-Sowińska, 2022). These legumes are important sources of protein and specific vitamins and minerals required for the healthy growth and development of animals. For example:

Alfalfa (*Medicago sativa*) is known as the "queen of forage crops" and is considered the gold standard of legume forage varieties (Rahmonov et al., 2020). It has a high protein content and is extremely nutrient rich, so it is an ideal feed for ruminants such as cattle and sheep. Alfalfa hay is also a primary feed for dairy cows, improving milk quality and yield.

Clover (*Trifolium* spp.), one of the most significant protein sources in animal forage diets, improves soil quality and reduces the use of synthetic nitrogen-fixing fertilizers (Płaza et al., 2022). Red (*Trifolium pratense*) and white clover (*Trifolium repens*) are commonly used legumes for ruminant and nonruminant feeds (Skøt et al., 2023).

Vetches (*Vicia* sp.), including common vetch (*Vicia sativa*), are another important legume used for forage (Piltz et al., 2021). In conventional rotations, vetch is an appealing source of feed for poultry, pigs, and other livestock because of its high carbohydrate and protein contents.

1.7.1.2 Protein Meals for Poultry, Cattle, and Aquaculture

The seeds of legumes, commonly soybeans, peas, and lentils, are used as ingredients in animal feeds. After the seeds are crushed to remove fat, they are fed to the animals. For these reasons, these animal foods are a significant source of protein, particularly for poultry, swine and aquaculture. For instance:

Soybean meal is one of the most common protein ingredients used by livestock. Soybean meal is the material remaining after extraction of oil from soybean flakes, with an approximately 40%

soy protein content (Toomer et al., 2024). It also contains a high concentration of essential amino acids that benefit animal growth and meat production.

Peas are also processed to make protein meal and are used to feed various animals. Pea seed protein meal (PPM), which is produced and utilized as a sustainable alternative to soybean meal (SBM) in animal nutrition and fish feed, has been identified as a nutritional and eco-friendly protein source (Hussain et al., 2024)).

In aquaculture, legume meals, especially soybean and pea protein, are external protein sources for the aquafarming of fish (Hekmatpour & Mozanzadeh, 2021). As wild fish populations decline, the demand for plant-based protein sources in aquaculture is increasing. Among legumes, soybeans have been recognized as an important component in aquaculture feeds of cultivated fish species, assisting in the optimum growth and health of these fish.

1.7.2 Use of legumes for Industrial Purposes

1.7.2.1 Biofuels and Bioplastics

The carbohydrate contents of legumes, soybeans, and peas, in particular, apply to industrial nonfood or feed uses. They are now being used more widely in biofuels and bioplastics, products derived from renewable resources that are generally regarded as more environmentally friendly than petroleum-based fuels and plastic items (Rajesh Banu & Godvin Sharmila, 2023). For example, beans such as soybeans can be used to produce biodiesel (da Silva César et al., 2019). Soy oil is also transformed into biodiesel, which burns cleaner than diesel does. Biodiesel produced from soy reduces greenhouse gas emissions and is biodegradable (Li et al., 2023). Legumes are also sources of bioplastics. Starch polymers from legumes, particularly peas, have been developed and have the potential to be used as biodegradable alternatives to conventional plastics (Teixeira-Costa & Andrade, 2021). They can be used in packaging, disposables, and other items as environmentally friendly alternatives to discarding plastic waste.

1.7.2.2 Legumes in the Pharmaceutical and Cosmetic Industries

In addition to the cultivation and production of biofuels, some companies use beans to manufacture drugs and cosmetics (Lee & Kwon, 2022). Research on the therapeutic effects of bioactive compounds found in legumes, such as isoflavones and saponins, is ongoing. Soybeans, in particular, are rich in isoflavones, which are plant compounds that have estrogen-like effects (Kim, 2021). Isoflavones are used in the pharmaceutical industry for hormone replacement therapy and are products for reducing menopausal symptoms in the food industry (Chen & Chen, 2021). Legumes such as soybeans and chickpeas have antioxidant properties and are used in skin-care products (Michalak, 2022). These legume compounds may do so through their antiinflammatory and skin-protective properties and, as a result, have already been incorporated into applications such as moisturizing creams, antiaging lotions, and hair care products (Cerulli et al., 2022; Lakshmikanthan et al., 2024).

1.7.2.3 Legumes in the Textile Industry

The textile industry uses fiber made of legumes, especially soybeans, to manufacture a variety of types of fabric, e.g., eco-friendly clothing (Tahir et al., 2024). The softness and easy dyeability of soybean fibers make them an attractive alternative to synthetic fibers in textiles (Asghar et al., 2024).

1.8 GLOBAL SIGNIFICANCE AND FUTURE OUTLOOK

Rising global concerns about food insecurity, hunger, climate change, and environmental sustainability have led many to consider legumes as strategically important crops. The future of food systems is based on the nutritional makeup, environmental benefits, and adaptability of legumes to diverse growing conditions. This approach is based on the idea that as the global population increases and the demand for sustainable food sources increases, legumes will be important in growing sustainable, resilient agricultural systems.

1.8.1 Role in Global Food Security

1.8.1.1 Addressing Food Insecurity

Substantial pressure will be placed on food systems to provide sufficient food availability for everybody when the world population reaches approximately 10 billion by 2050 (Ghosh et al., 2024). The consumption of legume solution (CLS) provides a sustainable way to address this challenge. They are nutritious, robust, and flexible in nature and may flourish in many environments, even those with poor or arid soils and challenging climatic conditions. In poor nations, where access to animal-derived protein is restricted or costly, it is vital for food security, reducing hunger and alleviating malnutrition. For example, beans and lentils are rich in protein, fiber, iron, and folate, which are vital for growth and development, especially in children and pregnant women (Begum, 2021; Parveen et al., 2022).

Climate change has caused more droughts and erratic weather patterns, and so the ability of legumes is vital for flourishing in poor conditions. Crops such as dry beans, lentils, and chickpeas are more drought resistant than many other staple foods, which makes them vital to diets in areas with more unpredictable weather conditions (Khatun et al., 2021). For at-risk demographic groups, legumes are a great source of necessary micronutrients, such as iron, zinc, and folate, which are often deficient in their diets. Malnutrition, including micronutrient shortages, remains a major worldwide problem affecting millions of people, who are suffering from inadequate consumption of vital elements. Given their great nutritional profile, legumes are set to help address this problem significantly. In countries that depend on staple crops such as rice and maize, which have low micronutrient contents, adding legumes to meals can help address deficiencies in essential vitamins and minerals. For example, beans and lentils contain substantial amounts of iron and folate, which are necessary to reduce the incidence of anemia (Yunus, 2020).

1.8.1.2 Environmental Sustainability and Climate Change Mitigation

Legumes have fewer environmental effects than other crops. They naturally fix nitrogen in the soil, reducing the doses of artificial fertilizer, which are costly and harmful to the environment (Jensen et al., 2020). The inclusion of legumes in agricultural crop rotations can increase soil health and reduce the reliance on chemical inputs, thereby increasing the sustainability of farming methods. Legumes also lessen the need for inorganic fertilizers, which are energy intensive to manufacture and cause soil acidification, water pollution, and the release of greenhouse gases (Saikia et al., 2025). Considering that agriculture is a significant source of greenhouse gas emissions worldwide, the decline in fertilizer consumption is remarkable. Several legumes are drought tolerant and need less water than grain crops, making them a viable option for areas with limited water. However, legumes in intercropping or crop rotation systems can help increase water use efficiency and strengthen farmers' climate change adaptation (Rahman et al., 2022). Legumes also contribute to the development of soil organic matter (SOM), enhancing soil health (Virk et al., 2021). SOM stores carbon in the soil, creating a carbon sink. By increasing soil carbon storage, legumes help reduce atmospheric CO_2 levels worldwide. Owing to their deep root systems, legume-based systems greatly improve soil carbon sequestration (Rashid et al., 2022). This carbon sequestration increases soil fertility and production and helps to resist climate change.

1.8.2 Future Outlook for Legumes

The legume market is projected to experience a steady rise in consumer product demand, given the growing desire for plant-based diets in recent years, particularly as a replacement for meat-based protein. The growing popularity of plant-based meat substitutes and the increase in vegan, vegetarian, and flexitarian diets are creating new possibilities for legumes (Alcorta et al., 2021). The market for vegan meat products, such as burgers, sausages, or nuggets made from legumes, is growing quickly (Bánáti, 2020). Companies such as Beyond Meat and Impossible Foods have developed ways to exploit the potential of legume proteins to produce meat-mimicking plant-based meat substitutes in terms of taste, texture, and appearance of animal-derived retail products (Fischer et al., 2023). This is set to increase even further as more people are searching for sustainable and ethical food choices.

The emergence of legume breeding, including genetic modification and CRISPR technology, can offer solutions for increasing yield, pest resistance, and climate-adaptive capacity in legumes (Bhowmik et al., 2021). Scientists are concentrating on breeding varieties that are more disease resistant, drought tolerant, and better able to tolerate extreme temperatures—all critical qualities—as climate change further disrupts crop production globally. Genetically modified (GM) soybeans, for example, have advanced considerably to increase yields and lower dependence on chemical pesticides because they are resistant to pests (Shelke et al., 2023). Legume breeding applications can help meet the growing global need for legumes and offer environmental benefits.

Governments, international organizations, and agricultural research institutions are becoming more aware of the possibilities of legumes for sustainability and food security. Increasing the usage and relevance of legumes in future food systems will depend on the implementation of encouraging policies for legume production, including subsidies for growth, market development, and research financing. Projects and plans encouraging legume-based agricultural systems, including crop rotation, intercropping, and sustainable farming practices, have the potential to reduce the carbon footprint of agriculture and improve food security under climate change. A global commitment is vital to invest in legume research, support smallholder farmers, and improve international commerce in legumes and legume-based products.

1.9 CONCLUSION

Legumes are also a cornerstone of sustainable agriculture, and they are more than just food; they have implications for many of the world's most critical issues. The excellent capacity of these materials to increase soil nitrogen and reduce the necessity of chemical fertilizers makes them an inseparable component of a sustainable farming system. The high protein, fiber, vitamin, and mineral contents of these products make them a critical food source in diets worldwide, especially in countries with low or expensive animal-based protein consumption. However, legumes are multitaskers, both in culinary and industrial uses, beyond improving soil health and human nutrition, from traditional recipes to trendy plant-based meat substitutes. Beans provide a cost-effective, nutrient-dense solution that fulfills consumers' desire for high-quality plant protein and addresses dietary and environmental demands in a world increasingly aware of the food system's impact on the environment. The need for legumes in the provision of food for the world and the search for sustainability worldwide will remain strong as legume products and breeding practices are advanced. Capable of thriving in most places and permitting stronger, healthier food systems, legumes represent the resilience and flexibility necessary in a world of warming climates, increasing populations, and deteriorating environments. Next-generation legumes will require additional research, legislative support, and global collaboration if they are to deliver their full potential for the benefit of future generations. Legumes are not only stakeholders in current food systems but also must be included in the future of agriculture and public health because of their potential to respond positively to adverse environmental food security and malnutrition.

REFERENCES

Affrifah, N. S., Uebersax, M. A., & Amin, S. (2023). Nutritional significance, value-added applications, and consumer perceptions of food legumes: A review. *Legume Science*, *5*(4), e192.

Ahmed, H. G. M. D., Naeem, M., Faisal, A., Fatima, N., Tariq, S., & Owais, M. (2023). Enriching the content of proteins and essential amino acids in legumes. In Nadeem, M.A. (Ed.), *Legumes biofortification* (pp. 417–447). Springer International Publishing.

Ahmad, M., Qureshi, S., Akbar, M. H., Siddiqui, S. A., Gani, A., Mushtaq, M., ... & Dhull, S. B. (2022). Plant-based meat alternatives: Compositional analysis, current development and challenges. *Applied Food Research*, *2*(2), 100154.

Alcorta, A., Porta, A., Tárrega, A., Alvarez, M. D., & Vaquero, M. P. (2021). Foods for plant-based diets: Challenges and innovations. *Foods*, *10*(2), 293.

Ali, T. M., Butt, N. A., & Hasnain, A. (2023). Global consumption and culinary trends in Lentil utilization. In *Lentils: Production, Processing Technologies, Products, and Nutritional Profile*, 393–414. Hoboken, New Jersey: John Wiley and Sons Ltd. https://doi.org/10.1002/9781119866923.ch16

Al-Khusaibi, M. (2019). Arab traditional foods: Preparation, processing and nutrition. In Al-Khusaibi, M., Al-Habsi, N., Shafiur Rahman, M. (eds) *Traditional Foods*. Food Engineering Series. Springer, Cham. https://doi.org/10.1007/978-3-030-24620-4_2.

Amin, S., & Borchgrevink, C. P. (2022). A Culinology® perspective of dry beans and other pulses. *Dry Beans and Pulses: Production, Processing, and Nutrition*, 453–480.

Amiri, S., Eyni-Nargeseh, H., Rahimi-Moghaddam, S., & Azizi, K. (2021). Water use efficiency of chickpea agro-ecosystems will be boosted by positive effects of CO2 and using suitable genotype× environment× management under climate change conditions. *Agricultural Water Management*, *252*, 106928.

Asghar, T., Javid, A., Zia, K. M., & Tabasum, S. (2024). Green synthesis of polyurethanes using soybean oil-based polyols for bioactive functional fabrics. *Arabian Journal for Science and Engineering*, *49*(1), 531–545.

Bakhsh, A., Lee, S. J., Lee, E. Y., Hwang, Y. H., & Joo, S. T. (2021). Traditional plant-based meat alternatives, current, and future perspective: A review. *Journal of Agricultural Life Sciences*, *55*(1), 1–11.

Bánáti, D. (2020). Veggie burgers, vegan meats?: The ruling of the European Parliament paved the way for meat substitutes with meat denominations. *Élelmiszervizsgálati Közlemények*, *66*(4), 3166–3174.

Banti, M., & Bajo, W. (2020). Review on nutritional importance and anti-nutritional factors of legumes. *International Journal of Food Sciences and Nutrition*, *9*(13), 8–49.

Basak, S., Chakraborty, S., & Singhal, R. S. (2023). Revisiting Indian traditional foods-A critical review of the engineering properties and process operations. *Food Control*, *143*, 109286.

Begum, A. A. (2021). *Pulses and legumes*. KAHM Unity Women's College, Manjeri.

Bhowmik, P., Konkin, D., Polowick, P., Hodgins, C. L., Subedi, M., Xiang, D., ... & Kagale, S. (2021). CRISPR/Cas9 gene editing in legume crops: Opportunities and challenges. *Legume Science*, *3*(3), e96.

Bodner, G., Mentler, A., & Keiblinger, K. (2021). Plant roots for sustainable soil structure management in cropping systems. In Z. Rengel & I. Djalovic (Eds.), *The Root Systems in Sustainable Agricultural Intensification* (pp. 45–90). John Wiley & Sons. https://doi.org/10.1002/9781119525417.CH3.

Cerulli, A., Masullo, M., Montoro, P., & Piacente, S. (2022). Licorice (Glycyrrhiza glabra, G. uralensis, and G. inflata) and their constituents as active cosmeceutical ingredients. *Cosmetics*, *9*, 7.

Chaturvedi, S., & Chakraborty, S. (2021). Review on potential nondairy synbiotic beverages: A preliminary approach using legumes. *International Journal of Food Science and Technology*, *56*(5), 2068–2077.

Chefonova, N. V., Ivanin, D. V., & Vitanov, O. D. (2023). Allelopatic compatibility of common bean and companion crops. *Vegetable and Melon Growing*, *73*, 59–71.

Chelladurai, V., & Erkinbaev, C. (2020). Lentils. In: Manickavasagan, A., Thirunathan, P. (eds) *Pulses*. Springer, Cham. https://doi.org/10.1007/978-3-030-41376-7_8.

Chen, L. R., & Chen, K. H. (2021). Utilization of isoflavones in soybeans for women with menopausal syndrome: An overview. *International Journal of Molecular Sciences*, *22*(6), 3212.

Chen, Y., Zhang, H., Liu, R., Mats, L., Zhu, H., Pauls, K. P., ... & Tsao, R. (2019). Antioxidant and anti-inflammatory polyphenols and peptides of common bean (Phaseolus vulga L.) milk and yogurt in Caco-2 and HT-29 cell models. *Journal of Functional Foods*, *53*, 125–135.

da Silva César, A., Conejero, M. A., Ribeiro, E. C. B., & Batalha, M. O. (2019). Competitiveness analysis of "social soybeans" in biodiesel production in Brazil. *Renewable Energy*, *133*, 1147–1157.

Damerau, K., Waha, K., & Herrero, M. (2019). The impact of nutrient-rich food choices on agricultural water-use efficiency. *Nature Sustainability*, *2*(3), 233–241.
Davy, P., & Vuong, Q. V. (2022). Soy milk byproduct: Its composition and utilization. *Food Reviews International*, *38*(sup1), 147–169.
Didinger, C., & Thompson, H. J. (2022). The role of pulses in improving human health: A review. *Legume Science*, *4*(4), e147.
Didinger, C., Foster, M. T., Bunning, M., & Thompson, H. J. (2022). Nutrition and human health benefits of dry beans and other pulses. In M. Siddiq and M.A. Uebersax (eds.,) *Dry Beans and Pulses*, 481–504. Hoboken, NJ: John Wiley & Sons. doi:10.1002/9781119776802.ch19.
Dingha, B. N., Jackai, L. E., Amoah, B. A., & Akotsen-Mensah, C. (2021). Pollinators on cowpea Vigna unguiculata: Implications for intercropping to enhance biodiversity. *Insects*, *12*(1), 54.
Dutta, A., Trivedi, A., Nath, C. P., Gupta, D. S., & Hazra, K. K. (2022). A comprehensive review on grain legumes as climate-smart crops: Challenges and prospects. *Environmental Challenges*, *7*, 100479.
Emir, A. A., Yildiz, E., & Sumnu, G. (2023). Utilization of lentils in different food products. In J. Ahmed, M. Siddiq, & M. A. Uebersax (Eds.), *Lentils: Production, Processing Technologies, Products, and Nutritional Profile*, (Chapter 10, pp. 237–259). John Wiley & Sons. https://doi.org/10.1002/9781119866923.ch10
EvergreenHealth. (2020, May 30). *Lentils are loaded with nutrition.* Well Together. www.welltogetherblog.com/well-together/nutrition-counseling/lentils-loaded-with-nutrition/
Febrianto, N. A., Sa'diyah, K., & Tejasari, T. (2016). Red kidney bean powder substituted milk in cinnamon herbal coffee: Consumer perception, sensory properties and nutrition content. *Pelita Perkebunan*, *32*(2), 109–119.
Fischer, A. R., Onwezen, M. C., & van der Meer, M. (2023). Consumer perceptions of different protein alternatives. In H. L. Meiselman & J. M. Lorenzo (Eds.), *Meat and Meat Replacements* (pp. 333–362). Woodhead Publishing. https://doi.org/10.1016/B978-0-323-85838-0.00005-5.
Gerliani, N., Hammami, R., & Aider, M. (2019). Production of functional beverage by using protein-carbohydrate extract obtained from soybean meal by electroactivation. *LWT*, *113*, 108259.
Ghosh, A., Kumar, A., & Biswas, G. (2024). Exponential population growth and global food security: Challenges and alternatives. In P. Kumar, A. L. Srivastav, V. Chaudhary, E. D. van Hullebusch, & R. Busquets (Eds.), *Bioremediation of emerging contaminants from soils* (pp. 1–20). Elsevier. https://doi.org/10.1016/B978-0-443-13993-2.00001-3.
Grdeń, P., & Jakubczyk, A. (2023). Health benefits of legume seeds. *Journal of the Science of Food and Agriculture*, *103*(11), 5213–5220.
Guo, F., Danielski, R., Santhiravel, S., & Shahidi, F. (2024). Unlocking the nutraceutical potential of legumes and their byproducts: Paving the way for the circular economy in the agri-food industry. *Antioxidants*, *13*(6), 636.
Heid, M. (2019, May 21). *Are veggie chips or straws healthier than potato chips?* TIME. https://time.com/5590711/veggie-chips-straws-vs-potato-chips/
Hekmatpour, F., & Mozanzadeh, M. T. (2021). Legumes, Sustainable alternative protein sources for aquafeeds. In J. C. Jimenez-Lopez & A. Clemente (Eds.), *Legumes Research - Volume 2* (doi: 10.5772/intechopen.99778). IntechOpen.
Hernandez-Rodriguez, R. (2021). *Food cultures of Mexico.*
Hernández-Ruiz, R. G., Olivares-Ochoa, X. C., Salinas-Varela, Y., Guajardo-Espinoza, D., Roldán-Flores, L. G., Rivera-Leon, E. A., & López-Quintero, A. (2025). Phenolic compounds and anthocyanins in legumes and their impact on inflammation, oxidative stress, and metabolism: Comprehensive review. *Molecules*, *30*(1), 174.
Hussain, S. M., Bano, A. A., Ali, S., Rizwan, M., Adrees, M., Zahoor, A. F., Sarker, P. K., Hussain, M., Zubair-ul-Hassan Arsalan, M., Yong, J. W. H., & Naeem, A. (2024). Substitution of fishmeal: Highlights of potential plant protein sources for aquaculture sustainability. *Heliyon*, *10*(4), e26573. https://doi.org/10.1016/j.heliyon.2024.e26573.
Islam, S. U., Ahmed, M. B., Ahsan, H., & Lee, Y. S. (2021). Recent molecular mechanisms and beneficial effects of phytochemicals and plant-based whole foods in reducing LDL-C and preventing cardiovascular disease. *Antioxidants*, *10*(5), 784.
Jennings, J., & Foster, J. (2020). Legume structure and morphology. *Forages: The Science of Grassland Agriculture*, *2*, 51–64.

Jensen, E. S., Carlsson, G., & Hauggaard-Nielsen, H. (2020). Intercropping of grain legumes and cereals improves the use of soil N resources and reduces the requirement for synthetic fertilizer N: A global-scale analysis. *Agronomy for Sustainable Development, 40*(1), 5.

Kaale, L. D., Siddiq, M., & Hooper, S. (2023). Lentil (Lens culinaris Medik) as nutrient-rich and versatile food legume: A review. *Legume Science, 5*(2), e169.

Kakaei, M., Rehman, F. U., & Fazeli, F. (2024). The effect of chickpeas metabolites on human diseases and the application of their valuable nutritional compounds suitable for human consumption. *Cellular, Molecular and Biomedical Reports, 4*(1), 30–42.

Kebede, E. (2021). Contribution, utilization, and improvement of legumes-driven biological nitrogen fixation in agricultural systems. *Frontiers in Sustainable Food Systems, 5*, 767998.

Khatun, M., Sarkar, S., Era, F. M., Islam, A. M., Anwar, M. P., Fahad, S., ... & Islam, A. A. (2021). Drought stress in grain legumes: Effects, tolerance mechanisms and management. *Agronomy, 11*(12), 2374.

Kim, I. S. (2021). Current perspectives on the beneficial effects of soybean isoflavones and their metabolites for humans. *Antioxidants, 10*(7), 1064.

Kocira, A., Staniak, M., Tomaszewska, M., Kornas, R., Cymerman, J., Panasiewicz, K., & Lipińska, H. (2020). Legume cover crops as one of the elements of strategic weed management and soil quality improvement. A review. *Agriculture, 10*(9), 394.

Kumar, S., & Pandey, G. (2020). Biofortification of pulses and legumes to enhance nutrition. *Heliyon, 6*(3), e03682. https://doi.org/10.1016/j.heliyon.2020.e03682.

Kumar, S., Bamboriya, S. D., Rani, K., Meena, R. S., Sheoran, S., Loyal, A., ... & Jhariya, M. K. (2022). Grain legumes: A diversified diet for sustainable livelihood, food, and nutritional security. In R. S. Meena & S. Kumar (Eds.), *Advances in legumes for sustainable intensification* (pp. 157–178). Academic Press. https://doi.org/10.1016/B978-0-323-85797-0.00007-0.

Kurek, M. A., Onopiuk, A., Pogorzelska-Nowicka, E., Szpicer, A., Zalewska, M., & Półtorak, A. (2022). Novel protein sources for applications in meat-alternative products—Insight and challenges. *Foods, 11*(7), 957.

Kyriakopoulou, K., Keppler, J. K., & van Der Goot, A. J. (2021). Functionality of ingredients and additives in plant-based meat analogues. *Foods, 10*(3), 600.

Lakshmikanthan, M., Muthu, S., Krishnan, K., Altemimi, A. B., Haider, N. N., Govindan, L., ... & Francis, Y. M. (2024). Journal of agriculture and food research. *Journal of Agriculture and Food Research, 17*, 101245.

Lee, J., & Kwon, K. H. (2022). Good ingredients from foods to vegan cosmetics after COVID-19 pandemic. *Journal of Cosmetic Dermatology, 21*(8), 3190–3199.

Li, Y., Xu, H., Northrup, D., & Wang, M. (2023). Effects of soybean varieties on life-cycle greenhouse gas emissions of biodiesel and renewable diesel. *Biofuels, Bioproducts and Biorefining, 17*(3), 449–462.

Liu, W. X., Wei, Y. X., Li, R. C., Chen, Z., Wang, H. D., Virk, A. L., ... & Zhang, H. L. (2022). Improving soil aggregates stability and soil organic carbon sequestration by no-till and legume-based crop rotations in the North China Plain. *Science of the Total Environment, 847*, 157518.

M. Tahat, M., M. Alananbeh, K., A. Othman, Y., & I. Leskovar, D. (2020). Soil health and sustainable agriculture. *Sustainability, 12*(12), 4859.

Maningat, C. C., Jeradechachai, T., & Buttshaw, M. R. (2022). Textured wheat and pea proteins for meat alternative applications. *Cereal Chemistry, 99*(1), 37–66.

Marazza, J. A., Nazareno, M. A., de Giori, G. S., & Garro, M. S. (2012). Enhancement of the antioxidant capacity of soymilk by fermentation with Lactobacillus rhamnosus. *Journal of Functional Foods, 4*(3), 594–601.

Medical News Today. (2024, December 9). *Everything you need to know about black beans*. www.medicalnewstoday.com/articles/289934

Michalak, M. (2022). Plant-derived antioxidants: Significance in skin health and the ageing process. *International Journal of Molecular Sciences, 23*(2), 585.

Modgil, R., Tanwar, B., Goyal, A., & Kumar, V. (2020). Soybean (glycine max). In B. Tanwar & A. Goyal (Eds.), *Oilseeds: Health attributes and food applications* (pp. 1–46). Springer Singapore.

Nadeem, F., Nawaz, A., & Farooq, M. (2019, August 28). Crop rotations, fallowing, and associated environmental benefits. In *Oxford Research Encyclopedia of Environmental Science*. Retrieved August 17, 2025, from https://oxfordre.com/environmentalscience/view/10.1093/acrefore/9780199389414.001.0001/acrefore-9780199389414-e-197.

Necheporuk, A. G., Tretyakova, E. N., Danilin, S. I., Toporkova, K. I., & Pershikova, A. G. (2021, November). Gluten-free products from chickpea flour. In *IOP Conference Series: Earth and Environmental Science* (Vol. 845, No. 1, p. 012077). IOP Publishing. https://doi.org/10.1088/1755-1315/845/1/012077

Okafor, G. I., Okpala, L. C., Uchegbu, N. N., & Anchang, M. M. (2024). Chapter 8 - Indigenous foods of South Eastern part of Nigeria. In O. C. Aworh, P. G. Owusu-Darko, H. Lelieveld, V. Andersen, V. Prakash, J. Prakash, & A. Baiçu (Eds.), *Nutritional and health aspects of food in Western Africa* (pp. 139–169). Academic Press. https://doi.org/10.1016/B978-0-443-27384-1.00009-7.

Oliveira, M., Castro, C., Coutinho, J., & Trindade, H. (2021). Grain legume-based cropping systems can mitigate greenhouse gas emissions from cereal under Mediterranean conditions. *Agriculture, Ecosystems & Environment*, *313*, 107406.

Parveen, S., Jamil, A., Pasha, I., & Ahmad, F. (2022). *Pulses: A potential source of valuable protein for human diet*. IntechOpen.

Pejčić, T., Zeković, M., Bumbaširević, U., Kalaba, M., Vovk, I., Bensa, M., ... & Tešić, Ž. (2023). The role of isoflavones in the prevention of breast cancer and prostate cancer. *Antioxidants*, *12*(2), 368.

Petre, A., & Ajmera, R. (2023, November 27). *The 18 best protein sources for vegans and vegetarians*. Healthline. www.healthline.com/nutrition/protein-for-vegans-vegetarians

Piltz, J. W., Rodham, C. A., Wilkins, J. F., & Hackney, B. F. (2021). A comparison of cereal and cereal/vetch crops for fodder conservation. *Agriculture*, *11*(5), 459.

Płaza, A., Niewiadomska, A., Górski, R., Rudziński, R., & Rzążewska, E. (2022). The effect of the nitrogen-fixing bacteria and companion red clover on the total protein content and yield of the grain of spring barley grown in a system of organic agriculture. *Agronomy*, *12*(7), 1522.

Poteraş, C. B., Culeţu, A., & Manolache, F. A. (2024). Nutritional importance of lentil, lupin, chickpea and Soy Legumes: A review. *Bulletin of University of Agricultural Sciences and Veterinary Medicine Cluj-Napoca. Food Science and Technology*, 81(2), 15-34

Rahman, M. M., Alam, M. S., Islam, M. M., Kamal, M. Z. U., Rahman, G. M., Haque, M. M., ... & Biswas, J. C. (2022). Potential of legume-based cropping systems for climate change adaptation and mitigation. In R. S. Meena & S. Kumar (Eds.), *Advances in legumes for sustainable intensification* (pp. 381–402). Academic Press. https://doi.org/10.1016/B978-0-323-85797-0.00030-6.

Rahman, V. (2021). *Simply plantb: Fabulous food for a healthy life*. Book Publishing Company.

Rahmonov, O., Zaurov, D. E., Islamov, B. S., & Eisenman, S. W. (2020). Resources along the Silk Road in Central Asia: Lagochilus inebrians Bunge (Turkestan Mint) and *Medicago sativa* L.(Alfalfa): Kazakhstan, Kyrgyzstan, Tajikistan, Turkmenistan, and Uzbekistan. In R. Cooper & J. J. Deakin (Eds.), *Natural products of silk road plants* (pp. 153–167). CRC Press. https://doi.org/10.1201/9780429061547

Rajesh Banu, J., & Godvin Sharmila, V. (2023). Review on food waste valorization for bioplastic production towards a circular economy: Sustainable approaches and biodegradability assessment. *Sustainable Energy & Fuels, 7*(14), 3165–3184. https://doi.org/10.1039/D3SE00500C

Ralat, J. R. (2020). *American Tacos: A history and guide*. University of Texas Press.

Rashid, M., Hussain, Q., Hayat, R., Ahmed, M., Riaz, M., Khan, K. S., ... & Khalid, R. (2022). Soil carbon and legumes. In R. S. Meena & S. Kumar (Eds.), *Advances in legumes for sustainable intensification* (pp. 329–344). Academic Press. https://doi.org/10.1016/B978-0-323-85797-0.00022-7

Saikia, N., Nithinkumar, K., Bagrecha, S., Ali, S. A., Jena, T., Kumari, M., ... & Babu, S. (2025). Crop diversification with legume for sustainable production and environmental gain. In Babu, S., Singh, R., Rathore, S.S., Das, A., Singh, V.K. (eds) Agricultural Diversification for Sustainable Food Production. *Sustainability Sciences in Asia and Africa*. Springer, Singapore. https://doi.org/10.1007/978-981-97-7517-0_9.

Samaranayaka, A., & Khazaei, H. (2024). Lentil: Revival of poor man's meat. In S. Nadathur, J. P. D. Wanasundara, & L. Scanlin (Eds.), *Sustainable protein sources* (2nd ed., pp. 201–217). Academic Press. https://doi.org/10.1016/B978-0-323-91652-3.00031-9.

Schmidt, H. D. O., & Oliveira, V. R. D. (2023). Overview of the incorporation of legumes into new food options: An approach on versatility, nutritional, technological, and sensory quality. *Foods*, *12*(13), 2586.

Shelke, D. B., Chambhare, M. R., Nikalje, G. C., & Nikam, T. D. (2023). Improvement of soybean crop for yield, stress tolerance, and value-added products using a transgenic approach. *Advances in Agriculture*, *2023*(1), 8166928.

Shevkani, K. (2023). Protein from land—legumes and pulses. In B. K. Tiwari & L. E. Healy (Eds.), *Future proteins* (pp. 35–68). Academic Press. https://doi.org/10.1016/B978-0-323-91739-1.00003-9.

Sidhu, J. S., Zafar, T., Benyathiar, P., & Nasir, M. (2022). Production, processing, and nutritional profile of chickpeas and lentils. In *Dry Beans and Pulses: Production, Processing, and Nutrition* (Chapter 15, pp. 383–407), edited by M. Siddiq & M. A. Uebersax. Hoboken, New Jersey, USA: John Wiley & Sons, Ltd. https://doi.org/10.1002/9781119776802.ch15

Skøt, L., Marley, C., Lloyd, D., Kingston-Smith, A., & Humphreys, M. (2023). Matching feed characteristics to animal requirements through plant breeding. Kyriazakis, I. (eds) *Smart Livestock Nutrition*. Smart Animal Production, vol 1. Springer, Cham. https://doi.org/10.1007/978-3-031-22584-0_2.

Sowiński, J., & Adamczewska-Sowińska, K. (2022). Forage legumes for human, animals, and environment. In R. S. Meena & S. Kumar (Eds.), *Advances in legumes for sustainable intensification* (pp. 205–226). Academic Press. https://doi.org/10.1016/B978-0-323-85797-0.00021-5.

Spalevic, V., Janmohammadi, M., Sabaghnia, N., & Kader, S. (2025). Adapting to climate change in semiarid regions via conservation measures: Climate-smart crop rotations of food legumes in cool seasons. *Turkish Journal of Agriculture and Forestry*, *49*(2), 242–259.

Tahir, M., Li, A., Moore, M., Ford, E., Theyson, T., & Seyam, A. F. M. (2024). Development of eco-friendly soy protein fibre: A comprehensive critical review and prospects. *Fibres*, *12*(4), 31.

Tang, J., Wan, Y., Zhao, M., Zhong, H., Zheng, J. S., & Feng, F. (2020). Legume and soy intake and risk of type 2 diabetes: A systematic review and meta-analysis of prospective cohort studies. *The American Journal of Clinical Nutrition*, *111*(3), 677–688.

Teixeira-Costa, B. E., & Andrade, C. T. (2021). Natural polymers used in edible food packaging—History, function and application trends as a sustainable alternative to synthetic plastic. *Polysaccharides*, *3*(1), 32–58.

TOI Lifestyle Desk. (2025, March 28). *5 iron-rich foods to eat to boost hair growth.* The Times of India. https://timesofindia.indiatimes.com/life-style/beauty/5-iron-rich-foods-to-eat-to-boost-hair-growth/amp_etphotostory/119423306.cms

Toomer, O. T., Oviedo-Rondón, E. O., Ali, M., Joseph, M., Vu, T., Fallen, B., & Mian, R. (2024). Full-fat Soybean meals as an alternative poultry feed ingredient—Feed processing methods and utilization—Review and perspective. *Animals*, *14*(16), 2366.

Tucker, L. A. (2023). Legume intake, body weight, and abdominal adiposity: 10-year weight change and cross-sectional results in 15,185 US adults. *Nutrients*, *15*(2), 460.

USDA (U.S. Department of Agriculture). (2022). *FoodData Central.* https://fdc.nal.usda.gov/

Virk, A. L., Yadav, G. S., Zhao, X., Kan, Z. R., Qi, J. Y., Ahmad, N., ... & Zhang, H. L. (2021). Role of legumes in managing soil organic matter and improving crop yield. In *Soil organic matter and feeding the future* (1st ed., pp. 259–278), edited by Rattan Lal. CRC Press. https://doi.org/10.1201/9781003102762.

Yanti, Violina, V., Putri, C. E., & Lay, B. W. (2023). Branched chain amino acid content and antioxidant activity of mung bean tempeh powder for developing oral nutrition supplements. *Foods*, *12*(14), 2789.

Yigezu, Y. A., El-Shater, T., Boughlala, M., Bishaw, Z., Niane, A. A., Maalouf, F., ... & Aw-Hassan, A. (2019). Legume-based rotations have clear economic advantages over cereal monocropping in dry areas. *Agronomy for Sustainable Development*, *39*(6), 58.

Yunus, F. (2020). *The effects of iron-fortified lentils on the iron (fe) status of adolescent girls in Bangladesh: A double-blind, community-based, cluster-randomized controlled trial* (Doctoral dissertation, University of Saskatchewan).

Zare, F., Boye, J. I., Champagne, C. P., Orsat, V., & Simpson, B. K. (2013). Probiotic milk supplementation with pea flour: Microbial and physical properties. *Food and Bioprocess Technology*, *6*, 1321–1331.

Zare, F., Boye, J. I., Orsat, V., Champagne, C., & Simpson, B. K. (2011). Microbial, physical and sensory properties of yogurt supplemented with lentil flour. *Food Research International*, *44*(8), 2482–2488.

2 Legumes
Need to Address Nutritional Security and Worldwide Hunger

Md. Parvez Kabir, Md. Taharat Al Tauhid, Md. Monoar Hossain, Md. Nasir Uddin, Arumoy Halder, and Akbar Hossain

2.1 INTRODUCTION

Hunger and malnutrition are among the greatest global challenges of our time. As reported by the US Food and Agriculture Organization (FAO), there are over 800 million chronically undernourished people globally, as well as an estimated 2 billion people suffering from some degree of micronutrient deficiency (FAO, 2022). Economic inequality, climate change, conflict, and poorly functioning agricultural systems that fail to provide access to enough safe and nutritious food are exacerbated by these issues.

There are no overnight solutions to these complicated issues, but legumes – a group of more than 19,000 plant species, including beans, lentils, chickpeas, and peas (Bennetau-Pelissero, 2019) – provide a potent but frequently overlooked answer. Leguminous crops have been consumed by humans for thousands of years and are commonly known for their low price, easy preparation, and effectiveness in meeting nutritional requirements in low-income and developing nations (Kumar et al., 2022). They are a high-quality source of plant-based protein, micronutrients, minerals, and dietary fiber and are a critical part of food systems that are designed to improve nutrition globally (Lisciani et al., 2024).

Legumes also offer a sustainable answer to hunger. They are highly adaptable to diverse climates and soils, can be produced in resource-poor settings, and require less water and land to produce than animal-based protein sources (Vadez et al., 2012). Another advantage is that legumes help improve soil health by fixing nitrogen in the soil, which not only reduces reliance on synthetic fertilizers but also encourages sustainable agricultural practices (Meena & Lal, 2018).

Despite their health benefits, legumes tend to be underconsumed worldwide (Magrini et al., 2023). They are often held back due to cultural preferences, market inefficiencies, or limitations of availability. Here, we address the nutritional and micronutrient potential of legumes, their contribution to global food security, and opportunities for food systems and policies to reflect the importance of legumes in diverse human diets. Overcoming these challenges can help realize the full potential of legumes, a primary solution not only to nutritional security but also to the burgeoning hunger crisis.

2.2 IMPORTANT LEGUMES THAT MEET FOOD AND NUTRITIONAL SECURITY REQUIREMENTS ACROSS THE GLOBE

Legumes are highly diverse groups of plants that have been domesticated in different regions of the world for thousands of years and are important components of human diets worldwide (Smýkal et al., 2015). Some leguminous plants play a vital role in contributing significantly to global food

DOI: 10.1201/9781003477235-3

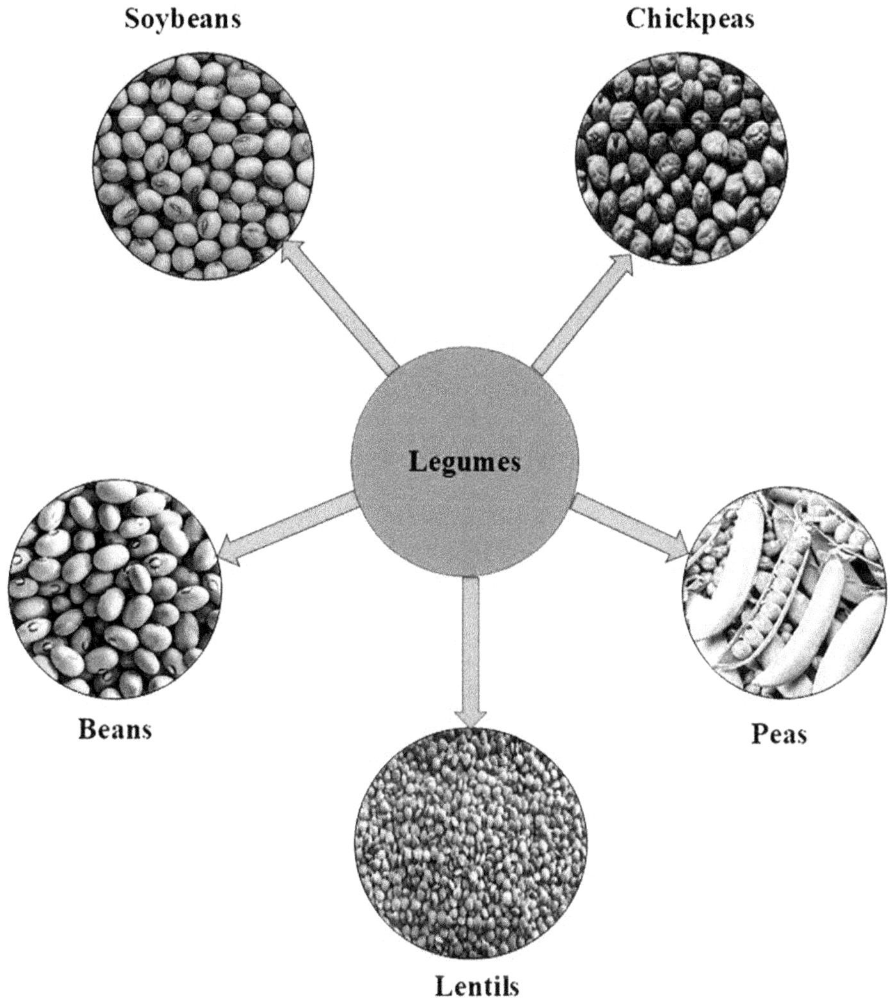

FIGURE 2.1 Some important legumes are grown worldwide. (Author's unpublished data.)

security and human nutrition, especially in developing countries, as well as sustainable agricultural systems (Foyer et al., 2016). Here, we look at some of the most important food legumes that have a major impact on food security worldwide (Figure 2.1).

2.2.1 Soybeans (*Glycine max*)

Soybeans originated in East Asia (specifically China) and are estimated to have been domesticated more than 6000 years ago (Sedivy et al., 2017). Today, they are widely grown in countries such as the United States, Brazil, Argentina, and China. These nations also are significant producers, and soybeans cover approximately 125 million hectares worldwide (FAO, n.d.). Soybeans grow in well-drained and fertile soil and require a temperate climate for optimum growth but can also be grown in tropical and subtropical regions. As long as they have adequate water and nutrients, they are resistant to climate change, as they have the ability to adjust to different climates. However, extreme weather events, such as droughts and floods, could hinder their production in some regions (Bijlwan et al., 2025). Worldwide, soybean production now exceeds 350 million metric tons per year (FAO, n.d.). The areas and production rates for the leading soybean-exporting countries are shown in Figure 2.2.

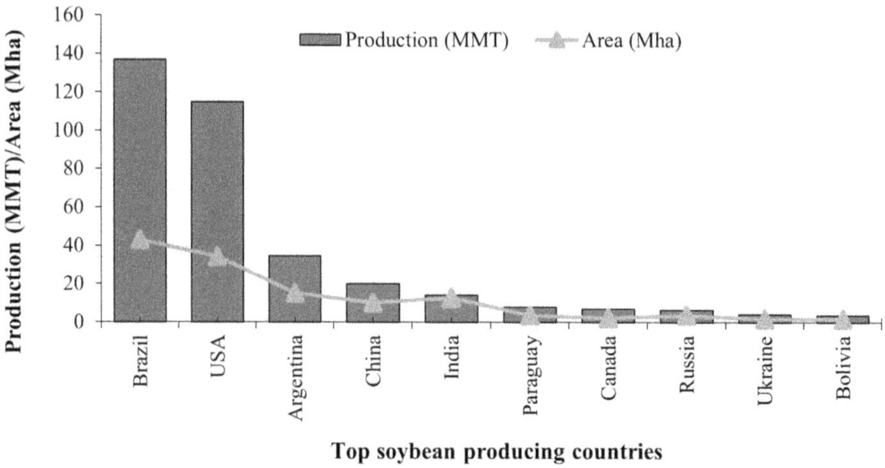

FIGURE 2.2 Top 10 soybean-producing countries worldwide. (FAO (2025) FAOSTAT. Food and Agriculture Organization of the United Nations. Retrieved from www.fao.org/faostat/en/#data/QCL/visualize (accessed March 15, 2025).)

Soybeans are affordable, high-protein food sources that are critical in communities with limited access to animal-based proteins. They are commonly consumed in vegetarian and vegan diets and provide an environmentally friendly alternative to combat protein malnutrition in developing countries. There are various uses of soybeans, such as tofu, soy milk, tempeh, soy sauce, edamame, soy protein, soy flour, soy lecithin, soy oil, and soy; miso, natto, and soy-based vegan products; and soy ice cream and soy cheese. Their byproducts also act as livestock feed, further contributing to animal protein production worldwide (Thrane et al., 2017).

From a nutritional perspective, soybeans are an excellent source of protein, containing 36.5 g per 100 g, with a complete set of all nine essential amino acids, making them a key item in plant-based diets. Soybeans also contain beneficial fats (20 g per 100 g), including omega-3 fatty acids, and essential minerals such as calcium (277 mg), iron (15.7 mg), magnesium (280 mg), and phosphorus (704 mg) (Gowda et al., 2014). Other nutritional and mineral compositions of soybean are provided in Table 2.1. Soybeans are also nitrogen fixers, which improve soil health and allow for sustainable farming practices (Coskan & Dogan, 2011).

2.2.2 Chickpea (*Cicer arietinum*)

Chickpeas, often called garbanzo beans, are native to the Middle East and Mediterranean regions, where they have been cultivated for more than 7000 years (Pratap & Kumar, 2011). India is the leading producer with almost 70% of global production, while Australia, Turkey, Pakistan, and Iran are also among its major cultivators (FAO, n.d.). Chickpeas are well adapted to arid and semiarid regions and are drought resistant, which means that they can be easily grown in areas with a limited water supply. They need well-drained soil and moderate temperatures. Additionally, dry areas are ideally suited to growing chickpeas, as they are able to survive prolonged periods without water, making them especially drought tolerant and desirable in the modern era of climate change (Rani et al., 2020). Globally, approximately 165 million metric tons of chickpea are produced every year, although this takes up nearly 106 million hectares of land (FAO, 2025). Figure 2.3 shows the areas and production rates of the top chickpea-producing countries.

Chickpea provides cost-effective nutrition where other sources of protein are scarce or costly. They are heavily utilized in vegetarian diets and provide balanced nutrition by offering protein, fiber,

TABLE 2.1
Nutritional and Mineral Composition of Important Grain Legumes

Composition	Soybean	Chickpea	Lentil	Bean	Field Pea
Carbohydrate (g/100 g)	30.2	60.7	60.1	60.3	60.4
Fat (g/100 g)	20.0	6.0	1.1	0.9	1.2
Protein (g/100 g)	36.5	19.3	25.8	23.4	24.6
Total dietary fiber (g/100 g)	9.3	17.4	30.5	15.2	25.5
Total sugars (g/100 g)	7.3	10.7	2.0	2.1	8.0
Ash (g/100 g)	4.9	2.5	2.7	4.2	2.7
Starch (g/100 g)	0.8–6.9	41.1	45.0	38.0	46.0-49.5
Energy (kcal/100 g)	446	364	353	333	341
Calcium (mg/100 g)	277	80.5–226.5	168–170	321–377	75–91
Iron (mg/100 g)	15.7	4.3–7.6	6.6–7.3	8.6–10.7	4.2–5.2
Magnesium (mg/100 g)	280	143.7–212.8	4.6–5.2	289–311	117–128
Phosphorus (mg/100 g)	704	276.2–828.8	301–341	185–210	283–319
Potassium (mg/100 g)	1797	816–1580	978–1024	1151–1319	888–953
Sodium (mg/100 g)	2	21–23	65–81	23–27	-
Zinc (mg/100 g)	4.9	3.5–3.6	4.2–4.3	2.5–2.8	2.4–3.7
Copper (mg/100 g)	1.7	1.2–1.3	4.6–6.1	0.5–0.7	0.6–0.8
Manganese (mg/100 g)	2.5	2.3–4.8	1.4–1.6	2.2–2.9	0.9–1.3
Selenium (mcg/100 g)	17.8	8.2	8.3	12.8	1.6

Source: Adapted and modified from Gowda et al. (2014).

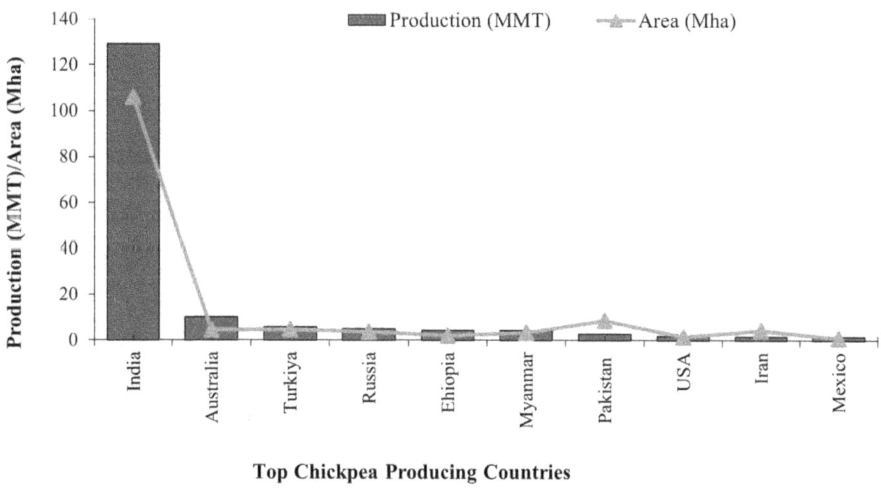

FIGURE 2.3 Top 10 chickpea-producing countries worldwide. (FAO (2025). FAOSTAT. Food and Agriculture Organization of the United Nations. Retrieved from www.fao.org/faostat/en/#data/QCL/visualize (accessed March 15, 2025).)

and micronutrients, especially in regions affected by malnutrition (Purewal et al., 2023). Chickpeas contain 19.3 g of protein per 100 g, as well as 17.4 g of fiber, and are also a source of essential micronutrients such as iron (4.3–7.6 mg), magnesium (143.7–212.8 mg), and folate, which are required for immune function and red blood cell production (Gowda et al., 2014). The detailed nutritional and mineral compositions of the chickpea are shown in Table 2.1. As nitrogen fixers, chickpeas increase soil health and reduce reliance on synthetic fertilizers (Sindhu et al., 2019).

2.2.3 Lentils (*Lens culinaris*)

Lentils are native to the Near East and Mediterranean regions and have been cultivated for more than 9000 years (Stefaniak & McPhee, 2015). Lentils are produced mainly in Canada, with significant cultivation in Australia, India, and Turkey. Lentils favor a cooler, dry climate and well-drained, loamy soils. Their short growing cycle (90–120 days) allows them to be grown in regions with shorter growing seasons. They are particularly suited to grow in climate change-affected areas, especially in drought-prone areas (Vadez et al., 2012). An estimated 65 million metric tons of lentils are produced each year. Lentils are cultivated over approximately 50 Mha worldwide (FAO, 2025). Figure 2.4 shows the areas and production rates of the leading lentil-producing countries.

In the context of high poverty, lentils are an important source of affordable protein and micronutrients. They can play a significant role in addressing malnutrition, especially iron-deficiency anemia, in regions where meat is not easily available (Kumar et al., 2022). One hundred grams of lentils contains 25.8 g of protein and 30.5 g of fiber and is among the best plant-based proteins we can obtain. They are also high in iron (6.6–7.3 mg), folate, and B vitamins, which are essential for energy production and immune health (Gowda et al., 2014). Other nutritional and mineral compositions of lentil are described in Table 2.1. Lentils can also fix nitrogen, enriching the soil and promoting sustainable agricultural practices (Montejano-Ramírez & Valencia-Cantero, 2024).

2.2.4 Beans (*Phaseolus vulgaris*)

Beans originated from Central and South America and have been cultivated for over 7000 years (Gepts, 2020). They are cultivated on a large scale in countries such as Latin America, Africa, Asia, and the United States, with major production in countries such as India, Brazil, Myanmar, China, and the United States. Beans prefer moderate temperatures and well-drained soils. They can withstand a range of climate conditions, from temperate to tropical environments, and are relatively

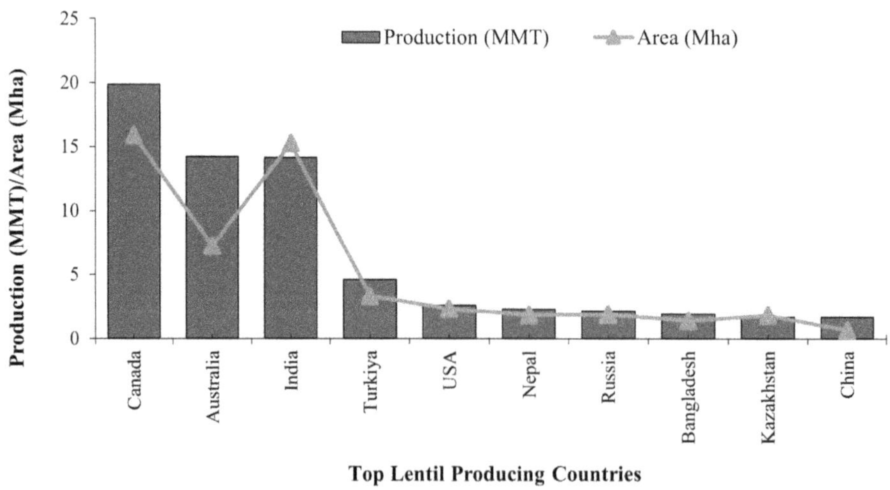

FIGURE 2.4 Top 10 lentil-producing countries worldwide. (FAO (2025). FAOSTAT. Food and Agriculture Organization of the United Nations. Retrieved from www.fao.org/faostat/en/#data/QCL/visualize (accessed March 15, 2025).)

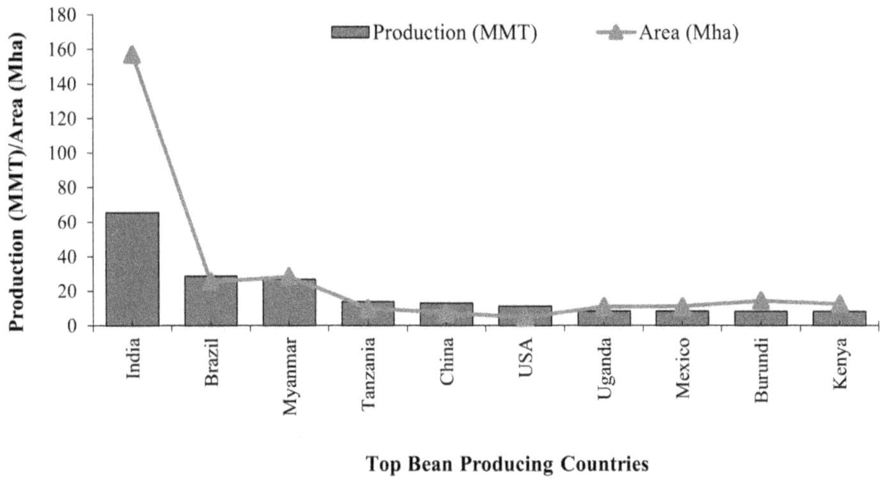

FIGURE 2.5 Top 10 bean-producing countries worldwide. (FAO (2025). FAOSTAT. Food and Agriculture Organization of the United Nations. Retrieved from www.fao.org/faostat/en/#data/QCL/visualize (accessed March 15, 2025).)

resilient to the effects of climate change. Beans are sensitive to frost, and also require proper conditions for optimal yield (Parker et al., 2022). Bean production is approximately 200 million metric tons worldwide per year, and more than 280 million hectares of beans are cultivated across the globe (FAO, 2025). Figure 2.5 shows the areas and production rates of the major bean-producing countries.

In many developing countries worldwide, beans are a staple food, offering a low-cost, nutritious source of protein. They complement other staple crops, such as rice and maize, helping to balance the diet and alleviate protein malnutrition (Temba et al., 2016). Beans contain 23.4 g of protein per 100 g and are high in fiber (15.2 g), iron (8.6–10.7 mg), potassium (1151–1319 mg), and magnesium (289–311 mg) (Gowda et al., 2014). They can be a good source of plant-based protein as well as micronutrients. Other nutritional and mineral compositions of beans are shown in more detail in Table 2.1. Like all legumes, beans fix nitrogen, add nutrients to soil, and are a part of sustainable agriculture (Kakraliya et al., 2018).

2.2.5 Peas (*Pisum sativum*)

Peas are native to the Mediterranean and Near Eastern regions and were first cultivated over 7000 years ago (Singh, 2020). Peas are now extensively grown in Europe, North America, and parts of Asia, with major production in Canada, Russia, and China. Cool weather and well-drained soils are ideal for growing peas. They need moderate amounts of rainfall and can be grown in areas where the growing season is short, meaning that they may be a good fit for regions affected by changing weather patterns. Peas are relatively tolerant of temperature extremes but are sensitive to dry heat and frost (Dhall, 2017). Peas are grown on an estimated total area of 60 million hectares, with approximately 120 million metric tons produced each year (FAO, 2025). Figure 2.6 shows the areas and production rates of the leading pea-producing countries.

Peas represent an important crop in regions with limited access to animal protein, providing an affordable, nutrient-dense food source. This is especially true in developing areas, where they are part of many traditional meals (Vilakazi et al., 2025). Peas (100 g) contain 24.6 g protein and 25.5 g fiber, as well as antioxidants, vitamin K, folate, iron, and magnesium, which support bone health and immune function (Gowda et al., 2014). In addition, Table 2.1 provides more details about the

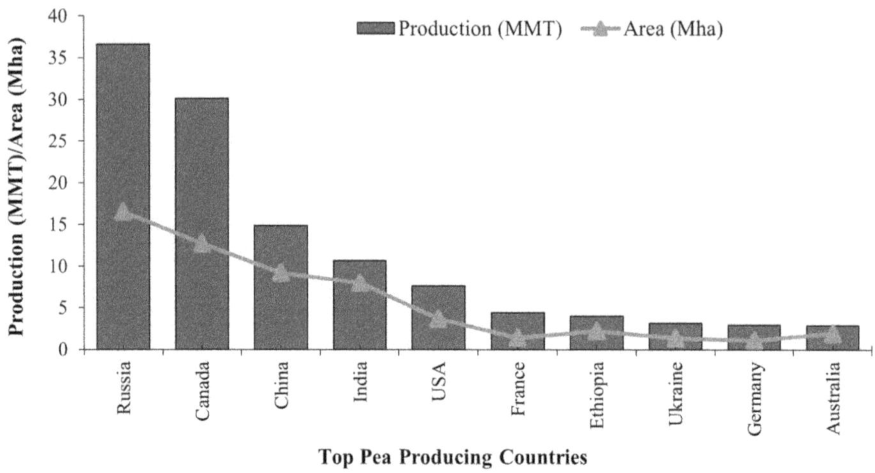

FIGURE 2.6 Top 10 pea-producing countries worldwide. (FAO (2025). FAOSTAT. Food and Agriculture Organization of the United Nations. Retrieved from www.fao.org/faostat/en/#data/QCL/visualize (accessed March 15, 2025).)

other nutritional and mineral contents of pea. As nitrogen fixers, peas also promote soil health and decrease reliance on chemical fertilizers, making them sustainable crops for agriculture.

2.3 THE ROLE OF LEGUMES IN NUTRITION SECURITY AND HEALTH

Legumes are among the most nutritious foods on Earth, providing various key nutrients at a low price (Maphosa & Jideani, 2017). They are primarily used as a source of food and represent an important strategy to combat protein malnutrition in developing countries (Temba et al., 2016). Legumes are also crucial for food security, and their nutritional profiles to provide a convenient means of addressing nutritional security, particularly for regions with limited access to animal-based protein sources (Ikhajiagbe et al., 2022). In an effort to emphasize the importance of legumes as a staple food group, Table 2.1 offers a detailed enumeration of the nutritional and mineral compositions of major grain legumes, indicating the contribution of such legumes to nutritional security.

Legumes, which include beans, lentils, chickpeas, peas, and soybeans, provide many scientifically proven health benefits. These nutrient-rich foods are a great source of high-quality protein, which provides amino acids that play a role in muscle repair, immune function, and general health. For example, one cup of cooked lentils contains approximately 18 grams of protein, making it an essential component of a balanced diet (Healthline, 2022). Legumes also support heart health; studies have shown that they can lower levels of low-density lipoprotein cholesterol, a major factor in cardiovascular disease (Flight & Clifton, 2006).

They are also rich in fiber, which helps with digestion, regularity, and a healthy gut microbiome (Ajay et al., 2024). A cup of cooked chickpea, for example, contains approximately 12.5 grams of fiber, promoting digestive health (Healthline, 2024). They also have a low glycemic index, meaning that they release sugar into the bloodstream slowly, which is particularly helpful for people regulating their blood sugar (Singh et al., 2021). Chickpeas, for example, have a glycemic index of 28, promoting stable blood glucose levels (Smith, 2019).

Additionally, legumes play a crucial role in weight maintenance, as they induce satiety (King et al., 2024). Studies have shown that those who eat legumes tend to lose more weight than those who do not (Jones, 2023). In addition to their fat-fighting potential, legumes are packed with critical vitamins and minerals such as folate, iron, magnesium, and potassium. The process of normalizing these nutrients, which is important for promoting heart health, bone density, and other bodily functions, is said to take place over a very short time frame (Kim et al., 2021). A cup of cooked lentils, for example, accounts for 90% of the daily recommended intake of folate (Jones, 2023) and 37% of the daily recommended intake of iron (Gunnars, 2023).

The magnesium and phosphorus contained in legumes also support stronger bones, as higher magnesium intake is associated with higher bone mineral density. In addition to their well-known health properties, legumes such as soybeans are excellent sources of essential fatty acids that contribute to both heart and brain health. For example, soybeans are a good source of omega-3 fatty acids, which have shown evidence that they help balance cholesterol levels and reduce inflammation (Messina et al., 2021).

Therefore, incorporating legumes into the diet can enhance overall health and help prevent chronic diseases while also contributing to better nutritional security and sustainable agricultural practices. By offering an array of health benefits, legumes are a crucial part of a balanced diet and an essential food source in the fight against global hunger and nutritional deficiencies.

2.4 BARRIERS TO THE GLOBAL ACCEPTANCE OF LEGUMES

Legumes have many advantages but face challenges as they become more accepted globally. This is due to a range of factors, such as cultural perceptions, dietary habits, and general ignorance of their nutritional value and versatility (Vaz Patto et al., 2015). In many cultures, legumes are not staple foods, and meat is often the primary protein source in traditional Western diets. Myths about vitamin deficiencies caused by plant-based proteins being inferior to those obtained from animal proteins and digestive issues such as bloating hinder their consumption. Accessibility issues, such as the lack of availability and higher costs in some regions, further hamper widespread adoption (Chibarabada et al., 2017). Economic factors, such as the prominence of meat and dairy in the food world, also lessen legume visibility and marketing (Goldstein and Reifen, 2022). Moreover, environmental and agricultural challenges, including monocropping and inadequate climatic conditions, limit the cultivation of legumes (Gebru, 2015). Addressing these obstacles requires education, accessibility, marketing, and innovations in convenience products so that legumes can become part of everyday diets and be appreciated for their health and environmental benefits.

2.5 CONCLUSION

Legumes are critical to global food security and nutrition worldwide, across different geographical and socioeconomic contexts, as they provide important protein, fiber, and micronutrients. They thrive in diverse climatic conditions and offer a potential route to combat human malnutrition and food insecurity as versatile foods for human and animal consumption. Leguminous crops also promote sustainable agriculture by naturally fixing nitrogen in the soil, which decreases the dependency on synthetic fertilizers. The growth and consumption of these crops support various agrarian economies, offering income prospects for farmers, particularly in developing nations. However, legume production stability is threatened by challenges such as climate change, water scarcity, and land degradation. A key barrier for smallholder farmers lies in market access, especially in low-income countries. Policies to address sustainable legume production, more efficient agricultural practices, and access to all people are critical to meeting these challenges. Key areas for collaboration among governments, international organizations, and private sector players include enhancing agricultural research, strengthening market linkages, and providing financial and technical support to farmers. Consumer education on the health benefits of legumes, along with their versatility, is also critical to increasing demand and encouraging consumption.

REFERENCES

Ajay, A., Gaur, S. S., Shams, R., Dash, K. K., Mukarram, S. A., & Kovács, B. (2024). Chickpeas and gut microbiome: Functional food implications for health. *Heliyon*, *10*(20), e39314. https://doi.org/10.1016/j.heliyon.2024.e39.

Bennetau-Pelissero, C. (2019). Plant proteins from legumes. *Bioactive Molecules in Food*, *1*, 223–266.

Bijlwan, A., Ranjan, R., Kumar, M., & Jha, A. (2025). Climate change: Projections and its possible impact on Soybean. In: Singh, K.P., Singh, N.K., T, A. (eds) *Soybean production technology: Physiology, production and processing* (pp. 19–44). Springer Nature Singapore.

Chibarabada, T. P., Modi, A. T., & Mabhaudhi, T. (2017). Expounding the value of grain legumes in the semiand arid tropics. *Sustainability*, *9*(1), 60.

Coskan, A., & Dogan, K. (2011). Symbiotic nitrogen fixation in soybean. *Soybean Physiology and Biochemistry*, *307*, 167–182.

Dhall, R. K. (2017). Pea cultivation. *Bulletin no PAU/2017/Elec/FB/E/29*. Punjab Agricultural University, Ludhiana.

FAO. (2025). *FAOSTAT*. Food and Agriculture Organization of the United Nations. Retrieved March 15, 2025 from www.fao.org/faostat/en/#data/QCL/visualize

Flight, I., & Clifton, P. (2006). Cereal grains and legumes in the prevention of coronary heart disease and stroke: A review of the literature. *European Journal of Clinical Nutrition*, *60*(10), 1145–1159.

Food and Agriculture Organization. (2022). *The state of food security and nutrition in the world 2022: Transforming food systems for affordable healthy diets*. Food and Agriculture Organization.

Foyer, C. H., Lam, H. M., Nguyen, H. T., Siddique, K. H., Varshney, R. K., Colmer, T. D., ... & Considine, M. J. (2016). Neglecting legumes has compromised human health and sustainable food production. *Nature Plants*, *2*(8), 1–10.

Gebru, H. (2015). A review on the comparative advantages of intercropping to mono-cropping system. *Journal of Biology, Agriculture and Healthcare*, *5*(9), 1–13.

Gepts, P. (2020). Beans: Origins and development. In: Smith, C. (ed) *Encyclopedia of global archaeology* (pp. 1362–1366). Springer, Cham. https://doi.org/10.1007/978-3-030-30018-0_2169.

Goldstein, N., & Reifen, R. (2022). The potential of legume-derived proteins in the food industry. *Grain & Oil Science and Technology*, *5*(4), 167–178.

Gowda, C. L. L., Jukanti, A. K., & Gaur, P. M. (2014). *Contribution of grain legumes in combating food and nutrition in-security in different regions of the world*. In *The Basics of Human Civilization: Food, Agriculture and Humanity* (Food). PNASF, Bangalore, pp. 469–500.

Gunnars, K. (2023, July 3). *12 healthy iron-rich foods*. www.healthline.com/nutrition/healthy-iron-rich-foods

Healthline. (2022, November 14). *Lentils: Nutrition, benefits, and types*. www.healthline.com/nutrition/lentils

Healthline. (2024, March 1). *Chickpeas: Nutrition, benefits, and more*. www.healthline.com/nutrition/chickpeas-nutrition-benefits

Ikhajiagbe, B., Ogwu, M. C., Ogochukwu, O. F., Odozi, E. B., Adekunle, I. J., & Omage, Z. E. (2022). The place of neglected and underutilized legumes in human nutrition and protein security in Nigeria. *Critical Reviews in Food Science and Nutrition*, *62*(14), 3930–3938.

Jones, M. (2023, October 25). *Foods high in folate (folic acid)*. www.healthline.com/nutrition/foods-high-in-folate-folic-acid

Jones, S. (2023, June 5). *20 Most weight-loss-friendly foods on earth — Backed by science*. Healthline. www.healthline.com/nutrition/128222

Kakraliya, S. K., Singh, U., Bohra, A., Choudhary, K. K., Kumar, S., Meena, R. S., & Jat, M. L. (2018). Nitrogen and legumes: A meta-analysis. In: Meena, R., Das, A., Yadav, G., Lal, R. (eds) *Legumes for Soil Health and Sustainable Management*. Springer, Singapore. https://doi.org/10.1007/978-981-13-0253-4_9

Kim, I. S., Kim, C. H., & Yang, W. S. (2021). Physiologically active molecules and functional properties of soybeans in human health—A current perspective. *International Journal of Molecular Sciences*, *22*(8), 4054.

King, J., Leong, S. Y., Alpos, M., Johnson, C., Mcleod, S., Peng, M., ... & Oey, I. (2024). Role of food processing and incorporating legumes in food products to increase protein intake and enhance satiety. *Trends in Food Science & Technology*, 147, 104466. https://doi.org/10.1016/j.tifs.2024.104466

Kumar, S., Bamboriya, S. D., Rani, K., Meena, R. S., Sheoran, S., Loyal, A., ... & Jhariya, M. K. (2022). Grain legumes: A diversified diet for sustainable livelihood, food, and nutritional security. In: Meena, R. S. & Kumar, S. (eds) *Advances in legumes for sustainable intensification* (pp. 157–178). Academic Press. https://doi.org/10.1016/B978-0-323-85797-0.00007-0

Lisciani, S., Marconi, S., Le Donne, C., Camilli, E., Aguzzi, A., Gabrielli, P., ... & Ferrari, M. (2024). Legumes and common beans in sustainable diets: Nutritional quality, environmental benefits, spread and use in food preparations. *Frontiers in Nutrition, 11*, 1385232.

Magrini, M. B., Salord, T., & Cabanac, G. (2023). The unbalanced development among legume species regarding sustainable and healthy agrifood systems in North-America and Europe: focus on food product innovations. *Food Security, 15*(1), 187–200.

Maphosa, Y., & Jideani, V. A. (2017). The role of legumes in human nutrition. *Functional Food-Improve Health through Adequate Food, 1*, 13.

Meena, R. S., & Lal, R. (2018). Legumes and sustainable use of soils. In: Meena, R., Das, A., Yadav, G., Lal, R. (eds) *Legumes for Soil Health and Sustainable Management*, (pp. 1–31). Springer, Singapore. https://doi.org/10.1007/978-981-13-0253-4_1

Messina, M., Shearer, G., & Petersen, K. (2021). Soybean oil lowers circulating cholesterol levels and coronary heart disease risk, and has no effect on markers of inflammation and oxidation. *Nutrition, 89*, 111343.

Montejano-Ramírez, V., & Valencia-Cantero, E. (2024). The importance of lentils: An overview. *Agriculture, 14*(1), 103.

Parker, T. A., Gallegos, J. A., Beaver, J., Brick, M., Brown, J. K., Cichy, K., ... & Gepts, P. (2022). Genetic resources and breeding priorities in Phaseolus beans: Vulnerability, resilience, and future challenges. *Plant Breeding Reviews, 46*, 289–420.

Pratap, A., & Kumar, J. (2011). History, origin and evolution. In: A. Pratap & J. Kumar (eds) *Biology and breeding of food legumes* (pp. 1–18). CABI.

Purewal, S. S., Kaur, P., & Salar, R. K. (Eds.). (2023). *Chickpea and cowpea: Nutritional profile, processing, health prospects and commercial uses*. CRC Press.

Rani, A., Devi, P., Jha, U. C., Sharma, K. D., Siddique, K. H., & Nayyar, H. (2020). Developing climate-resilient chickpea involving physiological and molecular approaches with a focus on temperature and drought stresses. *Frontiers in Plant Science, 10*, 1759.

Sedivy, E. J., Wu, F., & Hanzawa, Y. (2017). Soybean domestication: The origin, genetic architecture and molecular bases. *New Phytologist, 214*(2), 539–553.

Sindhu, S. S., Sharma, R., Sindhu, S., & Sehrawat, A. (2019). Soil fertility improvement by symbiotic rhizobia for sustainable agriculture. In: Panpatte, D., Jhala, Y. (eds) *Soil Fertility Management for Sustainable Development*. Springer, Singapore. https://doi.org/10.1007/978-981-13-5904-0_7

Singh, A. K. (2020). Early history of crop presence/introduction into India: VI. African and West and Central Asian Leguminous crops. *Asian Agri-History, 24*(1), 23–41.

Singh, M., Manickavasagan, A., Shobana, S., & Mohan, V. (2021). Glycemic index of pulses and pulse-based products: A review. *Critical Reviews in Food Science and Nutrition, 61*(9), 1567–1588.

Smith, J. (2019, March 10). *Best low-GI foods: What to know*. Medical News Today. www.medicalnewstoday.com/articles/324871

Smýkal, P., Coyne, C. J., Ambrose, M. J., Maxted, N., Schaefer, H., Blair, M. W., ... & Varshney, R. K. (2015). Legume crops phylogeny and genetic diversity for science and breeding. *Critical Reviews in Plant Sciences, 34*(1–3), 43–104.

Stefaniak, T. R., & McPhee, K. E. (2015). Lentil. In: De Ron, A. (eds) *Grain Legumes. Handbook of Plant Breeding*, vol 10. Springer, New York, NY. https://doi.org/10.1007/978-1-4939-2797-5_4.

Temba, M. C., Njobeh, P. B., Adebo, O. A., Olugbile, A. O., & Kayitesi, E. (2016). The role of compositing cereals with legumes to alleviate protein energy malnutrition in Africa. *International Journal of Food Science and Technology, 51*(3), 543–554.

Thrane, M., Paulsen, P. V., Orcutt, M. W., & Krieger, T. M. (2017). Soy protein: Impacts, production, and applications. In: S. R. Nadathur, J. P. D. Wanasundara, & L. Scanlin (eds.) *Sustainable protein sources* (pp. 23–45). Academic press. https://doi.org/10.1016/B978-0-12-802778-3.00002-0

Vadez, V., Berger, J. D., Warkentin, T., Asseng, S., Ratnakumar, P., Rao, K. P. C., ... & Zaman, M. A. (2012). Adaptation of grain legumes to climate change: A review. *Agronomy for Sustainable Development, 32*, 31–44.

Vaz Patto, M. C., Amarowicz, R., Aryee, A. N., Boye, J. I., Chung, H. J., Martin-Cabrejas, M. A., & Domoney, C. (2015). Achievements and challenges in improving the nutritional quality of food legumes. *Critical Reviews in Plant Sciences*, *34*(1–3), 105–143.

Vilakazi, B., Mafongoya, P. L., Odindo, A. O., & Phophi, M. M. (2025). The role of neglected grain legumes in Food and Nutrition Security and Human Health. *Sustainability*, *17*(1), 350.

3 Legume Production and Utilization in West Africa

Challenges and Opportunities

Tretty Ngobeni and Lucy Molatudi

3.1 INTRODUCTION

Legume crops, members of the Fabaceae family, are distinguished by their symbiotic association with nitrogen-fixing bacteria, enabling a steady supply of atmospheric nitrogen to the soil. Incorporating legumes into cropping systems provides numerous advantages, such as stimulation of beneficial soil microbial activity, enhanced nutrient availability, improved soil properties, and disruption of diseases and pest life cycles. Legume cultivation significantly reduces greenhouse gas emissions by replacing synthetic nitrogen fertilizers, thereby lowering nitrous oxide emissions (Piyush et al. 2024). Legumes are widely recognized for their nutritional properties, offering an affordable source of protein, carbohydrates, vitamins, and essential micronutrients for consumption (Santhiya et al., 2024). Table 3.1 highlights key leguminous crops commonly grown across the West African region, along with their respective nutrient compositions.

The protein level of legumes satisfies dietary needs while conforming to the increasing trend of minimizing reliance on animal-derived proteins (Ranamukhaarachchi & Nanayakkara, 2025). Legumes are regarded as one of the healthy foods for humans worldwide and are crucial for effectively addressing the global challenges of food and nutrition security (Yeremko et al., 2025).

Legumes, including cowpea, soybean, groundnut, and pigeon pea are widely grown across Africa due to their capacity for adaptation to diverse climatic conditions and their ability to improve soil nitrogen through biological nitrogen fixation (Kebede, 2021). West Africa, in particular, is a major supplier of legumes, with nations such as Nigeria, Ghana, Burkina Faso, and Senegal producing quite large areas of cowpea and groundnut across the African continent (Oso and Ashafa, 2020). Legumes are largely cultivated by smallholder farmers in Africa and play a crucial role in agricultural systems, economies, and social advantages (Isgren et al., 2020, Kebede, 2020). Despite their importance, legume production faces several kinds of issues, including extreme temperatures, droughts, pests and diseases, and soil acidity (Khan et al., 2025).

As Africa addresses food security issues and climate-related difficulties, legume farming remains a critical component of the continent's long-term agricultural growth. An increase in legume cultivation is driven by the rising need for plant-based protein, and the growing demand for substitutes to meat presents the chance to produce legume-based meat alternatives (Yanni et al., 2023). This chapter provides an overview of the current status of legume cultivation and use in West Africa, highlighting key crops, production practices, and utilization patterns.

DOI: 10.1201/9781003477235-4

TABLE 3.1
Nutrient Composition of Commonly Cultivated Legumes in West Africa

Legume	Protein (%)	Fat (%)	Carbohydrates (%)	Notable minerals (mg/100 g)	Reference
Bambara groundnut	18	6.5	65	Ca (3.50), Cu (30), Fe, Zn (139), Mg (3.9), P (17.40), K (31.50), Na (0.11)	Mazahib et al., 2013; Musah et al., 2021
Groundnut	20.7–25.3	31–46	21–37	Ca (245.85), K (218.08), Mg (46.04), Na (402.63), P (40.57), Cu (2.74), Fe (25.51), Zn (3.44), Mn (4.77)	Bonku & Yu, 2020; Udo et al., 2021
Cowpea	23–32		50–60	Zn (1.94–5.57), Cu (0.35–4.34), K (362.80–1150.67), Na (16.59–131.73), P (182.71–884.67), Mg (70.41–224.80), Fe (018–25.77), Ca (42.40–125), Mn (0.17–2.17)	Abebe & Alemayehu, 2022; Mbuma et al., 2021
African yam bean	24–28	1.5–2	74.10	Ca (244.5), K (166.2), Cu (2.44), Fe (13.2), Mn (3.35), Zn (6.05), Mg (40.56), P (28.07), Na (352.61)	Gbenga-Fabusiwa, 2021; Ndidi et al., 2014
Common bean	16.54–25.23	0.78–1.58	58.74–66.37	Ca (1295.50), Fe (73.69) Cu (4.64), Mg (1895.10), Na (695.81), K (16473.42), Zn (31) measured in mg/kg	Neupane et al., 2024
African locust bean	32.51–33.52	9.01–9.37	31.90–32.84	Ca (711.56–745.19), mg (64.63–83.13), Na (9.21–13.27), Cu (15.34–21.09), Zn (210.28–293.85)	Famuwagun & Taiwo, 2023

3.2 LEGUME PRODUCTION TRENDS IN WEST AFRICA

Legumes play a crucial role in sustainable agriculture across West and Central Africa, significantly enhancing food security, improving nutritional intake, and enriching soil fertility (Bationo et al., 2011). West African countries are major countries in Africa for legume cultivation (Nounagnon et al., 2024), serving as the foundation of traditional diets and cultivation methods (Paliwal et al., 2020; Vilakazi et al., 2025). Nigeria is one of the world's major producers of cowpeas, producing approximately 2.14 million metric tons annually (Aina, 2022). Another important legume in Africa is groundnuts (*Arachis hypogaea*), additionally referred to as peanuts, with Nigeria and Ghana being the major producers (Gelaye and Luo, 2024; Iddrisu et al., 2024). Soybeans (*Glycine max*) have also become a potential crop in Africa as a non-staple and non-native, notably in Nigeria and Ghana, where they are produced for human consumption, animal feed, and industrial uses (Sedibe et al., 2023). Additionally, Bambara groundnut (*Vigna subterranea*) is another important crop in Africa, highly valued for its drought resistance (Khan et al., 2022). Over the last few decades, farming production in West Africa has seen an outstanding increase, driven by growing population and expanded market possibilities (Eeswaran et al., 2022). According to FAOSTAT (2025), cowpea production in Nigeria increased from 1.64 million tons in 2011 to more than 4.29 million tons in 2023, resulting in an average annual yield of 861 kg/ha in 2023. Similarly, groundnut production in Senegal has steadily increased, achieving 1.49 million tons in 2023 (FAOSTAT, 2025).

Growing demand from local traders and national enterprises, which switched from cottonseed to soybean for oil production, has caused soybean farming to spread throughout central and north Benin since the cotton crisis (Hounhouigan et al., 2020). Soybean output in Benin climbed from 62,987 tons in 2010 to an estimated 520,928.9 tons in 2023, representing a more than 300% increase in production (FAOSTAT, 2025).

TABLE 3.2
Overview of the 2023 Production Trends of Key Legumes in Selected West African Countries

Legume crop	Country	Production (ton)	Area harvested (ha)	Yield (kg/ha)
Bamabara groundnut	Niger	60,982.17	107,531	567.1
	Burkina Faso	59,117.17	56,258	1050.8
Cowpea	Cameroon	182,040.42	224,457	811
	Mali	234,834	485,356	483.8
Soybean	Nigeria	1,350,000	1,200,000	1125
	Ghana	260,000	140,000	1857.1
Groundnut	Senegal	1,493,090.16	1,188,774	1256
	Nigeria	4,300,000	3,450,000	1246.4

Source: FAOSTAT (2025).

3.3 USES OF LEGUMES IN WEST AFRICA

3.3.1 Legumes as Staple and Processed Foods

Legumes are an important source of plant-based protein, making them necessary for human and animal nutrition, particularly in areas with limited availability of animal protein (Benevides et al., 2018). They are high in complex carbohydrates, notably starch, which provides sustained energy and aids metabolic activities (Obadi & Xu, 2024). Furthermore, several health programs and nutritional organizations promote consuming no fewer than three meals of legumes every week (Clemente & Olias, 2017). Legumes play a crucial role in many culinary traditions across the developing nations in Africa (Maphosa & Jideani, 2017). Legumes such as Bamabara groundnut, cowpea, and groundnut are used in a variety of ways in many West African countries, ranging from fresh to processed items, for example, the fresh pods of Bambara groundnut can be boiled with salt and pepper and consumed as a snack (Mayes et al., 2019).

Traditional legume-based dishes are integral to West African food culture. For example, cowpeas are used to make the well-known meal moimoi, which is mostly made in West Africa, where cowpea flour is blended with water, vegetable oil, and a few seasonings, then put in packaging materials like tin cans and aluminum containers and steam-cooked (Nnenna and Wisdom, 2022). Akara (bean cake) is a traditional native meal eaten by multiple ethnic groups in West African nations like Nigeria, Ghana, and Benin which is made from wet-milled cowpeas, which are thoroughly mashed and prepared with vital components like pepper, salt, and onion to enhance digestibility and nutritional quality of the meal (Opaleke et al., 2023).

Groundnuts are an important part of traditional cuisine in West African countries such as Ghana, where they can be found in savory dishes like maafe (meat stew), nkate nkwan (peanut butter soup), and gyorgyor, as well as beloved snacks and sweets like nkate cake (peanut candy) and kuli-kuli (deep-fried groundnut) (Naawe & Angyiereyiri, 2020). Additionally, fermented dish like dawadawa, made from African locust bean seeds (*Parkia biglobosa*), are used as a traditional condiment in soups and sauces, improving both the flavor and nutritional value (Okolo et al., 2023).

Legumes also provide a substantial contribution to the expanding food processing sector in Africa. According to reports, Bambara groundnut flour is used to make milk and yogurt, snacks and pastries, morning cereals, and pasta, offering alternative protein sources for poor communities (Quintieri et al., 2023). Moreover, the potential of legumes to be kept for extended periods while dried without considerable nutritional loss makes them a viable food source, particularly during times of scarcity (Singh et al., 2022).

3.3.2 Importance of Legumes in Combating Food Crisis in Africa

Legumes serve an important role in providing adequate nutrition in Africa, where malnutrition remains a major public health problem (Popoola et al., 2022). Legumes are an important source of protein for low-income households since they are affordable and easily accessible (Odeku et al., 2024). Their prolonged shelf-life and ability to be grown under varied climatic conditions further increase their impact on food security in households (Ahmed et al., 2023; Chettri et al., 2023; Sarkar et al., 2021).

Extreme weather events and climate variability such as rising temperatures, accelerated sea level rises, droughts, and shifting rainfall patterns driven by climate change are significantly impacting agriculture across Africa. Climate change reduces agricultural output and raises food prices, resulting in greater food insecurity and poverty (Acquaye, 2021).

Leguminous crops, such as Bambara groundnut and cowpea, are tolerant of drought and have the ability to grow in dry and semi-arid areas where other main crops struggle to thrive (Gao et al., 2023; Srivastava et al., 2023). This makes legumes especially important in the setting of climate change, as they provide an adequate supply of food even in harsh environments.

3.3.3 Medicinal and Pharmaceutical Uses of Legumes

Legumes and their components are being studied for their potential applications besides food, such as for pharmaceutical uses (Benevides et al., 2018). Legumes are rich sources of plant-based protein, dietary fiber, essential vitamins, and minerals, as well as beneficial bioactive compounds like polyphenols and flavonoids. These compounds offer numerous health-promoting effects, including antioxidant, antiinflammatory, and blood pressure-lowering properties. Regular consumption of legumes has been associated with various health benefits, such as reduced risk of heart disease, improved cholesterol levels, better blood sugar regulation, enhanced digestive health, and effective weight management (Zain et al., 2023; Olabanji et al., 2018). Additionally, legumes may play a role in preventing the onset and progression of certain cancers. In countries such as Cameroon, Ghana, Nigeria, and Uganda, consuming legumes like Bambara groundnuts, soybeans, cowpeas, and groundnuts has been associated with the prevention of metabolic disorders, including type II diabetes, hypertension, Alzheimer disease, Parkinson disease, obesity, and stroke (Hamadou et al., 2022; Irondi et al., 2019).

Additionally, Bambara groundnut has also been shown to contain beneficial antioxidants and antiinflammatory chemicals, with the ability to scavenge radicals that are harmful and promote a healthy body (Oyeyinka et al., 2021). Crushed Bambara groundnut grains are utilized as remedies for nausea and sight issues in Senegal, while black Bambara groundnut grains are blended in water to treat children's diarrhea and white grains are combined with guinea fowl meat to alleviate diarrhea in old people in Ghana. The Ibo people of Nigeria use Bambara groundnut to treat sexually transmitted diseases (Ramatsetse et al., 2023). Cowpea (*Vigna unguiculata*) is another legume widely utilized in traditional African medicine. Traditional healers use cowpea leaves and green pods to treat burns, adenitis, ulcers, and measles (Takaidza, 2023).

Another underutilized legume is the African locust bean, which is used for treatment of numerous health problems in Ghana (Nyadanu et al., 2017). The stem bark has been shown to have anti-snake venom properties, and in traditional medicine, an extract of the stem bark has been utilized as a warm mouthwash to ease toothaches, while a combination of the root and leaves is thought to be an efficient cure for eye irritation and dental cavities (Airaodion & Ogbuagu, 2020). A study found that consuming fermented African locust bean seeds significantly reduced blood pressure and heart rate, thus reducing hypertension in the Bogou people of Togo (Ognatan et al., 2011).

TABLE 3.3
Medicinal Uses of Selected Legumes in African Traditional Medicine

Legume	Medicinal use	Country	Reference
Bambara groundnut (Vigna subterranea)	Crushed grains used to treat nausea and vision problems	Senegal	Ramatsetse et al., 2023
	Black grains mixed with water to treat diarrhea in children	Ghana	Ramatsetse et al., 2023
	White grains blended with guinea fowl meat to treat diarrhea in the elderly	Ghana	Ramatsetse et al., 2023
	Used by the Ibo people to treat sexually transmitted diseases	Nigeria	Ramatsetse et al., 2023
African locust bean (Parkia biglobosa)	Used as anti-snake venom	Nigeria	Asuzu & Harvey, 2003
	Used as a warm mouthwash for toothaches	Ghana	Airaodion & Ogbuagu, 2020
	Root and leaves used to treat eye irritation and dental cavities	Ghana	Airaodion & Ogbuagu, 2020
	Fermented seeds lower blood pressure and heart rate, reducing hypertension	Togo	Ognatan et al., 2011
	Treating wounds, malaria, rheumatism, skin infections, headaches, and female sterility	Nigeria	Ayanrinde et al., 2023
Cowpea (Vigna unguiculata)	Traditional uses for the leaves and green pods include treating measles, adenitis, burns, and ulcers	General	Takaidza, 2023

3.3.4 Legumes in Livestock Feed and Fodder Production

Besides their importance in human nutrition, legumes serve as key components of feed for livestock. Cowpea hay is a crucial fodder supply for animal feed in crop–livestock systems in the Sahel area of West Africa, where feed shortages and seasonality are significant challenges to livestock productivity (Amole et at., 2022). Legumes including alfalfa, soybean, and cowpea are especially well known for their high protein content, which considerably improves livestock development and production (Kulkarni et al., 2018). Groundnut haulms, for example, are commonly utilized as supplemental feed for livestock in Nigeria, while Africans have long cultivated cowpea not only for their own consumption but also as a high-protein feed for their livestock (Abebe & Alemayehu, 2022; Adewuyi, 2021). Legume-based diets for ruminants can help lower methane (CH_4) emissions because they contain less fiber, which not only enhances livestock productivity but also benefits the environment (Ahmed et al., 2023). Legumes improve soil health using biological nitrogen fixation, which increases soil fertility and decreases the demand for synthetic fertilizers (Kebede, 2021). This, in turn, encourages sustainable livestock production techniques that reduce environmental deterioration.

3.3.5 Sustainable Agricultural Practices in Africa

Legumes such as groundnuts provide several kinds of environmental, social, and food choice advantages for rural families in many small-scale settings, such as the opportunity for women who have traditionally been excluded in managing farm resources and revenues (Paul et al., 2023). West African countries are actively practicing climate-smart agriculture strategies for legume cultivation (Agyekum et al., 2024). Given that legumes improve soil fertility by fixing atmospheric nitrogen through biological nitrogen fixation, they have become widely accepted ways to enhance soil fertility and improve yields (Yatogo, 2024). Farmers in Niger often utilize cowpea–millet intercropping,

TABLE 3.4
Sustainable Agricultural Practices using Legume Crops

Legume crop	Sustainable practice	Description	Benefits	References
Cowpea (Vigna unguiculata)	Intercropping with maize and millet	Cowpeas are grown alongside cereals to increase soil fertility and crop diversification	Enhances nitrogen fixation, minimizes soil degradation, and increases production	Namatsheve et al., 2020
Soybean (Glycine max)	Use of *Rhizobium* inoculants	The use of *Rhizobium* bacteria increases nitrogen fixation and crop development	Reduces dependence on synthetic fertilizers while increasing output	Kibido et al., 2020
Groundnut (Arachis hypogaea)	Use of *Rhizobium* inoculants	Groundnut is used as a cover crop to prevent soil erosion and fix nitrogen	Improves soil fertility, retains moisture, and reduces land deterioration	Hussainy et al., 2020
Pigeon pea (Cajanus cajan)	Agroforestry with nitrogen-fixing trees	Integrated into cropping systems with trees like *Gmelina* and mango to improve soil structure	Enhances soil organic matter, offers shade, productivity, and promotes biodiversity	Biswas et al., 2022
Bambara groundnut (Vigna subterranea)	Drought-resistant cultivation	Drought-tolerant Bambara groundnut is grown in semi-arid areas	Improves food security, helps smallholder farmers, and enhances resistance to climate change	Soumare et al., 2022
Lablab (Lablab purpureus)	Regenerative agriculture practices	Used as a rotating crop to revitalize deteriorated soil and control pests and diseases of other crops	Enhances soil nutrients, and provides feed for livestock	Aleme, 2022

which not only promotes land productivity but also improves nitrogen fixation, lowering the demand for synthetic fertilizers (Abdoul-Karim et al., 2022).

Farmers in Tanzania use conservation agricultural practices such as cropping systems, limited tillage, cover cropping, and intercropping lablabs and sweet-beans (Marambanyika, 2022). Cowpea cultivars, such as, Padi-tuya, Kirkhous Benga, Nketewade, and Zamzam, are tolerant of pest and disease infestations, mature early, and are suited for areas with unpredictable rainfall patterns (GNA, 2022). A study in Kenya demonstrated that adding *Rhizobium* inoculation agents to cowpea fields considerably increases production by enhancing root nodulation and biological nitrogen fixing (Wanja, 2022).

3.4 CHALLENGES FACING LEGUME CULTIVATION

Legumes are a key component of the agriculture, economies, and dietary traditions in West African nations. In a region largely dependent on smallholder and subsistence farming, crops like cowpea, groundnut, soybean, and Bambara groundnut play a vital role in enhancing food security, improving soil fertility, and supporting livelihoods. Despite their importance, the cultivation of legumes in West Africa faces many challenges, which include receiving little attention from researchers, policymakers, and the agricultural sector, which limits both profitability and output (Popoola et al., 2022; Vilakazi et al., 2025). Limited investment in research has resulted in a lack of high-quality seeds, leading to low productivity and restricted access to markets (Muhammad et al., 2020; Chongtham et al., 2022; Vilakazi et al., 2025). Policies often do not support efforts to raise awareness, promote dietary diversity, or encourage the incorporation of these legumes into everyday meals.

Legumes are often cultivated by vulnerable and marginalized communities lacking the necessary tools and technical expertise in marginal areas such as the Sahelian regions of Niger, Mali, and Burkina Faso which are characterized by poor soils, high temperatures, increased coastal flooding, drought, and unpredictable rainfall patterns, all of which highlight the visible impact of climate change in the region (ICRISAT, 2024, Vilakazi et al., 2025). Countries such as Ghana also lack infrastructure for processing, transporting, managing, and storing agricultural output, which results in substantial post-harvest losses (Sugri et al., 2021; Binge et al., 2023).

Another significant concern is pest and disease pressure. *Maruca vitrata* pod borer and aphids are important dangers to cowpea production, generating up to 80% yield losses in serious cases, with greater visible damage in Africa, and attempts to produce and deploy resistant cowpea varieties have not shown substantial benefits (Sodedji et al., 2020). Aflatoxin contamination also has an impact on groundnut production, restricting access to profitable markets, as well as affecting people that consume it (Ojiewo et al., 2020). Limited access to financial services hinders farmers' ability to adopt modern technologies and invest in the management of their farms. Additionally, limited market access and market information make it difficult for farmers to sell their produce at competitive prices.

Legume crops are generally grown as secondary crops, intercropped with cereals, or occasionally used in crop rotation systems (Bhatt et al., 2021; Singh, 2020; Singh et al., 2020; Vilakazi et al., 2025). The absence of agricultural extension support further restricts smallholder farmers from implementing improved farming practices. Furthermore, challenges such as social inequality, land ownership issues mostly affecting women and marginalized groups, and an aging farming population, compounded by the lack of youth interest in agriculture, result in a decline in innovation and modern techniques in the cultivation of traditional legumes (Vilakazi et al., 2025). Managing these difficulties would need collaborative efforts from governments, the commercial sector, and research institutions to increase input supply, extension services, and market access.

3.5 CONCLUSION

Legume cultivation is crucial to West Africa's agricultural systems, significantly enhancing food and nutrition security, environmental sustainability, and economic development. Crops such as cowpea, groundnut, soybean, and Bambara groundnut provide numerous benefits. They improve soil fertility through biological nitrogen fixation, offer essential nutrients for both human and animal consumption, and support sustainable farming practices. However, legume production faces several challenges. These include limited investment in research, poor access to quality seeds, lack of market integration, the impacts of climate change, and socioeconomic barriers such as issues related to land tenure, gender inequality, and limited youth engagement in agriculture. Furthermore, inadequate extension services and the absence of supportive policies also impede the adoption of legume cultivation.

To fully utilize the potential of legumes in addressing food insecurity, poverty, and environmental degradation, a collaborative approach involving multiple stakeholders is crucial. This approach should focus on increasing investment in research and development, enhancing infrastructure, and improving access to markets and financial services. Additionally, it should include supportive agricultural policies that promote awareness, value addition, and equitable participation among all farming communities.

To unlock the potential of legumes for sustainable development in West Africa, smallholder farmers, especially women and youth, must be empowered. This can be done through education, innovative practices, and better access to resources. These steps are essential for transforming farming and improving livelihoods in the region.

REFERENCES

Abdoul-Karim, T.D., Sanoussi, A., Soulé, M. and Yacoubou, B., 2022. Combined effect of fertilizer micro-dosing and intercropped millet/cowpea effect on agronomic and economic advantages in prone Sahel area, Niger. *Discover Sustainability*, *3*(1), 31. DOI: 10.1007/s43621-022-00099-2

Abebe, B.K. and Alemayehu, M.T., 2022. A review of the nutritional use of cowpea (Vigna unguiculata L. Walp) for human and animal diets. *Journal of Agriculture and Food Research*, *10*, 100383. DOI: 10.1007/s43621-022-00099-2

Acquaye, P., 2021. Impact of climate change on food security in Ghana. Available at: https://blog.bluecrest.edu.gh/2021/10/07/impact-of-climate-change-on-food-security-in-ghana/

Adewuyi, P.A., 2021. *Nutritive value of processed groundnut (Arachis hypogaea L.) haulm-based diets for West African Dwarf rams*. Ph.D. thesis, University of Ibadan, Nigeria.

Agyekum, T.P., Antwi-Agyei, P., Dougill, A.J. and Stringer, L.C., 2024. Benefits and barriers to the adoption of climate-smart agriculture practices in West Africa: A systematic review. *Climate Resilience and Sustainability*, *3*(3), e279. DOI: 10.1002/cli2.79

Ahmed, M., Sameen, A., Parveen, H., Ullah, M.I., Fahad, S. and Hayat, R., 2023. Climate change impacts on legume crop production and adaptation strategies. In Ahmed, M. (ed.) *Global Agricultural Production: Resilience to Climate Change* (pp. 149–181). Cham: Springer International Publishing. DOI: 10.1007/978-3-031-14973-3_5

Aina, F.B., 2022. *Evaluation of insecticide spraying times and weeding regime on Cowpea insect pests and grain yield in the Southern Guinea Savanna of Nigeria* (Master's thesis, Kwara State University).

Airaodion, A.I. and Ogbuagu, E.O., 2020. Ameliorative effect of Parkia biglobosa (African locust bean) against egg-yolk induced hypertension. *Hypertension*, *17*, 18. DOI: 10.5281/zenodo.7013664

Aleme, M., 2022. Potentials distribution and origin of lablab [Lablab purpureus (L.) sweet]: A review. *Agricultural Reviews*, *43*(3), 341–347. DOI: 10.18805/ag.RF-226.

Amole, T., Augustine, A., Balehegn M. and Adesogoan, A.T., 2022. Livestock feed resources in West African Sahel. *Agronomy Journal*, *114*, 26–45. DOI: 10.1002/agj2.20955

Asuzu, I.U. and Harvey, A.L., 2003. The antisnake venom activities of Parkia biglobosa (Mimosaceae) stem bark extract. *Toxicon*, *42*(7), 763–768. DOI: 10.1016/j.toxicon.2003.10.004

Ayalew, T. and Yoseph, T., 2022. Cowpea (Vigna unguiculata L. Walp.): A choice crop for sustainability during the climate change periods. *Journal of Applied Biology and Biotechnology*, *10*, 154–162. DOI: 10.7324/JABB.2022.100320

Ayanrinde, F.A., Oyewole, S.O., Ayanrinde, O.A. and Oyewole, A.L., 2023. Importance of Parkia biglobosa (African Locust Beans) to the women livelihoods in Ibarapa North and Ibarapa Central Local Government areas of Oyo State, Nigeria. *Journal of Applied Sciences and Environmental Management*, *27*(9), 2097–2100. DOI: 10.4314/jasem.v27i9.27

Bationo, A., Kimetu, J., Vanlauwe, B., Bagayoko, M., Koala, S. and Mokwunye, A.U., 2011. Comparative analysis of the current and potential role of legumes in integrated soil fertility management in West and Central Africa. In Bationo, A., Waswa, B., Okeyo, J., Maina, F., Kihara, J. and Mokwunye, U. (eds.) *Fighting Poverty in Sub-Saharan Africa: The Multiple Roles of Legumes in Integrated Soil Fertility Management*. Dordrecht: Springer. DOI: 10.1007/978-94-007-1536-3_6

Benevides, C.M., Trindade, B.A. and Lopes, M.V., 2018. Potentialities of legumes in the pharmaceutical industry. *Journal of Analytical & Pharmaceutical Research*, *7*(3), 369–373. DOI: 10.15406/japlr.2018.07.00253

Bhatt, K., Malav, P.K., Gore, P.G., Tripathi, K., Rathi, R., Tiwari, U.L. and Ahlawat, S.A., 2021. Note on distribution and potential of Japanese wild adzuki bean [Vigna angularis var. nipponensis (Ohwi) Ohwi and H. Ohashi] in India. *Genetic Resources and Crop Evolution*, *68*, 2157–2166.

Binge, B., Jalango, D. and Tesfaye, L., 2023. *Post-harvest losses management through climate smart innovations: A collaborative approach among value chain actors* (p. 53). AICCRA technical report. Accelerating Impacts of CGIAR Climate Research for Africa (AICCRA).

Biswas, B., Chakraborty, D., Timsina, J., Bhowmick, U.R., Dhara, P.K., Lkn, D.K.G., Sarkar, A., Mondal, M., Adhikary, S., Kanthal, S. and Patra, K., 2022. Agroforestry offers multiple ecosystem services in degraded lateritic soils. *Journal of Cleaner Production*, *365*, 132768. DOI: 10.1016/j.jclepro.2022.132768

Bonku, R. and Yu, J., 2020. Health aspects of peanuts as an outcome of its chemical composition. *Food Science and Human Wellness*, *9*(1), 21–30. DOI: 10.1016/j.fshw.2019.12.005

Chettri, S., Sharma, N. and Mohite, A.M., 2023. Utilization of lima bean starch as an edible coating base material for sapota fruit shelf-life enhancement. *Journal of Agriculture and Food Research, 12*, 100615. DOI: 10.1016/j.jafr.2023.100615

Chongtham, S.K., Devi, E.L., Samantara, K., Yasin, J.K., Wani, S.H., Mukherjee, S., Razzaq, A., Bhupenchandra, I., Jat, A.L. and Singh, L.K., 2022. Orphan legumes: Harnessing their potential for food, nutritional and health security through genetic approaches. *Planta, 256*, 24.

Clemente, A. and Olias, R., 2017. Beneficial effects of legumes in gut health. *Current Opinion in Food Science, 14*, 32–36.

Eeswaran, R., Nejadhashemi, A.P., Faye, A., Min, D., Prasad, P.V. and Ciampitti, I.A., 2022. Current and future challenges and opportunities for livestock farming in West Africa: Perspectives from the case of Senegal. *Agronomy, 12*(8), 1818. DOI: 10.3390/agronomy12081818

Famuwagun, A.A. and Taiwo, K.A., 2023. Drying characteristics, nutritional and anti-nutritional properties of locust bean seed. *Cogent Food & Agriculture, 9*(1), 2202276. DOI: 10.1080/23311932.2023.2202276

FAO, 2025. FAOSTAT. Crops production data. Available at: www.fao.org/faostat/ (Accessed: 14 March 2025).

Gao, X., Chai, H.H., Ho, W.K., Mayes, S. and Massawe, F., 2023. Deciphering the molecular basis for photosynthetic parameters in Bambara groundnut (Vigna subterranea L. Verdc) under drought stress. *BMC Plant Biology, 23*(1), 287.

Gbenga-Fabusiwa, F.J., 2021. African yam beans (Sphenostylis stenocarpa): A review of a novel tropical food plant for human nutrition, health and food security. *African Journal of Food Science, 15*(2), 33–47. DOI: 10.5897/AJFS2020.1961

Gelaye, Y. and Luo, H., 2024. Optimizing Peanut (Arachis hypogaea L.) production: Genetic insights, climate adaptation, and efficient management practices: Systematic review. *Plants, 13*(21), 2988. DOI: 10.3390/plants13212988

Ghana News Agency., 2022. *Farmers introduced to climate-resilient improved maize, cowpea varieties.* Available at: https://gna.org.gh/2022/08/farmers-introduced-to-climate-resilient-improved-maize-cowpea-varieties/

Hamadou, M, Alain, M.M.M., Obadias, F.V., Muhammad Zaffar Hashmi, M.Z., Başaran, B., Paul, B.J. and René, M.S., 2022. Consumption of underutilised grain legumes and the prevention of type II diabetes and cardiometabolic diseases: Evidence from field investigation and physicochemical analyses. *Environmental Challenges, 9*, 100621. DOI: 10.1016/j.envc.2022.100621

Henderson, P., 2023. Pigeon Pea (Cajanus cajan). In Chen, K. and Byrne, P. (eds.) *Understudied Indigenous Crops*. Fort Collins, Colorado: Colorado State University. https://colostate.pressbooks.pub/understudiedindigenouscrops/chapter/pigeon-pea/

Hounhouigan, M.H., Kounouewa, K.M., Ayesiga, C. and Ingenbleek, P.T., 2020. Sojagnon: shaping the Beninese soy system to meet the challenges of an emerging market. *International Food and Agribusiness Management Review, 23*(1), 143–156. DOI: 10.22434/IFAMR2019.0026

Hussainy, S.A.H., Brindavathy, R. and Vaidyanathan, R., 2020. Production potential of groundnut (Arachis hypogaea L.) under intercropping system-A review. *Crop Research, 55*(1–2), 36–47. DOI: 10.31830/2454-1761.2020.010

ICRISAT, 2024. *West and Central Africa Overview*. International Crops Research Institute for the Semi-Arid Tropics. Available at: www.icrisat.org/regions/west-and-central-africa/overview

Iddrisu, A., Adjei, E., Asomaning, S.K., Santo, K.G., Isaac, A.P. and Danson-Anokye, A., 2024. Effect of variety and plant spacing on growth and yield of groundnuts (Arachis hypogaea L.). *Agricultural Sciences, 15*(1), 54–70. DOI: 10.4236/as.2024.151004

Irondi, E.A., Ogunsanmi, A.O., Ahmad, R.S., Ajani, E.O., Adegoke, B.M. and Boligon, A.A., 2019. Effect of roasting on phenolics composition, enzymes inhibitory and antioxidant properties of cowpea pulses. *Journal of Food Measurement and Characterization, 13*, 1489–1496. DOI: 10.1007/s11694-019-00064-0

Isgren, E., Andersson, E. and Carton, W., 2020. New perennial grains in African smallholder agriculture from a farming systems perspective. A review. *Agronomy for Sustainable Development, 40*, 1–14. DOI: 10.1007/s13593-020-0609-8

Kabbashi, E.B., 2024. Major contaminants of peanut and its products and their methods of management. In Marc, R.A., Mureşan, C.C. and Postolache, A.N. (eds.) *Nut Consumption and its Usefulness in the Modern World.* London: IntechOpen. DOI: 10.5772/intechopen.1004630

Kebede, E., 2020. Grain legumes production and productivity in Ethiopian smallholder agricultural system, contribution to livelihoods and the way forward. *Cogent Food & Agriculture*, *6*(1), 1722353. DOI: 10.1080/23311932.2020.1722353

Kebede, E., 2021. Contribution, utilization, and improvement of legumes-driven biological nitrogen fixation in agricultural systems. *Frontiers in Sustainable Food Systems*, *5*, 767998. DOI: 10.3389/fsufs.2021.767998

Khan, M.M.H., Rafi, M.Y., Ramlee, S.I., Jusoh, M. and Mamun, A., 2022. Bambara groundnut (Vigna Subterranea L. Verdc): A model underutilized legume, its resiliency to drought and excellency of food & nutrient values-A review. *Malaysian Journal of Genetics*, *2*, 12.

Khan, M.M.H., Rafii, M.Y., Ramlee, S.I., Jusoh, M. and Al-Mamun, M., 2024. Bambara groundnut (Vigna subterranea L.): A climate-resilient crop to address food and nutritional security. *Journal of Underutilised Crops Research*, *3*(1), 2. DOI: 10.4102/jucr.v3i1.2

Khan, M.N., Wahab, S., Wahid, N., Shah, S.N., Ullah, B., Kaplan, A., Razzaq, A., Bibi, M., Suleman, F. and Ali, B., 2025. Impact of climate change on yield and quality of legumes. In Shah, F., Adnan, M., Munir, I., Lal, R., Nawaz, T. and Saud, S. (eds.) *Challenges and Solutions of Climate Impact on Agriculture* (pp. 85–111). Cambridge: Academic Press. DOI: 10.1016/B978-0-443-23707-2.00004-0.

Kibido, T., Kunert, K., Makgopa, M., Greve, M. and Vorster, J., 2020. Improvement of rhizobium-soybean symbiosis and nitrogen fixation under drought. *Food and Energy Security*, *9*(1), e177. DOI: 10.1002/fes3.177

Kulkarni, K.P., Tayade, R., Asekova, S., Song, J.T., Shannon, J.G. and Lee, J.D., 2018. Harnessing the potential of forage legumes, alfalfa, soybean, and cowpea for sustainable agriculture and global food security. *Frontiers in Plant Science*, *9*, 1314. DOI: 10.3389/fpls.2018.01314

Lai, H., Gao, F., Su, H., Zheng, P., Li, Y. and Yao, H., 2022. Nitrogen distribution and soil microbial community characteristics in a legume–cereal intercropping system: A review. *Agronomy*, *12*(8), 1900. DOI: 10.20944/preprints202207.0230.v1

Maphosa, Y. and Jideani, V.A., 2017. The role of legumes in human nutrition. *Functional Food-Improve Health Through Adequate Food*, *1*, 13. DOI: 10.5772/intechopen.69127

Marambanyika, V., 2022. *The role of Conservation Agriculture in improving smallholder farmers' crop yields in Malawi* (Master's thesis, Norwegian University of Life Sciences, Ås).

Mayes, S., Ho, W.K., Chai, H.H., Gao, X., Kundy, A.C., Mateva, K.I., Zahrulakmal, M., Hahiree, M.K.I.M., Kendabie, P., Licea, L.C. and Massawe, F., 2019. Bambara groundnut: an exemplar underutilised legume for resilience under climate change. *Planta*, *250*, 803–820. DOI: 10.1007/s00425-019-03191-6

Mazahib, A.M., Nuha, M.O., Salawa, I.S. and Babiker, E.E., 2013. Some nutritional attributes of bambara groundnut as influenced by domestic processing. *International Food Research Journal*, *20*(3), 1165.

Mbuma, N.W., Gerrano, A.S., Lebaka, N., Amoo, S., Mofokeng, A., Labuschagne, M., 2021. Variability in the concentration of mineral elements and phytochemical contents of cowpea genotypes for crop improvement. *Acta Agriculturae Scandinavica Section B, Soil Plant Science*, *71*, 132–144. DOI: 10.1080/09064710.2020.1859609

Muhammad, I., Rafii, M.Y., Ramlee, S.I., Nazli, M.H., Harun, A.R., Oladosu, Y., Musa, I., Arolu, F., Chukwu, S.C., Sani Haliru, B. and Silas Akos, I., 2020. Exploration of bambara groundnut (Vigna subterranea (L.) Verdc.), an underutilized crop, to aid global food security: Varietal improvement, genetic diversity and processing. *Agronomy*, *10*(6), 766. DOI: 10.3390/agronomy10060766

Musah, M., Azeh, Y., Mathew, J.T., Nwakife, N.C., Mohammed, A.I. and Saidu, F., 2021. Nutritional evaluation of bambara groundnut (Vigna subterranea (L.) Verdc) from Lapai, Nigeria. *African Journal of Agriculture and Food Science*, *4*(4), 32–39. DOI: 10.52589/AJAFS_SQI5U7CN

Naawe, E.K. and Angyiereyiri, E.D., 2020. Effect and infestation levels of groundnut pod borer (Elasmolomus sordidus) on groundnut (Arachis hypogaea) and farm control measures in Tedema, Builsa-North District of the Upper East Region, Ghana. *International Journal of Tropical Insect Science*, *40*, 687–696. DOI: 10.1007/s42690-020-00121-y

Namatsheve, T., Cardinael, R., Corbeels, M. and Chikowo, R., 2020. Productivity and biological N2-fixation in cereal-cowpea intercropping systems in sub-Saharan Africa. A review. *Agronomy for Sustainable Development*, *40*(4), 30. DOI: 10.1007/s13593-020-00629-0

Ndidi, U.S., Ndidi, C.U., Olagunju, A., Muhammad, A., Billy, F.G. and Okpe, O., 2014. Proximate, antinutrients and mineral composition of raw and processed (Boiled and Roasted) Sphenostylis stenocarpa seeds from Southern Kaduna, Northwest Nigeria. *International Scholarly Research Notices*, (1), 280837. DOI: 10.1155/2014/280837

Neupane, B.S., Olee, D., Shrestha, D.S., Kharel, G.P. and Koirala, N., 2024. "Nutritional and phytochemicals analysis of high-altitude common bean (Phaseolus vulgaris L.) cultivars of Nepal." *eFood*, *5*(5), e182. DOI: 10.1002/efd2.182

Nnenna, J.P. and Wisdom, J., 2022. Effects of wrapping materials on the microbial population of steam bean pudding (Moimoi). *Journal of Food Technology*, *9*(1), 80–93. DOI: 10.18488/jftr.v9i1.2955

Nounagnon, M., Roko, G., Agbodjato, N.A., Dah-Nouvlessounon, D., Babalola, O.O. and Baba-Moussa, L., 2024. Cowpea (Vigna unguiculata). In Chandora, R., Basavaraja, T. and Pratap, A. (eds.) *Potential Pulses: Genetic and Genomic Resources*. Wallingford: CABI. DOI: 10.1079/9781800624658.0004

Nyadanu, D., Adu Amoah, R., Obeng, B., Kwarteng, A.O., Akromah, R., Aboagye, L.M. and Adu-Dapaah, H., 2017. Ethnobotany and analysis of food components of African locust bean (Parkia biglobosa (Jacq.) Benth.) in the transitional zone of Ghana: implications for domestication, conservation and breeding of improved varieties. *Genetic Resources and Crop Evolution*, *64*, 1231–1240. DOI: 10.1007/s10722-016-0432-x

Obadi, M.O.M. and Xu, B., 2024. A review of the effects of physical processing techniques on the characteristics of legume starches and their application in low-glycemic index foods. *International Journal of Biological Macromolecules*, , *279*(1), 135124. DOI: 10.1016/j.ijbiomac.2024.135124

Odeku, O.A., Ogunniyi, Q.A., Ogbole, O.O. and Fettke, J., 2024. Forgotten gems: exploring the untapped benefits of underutilized legumes in agriculture, nutrition, and environmental sustainability. *Plants*, *13*(9), 1208. DOI: 10.3390/plants13091208

Ognatan, K., Adi, K., Lamboni, C., Damorou, J.M., Aklikokou, K.A., Gbeassor, M. and Guilland, J.C., 2011. Effect of dietary intake of fermented seeds of Parkia biglobosa (Jacq) Benth (African locust bean) on hypertension in Bogou and Goumou-kope areas of Togo. *Tropical Journal of Pharmaceutical Research*, *10*(5), 603–609. DOI: 10.4314/tjpr.v10i5.9

Ojiewo, C.O., Janila, P., Bhatnagar-Mathur, P., Pandey, M.K., Desmae, H., Okori, P., Mwololo, J., Ajeigbe, H., Njuguna-Mungai, E., Muricho, G. and Akpo, E., 2020. Advances in crop improvement and delivery research for nutritional quality and health benefits of groundnut (Arachis hypogaea L.). *Frontiers in Plant Science*, *11*, 29. DOI: 10.3389/fpls.2020.00029

Okolo, E., Simon, V.O. and Eze, M.E., 2023. Mycoflora of fermented locust beans (Parkia biglobosa) Seed,"Iru" Sold in Lokoja Markets, Kogi State, Nigeria. *FUDMA Journal of Sciences*, *7*(4), 76–81. DOI: 10.33003/fjs-2023-0704-1816

Olabanji, I.O., Ajayi, O.S., Oluyemi, E.A, Olawuni, I.J., Adeniji, A.O., Olasupo, O.F., Agboola, R.O. and Olusesi, I.M., 2018. Nutraceuticals in different varieties of cowpeas. *American Journal of Food Science and Technology*, *6*, 68–75. DOI: 10.12691/ajfst-6-2-2

Opaleke, D.O., Adebisi, T.T., Karim, O.R., Oluwadare, A.A., Ihensekhien, I. and Lawal, S.M., 2023. Preparation and characterization of akara and senke cakes made from cowpea flour paste. *Journal of Current Science and Technology*, *13*(2), 351–363. DOI: 10.59796/jcst.V13N2.2023.1750

Oso, A.A. and Ashafa, A.O., 2020. Insect pests preference for Cowpea: The dilemma of darmers in West Africa. *Tropical and Subtropical Agroecosystems*, *23*(1)), 1–11. DOI: 10.56369/tsaes.3025

Oyeyinka, S.A., Abdulsalam, A.O., Ahmed El-Imam, A.M., Oyeyinka, A.T., Olagunju, O.F., Kolawole, F.L., Arise, A.K., Adedeji, E.O. and Njobeh, P.B., 2021. Total phenolic content, antioxidant, anti-inflammatory and anti-microbial potentials of Bambara groundnut (Vigna subterranea L.) seed extract. *British Food Journal*, *123*(11), 3421–3435. DOI: 10.1108/BFJ-07-2020-0637

Paliwal, R., Abberton, M., Faloye, B. and Olaniyi, O., 2020. Developing the role of legumes in West Africa under climate change. *Current Opinion in Plant Biology*, *56*, 242–258. DOI: 10.1016/j.pbi.2020.05.002

Paul Jr, M., Aihounton, G.B. and Lokossou, J.C., 2023. Climate-smart agriculture and food security: Cross-country evidence from West Africa. *Global Environmental Change*, *81*, 102697. DOI: 10.1016/j.gloenvcha.2023.102697

Piyush, K.R, Kishan, K.R., Kholu, M., Raj, B. and Utkarsh, T., 2024. The role of legumes in climate-smart cropping systems: enhancing crop, soil, and environmental health security. In Kumar, G.S., Wanniang, S.K., Bahadur, R., Anbarasan, S., Raj, P. and Manu, M.S. (eds.) *From Soil to Harvest: Integrating Agronomy with Soil Science*. Delhi: Ninetales Publishing. www.researchgate.net/publication/383455081_The_role_of_legumes_in_climate_smart_cropping_systems_Enhancing_crop_soil_and_environmental_health_security

Popoola, J.O., Aworunse, O.S., Ojuederie, O.B., Adewale, B.D., Ajani, O.C., Oyatomi, O.A., Eruemulor, D.I., Adegboyega, T.T. and Obembe, O.O., 2022. The exploitation of orphan legumes for food, income, and nutrition security in Sub-Saharan Africa. *Frontiers in Plant Science*, *13*, 82140. DOI: 10.3389/fpls.2022.782140

Quintieri, L., Nitride, C., De Angelis, E., Lamonaca, A., Pilolli, R., Russo, F. and Monaci, L., 2023. Alternative protein sources and novel foods: benefits, food applications and safety issues. *Nutrients*, *15*(6), 1509. DOI: 10.3390/nu15061509

Ramatsetse, K.E., Ramashia, S.E. and Mashau, M.E., 2023. A review on health benefits, antimicrobial and antioxidant properties of Bambara groundnut (Vigna subterranean). *International Journal of Food Properties*, *26*(1), 91–107. DOI: 10.1080/10942912.2022.2153864

Ranamukhaarachchi, S. and Nanayakkara, D., 2025. Legumes: Cultivation, uses, and benefits. In Jimenez-Lopez, J.C. and Escudero-Feliu, J. (eds.) *Legume Crops for Food Security – Cultivation and Benefits*. London: IntechOpen. DOI: 10.5772/intechopen.1001792

Santhiya, S., Singh, A., Thakur, A., Dev, R., Hedau, N.K. and Kant, L., 2024. Importance of legume vegetables in nutritional and food security-An Indian perspective. *Journal of Food Legumes*, *37*(4), 359–366. DOI: 10.59797/jfl.v37.i4.219

Sarkar, S., Khatun, M., Era, F.M., Islam, A.M., Anwar, M.P., Danish, S., Datta, R. and Islam, A.A., 2021. Abiotic stresses: Alteration of composition and grain quality in food legumes. *Agronomy*, *11*(11), 2238. ttps://doi.org/10.3390/agronomy11112238

Sedibe, M.M., Mofokeng, A.M. and Masvodza, D.R., 2023. Soybean production, constraints, and future prospects in poorer countries: A review. In Hasanuzzaman, M, (eds.) *Production and Utilization of Legumes-Progress and Prospects*. London: IntechOpen. DOI: 10.5772/intechopen.109516

Shankar, S., Haque, E., Ahmed, T., Kiran, G.S., Hassan, S. and Selvin, J., 2021. Rhizobia–legume symbiosis during environmental stress. In Shrivastava, N., Mahajan, S. and Varma, A. (eds.) *Symbiotic Soil Microorganisms: Biology and Applications* (pp. 201–220). Cham: Springer International Publishing. DOI: 10.1007/978-3-030-51916-2_13

Singh, B., 2020. *Cowpea: The Food Legume of the 21st Century*. Hoboken, NJ, USA: John Wiley; p. 164.

Singh, M., Rundan, V. and Onte, S., 2020. Ricebean. *Indian Farming*, *70*, 27–31.

Singh, N., Jain, P., Ujinwal, M. and Langyan, S., 2022. Escalate protein plates from legumes for sustainable human nutrition. *Frontiers in Nutrition*, *9*, 977986.y. DOI: 10.3389/fnut.2022.977986

Sodedji, F.A.K., Agbahoungba, S., Nguetta, S.P.A., Agoyi, E.E., Ayenan, M.A.T., Sossou, S.H., Mamadou, C., Assogbadjo, A.E. and Kone, D., 2020. Resistance to legume pod borer (Maruca vitrata Fabricius) in cowpea: Genetic advances, challenges, and future prospects. *Journal of Crop Improvement*, *34*(2), 238–267. DOI: 10.1080/15427528.2019.1680471

Soumare, A., Diedhiou, A.G. and Kane, A., 2022. Bambara groundnut: A neglected and underutilized climate-resilient crop with great potential to alleviate food insecurity in sub-Saharan Africa. *Journal of Crop Improvement*, *36*(5), 747–767. DOI: 10.1080/15427528.2021.2000908

Srivastava, R., Kobayashi, Y., Koyama, H. and Sahoo, L., 2023. Cowpea NAC1/NAC2 transcription factors improve growth and tolerance to drought and heat in transgenic cowpea through combined activation of photosynthetic and antioxidant mechanisms. *Journal of Integrative Plant Biology*, *65*(1), 25–44. DOI: 10.1111/jipb.13365

Sugri, I., Abubakari, M., Owusu, R.K. and Bidzakin, J.K., 2021. Postharvest losses and mitigating technologies: evidence from Upper East Region of Ghana. *Sustainable Futures*, *3*, 100048. DOI: 10.1016/j.sftr.2021.100048

Takaidza, S., 2023. Indigenous South African food: Nutrition and health benefits. In Soto-Hernández, M., Aguirre-Hernández, E. and Palma-Tenango, M. (eds.) *Phytochemicals in Agriculture and Food*. London: IntechOpen. DOI: 10.5772/intechopen.110732

Udo, N.N., Effiong, O.O., George, U.E. and Okon, E.J., 2021. Effects of processing on the mineral content and proximate composition of Arachis hypogeae (Groundnut) seeds. *World Journal of Advanced Research and Reviews*, *12*(1), 387–395. DOI: 10.30574/wjarr.2021.12.1.0379

Vilakazi, B., Mafongoya, P.L., Odindo, A.O. and Phophi, M.M., 2025. The role of neglected grain legumes in Food and Nutrition Security and Human Health. *Sustainability*, *17*(1), 350. DOI: 10.3390/su17010350

Wanja, N.J., 2022. *Potential of Rhizobia and Arbuscular mycorrhizal fungi to enhance nitrogen fixation and growth of Cowpea genotypes in Eastern Kenya* (Doctoral dissertation, Kenyatta University).

Yanni, A.E., Iakovidi, S., Vasilikopoulou, E. and Karathanos, V.T., 2023. Legumes: A vehicle for transition to sustainability. *Nutrients*, *16*(1), 98. DOI: 10.3390/nu16010098

Yatogo, H., 2024. *Assessment of maize-legume intercropping as a way for sustainable intensification in mixed farming systems for smallholder farmers in Jimma, Ethiopia* (Doctoral dissertation, International Maize and Wheat Improvement Center).

Yeremko, L., Czopek, K., Staniak, M., Marenych, M. and Hanhur, V., 2025. Role of environmental factors in legume-rhizobium symbiosis: A review. *Biomolecules*, *15*, 118. DOI: 10.3390/biom15010118

Zain, M., Despal, Tanuwiria, U.H., Pazla, R., Putri, E.M. and Amanah, U., 2023. Evaluation of legumes, roughages, and concentrates based on chemical composition, rumen degradable and undegradable proteins by in vitro method. *International Journal of Veterinary Science*, *12*(5), 28–38. DOI: 10.47278/journal.ijvs/2022.218

4 Chickpea
Genetic and Genomic Resources and Their Utilization

Sarvjeet Kukreja, Samiksha, and Umesh Goutam

4.1 INTRODUCTION

Chickpea (*Cicer arietinum* L.) is one of the most important pulse crops globally, playing a crucial role in food security, nutrition, and sustainable agriculture. As a rich source of protein, dietary fiber, vitamins, and minerals, chickpea significantly contributes to human health and well-being (Kumar et al., 2023; Ajay et al., 2024). The crop is primarily cultivated in arid and semi-arid regions, with major production hubs in South Asia, the Middle East, Africa, and the Mediterranean region. Despite its agronomic and economic significance, chickpea production is frequently challenged by biotic and abiotic stresses, including diseases, pests, drought, and heat stress, which often lead to substantial yield losses (Khan et al., 2025; Naveed et al., 2025).

To enhance chickpea productivity and adaptability, substantial efforts have been made in the exploration and utilization of its genetic and genomic resources. Traditional breeding programs have contributed significantly to the development of improved chickpea varieties with desirable agronomic traits, including high yield potential, stress tolerance, and disease resistance (Singh et al., 2022; Kumar et al., 2024). However, the advent of modern genomic tools and biotechnological interventions has revolutionized chickpea improvement strategies. Advances in molecular genetics, whole-genome sequencing, and functional genomics have facilitated a deeper understanding of the genetic architecture of key traits, enabling targeted breeding approaches such as marker-assisted selection (MAS) and genomic selection (GS) (Singh et al., 2023; Altaf et al., 2024).

The availability of diverse genetic resources, including landraces, wild relatives, and germplasm collections, forms the foundation for chickpea improvement. These resources harbor valuable alleles that can be harnessed for enhancing resilience to environmental challenges (Jha et al., 2024; Kumar et al., 2024). The integration of genomics-assisted breeding, gene-editing technologies, and transcriptomic analyses has further accelerated the development of superior chickpea varieties tailored to specific agro-climatic conditions (Kumar et al., 2024; Pandey et al., 2024).

This chapter provides an in-depth overview of the genetic and genomic resources available for chickpea improvement, highlighting their significance in crop enhancement programs. It also discusses the latest advancements in chickpea genomics, biotechnological tools, and breeding approaches aimed at addressing global food security challenges. The efficient utilization of these resources will be instrumental in meeting the growing demand for chickpea while ensuring sustainable agricultural practices.

4.2 OVERVIEW OF CHICKPEA AS A LEGUME CROP

Chickpeas (*Cicer arietinum* L.) are one of the most significant leguminous crops grown globally. It is a member of the *Fabaceae* family and provides an important source of plant-based protein, dietary fiber, vitamins, and minerals. Chickpea, a cool-season grain legume, thrives in semi-arid and

DOI: 10.1201/9781003477235-5

Mediterranean climates, making an important contribution to world food security and sustainable agriculture (Zhang et al., 2024; Dave et al., 2024).

4.2.1 Importance in Global Agriculture and Production

Chickpeas play an important role in worldwide agriculture due to their high nutritional content and capacity to boost soil fertility through biological nitrogen fixation. It is a staple protein for millions, mainly in South Asia, the Middle East, and portions of Africa (Kafeel et al., 2023; Zhang et al., 2024). With the growing popularity of plant-based diets, chickpeas have attracted attention as a functional food with possible health advantages such as glycemic management, cardiovascular health, and gut microbiota modification (Bose et al., 2024; Chehade et al., 2024).

4.2.2 Historical Context and Cultivation Regions

Chickpeas were one of the first domesticated crops, with archeological evidence dating back over 7000 years to the Fertile Crescent. Historically, its cultivation spread over the Mediterranean and South Asia, eventually spreading to Australia, North America, and parts of South America. There are two primary varieties of chickpeas: Desi and Kabuli chickpea. Desi chickpea is a small, dark-colored seed with a rough covering that is grown mostly in South Asia, Ethiopia, and Australia. However, Kabuli chickpea, having larger, beige-colored seeds with a smooth coat, is widely grown in the Mediterranean, North America, and West Asia (Rocha et al., 2023; Wrigley & Taylor, 2023).

Today, India is the world's greatest producer and consumer of chickpeas, accounting for more than 70% of worldwide output. Other important chickpea-growing nations include Pakistan, Turkey, Australia, Canada, and Ethiopia. Chickpea farming is expanding into non-traditional places due to the crop's flexibility, better breeding efforts, and rising market demand for plant-based protein sources (Zhang et al., 2024).

Advances in genomic research, genetic diversity exploration, and breeding tactics continue to improve chickpea yield and tolerance to environmental challenges, guaranteeing its critical position in global food security and sustainable agricultural systems.

4.3 GENETIC RESOURCES OF CHICKPEA

Chickpea (*Cicer arietinum* L.) is an important legume crop with high genetic diversity, contributing significantly to world food security and agricultural sustainability (Jha et al., 2022; Sharma et al., 2024). Its genetic resources include a wide range of genotypes, from landraces to wild cousins to recent commercial cultivars. These materials serve as the foundation for breeding initiatives that aim to improve agronomic performance, nutritional value, and resistance to environmental stress (Thakur et al., 2024; Srivastava & Manojkumar, 2024).

4.3.1 Variations in Seed Traits and Nutritional Profiles

Chickpea genotypes have a diverse variety of seed properties, including size, shape, color, and biochemical content. Variations in protein, fiber, and vitamin content between genotypes have been observed, emphasizing their importance in treating malnutrition and dietary diversity (Eker et al., 2022; Sivakumar et al., 2024). Selection of genotypes with higher nutritional profiles is a significant aim in the chickpea breeding program.

4.3.2 Landraces vs. Common Varieties

Landraces, which have evolved over centuries through natural selection and traditional farming practices, exhibit greater genetic diversity compared to commercial cultivars. These landraces are often more resilient to environmental fluctuations and possess unique agronomic traits that can be harnessed for crop improvement (Altaf et al., 2024; Chehade et al., 2024). In contrast, commercial varieties are bred for specific traits such as high yield, disease resistance, and uniformity, often at the expense of genetic diversity. The integration of landraces into breeding programs can enhance the adaptability and resilience of chickpea cultivars (Mohanty et al., 2024; Sharma et al., 2024).

4.3.3 Genetic Divergence and Clustering

The study of genetic divergence in chickpea populations is critical for understanding the evolutionary links and breeding potential of various genotypes (Shahnaz et al., 2025). Various multivariate techniques, including principal component analysis (PCA) and cluster analysis, have been used to categorize chickpea genotypes based on morphological and molecular characteristics (Jameel et al., 2024; Parveen et al., 2025). These strategies help to identify genetically unique clusters, making it easier to select various parental lines for hybridization and genetic improvement.

Chickpea genetic resources must be conserved and used in order to generate cultivars that are both hardy and nutritious (Sharma et al., 2024). Advances in genomic technology, along with conventional breeding methods, continue to unleash the promise of chickpea genetic resources for sustainable agriculture and food security.

4.4 MULTIVARIATE APPROACHES TO GENETIC ANALYSIS

The use of multivariate statistical approaches has substantially enhanced the genetic study of chickpea (*Cicer arietinum* L.). These methods provide a thorough assessment of genetic diversity, making genotype categorization easier when numerous features are considered at the same time (Mahmood et al., 2024). Multivariate approaches, which integrate numerous phenotypic and molecular markers, give greater insights into the genetic structure and divergence of chickpea populations, assisting in efficient breeding tactics (Azam et al., 2024; Bhanu et al., 2024).

4.4.1 Principal Component Analysis

Principal component analysis (PCA) is commonly employed in chickpea genetic research to minimize dimensionality while retaining the most valuable data (Asadullah et al., 2024). PCA efficiently separates genotypes based on essential agronomic and nutritional characteristics by converting correlated parameters into uncorrelated principal components (Safdar et al., 2021; Kakaei et al., 2024). Recent research has used PCA to detect patterns of variation in seed characteristics, protein content, and stress tolerance mechanisms, which helps to categorize landraces and improve cultivars into separate genetic clusters (Jameel et al., 2024).

4.4.2 Cluster Analysis

Cluster analysis is another important approach for determining genetic links among chickpea genotypes. Clustering approaches, both hierarchical and non-hierarchical, divide genotypes into groups based on phenotypic and genotypic similarities and differences (Philanim et al., 2024). The use of clustering approaches in chickpea breeding has resulted in the identification of genetically varied lines, which are critical for hybridization programs aiming at increasing yield, disease resistance, and nutritional value (Jameel et al., 2024; Tunc et al., 2025).

4.4.3 Discriminant Function Analysis

Discriminant function analysis (DFA) is employed to validate the classification of genotypes into predefined groups. This method helps distinguish between landraces and commercial varieties based on morphological, biochemical, and molecular markers (Yakubu et al., 2022). It has been demonstrated that the effectiveness of DFA in differentiating chickpea genotypes based on adaptive traits ensures accurate selection in breeding programs targeting climate resilience and enhanced productivity (Jameel et al., 2024).

4.4.4 Canonical Correspondence Analysis

Canonical correspondence analysis (CCA) combines environmental and genetic data to evaluate genotype–environment relationships (Ghafoor et al., 2024). This method is very effective in chickpea breeding for stress adaptation, since it allows researchers to identify which genetic features contribute to performance under distinct climatic circumstances. Studies have demonstrated that CCA may efficiently select genotypes with higher drought and salinity tolerance, providing useful insights for breeding programs in arid and semi-arid settings (Panigrahi et al., 2024; Jameel et al., 2024).

4.4.5 Genetic Distance and Molecular Marker Analysis

Multivariate approaches are also useful in estimating genetic distances using molecular markers like simple sequence repeats (SSRs) and single-nucleotide polymorphisms (SNPs) (Mishra et al., 2023). Researchers can use tools such as Nei's genetic distance and Mantel tests to visualize population patterns and infer evolutionary links among chickpea accessions. Some researches have emphasized the need of combining these studies with multivariate models to improve the accuracy of genetic diversity evaluations (Jameel et al. 2024; Pipan et al., 2024).

4.5 GENOMIC ADVANCES IN CHICKPEA

4.5.1 Genome-wide Association Studies

Genome-wide association studies (GWASs) have emerged as an effective technique for finding genetic variants linked to important agronomic and nutritional aspects in chickpea (Altaf et al., 2024). Researchers can use high-density single-nucleotide polymorphism (SNP) markers to pinpoint loci associated with features including seed size, protein content, drought tolerance, and disease resistance. The discovery of SNPs and candidate genes for nutritional characteristics has been a major focus of chickpea genomics (Sahito et al., 2024; Sari et al., 2024). GWASs have identified new alleles that contribute to increased micronutrient accumulation, such as iron and zinc, allowing for marker-assisted selection (MAS) of biofortified chickpea types (Ahmad et al., 2023).

4.5.2 Proteome Profiling

Advancements in proteomics have provided insights into genotype-dependent protein variations in chickpea. Proteome profiling involves the large-scale analysis of protein expression patterns, offering a comprehensive understanding of how genetic and environmental factors influence seed protein composition (Jha et al., 2022; Kumar et al., 2024). Recent studies have highlighted differentially expressed proteins (DEPs) between landraces and commercial varieties, revealing candidate proteins linked to stress resilience and nutritional quality (Bose et al., 2024). These insights contribute to targeted breeding strategies aimed at improving protein content and digestibility in chickpea.

4.5.3 Genomic Selection and Marker-assisted Breeding

The combination of genomic selection (GS) and marker-assisted breeding (MAB) has transformed chickpea improvement efforts. GS uses genome-wide markers to forecast the genetic potential of breeding lines, which improves selection efficiency for complicated characteristics like yield stability and drought tolerance (Altaf et al., 2024; Guo, 2024). MAB, on the other hand, allows for precision introgression of desired alleles that govern certain agronomic and nutritional characteristics. The use of these genetic technologies speeds the creation of high-yielding, nutritionally dense chickpea cultivars, therefore addressing global food security concerns (Kamaluddin et al., 2022).

4.5.4 Functional Genomics and Gene Editing

Functional genomics technologies, including transcriptomics and CRISPR/Cas9-mediated gene editing, have increased the potential for chickpea genetic modification. Transcriptome sequencing has allowed the discovery of genes that respond to biotic and abiotic stressors, paving the way for genetic therapies targeted at improving stress tolerance. The use of CRISPR/Cas9 technology has enabled targeted alterations in genes related to seed quality and disease resistance, providing a precise and efficient technique for crop development (Singh et al., 2024).

4.6 INSIGHTS INTO GENOTYPE-DEPENDENT PROTEIN VARIATIONS

The protein content and composition of chickpea seeds vary significantly among genotypes, making genotype-dependent protein profiling a critical component of chickpea genetic research. Recent breakthroughs in proteomic research have shed light on the molecular processes driving protein accumulation, content, and functioning in chickpea seeds (Bose et al., 2024).

4.6.1 Proteomic Approaches for Chickpea Genotype Characterization

High-throughput proteomic tools, including two-dimensional gel electrophoresis (2D-GE), mass spectrometry (MS), and liquid chromatography (LC), have made it easier to analyze genotype-dependent protein changes in chickpeas. These approaches allow for the detection and measurement of variable protein expression patterns across various chickpea genotypes, providing molecular-level knowledge of genetic impacts on protein production and accumulation (Pandya, 2023; Bose et al., 2024).

4.6.2 Genotypic Variation in Seed Storage

Chickpea's principal protein components are seed storage proteins, specifically globulins and albumins. Variations in these proteins between genotypes have been linked to changes in seed quality, digestibility, and nutritional value (Shi et al., 2024). Bose et al. (2024) discovered substantial genotype-dependent changes in the expression levels of the 7S vicilin and 11S legumin proteins, which are essential for seed development and human feeding. Furthermore, the amount of sulfur-rich amino acids, such as methionine and cysteine, differs between genotypes, altering the total protein quality of chickpea varieties (Ahmad et al., 2023; Sultan et al., 2023).

4.6.3 Stress-responsive Proteins and Adaptive Mechanisms

Drought and heat stress have varying effects on the protein profiles of chickpea seeds among genotypes (Devi et al., 2023). Proteome profiling has demonstrated that key stress-responsive proteins, such as late embryogenesis abundant (LEA) proteins, heat shock proteins (HSPs), and antioxidant enzymes, are differently expressed in tolerant and sensitive genotypes. These proteins help to preserve seed viability, increase stress tolerance, and improve yield stability in severe environmental circumstances (Yadav et al., 2022; Devi et al., 2023).

4.6.4 Implications for Nutritional Importance Ad Breeding

Understanding genotype-dependent protein differences allows for focused breeding strategies to improve chickpea nutritional value. Breeders can create new cultivars with greater functional and health-promoting qualities by finding genotypes with higher protein content and more favorable amino acid profiles. Furthermore, combining proteomic data with genomic and transcriptome resources allows for the discovery of candidate genes and molecular markers associated with desired protein characteristics, hence aiding marker-assisted selection (MAS) in chickpea breeding (Dwivedi et al., 2022; Altaf et al., 2024).

4.7 UTILIZATION IN BREEDING PROGRAMS

Chickpea (*Cicer arietinum* L.) breeding initiatives attempt to improve the crop's nutritional value, agronomic qualities, and general robustness. The use of genetic and genomic resources in breeding procedures has considerably accelerated chickpea improvement by addressing issues such as nutritional shortages, climate change, and yield instability (Yadav et al., 2023; Hassen et al., 2024).

4.7.1 Breeding for Nutritional Enhancement

Chickpea breeding aims to increase nutrition by boosting protein content and micronutrient levels like iron and zinc. Genome-wide association studies (GWASs) and quantitative trait locus (QTL) mapping have revealed critical single-nucleotide polymorphisms (SNPs) and candidate genes linked to these attributes (Hassen et al., 2024; Sari et al., 2024). Marker-assisted selection (MAS) is increasingly being used to introduce high-nutrient genotypes into commercial cultivars, resulting in improved nutritional advantages for consumers, particularly in areas with prevalent malnutrition (Jha et al., 2024).

Recent developments in biofortification have also resulted in the production of chickpea lines with enhanced amino acid profiles, which increases their nutritional value. Furthermore, research on seed coat composition is being conducted to eliminate antinutritional elements such as phytic acid and tannins, making chickpeas more digestible and good for human health (Hassen et al. 2024; Jha et al., 2024).

4.7.2 Agronomic Traits and Yield Improvement

Enhancing agronomic features such as yield, drought tolerance, and disease resistance are critical to guaranteeing chickpea output and sustainability. Breeding efforts use a variety of genetic resources, including landraces and wild cousins, to incorporate resilience qualities into cultivated cultivars (Tiwari et al., 2023; Sichkar et al., 2024).

Pod number, seed weight, and biomass buildup are among the selection parameters used to enhance yield (Richards et al., 2022; Tiwari et al., 2023). Genomic selection, along with high-throughput phenotyping, has made it easier to identify superior genotypes with higher output in a variety of climatic circumstances. Targeted breeding efforts have enhanced resistance to biotic stressors like as *Ascochyta* blight and *Fusarium* wilt by integrating resistance genes from wild relatives and using genetic markers to discover resistant lines (Sinha et al., 2023; Chauhan et al., 2024).

4.8 APPLICATIONS IN SUSTAINABLE AGRICULTURE

The role of chickpea in sustainable agricultural systems has gained prominence due to its nitrogen-fixing ability and adaptability to diverse agroecological conditions. Breeding programs are now focused on developing climate-resilient chickpea varieties that can withstand abiotic stresses such as heat and drought while maintaining nutritional integrity (Chehade et al., 2024; Kumar et al., 2025).

Additionally, integrating chickpea into crop rotation systems enhances soil fertility and reduces the need for synthetic fertilizers, promoting environmentally friendly farming practices. Researchers are also exploring genotypes with enhanced water-use efficiency and minimal input requirements, ensuring economic viability for farmers while supporting sustainable food production systems (Gupta et al., 2024; Zou et al., 2024).

4.8.1 Role in Agroecological Transition

Agroecology promotes biodiversity, ecological resilience, and resource efficiency, all of which are compatible with chickpea agriculture. Chickpea, as a legume crop, contributes to sustainable agriculture by increasing soil fertility, lowering reliance on synthetic fertilizers, and supporting diverse cropping systems (Xing & Wang, 2024).

4.8.1.1 Contributions to Sustainable Food Systems

Chickpea contributes significantly to sustainable food systems due to its adaptability to diverse agro-climatic conditions and its role in regenerative farming practices. Key contributions include:

i. **Nitrogen Fixation and Soil Health:** Chickpea establishes symbiotic relationships with rhizobia, enhancing soil nitrogen content and reducing the need for chemical fertilizers. This contributes to long-term soil fertility and lower greenhouse gas emissions.
ii. **Climate Resilience:** Chickpea is drought-tolerant and can be grown in semi-arid and arid regions, supporting food security in areas vulnerable to climate change.
iii. **Diversified Cropping Systems:** Inclusion of chickpea in crop rotations improves soil microbial diversity, disrupts pest and disease cycles, and enhances farm resilience.
iv. **Resource Efficiency:** Compared to cereal crops, chickpea requires less water and has a lower carbon footprint, making it a viable option for climate-smart agriculture.
v. **Socioeconomic Benefits:** Sustainable chickpea production supports smallholder farmers by providing a low-input, high-nutrition crop, thereby contributing to rural livelihoods and reducing food insecurity.

By integrating chickpea into sustainable agricultural models, global food systems can be enhanced to meet environmental and nutritional demands while addressing the challenges posed by climate change.

4.8.2 Nutritional and Antinutritional Factors

Chickpeas are recognized for their high protein content, essential amino acids, dietary fiber, and micronutrients, making them an important part of plant-based diets. However, antinutritional substances such as phytates, tannins, and enzyme inhibitors can reduce nutrient absorption. Understanding these characteristics is critical for producing chickpea-based food items with improved health results.

4.8.2.1 Implications for Health-promoting Food Products

The growing interest in functional foods has highlighted chickpea's potential in health-oriented product development. Key aspects include:

i. **High-Quality Plant Protein:** Chickpea serves as an excellent protein source, especially in vegetarian and vegan diets, aiding in muscle development and overall health.
ii. **Gut Health and Digestive Benefits:** The high fiber content in chickpea promotes gut microbiota diversity, enhances digestion, and helps regulate blood sugar levels, making it beneficial for individuals with diabetes.

iii. **Reduction of Antinutritional Factors:** Food processing techniques such as soaking, sprouting, fermentation, and heat treatments can reduce phytates and tannins, improving mineral bioavailability and digestibility.
iv. **Potential in Functional Foods:** Chickpea-derived ingredients, such as flour, protein isolates, and fermented products, are increasingly used in gluten-free and protein-enriched foods. These innovations contribute to the development of health-promoting food options, particularly in addressing malnutrition and non-communicable diseases.

The application of chickpea in sustainable agriculture and health-focused food innovations underscores its importance in future food systems. Further research and breeding efforts aimed at improving both nutritional profiles and agronomic performance will enhance its contribution to global food security and sustainability.

4.9 CHALLENGES AND FUTURE DIRECTIONS

Despite its many benefits, chickpea cultivation confronts a number of obstacles, including vulnerability to abiotic and biotic stressors, minimal genetic diversity in commercial cultivars, and limited genomic resources for breeding (Dwivedi et al., 2023). Climate change-related pressures, such as droughts and heat waves, endanger chickpea yields and worldwide production stability (Gupta et al., 2024). Furthermore, while genomic discoveries have hastened breeding efforts, the actual application of these improvements in farmers' fields is hampered by financial and technological restrictions.

4.9.1 Addressing Climate Change and Food Security

Climate change poses a significant danger to global food security, particularly in areas where chickpeas are a key crop. Rising temperatures, irregular rainfall patterns, and a rise in pests and illnesses have created the need for climate-smart chickpea cultivars that can tolerate these challenges while retaining high nutritional value and yield (Mahto et al., 2022; Dave et al., 2024).

4.9.1.1 Genomics-assisted Breeding for Resilience

Genomics-assisted breeding has emerged as an effective approach for creating climate-resistant chickpea cultivars. Advances in genome-wide association studies (GWASs) and high-throughput sequencing technology have enabled the discovery of important genetic loci linked to drought tolerance, heat resistance, and disease resistance. Researchers can speed up the production of superior chickpea varieties with greater tolerance to climatic stresses by adding these genetic markers into breeding procedures (Ma et al., 2024; Zhang et al., 2024).

Furthermore, transcriptomic and proteomic investigations have shed light on the molecular pathways underpinning stress tolerance, assisting in the identification of prospective genes for targeted genetic alteration (Satrio et al., 2024). The combination of marker-assisted selection (MAS) and genomic selection (GS) improves breeding efficiency by allowing for the early selection of superior genotypes, shortened breeding cycles, and faster variety release (Ceran et al., 2024; Tyagi et al., 2024).

4.9.2 Potential for Innovation in Chickpea Cultivation

Chickpea production can benefit from innovative methods that improve productivity, nutritional quality, and flexibility to changing climates. Precision agriculture, digital phenotyping, and AI-based prediction models are emerging as revolutionary technologies for optimizing chickpea farming techniques and increasing yields (Gupta et al., 2024; Kaya, 2025).

4.9.2.1 Opportunities for New Breeding Programs

Novel breeding methods adapted to regional conditions can improve chickpea output and sustainability. Targeted breeding techniques that focus on early-maturing, high-yielding, and climate-resilient genotypes can help to alleviate production restrictions in a variety of agricultural zones. The study of wild chickpea cousins and landraces as genetic reservoirs provides an important opportunity to widen the genetic basis of farmed chickpea (Altaf et al., 2024; Hassen et al., 2024).

Furthermore, using biofortification techniques in breeding programs can increase the nutritional value of chickpea by increasing its protein, iron, and zinc contents (Jha et al., 2024). This can make a substantial contribution to global efforts to address hunger and increase food security, especially in underdeveloped countries (Kundu & Gupta, 2024).

Overall, continuous investment in creative breeding efforts, together with advances in genomic research, can unleash the full potential of chickpea as a resilient, nutritious, and sustainable crop for future agricultural systems.

4.10 CONCLUSION

Chickpeas remain an important crop in worldwide agriculture due to their nutritional value, flexibility, and contribution to sustainable agricultural methods. The high genetic variety of chickpea genotypes presents substantial opportunities for crop enhancement through focused breeding efforts. Advances in genomic research, such as GWAS and proteomics, have enabled the discovery of critical genetic markers for improving nutritional and agronomic features. The use of these genetic technologies in breeding tactics has resulted in the creation of high-yielding, nutrient-dense, and climate-resistant chickpea cultivars. Despite these advances, concerns such as climate change, disease susceptibility, and the need for sustainable production techniques remain. Future research should concentrate on leveraging genetic resources, adopting precision breeding, and investigating novel farming techniques to assure food security and environmental sustainability. The ongoing study of chickpea genetics and genomics will be critical in satisfying the growing worldwide need for healthy and robust crops.

REFERENCES

Ahmad HM, Zahra S, Oranab S, Arif S, Zakia S, Raina A, Khan MZ, Shimira F, Zahid G, Bano S. Contribution of conventional breeding approaches in legumes biofortification. In Fatima H, Kanwal A, Afzal S, Iqbal N (eds.) *Legumes biofortification* 2023 Nov 9 (pp. 111–129). Cham: Springer International Publishing.

Ajay A, Gaur SS, Shams R, Dash KK, Mukarram SA, Kovács B. Chickpeas and gut microbiome: functional food implications for health. *Heliyon*. 2024 Oct 30;10(20).

Altaf MT, Liaqat W, Ali A, Jamil A, Fahad M, Rahman MA, Baloch FS, Mohamed HI. Advancing chickpea breeding: omics insights for targeted abiotic stress mitigation and genetic enhancement. *Biochemical Genetics*. 2024 Nov 12;3:1–53.

Asadullah HM, Malik SR, Khalid R, Ajmal SA, Faisal S, Abdullah M, Ghuffar S. Genetic analysis and breeding strategies for Mungbean (Vigna radiata L.): insights from correlation, principal component, and cluster analyses. *Planta Animalia*. 2024 Oct 30;3(1):57–64.

Azam MG, Sarker U, Hossain MA, Mahabubul Alam AK, Islam MR, Hossain N, Alamri S. Phenotypic diversity in qualitative and quantitative traits for selection of high yield potential field pea genotypes. *Scientific Reports*. 2024 Aug 9;14(1):18561.

Bhanu AN, Sharma P, Singh D, Singh S. Multivariate analysis reveals genetic diversity in chickpea (Cicer arietinum L.). *Journal of Genetics*. 2024;103(2):e120. https://doi.org/10.1007/s12041-024-01420-5

Bose U, Buck S, Sirault X, Bahmani M, Byrne K, Stockwell S, McWilliam S, Colgrave M, Juhász A, Ral JP. Chickpea proteome analysis reveals genotype-dependent variations associated with seed traits. *Journal of Agricultural and Food Chemistry*. 2024 Nov 21;72(48):27030–42.

Ćeran M, Miladinović D, Đorđević V, Trkulja D, Radanović A, Glogovac S, Kondić-Špika A. Genomics-assisted speed breeding for crop improvement: present and future. *Frontiers in Sustainable Food Systems.* 2024 Apr 8;8:1383302. https://doi.org/10.3389/fsufs.2024.1383302

Chauhan N, Makwana T, Patel D, Radadiya C. *Recent developments in breeding for biotic resistance. Amalgamation of recent efforts in plant breeding and biotechnology.* 1st ed. Maharashtra, India: Bhumi; 2024 Oct: 13–39.

Chehade A, El-Sayed, A, Bitar, A, Hariri, A, Zein, E. Chickpea-based diets and cardiometabolic health: *A Systematic Review. Nutrients.* 2024;16(7):1432. https://doi.org/10.3390/nu16071432

Dave K, Kumar A, Dave N, Jain M, Dhanda PS, Yadav A, Kaushik P. Climate change impacts on legume physiology and ecosystem dynamics: A multifaceted perspective. *Sustainability.* 2024 Jul 15;16(14):6026.

Devi P, Awasthi R, Jha U, Sharma KD, Prasad PV, Siddique KH, Roorkiwal M, Nayyar H. Understanding the effect of heat stress during seed filling on nutritional composition and seed yield in chickpea (Cicer arietinum L.). *Scientific Reports.* 2023 Sep 18;13(1):15450.

Dwivedi SL, Mattoo AK, Garg M, Dutt S, Singh B, Ortiz R. Developing germplasm and promoting consumption of anthocyanin-rich grains for health benefits. *Frontiers in Sustainable Food Systems.* 2022 Apr 25;6:867897.

Eker T, Sari D, Sari H, Tosun HS, Toker C. A kabuli chickpea ideotype. *Scientific Reports.* 2022 Jan 31;12(1):1611.

Ghafoor AZ, Derejko A, Studnicki M. Identification of plant and soil characteristics affecting stability of winter wheat cultivar in temperate climates. *Agronomy.* 2024 Apr 10;14(4):779.

Guo T. Genetic improvement of chickpeas: traits, targets, and technology. *Legume Genomics and Genetics.* 2024 Oct 15;15.

Gupta DK, Pagani A, Zamboni P, Singh AK. AI-powered revolution in plant sciences: advancements, applications, and challenges for sustainable agriculture and food security. *Exploration of Foods and Foodomics.* 2024 Aug 6;2(5):443–59.

Gupta S, Ranjan R, Bijlwan A, Jha A, Jena RK. Enhancing water use efficiency through trait-based approaches. In Thind HS, Bala A, Singh R, Gupta SK (eds.) *Plant functional traits for improving productivity* 2024 Apr 23 (pp. 133–147). Singapore: Springer Nature Singapore.

Hassen A, Mekonnen F, Zander KK, Seid N, Mekonen A, Ibrahim S, Abebe S. Recent advances in molecular breeding and key insights into major pulse improvement efforts for sustainable food security. *Gene Reports.* 2024 Sep 1;36:101997. https://doi.org/10.1016/j.genrep.2024.101997

Jameel S, Hameed A, Shah TM, Coyne CJ. Demystifying the nutritional and anti-nutritional genetic divergence of Pakistani chickpea (Cicer arietinum L.) genetic resource via multivariate approaches. *Frontiers in Nutrition.* 2024 Oct 4;11:1407096.

Jha UC, Nayyar H, Parida SK, Deshmukh R, von Wettberg EJ, Siddique KH. Ensuring global food security by improving protein content in major grain legumes using breeding and 'Omics' tools. *International Journal of Molecular Sciences.* 2022 Jul 12;23(14):7710.

Jha UC, Nayyar H, Sharma KD, Von Wettberg EJ, Siddique KH, editors. *Legume crop wild relatives: Their role in improving climate resilient legumes.* Boca Raton, Florida: CRC Press; 2024 Nov 4.

Kafeel U, Jahan U, Khan FA. Role of mineral nutrients in biological nitrogen fixation. *InSustainable Plant Nutrition* 2023 Jan 1 (pp. 87–106). Academic Press.

Kakaei M, Mirmazloum SI. Evaluation of the genetic diversity and agro-morphophysiological traits of bread wheat varieties under postflowering drought stress using different statistical methods. *Journal of Genetic Resources.* 2024 Feb 1;10(1):46–56.

Kamaluddin, Sonkar P, Sharma V, Kumar H, Kumar M, Negi HS, Kiran U, Abdin MZ, Choubey AK. Marker-assisted selection for value addition in crop plants. *InTechnologies in Plant Biotechnology and breeding of field crops* 2022 Oct 6 (pp. 23–39). Singapore: Springer Nature Singapore.

Kaya C. Optimizing crop production with plant phenomics through high-throughput phenotyping and AI in controlled environments. *Food and Energy Security.* 2025 Jan;14(1):e70050.

Khan SA, Zubair H, Rathnayake I, Abeysekara I, Korai PK, Mumtaz H, Kamran M. Pulses based cropping system under changing climate in Pakistan. *Dialogue Social Science Review* (DSSR). 2025 Feb 7;3(2):342–77.

Kumar N, Kancheti M, Hazra KK, Nath CP, Hashim M, Deo MM. Enhancing nutrient use efficiency in pulses under diversified Farming through Agronomic Manipulations. In Babu S, Singh R, Rathore SS, Singh

VK (eds.) *Agricultural diversification for sustainable food production* 2025 Jan 24 (pp. 197–213). Singapore: Springer Nature Singapore.

Kumar S, Gopinath KA, Sheoran S, Meena RS, Srinivasarao C, Bedwal S, Jangir CK, Mrunalini K, Jat R, Praharaj CS. Pulse-based cropping systems for soil health restoration, resources conservation, and nutritional and environmental security in rainfed agroecosystems. *Frontiers in Microbiology*. 2023 Feb 3;13:1041124.

Kumar T, Tiwari N, Saxena DR, Istanbuli T, Hamwieh A. Genetic and genomic resources for improving biotic stress tolerance in Chickpea. In *Genomics-aided breeding strategies for biotic stress in grain legumes* 2024 Aug 14 (pp. 13–48). Singapore: Springer Nature Singapore.

Kumar T, Tiwari N, Saxena DR, Istanbuli T, Hamwieh A. Genetic and genomic resources for improving biotic stress tolerance in chickpea. In *Genomics-aided breeding strategies for biotic stress in grain legumes* 2024 Aug 14 (pp. 13–48). Singapore: Springer Nature Singapore.

Kundu D, Gupta D. Building a fairer future: joint actions for poverty, hunger and inequality reduction by G20 nations. *Revista Tempo do Mundo*. 2024 Apr 30(34):95–118.

Ma Y, Li Y. The research progress on genomic selection and breeding for drought resistance in wheat: From genetic analysis to smart breeding applications. *Geographical Research Bulletin*. 2024 Dec 19;3(4):625–46.

Mahmood MT, Hussain K, Amin A, Batool A, Aziz A, Kaukab S, Hussain A, Rasool I, Kashif M, Waseem M, Zubair HK. Genetic variability assessment among elite chickpea (Cicer arientinum) genotypes under water deficit conditions. *Pak-Euro Journal of Medical and Life Sciences*. 2024 Mar 31;7(1):31–8.

Mahto RK, Ambika, Singh C, Chandana BS, Singh RK, Verma S, Gahlaut V, Manohar M, Yadav N, Kumar R. Chickpea biofortification for cytokinin dehydrogenase via genome editing to enhance abiotic-biotic stress tolerance and food security. *Frontiers in Genetics*. 2022 May 20;13:900324.

Mishra G, Meena RK, Kant R, Pandey S, Ginwal HS, Bhandari MS. Genome-wide characterization leading to simple sequence repeat (SSR) markers development in Shorea robusta. *Functional & Integrative Genomics*. 2023 Mar;23(1):51.

Mohanty TA, Mandal P, Kumaresan D, Swaminathan M. Fast-forward breeding in rice. In Singh A, Singh SK, Shrestha J (eds.) *Climate-smart rice breeding* 2024 Nov 16 (pp. 301–322). Singapore: Springer Nature Singapore.

Naveed M, Aslam M, Ahmed SR, Tan DK, De Mastro F, Tariq MS, Sakhawat A, Asad MA, Liu Y. An overview of heat stress in Chickpea (Cicer arietinum L.): effects, mechanisms and diverse molecular breeding approaches for enhancing resilience and productivity. *Molecular Breeding*. 2025 Feb;45(2):18.

Pandey S, Anand A, Rathore A, Taddi S, Kole C. Prospects and potentials of allele mining in chickpea for qualitative and quantitative traits. In Kole C, Bharadwaj C, Sarkar A (eds.) *Allele mining for genomic designing of grain legume crops* 2024 (pp. 50–70). Boca Raton: CRC Press.

Pandya AK. Genotype-dependent phenotypic variation for agronomic traits, seed composition, and root architecture of Chilean Quinoa (Chenopodium quinoa Willd) . (Doctoral dissertation, Universitäts-und Landesbibliothek Bonn).

Panigrahi S, Kumar U, Swami S, Singh Y, Balyan P, Singh KP, Dhankher OP, Varshney RK, Roorkiwal M, Amiri KM, Mir RR. Meta QTL analysis for dissecting abiotic stress tolerance in chickpea. *BMC Genomics*. 2024 May 2;25(1):439.

Parveen N, Umer S, Tan C, Jabbar A, Kanwal B, Haider I, Raza W, Usma A, Mehmood A, Junaid MB, Abbasi SH. Multivariate and association analyses of various seed yield contributing traits divulge genetic diversity among Pisum sativum L. *genotypes. Plant Molecular Biology Reporter*. 2025 Jan 21;43:1–9.

Philanim WS, Bharadwaj C, Kumar N, Ngangkham U, Iangrai B, Singh BK. Genetic diversity, population structure and marker-trait association analysis for yield-related traits in chickpea (Cicer arietinum L.). *Genetic Resources and Crop Evolution*. 2024 Sep 29;71:1–9.

Pipan B, Neji M, Meglič V, Sinkovič L. Genetic diversity of kale (Brassica oleracea L. var acephala) using agro-morphological and simple sequence repeat (SSR) markers. *Genetic Resources and Crop Evolution*. 2024 Mar;71(3):1221–39.

Richards MF, Maphosa L, Preston AL. Impact of sowing time on chickpea (Cicer arietinum L.) biomass accumulation and yield. *Agronomy*. 2022 Jan 10;12(1):160.

Rocha FS, Sharma M, Tarafdar A, Chen W, Azevedo DM, Castillo P, Costa CA, Chobe DR. Diseases of chickpea. In Elmer WE, McGrath M, McGovern RJ (eds.) *Handbook of vegetable and herb diseases* 2023 Dec 4 (pp. 1–44). Cham: Springer International Publishing.

Safdar LB, Umer MJ, Almas F, Uddin S, Blighe K, Quraishi UM. Identification of genetic factors controlling phosphorus utilization efficiency in wheat by genome-wide association study with principal component analysis. *Gene*. 2021 Feb 5;768:145301.

Sahito JH, Zhang H, Gishkori ZG, Ma C, Wang Z, Ding D, Zhang X, Tang J. Advancements and prospects of genome-wide association studies (GWAS) in maize. *International Journal of Molecular Sciences*. 2024 Feb 5;25(3):1918.

Sari H, Uhdre R, Wallace L, Coyne CJ, Bourland B, Zhang Z, Russo M, Kiszonas A, Warburton ML. Genome-wide association study in Chickpea (Cicer arietinum L.) for yield and nutritional components. *Euphytica*. 2024 Jun;220(6):84.

Satrio RD, Fendiyanto MH, Miftahudin M. Tools and techniques used at global scale through genomics, transcriptomics, proteomics, and metabolomics to investigate plant stress responses at the molecular level. In Shahid M, Gaur R (eds.) *Molecular dynamics of plant stress and its management* 2024 Jun 26 (pp. 555–607). Singapore: Springer Nature Singapore. https://doi.org/10.1007/978-981-99-2097-1_19

Shahnaz Z, Shahnaz A, Ismail A, Manzoor Z, Iqbal R, Zia MA. Chickpea (Cicer arietinum) genetic diversity: genetic resources, breeding progress, and future prospects. *Crop and Pasture Science*. 2025 Jan 17;76(1).

Sharma P, Tailor A, Joshi A, Bhoi TK. Genetic diversity of grain legumes for food and nutritional security. In Al Khayri AM, Jain SM, Penna S (eds.) *Sustainable utilization and conservation of plant genetic diversity* 2024 Mar 29 (pp. 63–105). Singapore: Springer Nature Singapore.

Sharma V, Thakur M, Maan SS, Verma K, Thakur A, Penna S. In vitro mutagenesis: a non-invasive technology for effective crop improvement to assure food and nutritional security—current trends, advancements and future perspectives. *Journal of Plant Growth Regulation*. 2024 Sep 16; 44(2):1–24.

Shi Y, Li H, Zhou J, Wang R, Khan A, Zhang L. Proteomic insights into seed storage protein variation in chickpea (Cicer arietinum L.). *Food and Bioprocess Technology*. 2024; 17(10):3253–3266. https://doi.org/10.1007/s11947-024-03190-y

Singh C, Kumar R, Sehgal H, Bhati S, Singhal T, Gayacharan, Nimmy MS, Yadav R, Gupta SK, Abdallah NA, Hamwieh A. Unclasping potentials of genomics and gene editing in chickpea to fight climate change and global hunger threat. *Frontiers in Genetics*. 2023 Apr 18;14:1085024.

Singh RK, Singh C, Ambika, Chandana BS, Mahto RK, Patial R, Gupta A, Gahlaut V, Gayacharan, Hamwieh A, Upadhyaya HD. Exploring chickpea germplasm diversity for broadening the genetic base utilizing genomic resourses. *Frontiers in Genetics*. 2022 Aug 4;13:905771.

Singh S, Mishra V, Riyazuddin R, Chugh V, Upadhyay SK. CRISPR/Cas9-mediated genome editing for trait improvement and stress tolerance in leguminosae (Legume Family). *Plant Breeding*. 2024;3:149–163.

Sinha D, Maurya AK, Abdi G, Majeed M, Agarwal R, Mukherjee R, Ganguly S, Aziz R, Bhatia M, Majgaonkar A, Seal S. Integrated genomic selection for accelerating breeding programs of climate-smart cereals. *Genes*. 2023 Jul 21;14(7):1484.

Sivakumar KB, Gautam A, Singh S, Panwar RK, Arora A, Verma SK. A pragmatic study on seed shape classification and its association among seed quality attributes in chickpea (Cicer arietinum L.). *Legume Research*. 2024;47(4):597–602.

Srivastava M, Manojkumar HG. Genetic diversity and crop improvement. *Genetics and Plant Breeding*. 70:131.

Sultan A, Al Tawaha AR, Altawaha AR. Enriching legume protein contents. In Nadeem MA, Baloch FS, Fiaz S, Aasim M, Habyarimana E, Sönmez O, Zencirci N (eds.) *Legumes biofortification* 2023 Nov 9 (pp. 449–463). Cham: Springer International Publishing.

Thakur NR, Ingle KP, Sargar PR, Baraskar SS, Kasanaboina K, Awio B, Pranati J, Abdi G. Sustainable utilization of wild germplasm Resources. In Kumar S, Singh VP, Kumar D (eds.) *Sustainable utilization and conservation of plant genetic diversity* 2024 Mar 29 (pp. 551–590). Singapore: Springer Nature Singapore.

Tiwari PN, Tiwari S, Sapre S, Tripathi N, Payasi DK, Singh M, Thakur S, Sharma M, Tiwari S, Tripathi MK. Prioritization of physio-biochemical selection indices and yield-attributing traits toward the acquisition of drought tolerance in chickpea (Cicer arietinum L.). *Plants*. 2023 Sep 5;12(18):3175.

Tunc M, Rufaioglu SB, Ipekesen S, Yakar M, Yorulmaz L, Bicer BT. Evaluation of the possibility of using fuzzy C-means clustering, AMMI analysis and GGE biplot methods to predict the yield of chickpea genotypes cultivated in different environments. *Agronomy*. 2025 Jan 25;15(2):300.

Tyagi A, Mir ZA, Almalki MA, Deshmukh R, Ali S. Genomics-assisted breeding: a powerful breeding approach for improving plant growth and stress resilience. *Agronomy*. 2024 May 25;14(6):1128.

Wrigley CW, Taylor JR. Origin, evolution, production, and utilization of cereals. In Köksel H, Álvarez I (eds.) *ICC Handbook of 21st Century Cereal Science and Technology* 2023 Jan 1 (pp. 1–15). London, UK: Academic Press.

Xing Y, Wang X. Precision agriculture and water conservation strategies for sustainable crop production in arid regions. *Plants*. 2024 Nov 13;13(22):3184.

Yadav RK, Tripathi MK, Tiwari S, Tripathi N, Asati R, Patel V, Sikarwar RS, Payasi DK. Breeding and genomic approaches towards development of fusarium wilt resistance in chickpea. *Life*. 2023 Apr 11;13(4):988.

Yakubu HG, Kovacs Z, Toth T, Bazar G. Trends in artificial aroma sensing by means of electronic nose technologies to advance dairy production–a review. *Critical reviews in food science and nutrition*. 2022 Dec 8;63(2):234–48.

Ye J, Shi N, Rozi P, Kong L, Zhou J, Yang H. A comparative study of the structural and functional properties of chickpea albumin and globulin protein fractions. *Food and Bioprocess Technology*. 2024 Oct;17(10):3253–66.

Zhang J, Wang J, Zhu C, Singh RP, Chen W. Chickpea: its origin, distribution, nutrition, benefits, breeding, and symbiotic relationship with Mesorhizobium species. *Plants*. 2024 Feb 1;13(3):429.

Zou Y, Liu Z, Chen Y, Wang Y, Feng S. Crop rotation and diversification in China: Enhancing sustainable agriculture and resilience. *Agriculture*. 2024 Aug 28;14(9):1465.

5 Heightening Legume Production in the Wake of Changing Climate and Increasing Environmental Stressors

Julian Wanja Nyaga, Morris Muthini, Priscillah Wakarera, Charles Kibet Kirui, Kipkorir Koech, Grace Ng'endo Kanyita, and Ezekiel Mugendi Njeru

5.1 INTRODUCTION

Legumes are the largest source of plant-based proteins in human diets (Yanni et al., 2024). With the global food demands likely to double by 2050, there is need to enhance the agricultural productivity of legumes to meet demand (Falcon et al., 2022). This is a challenging issue owing to global climate change, and biotic and abiotic stresses. Biotic and abiotic stresses are the leading causes of declining legume production, as they influence legume crop distribution, adaptation, and productivity in the different environments. Depending on the severity, biotic and abiotic stresses cause 30–100% losses in legume yields (Kide Mengistu Habtamu, 2022). These stresses exacerbate legume production on the global scale, causing production irregularities and declines.

Microbial mitigation strategies such as biocontrols and biofertilizers provide eco-friendly and sustainable practices to ameliorate losses and enhance legume tolerance to biotic and abiotic stresses (Anuar et al., 2023). In addition, establishing the genetic diversity of the cultivated legume crops to broaden their narrow genetic base is key to the breeding of high-yielding varieties adapted to the prevailing conditions (Nair et al., 2019). Specifically, the development of short-duration, disease-resistant, pest-tolerant, and drought-tolerant varieties is key to enhancing legume productivity. Together, microbial mitigation and genetic modifications provide the most viable and sustainable strategies to mitigate declining legume production.

In this context, we discuss the biotic and abiotic stresses affecting legume production and the strategies to counteract them. In addition, the molecular perspectives in strategy development and breeding legume crops well adapted to various stress conditions are discussed in this chapter.

5.2 BIOTIC STRESSES

Biotic stress results from damage caused by living macro- and microorganisms such as fungi, bacteria, viruses, nematodes, insects, herbivores, and weeds. These agents take in nutrients from the plants directly or indirectly, depriving them of essential elements. This ultimately cause damages or even death of the plant, leading to significant yield losses (Haldar et al., 2022). Biotic stress damage is seen through fruit and seed rot, loss of foliage, wilting, and leaf spots (Maiti et al., 2021).

DOI: 10.1201/9781003477235-6

5.2.1 Fungal Pathogens

Aerial fungal pathogens are some of the major causes of plant diseases leading to great losses in legume production globally. These diseases include rusts, downy mildew, and powdery mildew, which significantly limit legume production (Mahmoud, 2021). Rust species belonging to the genus *Uromyces* have been reported to affect common beans, chickpeas, faba beans, and lentils by extracting nutrients from the host (Osuna-Caballero et al., 2024). Powdery mildew and rust infections cause the formation of a nutrient sink as the fungal pathogens produce cytokinins, causing the diversion of nutrients from other parts of the plant to the infected site (Julian et al., 2023).

In addition, rust species of other genera cause legume yield losses; for example, Asian rust (*Phakopsora pachyrhizi*) results in soybean yield losses as it affects the leaves, forming lesions that reduce the chlorophyll area and thus affecting the general growth of the plant (Meira et al., 2020) . Powdery mildew in legumes is caused by *Erysiphe pisi* (Sharma et al., 2019). Severe infection can result in yield losses of up to 70% in field pea (Devi et al., 2022). Thus, fungal pathogens result in huge loses of legume crops in the field and subsequently threatening food sources for smallholder farmers.

In addition, necrotrophic fungal infections including *Ascochyta* blight, chocolate spot, and anthracnose also cause a significant reduction in growth and production in legumes (Mahadevakumar & Sridhar, 2021). *Ascochyta* blight (*Ascochyta rabei*) affects chickpea and leads to over 25% losses, as noted in the semi-arid regions of Australia (Reen et al., 2019). Under wet and cool weather, *Ascochyta fabae* can cause up to 90% yield losses in faba beans where it affects the leaves, stems, pods, and seeds. In Kenya, anthracnose (*Colletotrichum lindemuthianum*) is responsible for up to 90–100% losses in yield in common beans (Kiptoo et al., 2020). Thus, necrotrophic fungal infections are a threat to chickpeas, faba beans, and common beans.

Soil harbors many organisms that form a complex environment for plant growth. According to Naseri et al. (2021), soil-borne infections that affect legumes at the seedling stage include damping-off. *Rhizoctonia solani* is an example of a damping-off disease that can lead to massive losses in legumes. Moparthi et al. (2021) also reported that *Fusarium* root rot infects legumes at a young age and can lead to severe losses. *Fusarium* wilt (*Fusarium oxysporum*) has led to crop losses in sub-Saharan Africa, especially in chickpeas and lentils (Bedine Boat et al., 2020). These soil-borne infections result in losses as they hasten stem and root rots, which eventually impact the general health and cause the death of the plant leading to losses.

5.2.2 Bacterial Pathogens

Bacterial pathogens form part of the main biotic factors hindering the growth and production of legumes (Nair et al., 2019). Economically significant bacterial diseases affecting legumes include bacterial wilt, common blight, and halo blight (Mawar et al., 2021). In the field, infected seeds and previous plant debris are the primary sources of the disease inoculum. Humid and warm temperatures favor the growth and transmission of bacterial pathogens through windblown rains and irrigation water (Conner et al., 2021).

For example, halo blight, caused by *Pseudomonas syringae* pv. phaseolicola is favored by the same conditions as the common blight and can cause losses under severe conditions (Cooper et al., 2020). Bacterial brown spot is caused by *Pseudomonas syringae* pv. syringae (Besfat et al., 2020). Many smallholder farmers lack specialized tools and crop rotation programs; thereby, the disease can spread to the next season as it overwinters in the infected crop remains and weeds (Besfat et al., 2020). Thus, bacterial pathogens cause significant losses as they affect the leaves, reduce chlorophyll area, and affect the plant's general health.

5.2.3 Viral Pathogens

Viral pathogens cause significant legume losses at different stages of growth. Some common viral diseases include necrotic yellow and common mosaic viruses (Maiti et al., 2021). López et al. (2017) reported that viral pathogens, especially mosaic virus, are the main constraints to legume production in the tropics. They further noted that most cultivated food legumes are susceptible to over 100 viruses of different genera.

Viral infections are characterized by yellowing and deformation of leaves, stunted growth, and reduced yield. In severe outbreaks, viral diseases in legumes can result in between 50–100% losses depending on the cultivar, field conditions, location, and duration of infection (Rao & Reddy, 2020). Most viral pathogens, especially mosaic viruses, are transmitted by sap-sucking pests such as aphids (Mahatma & Shukla, 2024). The availability of data, showing loses of up to 100%, illustrates that viral pathogens are a threat to legume growth and production.

5.2.4 Insects

Insect infestation causes direct damage by destroying plant tissues or indirect damage through the spread of diseases. Common widespread insects affecting legumes include aphids, pod bugs, stem flies, bean mites, whitefly, bruchid, thrips, and nematodes (Moorthy et al., 2022). Researchers have noted that stem maggots, *Ophiomyia* spp., are the main field pests that hinder growth and subsequently yield in legumes (Moorthy et al., 2022). Their larvae are the most destructive, especially to the seedlings, as they feed on the stem leading to yellowing and cracking. Affected plants turn yellowish, wilt, and die due to low water uptake (Reena & Prasad, 2022). Thus, these larvae hinder the growth and production of legumes as seedlings are destroyed before they mature.

In addition, root rot nematodes (*Meloigedogyne* spp.) and cysts (*Globodera* and *Heterodora* spp.), which are sedentary endoparasites, affect the roots of many leguminous plants throughout the world (Otieno, 2023). In cowpea, parasitic nematodes are responsible for up to 14% yield loss globally (Ghosh et al., 2024). Root-knot nematodes feed on giant multinucleated cells on root tissues which they induce by modifying the gene expression and metabolism of the host causing root galls (Costa et al., 2021).

Cysts on the other hand develop syncytium which is a multinucleate feeding cell (Gupta et al., 2023). These nematodes lay eggs that can persist in soil for many years as a survival strategy. Other nematodes which have been reported to affect legumes include the reniform nematodes (*Rotylenchulus*, *Naccobus*, and *Xiphinema* spp.) reported in soybean (Khan, 2023) and root-lesion nematode (*Pratylenchus* spp.) which has been reported in chickpea (Ayaz et al., 2021). Given that nematodes infect and destroy the root system, they pose a serious threat to the growth and production of legumes.

The magnitude of reduced yields following biotic stresses is dependent on the severity of the stress, stage of plant growth at the time of infection, and prevailing environmental conditions (Nagaraja et al., 2021). One effect of biotic stress is reduced photosynthesis in infected plants. Viral, bacterial, and fungal diseases cause lesions and damage to leaves. At the same time, insects chew up the leaves, thereby reducing their surface area and ultimately reducing productivity.

Various mitigation measures are currently in place to guarantee the yield of legumes through the control of diseases and pests. The most common method is through the use of chemicals such as fungicides, pesticides, and herbicides (Ahmad et al., 2022). However, using these chemicals is associated with negative impacts on the environment, including harm to non-targeted organisms and pollution (Richard et al., 2022). Thus, to ensure better legume growth and production, there is need to implement the holistic control of biotic plant stresses.

5.3 ABIOTIC STRESSES

5.3.1 Drought Stress

Drought stress refers to a water-deficit condition where rainfall or irrigation water is insufficient to meet the requirements of a plant. Sofi et al. (2021) posit that most field crops suffer from drought stress. In semi-arid tropics, drought is cited as one of the significant challenges faced by legume farmers due to erratic rains (Ben Gaied et al., 2024). Drought conditions lead to a reduced quantity of water in the soil. Thus, plants lose a lot of water through evapotranspiration. Legumes, just like other field crops, require water for survival, mainly for the transport of nutrients up from the roots to the rest of the plant parts (Yadav et al., 2020). Thus, inadequate availability of water results in reduced vitality of plants.

Drought conditions can occur in different ways; they can either be intermittent, arising as episodes throughout the growing period, or terminal, where the drought progresses to a severe level due to lack of adequate soil moisture. In legumes, the effects of drought stress vary according to the intensity and duration of drought and the plant development stage (Nkomo et al., 2021).

At the vegetative stage of cowpea growth, drought stress results in reduced leaf growth owing to low nutrient absorption, stomatal resistance, and poor gaseous exchange (Boukar et al., 2019). This is followed by decreased turgor in plant cells and subsequent wilting of plants (Nadeem et al., 2019). In soybeans, water deprivation after the anthesis growth stage results in the loss of pods (Moura et al., 2023). On the same note, Bhattacharya (2021) noted that drought stress could lower the number of pods, seeds, dry seed weight, total biomass, and yield per hectare.

Drought stress in legumes induces changes in the root phenome, including reduced root lengths and biomass (Khatun et al., 2021). In soybeans, insufficient root mass and changes in root architecture lead to poor development of lateral root systems and xylem, which adversely reduces the surface area of roots and the amount of water taken up (Xiong et al., 2021). Furthermore, symbiotic nitrogen fixation is rapidly reduced by drought stress due to the heightened accumulation of ureides in nodules (Thu et al., 2020). Several reports indicate that nitrogen fixation is reduced due to low demand for N and reduced translocation due to low transpiration rates (Dubey et al., 2021; Islam et al., 2021). Oxidative damage is induced by drought conditions resulting in nodule damages (Chammakhi et al., 2022).

5.3.2 Salinity Stress

Salinity occurs due to increased salt content in water and soil environments. It hinders germination, plant vigor, and total legume yields in arid and semi-arid regions (Kirova & Kocheva, 2021). High salinity affects the plant by inducing water stress, ion toxicity, and reduction of cell division, and also alters metabolic processes (Balasubramaniam et al., 2023). A combination of these effects results in hindered growth and development of plants. Growth is affected in terms of radicle length, dry weight, shoot length, plant biomass, nodulation, pod numbers, and grain yield.

Most legumes are very sensitive to saline conditions (Elgharably & Benes, 2021), for example, barrel clover (*Medicago truncatula*) and pea (*Pisum sativum*), while others are more salt-tolerant, as seen in green beans (*Phaseolus vulgaris*) (Kirova & Kocheva, 2021). Many reports indicate that salt stress reduces shoots in soybeans, chickpeas, peas, and mung bean plants (Bouzroud et al., 2023; Joshi et al., 2021). Other ways in which saline conditions hinder the growth and production of legumes include damaging the photosynthetic machinery, accumulation of proline, and accretion of hydrogen peroxide in the roots and leaves of legume plants (Rane et al., 2021).

5.3.3 Soil Acidity Stress

Within the tropical regions of Africa, human activities and weather changes have contributed to soil acidity in arable land. Gurmessa (2021) reported that sub-Saharan African smallholder farmers engage in frequent tillage of their lands and remove crop residues to feed their animals, thus making their soil poor as there is lowered affinity for hydrogen ion. Zhu et al. (2020) added that soil acidity had been largely contributed by frequent use of fertilizers, such as urea which enhances the release of hydrogen ion.

Besides human activities, high sporadic rainfalls have also contributed to leaching basic cations, leading to increased soil acidity (Desta et al., 2021). Soil acidity limits the availability of mineral nutrients, including phosphorus and calcium as they form insoluble complexes (Penn & Camberato, 2019). Given the role of phosphorus in the growth and production of legumes, its deficiency due to acidity reduces the production of legumes (Bechtaoui et al., 2021). Nitrogen fixation is a high energy-consuming process, which requires high availability of phosphorus in the soil (Penn & Camberato, 2019). Phosphorus deficiency will therefore hinder nitrogen fixation.

Soil acidity is also associated with the availability of toxic forms of aluminum and manganese (Tadele, 2020). Aluminum in acidic soil promotes reactive oxygen species and lipid peroxidation, making the soil unsuitable for the normal growth of plants, leading to a reduction in yield (Ranjan et al., 2021). The availability of toxic soluble micronutrients hinders legume production and negatively affects chemical and biological processes such as the uptake of water and nitrogen fixation (Khatun et al., 2021).

5.3.4 Temperature Stress

Temperature is a crucial element affecting legume production (Al-Tawaha et al., 2022). In East and Central Africa, global warming has led to rising temperatures and changes in rainfall patterns in the region (Haile et al., 2020; Ongoma et al., 2018). High temperatures affect legume production, especially when temperature and rainfall are at extremes beyond the crop's minimum and maximum limits (Parthasarathi et al., 2022). A slight temperature increase above the threshold limit is described as heat stress in legumes. Heat stress can be categorized into either heat shock or moderate heat, depending on the duration of interaction and temperature (Kan et al., 2023).

The optimum temperature for legume growth varies among the genera, species, and genotypes. Heat stress for the majority of subtropical and tropical crops occurs when temperatures rise above 32–35 °C and >25 °C for cool-season crops (Kumari et al., 2021). Studies have also reported decreased yields at day temperatures > 30 °C and >20 °C at night (S. Basu et al., 2022).

Heat stress has harmful effects on legumes' vegetative and reproductive growth phases. During reproductive growth, heat stress affects gamete formation, flowering, and podding. Flowers are the most susceptible organs (Qian et al., 2025). The high temperatures shorten flowering and pod-filling durations, causing losses in yields (Ncama et al., 2021). Other effects of heat stress include withering, burning of lower leaves, dehydration of underdeveloped plants, flower and pod abortion, reduced root nodule formation, nitrogen fixation, and low seed yield (Ncama et al., 2021).

Stress related to low temperature can be described as either chilling (0–12 °C) or freezing (below 0 °C) (Bhattacharya, 2022). The optimum temperature required by warm-season legumes is 25–35 °C while that of cool-season legumes is 15–25 °C. Legumes such as soybean, groundnut, common bean, cowpea, pigeon pea, and lima bean grow during warm seasons and are more susceptible to temperatures below 10 °C (Bhat et al., 2022). Cold stress generally harms the vegetative growth and reproduction of legumes.

Low-temperature effects during the reproductive phase include shedding of flowers, ovule abortion, pollen tube deformation, reduced pod filling, production of sterile flowers, and overall reduced yields (Gogoi et al., 2018). Chilling decreases the membrane stability of cells by modifying lipids and proteins. The cell membrane loses its integrity due to increased solutes such as potassium ions, amino acids, carbohydrates, and membrane-changing chloroplast function (Bhandari et al., 2017). The overall result is a slower growth rate, higher chlorosis and necrosis of older leaves, and delays in germination and emergence (Gogoi et al., 2018).

5.3.5 Waterlogging

Waterlogging is the saturation of the soil surrounding the root with water. The deleterious effects of waterlogging in crops result from the slow diffusion rates of gases and poor O_2 solubility in water. Oxygen deficiency inhibits root respiration resulting in reduced nutrient uptake and root growth (Agualongo et al., 2022). Legume species vary in their tolerance and susceptibility to temporary waterlogging. Legumes prone to waterlogging during vegetative growth include lupin, lentil, chickpea, and field pea, while faba bean is tolerant (Zaman et al., 2024). Waterlogging generally decreases germination rates, seedling emergence, root and shoot growth, and plant biomass by up to 80% (Agualongo et al., 2022).

In temporarily waterlogged soils, legume yields decrease due to limited nitrogen fixation owing to reduced oxygen supply to the nodules (Senthamil et al., 2025). This effect is usually worsened by soil denitrification. Waterlogging also causes chlorosis and reduced crop growth vigor (Majeed et al., 2023). Prolonged waterlogging eventually results in anaerobic metabolism, which may cause severe growth retardation and even death (Sharma et al., 2021). Genetic screening to select for waterlogging-tolerant species and disease resistance is ideal for managing this stress (Langan et al., 2022).

5.3.6 Nutrient Deficiencies

5.3.6.1 Macronutrient Deficiencies

Macronutrient deficiencies in legumes include phosphorus, potassium, sulfur, calcium, and magnesium. Approximately 50% of bean production experiences phosphorus deficiency (Smith et al., 2022). Phosphorus deficiency limits legume production, particularly in tropical soils. Root morphology largely influences the phosphorus acquisition efficiency in legumes, with variations observed among breeding lines (Fathi & Mehdiniya, 2023).

Phosphorus is an essential requirement for nitrogen fixation and regulation of enzymatic activities that promote biological nitrogen fixation (Li et al., 2021). Phosphorus deficiency leads to poor growth and development of leguminous plants, with harmful effects on biological nitrogen fixation, nodule formation, and photosynthesis (Fathi & Mehdiniya, 2023).

Potassium is also deficient in many soils, but less research focuses on this nutrient (Mostofa et al., 2022). Sulfur deficiency in agricultural soils is rare, but due to its depletion in some soils, increasing reports of crop sulfur deficiency have been noted (Saggar & Bolan, 2024). Legumes that rely on biological nitrogen fixation to acquire nitrogen require higher phosphorus, potassium, and sulfur concentrations than those that obtain nitrogen directly from the soil (Kebede, 2021). These nutrients directly affect biological nitrogen fixation by regulating nodule growth, formation, and functioning, while indirect effects include a reduction in host plant growth (Kafeel et al., 2022).

5.3.6.2 Micronutrient Deficiencies

Micronutrient deficiencies affecting legume production include copper, zinc, iron, manganese, boron, and molybdenum (Abbasi et al., 2023). Large quantities of micronutrients such as cobalt and molybdenum are required for effective symbiotic nitrogen fixation (Li et al., 2024). Nutrient deficiencies have been reported to cause yield losses of varying degrees in legumes. Iron deficiency in legumes causes chlorosis in younger leaves (Banerjee & Nath, 2022). Insufficient molybdenum results in leaf wilting and chlorosis of older plants (Merry et al., 2022).

5.4 STRATEGIES USED TO OVERCOME BIOTIC AND ABIOTIC STRESSES IN LEGUME PRODUCTION

In recent decades, agriculture has developed the need for eco-friendly and sustainable farming management practices. A significant shift has been witnessed from the use of chemical inputs to

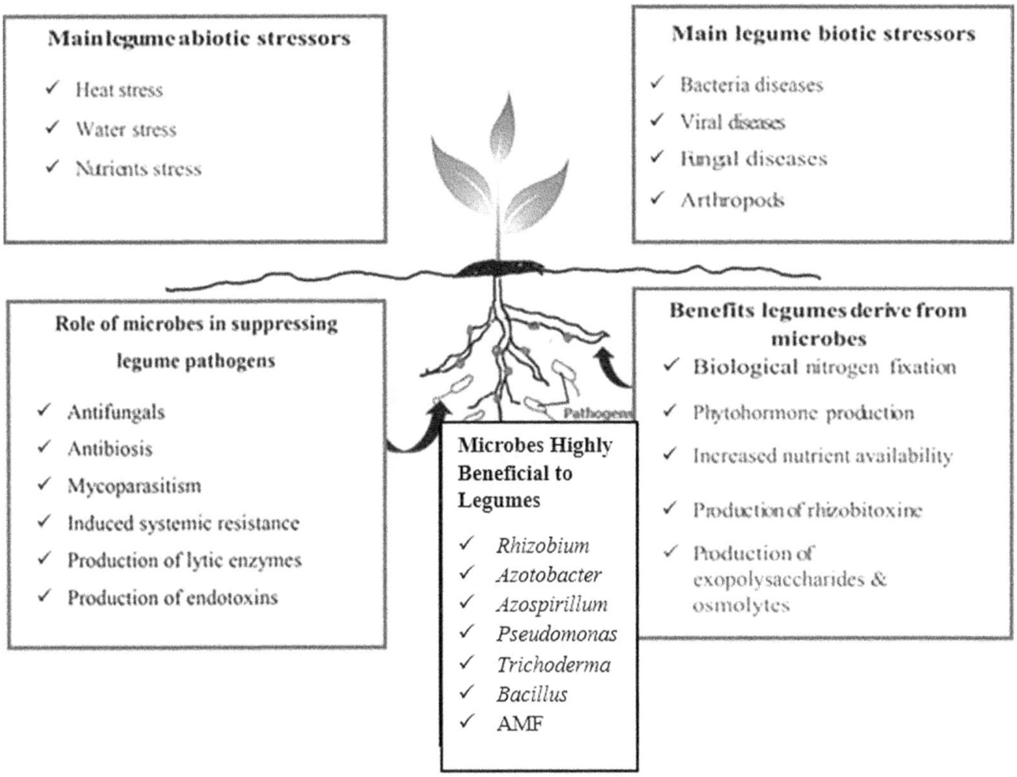

FIGURE 5.1 Different ways in which soil microbes enhance legume growth, production, and tolerance to biotic and abiotic stresses.

more affordable, environment-conscious alternatives that have included biological control agents, biofertilizers, and bio-enhancers. Biological control involves the utilization of specific natural organisms to control the population density of other organisms, like weeds, pests, and diseases that would negatively affect the production of a target crop.

Biofertilizers such as rhizobia enrich and increase the supply of essential nutrients to the plants. At the same time, bio-enhancers such as arbuscular mycorrhizal fungi enhance access to essential nutrients. In addition, biofertilizers and bio-enhancers improve plants' tolerance to abiotic stresses such as drought, acidity, salinity, and temperature.

5.4.1 Biocontrol Microbial Inoculants

Biotic stressors significantly decimate legume production (Abbey et al., 2019; Harrison et al., 2019). Legumes are susceptible to pests and phytopathogens at all stages of their development (Srivastava et al., 2020). Inorganic pesticides are deleterious to the environment and cause considerable harm to farmers due to their well-documented carcinogenic and teratogenic effects (Pallangyo et al., 2019). Biological control agents (BCAs) or BCA-based formulations have been developed for plant protection as biopesticides.

Several bacterial and fungal genera have been shown to have high biocontrol efficacy against pests and pathogens, increasing legume productivity (Srivastava et al., 2020). *Bacillus thuringiensis*, for instance, is the leading endotoxin producer known to significantly control many entomopathogens affecting legume production (Alves et al., 2024; Gemmeda et al., 2024; Vimala Devi et al., 2020).

Volatiles produced by these bacteria have been established to have larvicidal properties, increasing their biocontrol efficacy against Lepidopterian and Odopterian insect pests (Gemmeda et al., 2024). *B. thuringiensis* is the most potent bacterial biocontrol agent owing to its ability to be effectively integrated with other BCAs in pest management strategies.

In addition to *Bacillus* spp. that are highly effective in controlling entomopathogens, several *Pseudomonas* spp. have been made into formulations that are vital in controlling soil-borne pathogens affecting legume productivity (Gaurkhede et al., 2015; Sangiogo et al., 2018; Sunyar et al., 2024). These bacterial species are therefore exploited for their biological control ability.

A number of fungal genera such as *Trichoderma* have also been applied in biopesticide formulations (Table 5.1). *Beauveria* spp. has shown high efficacy in the biocontrol of entomopathogens that

TABLE 5.1
Strategies Involving Beneficial Soil Microbes which are used to Overcome Biotic and Abiotic Stresses in Legumes

Soil microbes	Legume crop	Stress managed	Strategy	Reference
Bacillus thuringiensis	Cowpea, Soybean Chickpea	Lepidopterian insect pests	Expression of insecticidal proteins	(Gemmeda et al., 2024) (Alves et al., 2024) (Vimala Devi et al., 2020)
Bacillus subtilis	Pea	Fungal pathogen (*Fusarium oxysporum*)	Production of volatiles, chitinase, and antifungal molecules	(Klimek-Kopyra et al., 2023)
Trichoderma spp.	*Phaseolus vulgaris* L.	*Rhizoctonia solani*	Induction of the expression of genes involved in plant defense responses	(Salman & Boyraz, 2023) (Abbas et al., 2017)
Pseudomonas fluorescens	*Phaseolus vulgaris*	Leaf spot Blight	Antagonistic behavior against the pathogen	(Sangiogo et al., 2018) (Sunyar et al., 2024)
Pseudomonas fluorescens	Chick pea	Collar rot disease	Antagonistic behavior against the pathogen	(Gaurkhede et al., 2015)
Trichoderma spp. and *Rhizobium* spp.	Chick pea Faba bean	Damping off Root rot	Antagonistic behavior against the pathogen	(Nirmalkar et al., 2017) (El-Khateeb, 2014)
Rhizobia	Cowpea Climbing bean	Nitrogen deficiency	Increased nodulation and nitrogen fixation	(Nyaga & Njeru, 2020) (Koskey et al., 2017)
Mesorhizobium mediterraneum	Chickpea	Phosphorus deficiency	Increased phosphorus solubilization	(Imen et al., 2015)
AMF and *Bradyrhizobium*	Green gram	Water stress	Increased absorption of nutrients and CO_2 assimilation	(Musyoka et al., 2020)
AMF	Chickpea	Drought stress	Enhancement of the relative water content, membrane stability index, and photosynthetic efficiency	(Hashem et al., 2019)
AMF (*Rhizophagus clarus*)	Cowpea	Soil salinity stress	Increased uptake of limiting soil nutrients such as phosphorus	(Tavares et al., 2024)
Pantoea sp. *Enterococcus*	Mungbean	Salinity stress	Increasing ACC deaminase activity and plant growth promotion	(Panwar et al., 2016)

affect legumes (McGuire & Northfield, 2020; Singh et al., 2016). *Metarhizium* and *Verticillium* are other genera with the capability to control legume pathogens (McGuire & Northfield, 2020). Arbuscular mycorrhizal fungi are also effective in the control of phytopathogens due to their high competitiveness.

The most potent and widely used fungal biological control agent is *Trichoderma* (Abbas et al., 2017; Nirmalkar et al., 2017; Salman & Boyraz, 2023). Species of *Trichoderma* are highly aggressive and hyper-mycoparasitic against a wide range of fungal phytopathogens. Species such as *Trichoderma harzianum*, *Trichoderma viride*, *Trichoderma virens*, and *Trichoderma asperellum* are well-known antibiotic producers. These species also produce antifungal compounds that act against many fungal pathogens affecting legumes (Bahadur & Dutta, 2022; Stracquadanio et al., 2020). *Trichoderma* are also producers of hydrolytic enzymes such as chitinases and cellulases that lyse pathogen walls, thus killing them.

Trichoderma species are highly competitive and quickly outgrow pathogens, hindering their manifestation. Several species have also been determined to induce systemic resistance in plants. What makes *Trichoderma* the model fungus of the green agriculture revolution is its plant growth-promoting potential in addition to its biocontrol efficacy (Mukhopadhyay & Kumar, 2020).

The BCAs are more advantageous than chemical formulated pesticides because they are less toxic, only affect the target organism, decompose quickly, and are effective in small quantities.

However, their high specificity requirement hinders their commercial production (Kumar et al., 2023). Several mechanisms of biological control have been explored in legume production. These include mycoparasitism, antibiosis, induced resistance, and competition (Figure 5.1).

5.4.1.1 Mycoparasitism

Mycoparasites are fungi that exhibit a parasitic relationship with other forms of fungi. Fungus-based biological control agents function through two methods, either by parasitism or hyperparasitism (Mishra et al., 2018). Parasitism involves recognizing a host fungus by the mycoparasite triggering the release of hydrolytic enzymes and antibiotics, whereas hyperparasitism occurs by direct lysis of the host's "fungus" mycelium by the mycoparasite (Köhl et al., 2019). The most studied fungus in controlling pathogenic fungi and parasitic nematodes affecting legume production is *Trichoderma* spp.

Trichoderma has been proven to be very efficient in the control of plant diseases and soil-borne pathogens. *Fusarium* wilt caused by the soil fungus *Fusarium oxysporum* is the major cause of early wilting and plant death in *Phaseolus vulgaris* L. (common bean). Nigam and Singh (2022) reported on the antagonistic activity of *Trichoderma* against *Fusarium oxysporum* in lentils. *Trichoderma* is a broad-spectrum biological control agent due to its wide host range and ability to survive in different ecologies and eliminate pathogens and legume-associated fungi.

5.4.1.2 Antibiosis

The production of metabolic substances by an organism to inhibit the development of another organism is referred to as antibiosis (Lacerda et al., 2022). These substances can be isolated from cultures of the microorganism, or the microorganism can be introduced into the environment (Ali Nusaibah & Musa, 2019). Vimala Devi et al. (2020) found that the use of *Bacillus thuringiensis* as a pesticide reduced the damage of *Helicoverpa armigera* (bollworm), which is a notorious pod borer, when sprayed on chickpeas. Macrocyclic lactones derived from cultured actinomycetes have also been recorded to be highly effective against *H. armigera* in in beans (Diab et al., 2024). *Trichoderma harzianum* produces a cell-wall-degrading enzyme in the soil that kills pathogens in the rhizosphere. El-Hassan et al. (2013) noted that *Trichoderma hamatum* and *Pseudomonas fluorescence* are effective against *Fusarium oxysporum*, responsible for lentil wilt disease, which significantly affects the yield and development of lentils.

5.4.1.3 Induced Resistance

Biological control agents can induce resistance to infections or attacks by parasites in plants. This form of resistance can be local or systemic and is a form of indirect antagonism (Kunda et al., 2020). Several biological control agents cause the thickening of the plant cell wall, lignification, callus formation, accumulation of phytoalexins, and triggering the synthesis of pathogen-related proteins (Srinivasan et al., 2019). Gupta et al. (2021) reported inhibition of the soil-borne pathogen *Rhizoctonia bataticola* in pulses through the production of siderophores, hydrogen cyanide, and antibiotics by *Bacillus* and *Paenibacillus* endophytes (Kunda et al., 2020). Root-associated mutualistic microorganisms such as Mycorrhiza, *Pseudomonas*, *Bacillus*. and *Trichoderma* species prime the plant response system, activating a defense against pathogenic microorganisms in the rhizosphere (Ansari et al., 2024).

5.4.1.4 Competition

Competition for the limiting nutrients in the rhizosphere inhibits the growth of some microbes that do not express dominance. More dominant and competitive strains of microorganisms can be utilized as biopesticides to ensure the exclusion of the pathogen from that environment (Köhl et al., 2019). Mycorrhizae colonize the surface of the roots of legumes before pathogens can, thereby limiting damage to plant roots (Kumar & Verma, 2018). Other species like *Pythium* and *Fusarium* that infect plants by attaching and infecting the roots through mycelial contact are more prone to be outcompeted than microorganisms that grow and develop on the surface of the plant (Zheng et al., 2017). The production of siderophores by plant growth-promoting rhizobacteria like *Azotobacter* limits the availability of iron for other microorganisms (Kour et al., 2019).

5.4.2 Use of Biofertilizers

With close to three million reported cases of chemical poisoning by the World Health Organization in many agroecosystems across the globe, the search for safer but effective alternatives is a need of the hour (Manzoor, 2020; Thompson & Raizada, 2018). The effect of the inorganic farming additives is more pronounced in the smallholder farming units, where 90% of the poisoning occurs (Pallangyo et al., 2019; Thompson & Raizada, 2018). In addition, to ensure sustainable agriculture with minimalist disturbance to the soil microecology, and therefore essential for green farming technologies, microbial inoculants have been used to beef up agricultural production in many ways. Microbial inoculants provide a means of transferring the benefits conferred by a living microorganism to a living plant in an economical and easier-to-use form (Koskey et al., 2021). They can broadly be classified as biofertilizers or biopesticides. In most cases, some microbes offer more than one essential benefit, hence they can possess plant growth-promoting attributes and biocontrol ability. Microbial inoculants can also be bacterial or fungal.

5.4.2.1 Bacterial Biofertilizers

For a long time, most bioinoculants in legume production were composed of rhizobia as the main bioactive component (Santos et al., 2019). A lot of effort has been placed toward incorporating other beneficial microorganisms due to the different beneficial ecological services they confer to the legumes (Santos et al., 2019). Determining the co-inoculants that can enable legumes to resist the environmental stressors and promote nutrient uptake is the critical factor in establishing the right choice of inoculants (Korir et al., 2017). While the symbiotic relationship between legumes and rhizobia has been elaborated since time immemorial, the increasing environmental stressors occasioned by climate change are making this classical symbiosis less efficient in overcoming most of the plant growth stressors. Intensified bioprospecting for rhizobia strains with greater nitrogen-fixing ability has yielded better bioinoculants of rhizobia essential in boosting nitrogen-fixing ability and yields of many legumes (Checcucci et al., 2017; Muleta et al., 2017). This significantly reduces the need

FIGURE 5.2 Cowpea plants inoculated with rhizobia in Tharaka Nithi County in Kenya during the vegetative and yielding stages.

for nitrogenous fertilizers since biological nitrogen fixation provides a renewable and sustainable nitrogen source for legumes (Elgharably & Benes, 2021). As a result, legumes inoculated with rhizobia are able to achieve optimum growth and production in poor soils (Figure 5.2). However, most legumes can only nodulate with specific rhizobia strains; hence species compatibility should be considered for maximum gains (Mburu et al., 2021).

Rhizobia have been documented in several studies to have the capacity to alleviate nutrient deficiency by promoting the uptake of nutrients such as nitrogen and phosphorus (Santos et al., 2019). Other than rhizobia, other rhizobacteria present in soil promote legume growth in several ways. This occurs primarily through phytohormone production (Janati et al., 2021). This has been documented in many bacteria genera such as *Bacillus*, *Pseudomonas*, *Azospirillum*, *Aeromonas*, and *Serratia* (Roriz et al., 2020). These plant-growth-promoting bacteria significantly influence the availability of the nutrients in the plant rhizosphere through siderophore production and nutrient solubilization (Ahmed et al., 2020; Andrić et al., 2020). Phosphorus, after nitrogen, is a macronutrient required by plants for normal metabolism but is most often available in insoluble forms that plants are often unable to exploit (Wang et al., 2020). Co-inoculating legumes with phosphate solubilizers has been shown to significantly increase legume production (Nosheen et al., 2021; Zaidi et al., 2017). Co-inoculation of legumes with rhizobia and *Azospirillum* has been attributed to increased nodulation in many legumes. Mixed cultures with rhizobia and *Azospirillum* produce elevated levels of phytohormones that can significantly benefit legume production (Swarnalakshmi et al., 2020). Sulfur-oxidizing bacteria such as *Thiobacillus* spp. have also been documented to ameliorate soil alkalinity contributing significantly to improved legume productivity (Raimi et al., 2017). Co-inoculation of rhizobia and *Thiobacillus* spp. strain LCH significantly improved nodulation and legume biomass in vivo (van Heerwaarden et al., 2018). The incorporation of sulfur-oxidizing bacteria has been associated with an increased oil content of peanuts.

5.4.2.2 Fungal Biofertilizers and Bioenhancers

While legume–rhizobia symbiosis is the most documented and studied, fungi have also been established to form highly beneficial associations with legumes. Effective fungal bioinoculants include arbuscular mycorrhizal fungi, *Beauveria*, *Metarhizium*, and *Trichoderma* (Abbey et al., 2019; McGuire & Northfield, 2020). Arbuscular mycorrhizal fungi have been well documented to promote nutrient uptake by legumes by siderophore production and phosphate solubilization (Njeru et al., 2020). In addition, arbuscular mycorrhizal fungi strains have also been associated with improved uptake of diffusion-limited mineral ions by legumes (Devi et al., 2022). While co-inoculation of different microorganisms is not often associated with an improved effect on overall productivity, arbuscular mycorrhizal fungi have been documented to have synergistic effects with many rhizobia strains and other phosphobacters (Ansari et al., 2024; Njeru et al., 2020). Arbuscular mycorrhizal fungi and *Trichoderma* are well known to promote plant tolerance to environmental abiotic stresses such as drought, salinity, and thermal stress (Abbey et al., 2019).

Trichoderma colonize legume crops root forming *Trichoderma*–root interactions. Such interactions not only confer tolerance to biotic stresses in the root but also to abiotic stresses such as drought and salinity stress tolerance. This is because *Trichoderma* root colonization in legumes increases root growth, plant growth, and nutrient uptake, as well as tolerance against oxidative stress (Mukhopadhyay & Kumar, 2020).

5.5 MOLECULAR APPROACHES IN MANAGING BIOTIC AND ABIOTIC STRESSES

Biotic and abiotic stresses are responsible for the evolution of plants through the development of adaptive mechanisms pre-domestication (Balasubramaniam et al., 2023; Dresselhaus & Hückelhoven, 2018). Plant breeding during crop domestication led to the loss of diversity in the gene pool as man improved yield production at the expense of disease resistance and stress tolerance traits (Wu et al., 2019). However domestication is also important for fixing suitable traits and enhancing variation of crop plants (Dev et al., 2024). Even in the advent of Mendelian crop breeding, the recombination of alleles wasn't well understood and attempts to improve resistance to pests and diseases through crossing with resistant plant species was challenging (Ojiewo et al., 2018). This challenge was attributable to abnormalities in chromosomal pairing resulting in infertile hybrids and very low efficiency in the fixation of desired traits (Manzoor, 2020).

Developments in the study of genetics and the molecular structure of genes have significantly impacted plant breeding (Abdul Aziz & Masmoudi, 2025). These advances led to the introduction of genetic engineering of plant genes to improve yield and alleviate biotic and abiotic stress (Chen et al., 2024). The utilization of biotechnology requires in-depth knowledge of the plant genetics, transcriptomics, proteomics, as well as the mechanisms by which the resistance or tolerance to biotic and abiotic stresses is mediated (Varadharajan et al., 2025).

Several studies have delved into molecular approaches within which these stresses can be alleviated in legumes. However, their polyploidy nature and large genome size have posed a major challenge (Smýkal et al., 2020). Therefore, two research model plants have been utilized to study resistance to stresses as well as plant microbe interactions, these include *Lotus japonicas* and *Medicago truncatula* (Ará Ujo et al., 2015; Zhao et al., 2023). These model plants have small diploid genomes, produce many seeds, and are autogamous in nature (Meena et al., 2017), making them ideal test models.

Advances in genome-sequencing techniques have resulted in whole-genome sequencing of most legumes; the availability of these data has improved their study and potential improvement (Khandelwal & Mukesh, 2018). Studies have found that genetic engineering of plants requires strategies that cover varying stress tolerance pathways; this is because plants are affected by different stresses at the same time and their responses are complex, involving activity at the physiological, cellular, and transcriptome levels (Lohani et al., 2022). Therefore, genes that confer tolerance through multiple pathways are preferred for legume improvement.

5.5.1 Genetic Transformation

The development of transgenic legume crops has significantly improved stress tolerance to biotic stress. New methods of pest control have been studied, including the modification of plants to enhance the expression of naturally occurring toxic defense proteins, for instance, chitinases, lectins, and amylase and protease inhibitors; and metabolite modulation and tissue-specific gene expression (Kocyigit et al., 2023). These methods are specifically tailored to combat commonly occurring pests such as aphids and white flies (Mangena & Sehaole, 2020). Legume production is greatly affected by aphids and white flies, leading to reduced yields.

The incorporation of resistance-conferring genes like the cry protein toxin produced by *Bacillus thuringiensis* for pest resistance has tremendously advanced studies in pest control (Hu et al., 2025). They are divided into four categories according to their insecticidal specificity, these include anti-Coleopteran, anti-Diptera, anti-Lepidoptera, and anti-nematode (Qin et al., 2019).

Soybean (*Glycine max* L. Merr) is among the most important food legumes cultivated by farmers. It suffers considerable losses in yield due to attacks by the dark black chaffer (*Holotrichiaparallela,* a Coleoptera) pest (Jiang et al., 2017). The transformation of soybean mediated by *Agrobacterium tumefaciens* has led to the expression of Cry-8-like proteins conferring resistance to the soybean (Qin et al., 2019). The exploitation of *Agrobacterium tumefaciens* as a vector for the transformation of legume crops was revolutionary in genetic engineering of crops. There is limited understanding of the mechanisms used by plants to overcome abiotic stress; this acts as a hindrance to the utilization of genetic transformation as a control strategy (Dita et al., 2006; Ojiewo et al., 2018). This is because abiotic stresses involve complex metabolic mechanisms and there is very limited documentation of the natural sources of tolerance (Meena et al., 2017).

5.5.2 Molecular Marker-assisted Breeding and Gene Pyramiding

Breeding strategies for crop improvement rely on genetic and genomic analyses to identify regions in the DNA that are linked to agronomic traits; these regions are known as molecular markers (Singh et al., 2020). The identification of these markers in legumes reduces the amount of time spent in traditional breeding practices like screening in a greenhouse (Ojiewo et al., 2018). Genetic markers such as simple sequence repeat markers and single-nucleotide polymorphisms are instrumental in genetic mapping and identification of quantitative trait loci responsible for economically important traits (Rani et al., 2020). The development of these markers has advanced the identification of key quantitative trait loci conferring tolerance to abiotic stress. Studies on single-nucleotide polymorphisms in cowpeas have sought to identify resistance markers to foliar thrips (Lucas et al., 2013). Mapping different quantitative trait loci responsible for agronomic traits allows for the understanding and introduction of several quantitative trait loci into improved cultivars to confer resistance traits (Khandelwal & Mukesh, 2018). This enhances breeding techniques through marker-assisted selection for tolerance traits in biotic and abiotic stress.

Stacking together different tolerance and resistance genes in a genotype is referred to as gene pyramiding (Ferreira et al., 2012). Research in common beans has led to marker-assisted selection and gene pyramiding to confer resistance to rust and this has been combined with disease resistance to bacterial blight and anthracnose (Gonçalves-Vidigal et al., 2020). The enhancement of chickpea and pigeon pea for drought tolerance and *Fusarium* wilt resistance have been advanced using marker-assisted selection and gene pyramiding (Mir et al., 2017).

5.5.3 Somaclonal Variation and in Vitro Mutagenesis

Somaclonal variation arising from callus formation during tissue culture leads to genetic variation in plants. The development of these genetic variants when culturing legumes can be exploited to produce agriculturally useful soma clones; this has been reported in peas and pigeon peas (Adero, 2016). Somaclonal variants resistant to *Fusarium* wilt and *H. armigera* have been isolated through in vitro regeneration of pigeon pea (Anil et al., 2018). Different in vitro mutagenesis strategies including fast neutron radiation and treatment with ethyl-methane-sulfonate to induce mutations have been utilized in plant breeding.

5.6 THE FUTURE OUTLOOK

Though microorganisms present in the rhizosphere have been proven to enhance legume tolerance to biotic and abiotic stresses, the effectiveness of the microbial species is highly variable depending on the agro-ecological conditions. Further identification of the specific microbial inoculants highly effective in different conditions is needed. This also applies to the newly bred legume varieties, whereby the microbe–varietal compatibilities need to be evaluated. The outlook here is to have tolerant varieties that are compatible with highly effective microbial inoculants, enhancing these varieties' resilience to prevailing biotic and abiotic stresses.

REFERENCES

Abbas, A., Jiang, D., & Fu, Y. (2017). Trichoderma Spp. as Antagonist of Rhizoctonia solani. *Journal of Plant Pathology & Microbiology*, *8*(3), 3. https://doi.org/10.4172/2157-7471.1000402

Abbasi, A., Hina, A., Shahid, S., Imtiaz, S., Intisar, A., & Khan, M. T. (2023). Micronutrients enrichments in legumes through agronomic and cultural practices. *Legumes Biofortification*, *2023*, 47–71. https://doi.org/10.1007/978-3-031-33957-8_3

Abbey, J. A., Percival, D., Abbey, Lord, Asiedu, S. K., Prithiviraj, B., & Schilder, A. (2019). Biofungicides as alternative to synthetic fungicide control of grey mould (Botrytis cinerea)–prospects and challenges. *Biocontrol Science and Technology*, *29*(3), 241–262. https://doi.org/10.1080/09583157.2018.1548574

Abdul Aziz, M., & Masmoudi, K. (2025). Molecular breakthroughs in modern plant breeding techniques. *Horticultural Plant Journal*, *11*(1), 15–41. https://doi.org/10.1016/j.hpj.2024.01.004

Adero, M. (2016). In vitro regeneration of pigeon pea using leaf explants. *African Crop Science Journal*, *24*, 191–201. https://doi.org/10.4314/acsj.v24i2.7

Agualongo, D. A. P., Da-Silva, C. J., Garcia, N., De Oliveira, F. K., Shimoia, E. P., Posso, D. A., De Oliveira, A. C. B., De Oliveira, D. D. S. C., & Do Amarante, L. (2022). Waterlogging priming alleviates the oxidative damage, carbohydrate consumption, and yield loss in soybean (Glycine max) plants exposed to waterlogging. *Functional Plant Biology*, *49*(12), 1029–1042. https://doi.org/10.1071/FP22030

Ahmad, Z., Tariq, R. M. S., Ramzan, M., Bukhari, M. A., Raza, A., Iqbal, M. A., Meena, R. S., Islam, M. S., Sytar, O., Godswill, N. N., Wasaya, A., Singh, K., Hossain, A., Raza, M. A., Hasanuzzaman, M., Soysal, S., Erman, M., Cig, F., Ceritoğlu, M., … & Sabagh, A. EL. (2022). Biological nitrogen fixation: An analysis of intoxicating tribulations from pesticides for sustainable legume production. *Managing Plant Production under Changing Environment*, *2022*, 351–374. https://doi.org/10.1007/978-981-16-5059-8_14

Ahmed, A., Sultan, T., Qadir, G., Afzal, O., Ahmed, M., Shah, S.-U.-S., Asif, M., Ali, S., & Mehmood, M. Z. (2020). Impact assessment of plant growth promoting rhizobacteria on growth and nutrient uptake of maize (Zea mays). *Pakistan Journal of Agricultural Research*, *33*(2), 234–246. https://doi.org/10.17582/journal.pjar/2020/33.2.234.246

Al-Tawaha, A. R. M., McNeil, D., Othman, Y. A., AlRawashdeh, I. M., Amanullah, Imran, Adnan, M., Zamin, M., Qaisi, A. M., Al-Tawaha, A., Jahan, N., Shah, M. A., Khalid, S., Sami, R., Rauf, A., Thangadurai, D., Sangeetha, J., Fahad, S., Youssef, R. A., … & Al-Taey, D. K. A. (2022). Legume Production and Climate Change. *Improvement of Plant Production in the Era of Climate Change*, *2022*, 221–248. https://doi.org/10.1201/9781003286417-11

Ali Nusaibah, S., & Musa, H. (2019). A review report on the mechanism of trichoderma spp. as biological control agent of the Basal Stem Rot (BSR) disease of Elaeis guineensis . *Trichoderma – The Most Widely Used Fungicide*. *2018*, 1–12. https://doi.org/10.5772/intechopen.84469

Alves, G. B., Lemes, T. S. O., Pereira, E. J. G., Jurat-Fuentes, J. L., Smagghe, G., Santos, G. R., Haddi, K., Corrêa, R. F. T., Melo, F. L., Jumbo, L. O. V., Oliveira, E. E., Peron, A. J., Ribeiro, B. M., & Aguiar, R. W. S. (2024). Draft genome of neotropical Bacillus thuringiensis UFT038 and its potential against lepidopteran soybean pests. *Folia Microbiologica*, *69*(1), 91–99. https://doi.org/10.1007/s12223-023-01114-3

Andrić, S., Meyer, T., & Ongena, M. (2020). Bacillus responses to plant-associated fungal and bacterial communities. *Frontiers in Microbiology*, *11,* 1–9. https://doi.org/10.3389/fmicb.2020.01350

Anil, V. S., Bennur, S., & Lobo, S. (2018). Somaclonal variations for crop improvement: Selection for disease resistant variants in vitro. *Plant Science Today*, *5*(2), 44–54. https://doi.org/10.14719/pst.2018.5.2.382

Ansari, M. M., Bisht, N., Singh, T., & Chauhan, P. S. (2024). Symphony of survival: Insights into cross-talk mechanisms in plants, bacteria, and fungi for strengthening plant immune responses. *Microbiological Research*, *285,* 1–11. https://doi.org/10.1016/j.micres.2024.127762

Anuar, M. S. K., Hashim, A. M., Ho, C. L., Wong, M. Y., Sundram, S., Saidi, N. B., & Yusof, M. T. (2023). Synergism: Biocontrol agents and biostimulants in reducing abiotic and biotic stresses in crop. *World Journal of Microbiology and Biotechnology*, *39*(5), 1–21. https://doi.org/10.1007/s11274-023-03579-3

Ará Ujo, S. S., Beebe, S., Crespi, M., Delbreil, B., González, E. M., Gruber, V., Lejeune-Henaut, I., Link, W., Monteros, M. J., Prats, E., Rao, I., Vadez, V., Patto, M. C. V. V., Araújo, S. S., Beebe, S., Crespi, M., Delbreil, B., González, E. M., Gruber, V., … & Patto, M. C. V. V. (2015). Abiotic stress responses in legumes: Strategies used to cope with environmental challenges. *Critical Reviews in Plant Sciences*, *34*, 237–280. https://doi.org/10.1080/07352689.2014.898450

Ayaz, İ., Kasapoğlu Uludamar, E. B., Behmand, T., & ElekciOğlu, İ. H. (2021). Investigation of the development of root lesion nematodes, Pratylenchus spp. (Tylenchida: Pratylenchidae) in three chickpea cultivars. *Turkiye Entomoloji Dergisi*, *45*(1), 23–31. https://doi.org/10.16970/entoted.753614

Bahadur, A., & Dutta, P. (2022). Trchoderma Spp.: Their impact in crops diseases management. *Trichoderma – Technology and Uses*. *2022*, 1–21 https://doi.org/10.5772/intechopen.101846

Balasubramaniam, T., Shen, G., Esmaeili, N., & Zhang, H. (2023). Plants' response mechanisms to salinity stress. *Plants*, *12*(12), 2253. https://doi.org/10.3390/plants12122253

Banerjee, P., & Nath, R. (2022). Prospects of molybdenum fertilization in grain legumes-A review. *Journal of Plant Nutrition*, *45*(9), 1425–1440. https://doi.org/10.1080/01904167.2021.2020831

Bechtaoui, N., Rabiu, M. K., Raklami, A., Oufdou, K., Hafidi, M., & Jemo, M. (2021). Phosphate-dependent regulation of growth and stresses management in plants. *Frontiers in Plant Science*, *12*, 679916. https://doi.org/10.3389/fpls.2021.679916

Bedine Boat, M. A., Sameza, M. L., Iacomi, B., Tchameni, S. N., & Boyom, F. F. (2020). Screening, identification and evaluation of Trichoderma spp. for biocontrol potential of common bean damping-off pathogens. *Biocontrol Science and Technology*, *30*(3), 228–242. https://doi.org/10.1080/09583157.2019.1700909

Ben Gaied, R., Brígido, C., Sbissi, I., & Tarhouni, M. (2024). Sustainable strategy to boost legumes growth under salinity and drought stress in semi-arid and arid regions. *Soil Systems*, *8*(3), 84. https://doi.org/10.3390/soilsystems8030084

Besfat, B. T., Merkuz, A. A., & Gezahegne, G. D. (2020). Assessment and characterization of mung bean (Vigna radiata) bacterial brown spot in Eastern Amhara, Ethiopia. *African Journal of Agricultural Research*, *16*(5), 606–621. https://doi.org/10.5897/ajar2019.14681

Bhandari, K., Sharma, K. D., Hanumantha Rao, B., Siddique, K. H. M., Gaur, P., Agrawal, S. K., Nair, R. M., & Nayyar, H. (2017). Temperature sensitivity of food legumes: a physiological insight. *Acta Physiologiae Plantarum*, *39*(3), 1–22. https://doi.org/10.1007/s11738-017-2361-5

Bhat, K. A., Mahajan, R., Pakhtoon, M. M., Urwat, U., Bashir, Z., Shah, A. A., Agrawal, A., Bhat, B., Sofi, P. A., Masi, A., & Zargar, S. M. (2022). Low temperature stress tolerance: An insight into the omics approaches for legume crops. *Frontiers in Plant Science*, *13*, 888710. https://doi.org/10.3389/fpls.2022.888710

Bhattacharya, A. (2021). Effect of soil water deficit on growth and development of plants: A review. *Soil Water Deficit and Physiological Issues in Plants*, *2022*, 393–488. https://doi.org/10.1007/978-981-33-6276-5_5

Bhattacharya, A. (2022). Effect of low temperature stress on photosynthesis and allied traits: A review. *Physiological Processes in Plants Under Low Temperature Stress*, *2022*, 199–297. https://doi.org/10.1007/978-981-16-9037-2_3

Boukar, O., Belko, N., Chamarthi, S., Togola, A., Batieno, J., Owusu, E., Haruna, M., Diallo, S., Umar, M. L., Olufajo, O., & Fatokun, C. (2019). Cowpea (Vigna unguiculata): Genetics, genomics and breeding. *Plant Breeding*, *138*(4), 415–424. https://doi.org/10.1111/pbr.12589

Bouzroud, S., Henkrar, F., Fahr, M., & Smouni, A. (2023). Salt stress responses and alleviation strategies in legumes: A review of the current knowledge. *3 Biotech*, *13*(8), 1–37. https://doi.org/10.1007/s13205-023-03643-7

Chammakhi, C., Boscari, A., Pacoud, M., Aubert, G., Mhadhbi, H., & Brouquisse, R. (2022). Nitric oxide metabolic pathway in drought-stressed nodules of Faba Bean (Vicia faba L.). *International Journal of Molecular Sciences*, *23*(21), 13057. https://doi.org/10.3390/ijms232113057

Checcucci, A., DiCenzo, G. C., Bazzicalupo, M., & Mengoni, A. (2017). Trade, diplomacy, and warfare: The Quest for elite rhizobia inoculant strains. *Frontiers in Microbiology*, *8*(NOV), 1–7. https://doi.org/10.3389/fmicb.2017.02207

Chen, F., Chen, L., Yan, Z., Xu, J., Feng, L., He, N., Guo, M., Zhao, J., Chen, Z., Chen, H., Yao, G., & Liu, C. (2024). Recent advances of CRISPR-based genome editing for enhancing staple crops. *Frontiers in Plant Science*, *15*, 01–21. https://doi.org/10.3389/fpls.2024.1478398

Conner, R. L., McRae, K. B., Hwang, S. F., Strelkov, S., Sager, S., Hou, A., Penner, W. C., Stoesz, D. B., & Turnbull, G. D. (2021). Effects of seed infection and hydration on the buildup of common bacterial blight and its impact on the yield of dry beans. *Canadian Journal of Plant Science*, *101*(2), 249–261. https://doi.org/10.1139/cjps-2020-0132

Cooper, B., Campbell, K. B., Beard, H. S., Garrett, W. M., & Ferreira, M. E. (2020). The proteomics of resistance to halo blight in common bean. *Molecular Plant-Microbe Interactions*, *33*(9), 1161–1175. https://doi.org/10.1094/MPMI-05-20-0112-R

Costa, S. R., Ng, J. L. P., & Mathesius, U. (2021). Interaction of symbiotic rhizobia and parasitic root-knot nematodes in legume roots: From molecular regulation to field application. *Molecular Plant-Microbe Interactions®*, *34*(5), 470–490. MPMI-12-20-0350. https://doi.org/10.1094/mpmi-12-20-0350-fi

Desta, G., Kassawmar, T., Tadesse, M., & Zeleke, G. (2021). Extent and distribution of surface soil acidity in the rainfed areas of Ethiopia. *Land Degradation and Development*, *32*(18), 5348–5359. https://doi.org/10.1002/ldr.4113

Dev, D., Antre, S. H., Nandni, S., Pandey, S., Singh, M. K., & Singh, A. (2024). Accelerated crop domestication: Candidates, advanced tools and prospects. *South African Journal of Botany*, *171*, 237–244. https://doi.org/10.1016/j.sajb.2024.06.011

Devi, J., Mishra, G. P., Sagar, V., Kaswan, V., Dubey, R. K., Singh, P. M., Sharma, S. K., & Behera, T. K. (2022). Gene-based resistance to Erysiphe species causing powdery mildew disease in peas (Pisum sativum L.). *Genes*, *13*(2), 316. https://doi.org/10.3390/genes13020316

Devi, N. O., Devi, R. K. T., Debbarma, M., Hajong, M., & Thokchom, S. (2022). Effect of endophytic Bacillus and arbuscular mycorrhiza fungi (AMF) against Fusarium wilt of tomato caused by Fusarium oxysporum f. sp. lycopersici. *Springer*, *32*(1), 1. https://doi.org/10.1186/s41938-021-00499-y

Diab, M. K., Mead, H. M., Ahmad Khedr, M. M., Abu-Elsaoud, A. M., & El-Shatoury, S. A. (2024). Actinomycetes are a natural resource for sustainable pest control and safeguarding agriculture. *Archives of Microbiology*, *206*(6), 1–27. https://doi.org/10.1007/s00203-024-03975-9

Dita, M., Rispail, N., Prats, E., Rubiales, D., & Singh, K. (2006). Biotechnology approaches to overcome biotic and abiotic stress constraints in legumes. *Euphytica*, *147*, 1–24. https://doi.org/10.1007/s10681-006-6156-9

Dresselhaus, T., & Hückelhoven, R. (2018). Biotic and abiotic stress responses in crop plants. *Agronomy*, *8*(11), 8–13. https://doi.org/10.3390/agronomy8110267

Dubey, R. S., Srivastava, R. K., & Pessarakli, M. (2021). Physiological mechanisms of nitrogen absorption and assimilation in plants under stressful conditions. *Handbook of Plant and Crop Physiology*, *2021*, 579–616. https://doi.org/10.1201/9781003093640-36

El-Hassan, S. A., Gowen, S. R., & Pembroke, B. (2013). Use of Trichoderma hamatum for biocontrol of lentil vascular wilt disease: efficacy, mechanisms of interaction and future prospects. *Journal of Plant Protection Research*, *53*(1), 12–26. https://doi.org/10.2478/jppr-2013-0002

El-Khateeb, N. M. M. (2014). Influence of rhizobium sp. Combined with trichoderma spp. on damping-off disease and growth parameters of faba bean plants. *Egyptian Journal of Biological Pest Control*, *24*(1), 139–149. https://search.ebscohost.com/login.aspx?direct=true&profile=ehost&scope=site&authtype=

crawler&jrnl=11101768&AN=97137195&h=maFFyWz0OVd9bcgborfFOj5XDfBvnX315Wlm1Uoaq FdvH5GxCTsd8mTaAPI0d2JUZK0GUINXdSHaWJIL8pIkMA%3D%3D&crl=c

Elgharably, A., & Benes, S. (2021). Alfalfa biomass yield and nitrogen fixation in response to applied mineral nitrogen under saline soil conditions. *Journal of Soil Science and Plant Nutrition*, *21*(1), 744–755. https://doi.org/10.1007/s42729-020-00397-6

Falcon, W. P., Naylor, R. L., & Shankar, N. D. (2022). Rethinking global food demand for 2050. *Population and Development Review*, *48*(4), 921–957. https://doi.org/10.1111/padr.12508

Fathi, A., & Mehdiniya, J. (2023). Plant growth and development in relation to phosphorus: A review. *Bulletin of University of Agricultural Sciences and Veterinary Medicine Cluj-Napoca. Agriculture*, *80*(1), 1–7. https://doi.org/10.15835/buasvmcn-agr:2022.0012

Ferreira, J. J., Campa, A., Pérez-Vega, E., Rodríguez-Suárez, C., & Giraldez, R. (2012). Introgression and pyramiding into common bean market class fabada of genes conferring resistance to anthracnose and potyvirus. *Theoretical and Applied Genetics*, *124*(4), 777–788. https://doi.org/10.1007/s00122-011-1746-x

Gaurkhede, J., Gupta, O. M., & Patil, M. (2015). Management of collar rot of chickpea by Pseudomonas fluorescens and identification of sources of resistance. *Indianjournals.Com*, *28*(2), 51–54. www.indianjournals.com/ijor.aspx?target=ijor:jfl&volume=28&issue=2&article=011

Gemmeda, L., Getu, E., & Muleta, D. (2024). Native Bacillus thuringiensis isolates against the old-world bollworm, Helicoverpa armigera (Hübner) (Lepidoptera: Noctuidae) in Ethiopia. *Biocontrol Science and Technology*, *34*(7), 616–629. https://doi.org/10.1080/09583157.2024.2362315

Ghosh, S. M., Debbarma, K., & Chakraborty, G. (2024). Estimation of yield loss of Cowpea, [Vigna unguiculata (L.) Walp.] with reaction response of few genotypes against root knot nematode, meloidogyne incognita (Kofoid and White), Chitwood. *Legume Research*, *47*(3), 484–489. https://doi.org/10.18805/LR-4514

Gogoi, N., Farooq, M., Barthakur, S., Baroowa, B., Paul, S., Bharadwaj, N., & Ramanjulu, S. (2018). Thermal stress impacts on reproductive development and grain yield in grain legumes. *Journal of Plant Biology*, *61*(5), 265–291. https://doi.org/10.1007/s12374-018-0130-7

Gonçalves-Vidigal, M. C., Gilio, T. A. S., Valentini, G., Vaz-Bisneta, M., Vidigal Filho, P. S., Song, Q., Oblessuc, P. R., & Melotto, M. (2020). New Andean source of resistance to anthracnose and angular leaf spot: Fine-mapping of disease-resistance genes in California Dark Red Kidney common bean cultivar. *PLOS ONE*, *15*(6), e0235215.

Gupta, J., Singh, U. N., Akram, M., Naimuddin, & Mishra, R. K. (2021). Potential of biological control agents for the management of soil-borne pathogens in pulse crops. *Journal of Food Legumes*, *34*(3), 149–165. www.indianjournals.com/ijor.aspx?target=ijor:jfl&volume=34&issue=3&article=002

Gupta, R., Mfarrej, M. F. B., Elnour, R. O., Hashem, M., & Ahmad, F. (2023). Defence response of host plants for cyst nematode: A review on parasitism and defence. *Journal of King Saud University – Science*, *35*(7), 102829. https://doi.org/10.1016/j.jksus.2023.102829

Gurmessa, B. (2021). Soil acidity challenges and the significance of liming and organic amendments in tropical agricultural lands with reference to Ethiopia. *Environment, Development and Sustainability*, *23*(1), 77–99. https://doi.org/10.1007/s10668-020-00615-2

Haile, G. G., Tang, Q., Hosseini-Moghari, S. M., Liu, X., Gebremicael, T. G., Leng, G., Kebede, A., Xu, X., & Yun, X. (2020). Projected impacts of climate change on drought patterns over East Africa. *Earth's Future*, *8*(7), e2020EF001502. https://doi.org/10.1029/2020EF001502

Haldar, S., Kumari, A., Ghosh, A., & Ghosh, A. (2022). Influence of biotic and abiotic stresses on Quinoa cultivation: Insights into microbe-assisted stress tolerance. *Biology and Biotechnology of Quinoa: Super Grain for Food Security*, *2021*, 167–194. https://doi.org/10.1007/978-981-16-3832-9_9

Harrison, R. D., Thierfelder, C., Baudron, F., Chinwada, P., Midega, C., Schaffner, U., & van den Berg, J. (2019). Agro-ecological options for fall armyworm (Spodoptera frugiperda JE Smith)management: Providing low-cost, smallholder friendly solutions to an invasive pest. *Journal of Environmental Management*, *243*, 318–330. https://doi.org/10.1016/j.jenvman.2019.05.011

Hashem, A., Kumar, A., Al-Dbass, A. M., Alqarawi, A. A., Al-Arjani, A. B. F., Singh, G., Farooq, M., & Abd_Allah, E. F. (2019). Arbuscular mycorrhizal fungi and biochar improves drought tolerance in chickpea. *Saudi Journal of Biological Sciences*, *26*(3), 614–624. https://doi.org/10.1016/j.sjbs.2018.11.005

Hu, D., Wang, D., Pan, H., & Liu, X. (2025). Molecular mechanisms underlying resistance to bacillus thuringiensis cry toxins in Lepidopteran pests: An updated research perspective. *Agronomy*, *15*(1), 1–21. https://doi.org/10.3390/agronomy15010155

Imen, H., Neila, A., Adnane, B., Manel, B., Mabrouk, Y., Saidi, M., & Bouaziz, S. (2015). Inoculation with phosphate solubilizing mesorhizobium strains improves the performance of Chickpea (Cicer aritenium L.) under phosphorus deficiency. *Journal of Plant Nutrition*, *38*(11), 1656–1671. https://doi.org/10.1080/01904167.2015.1061543

Islam, M. S., Fahad, S., Hossain, A., Chowdhury, M. K., Iqbal, M. A., Dubey, A., Kumar, A., Rajendran, K., Danish, S., Rahman, M. H. U., Raza, M. A., Arif, M., Saud, S., Hossain, M. A., Waraic, E., Ahmad, Z., Hussain, S., Çïg, A., Erman, M., … EL Sabagh, A. (2021). Legumes under drought stress: Plant responses, adaptive mechanisms, and management strategies in relation to nitrogen fixation. *Engineering Tolerance in Crop Plants Against Abiotic Stress*, *2021*, 179–207. https://doi.org/10.1201/9781003160717-9

Janati, W., Benmrid, B., Elhaissoufi, W., Zeroual, Y., Nasielski, J., & Bargaz, A. (2021). Will phosphate biosolubilization stimulate biological nitrogen fixation in grain legumes? *Frontiers in Agronomy*, *3*(March), 1–29. https://doi.org/10.3389/fagro.2021.637196

Jiang, J., Huang, Y., Shu, C., Soberon, M. A. B., C. L., & Al, E. (2017). Holotrichia oblita midgut proteins that bind to Bacillus thuringiensis cry8-like toxin and assembly of the H. oblita midgut tissue transcriptome. *Applied Environmental Microbiology*, *83,* 1–11. 10.1128/AEM.00541-17

Joshi, R., Ramawat, N., Jha, J., Durgesh, K., Singh, M., Talukdar, A., Tomar, S. M. S., & Singh, D. (2021). Salt stress in pulses: A learning from global research on salinity in crop plants. *Indian Journal of Genetics and Plant Breeding*, *81*(2), 159–185. https://doi.org/10.31742/IJGPB.81.2.1

Julian, N., Francioli, D., Mang, M., & Ludewig, U. (2023). Role of mineral nitrogen nutrition in fungal plant diseases of cereal crops. *Critical Reviews in Plant Sciences*, *42*(3), 93–123. https://doi.org/10.1080/07352689.2023.2196100

Kafeel, U., Jahan, U., & Khan, F. A. (2022). Role of mineral nutrients in biological nitrogen fixation. *Sustainable Plant Nutrition: Molecular Interventions and Advancements for Crop Improvement*, *2022*, 87–106. https://doi.org/10.1016/B978-0-443-18675-2.00004-3

Kan, Y., Mu, X. R., Gao, J., Lin, H. X., & Lin, Y. (2023). The molecular basis of heat stress responses in plants. *Molecular Plant*, *16*(10), 1612–1634. https://doi.org/10.1016/j.molp.2023.09.013

Kebede, E. (2021). Contribution, utilization, and improvement of legumes-driven biological nitrogen fixation in agricultural systems. *Frontiers in Sustainable Food Systems*, *5*, 767998. https://doi.org/10.3389/fsufs.2021.767998

Khan, M. R. (2023). Nematode pests of agricultural crops, a global overview. *Novel Biological and Biotechnological Applications in Plant Nematode Management*, 3–45. https://doi.org/10.1007/978-981-99-2893-4_1

Khandelwal, R., & Mukesh, J. (2018). Genome engineering tools for functional genomics and crop improvement in legumes. In Shabir Hussain Wani and Mukesh Jain (Eds) *Pulse Improvement: Physiological, Molecular and Genetic Perspectives* (pp. 219–235). Springer. https://doi.org/10.1007/978-3-030-01743-9

Khatun, M., Sarkar, S., Era, F. M., Islam, A. K. M. M., Anwar, M. P., Fahad, S., Datta, R., & Islam, A. K. M. A. (2021). Drought stress in grain legumes: Effects, tolerance mechanisms and management. *Agronomy*, *11*(12), 2374. https://doi.org/10.3390/agronomy11122374

Kide Mengistu Habtamu. (2022). Abiotic and biotic stress factors affecting storage of legumes in tropics. *Legumes Research – Volume 1*, 1–17. https://doi.org/10.5772/intechopen.99413

Kiptoo, G. J., Kinyua, M. G., Kiplagat, O. K., & Matasyoh, L. G. (2020). Incidence and severity of anthracnose (Colletotrichum lindemuthianum) on selected common bean (Phaseolus vulgaris L.) Genotypes. *American Journal of Agricultural Research*, *84,* 1–12. https://doi.org/10.28933/ajar-2019-09-2705

Kirova, E., & Kocheva, K. (2021). Physiological effects of salinity on nitrogen fixation in legumes – A review. *Journal of Plant Nutrition*, *44*(17), 1–10. https://doi.org/10.1080/01904167.2021.1921204

Klimek-Kopyra, A., Dłużniewska, J., & Sikora, A. (2023). Influence of biofungicides containing microorganisms such as ythium oligandrum and Bacillus subtilis on yield, morphological parameters, and pathogen suppression in ix Winter Pea Cultivars. *Agriculture (Switzerland)*, *13*(6), 1170. https://doi.org/10.3390/agriculture13061170

Kocyigit, E., Kocaadam-Bozkurt, B., Bozkurt, O., Ağagündüz, D., & Capasso, R. (2023). Plant toxic proteins: Their biological activities, mechanism of action and removal strategies. *Toxins*, *15*(6), 1–19. https://doi.org/10.3390/toxins15060356

Köhl, J., Kolnaar, R., & Ravensberg, W. J. (2019). Mode of action of microbial biological control agents against plant diseases: Relevance beyond efficacy. *Frontiers in Plant Science*, *10*, 1–19. https://doi.org/10.3389/fpls.2019.00845

Korir, H., Mungai, N. W., Thuita, M., Hamba, Y., & Masso, C. (2017). Co-inoculation effect of rhizobia and plant growth promoting rhizobacteria on common bean growth in a low phosphorus soil. *Frontiers in Plant Science*, *8*(FEBRUARY), 1–10. https://doi.org/10.3389/fpls.2017.00141

Koskey, G., Mburu, S. W., Awino, R., Njeru, E. M., & Maingi, J. M. (2021). Potential use of beneficial microorganisms for soil amelioration, phytopathogen biocontrol, and sustainable crop production in Smallholder agroecosystems. *Frontiers in Sustainable Food Systems*, *5*(April), 1–20. https://doi.org/10.3389/fsufs.2021.606308

Koskey, G., Mburu, S. W., Njeru, E. M., Kimiti, J. M., Ombori, O., & Maingi, J. M. (2017). Potential of native rhizobia in enhancing nitrogen fixation and yields of climbing beans (phaseolus vulgaris L.) in contrasting environments of eastern Kenya. *Frontiers in Plant Science*, *8*, 248296. https://doi.org/10.3389/fpls.2017.00443

Kour, D., Rana, K. L., Yadav, N., Yadav, A. N., Kumar, A., Meena, V. S., Singh, B., Chauhan, V. S., Dhaliwal, H. S., & Saxena, A. K. (2019). Rhizospheric microbiomes: Biodiversity, mechanisms of plant growth promotion, and biotechnological applications for sustainable agriculture. In Ashok Kumar and Vijay Singh Meena (Eds), *Plant Growth Promoting Rhizobacteria for Agricultural Sustainability* (pp. 19–65). Springer Singapore. https://doi.org/10.1007/978-981-13-7553-8_2

Kumar, A., & Verma, J. P. (2018). Does plant—Microbe interaction confer stress tolerance in plants: A review? *Microbiological Research*, *207*, 41–52. https://doi.org/10.1016/j.micres.2017.11.004

Kumar, V., Koul, B., Taak, P., Yadav, D., & Song, M. (2023). Journey of trichoderma from pilot scale to mass production: A review. *Agriculture (Switzerland)*, *13*(10), 1–37. https://doi.org/10.3390/agriculture13102022

Kumari, V. V., Roy, A., Vijayan, R., Banerjee, P., Verma, V. C., Nalia, A., Pramanik, M., Mukherjee, B., Ghosh, A., Reja, M. H., Chandran, M. A. S., Nath, R., Skalicky, M., Brestic, M., & Hossain, A. (2021). Drought and heat stress in cool-season food legumes in sub-tropical regions: Consequences, adaptation, and mitigation strategies. *Plants*, *10*(6), 1038. https://doi.org/10.3390/plants10061038

Kunda, P., Mukherjee, A., & Dhal, P. K. (2020). Bacterial biological control agents for soilborne diseases management in pulses: Present status and future prospects. *Microbial Mitigation of Stress Response of Food Legumes*, *2022*, 231–243. https://doi.org/10.1201/9781003028413-22

Lacerda, Í. C. dos S., Polonio, J. C., & Golias, H. C. (2022). Endophytic fungi as a source of antiviral compounds – A review. *Chemistry and Biodiversity*, *19*(6), e202100971. https://doi.org/10.1002/cbdv.202100971

Langan, P., Bernád, V., Walsh, J., Henchy, J., Khodaeiaminjan, M., Mangina, E., & Negrão, S. (2022). Phenotyping for waterlogging tolerance in crops: Current trends and future prospects. *Journal of Experimental Botany*, *73*(15), 5149–5169. https://doi.org/10.1093/jxb/erac243

Li, H., Wang, X., Liang, Q., Lyu, X., Li, S., Gong, Z., Dong, S., Yan, C., & Ma, C. (2021). Regulation of phosphorus supply on nodulation and nitrogen fixation in oybean plants with dual-root systems. *Agronomy*, *11*(11), 2354. https://doi.org/10.3390/agronomy11112354

Li, Y., Liu, Q., Zhang, D. X., Zhang, Z. Y., Xu, A., Jiang, Y. L., & Chen, Z. C. (2024). Metal nutrition and transport in the process of symbiotic nitrogen fixation. *Plant Communications*, *5*(4), 100829. https://doi.org/10.1016/j.xplc.2024.100829

Lohani, N., Singh, M. B., & Bhalla, P. L. (2022). Biological parts for engineering abiotic stress tolerance in plants. *BioDesign Research*, *2022*, 1–41 https://doi.org/10.34133/2022/9819314

López, M., Muñoz, N., Lascano, H. R., & Izaguirre-Mayoral, M. L. (2017). The seed-borne Southern bean mosaic virus hinders the early events of nodulation and growth in Rhizobium-inoculated Phaseolus vulgaris L. *Functional Plant Biology*, *44*(2), 208–218. https://doi.org/10.1071/FP16180

Lucas, M. R., Ehlers, J. D., Huynh, B. L., Diop, N. N., Roberts, P. A., & Close, T. J. (2013). Markers for breeding heat-tolerant cowpea. *Molecular Breeding*, *31*(3), 529–536. https://doi.org/10.1007/s11032-012-9810-z

Mahadevakumar, S., & Sridhar, K. R. (2021). *Diversity of Pathogenic Fungi in Agricultural Crops* (pp. 101–149), Springer, Singapore. https://doi.org/10.1007/978-981-16-3364-5_6

Mahatma, L., & Shukla, A. (2024). Diversity of arthropod vectors of plant viruses. *New and Emerging Plant Viruses: The Threat to Food Security, 2024*, 169–208. https://doi.org/10.1201/9781003497486-6

Mahmoud, G. A.-E. (2021). *Biotic Stress to Legumes: Fungal Diseases as Major Biotic Stress Factor* (pp. 181–212), Springer, Cham. https://doi.org/10.1007/978-3-030-68828-8_7

Maiti, R., Rodríguez, H. G., Kumari, C. A., & Begum, S. (2021). Biotic stresses affecting crop productivity. *Advances in Sorghum Science, 2021*, 139–160. https://doi.org/10.1201/9781003127628-8

Majeed, J. A., Mahmood, A., Bibi, S., Jabeen, A., Javaid, M. M., Ahmad, H. B., & Nargis, J. (2023). Waterlogging and crop productivity. *Climate-Resilient Agriculture, 1*, 237–256. https://doi.org/10.1007/978-3-031-37424-1_11

Mangena, P., & Sehaole, E. (2020). *Transgenic Grain Legumes* (pp. 148–172). Bentham Science Publisher. https://doi.org/10.2174/97898114796251200100 11

Manzoor, M. M. (2020). Environmental biotechnology: For sustainable future. *Bioremediation and Biotechnology, Vol 2: Degradation of Pesticides and Heavy Metals, 2*, 241–258. https://doi.org/10.1007/978-3-030-40333-1_14

Mawar, R., Mathur, M., & Tomer, A. S. (2021). Diseases of arid legumes: Extent and management. *Diseases of Pulse Crops and Their Management, 381*, 1–10. www.researchgate.net/profile/Ritu-Mawar/publication/350513560_Disease_of_Arid_Legumes_Extent_and_Management/links/645c9045fbaf5b27a4bbfef5/Disease-of-Arid-Legumes-Extent-and-Management.pdf#page=398

Mburu, S. W., Koskey, G., Njeru, E. M., & Maingi, J. M. (2021). Revitalization of bacterial endophytes and rhizobacteria for nutrients bioavailability in degraded soils to promote crop production. *AIMS Agriculture and Food, 6*(2), 496–524. https://doi.org/10.3934/agrfood.2021029

McGuire, A. V., & Northfield, T. D. (2020). Tropical occurrence and agricultural importance of Beauveria bassiana and Metarhizium anisopliae. *Frontiers in Sustainable Food Systems, 4,* 1–8. https://doi.org/10.3389/fsufs.2020.00006

Meena, K. K., Sorty, A. M., Bitla, U. M., Choudhary, K., Gupta, P., Pareek, A., Singh, D. P., Prabha, R., Sahu, P. K., Gupta, V. K., Singh, H. B., Krishanani, K. K., & Minhas, P. S. (2017). Abiotic stress responses and microbe-mediated mitigation in plants: The Omics strategies. *Frontiers in Plant Science, 8*, 172. https://doi.org/10.3389/fpls.2017.00172

Meira, D., Woyann, L. G., Bozi, A. H., Milioli, A. S., Beche, E., Panho, M. C., Madella, L. A., Barrionuevo, F., Marchioro, V. S., & Benin, G. (2020). Asian soybean rust: A scientometric approach of Phakopsora pachyrhizi studies. In Herman van Eck (Ed.) *Euphytica* (Vol. 216, Issue 8, pp. 1–12). Springer. https://doi.org/10.1007/s10681-020-02667-x

Merry, R., Dobbels, A. A., Sadok, W., Naeve, S., Stupar, R. M., & Lorenz, A. J. (2022). Iron deficiency in soybean. *Crop Science, 62*(1), 36–52. https://doi.org/10.1002/csc2.20661

Mir, R. R., Rather, I. A., Bhat, M. A., Parray, G. A., & Varshney, R. K. (2017). *Molecular Mapping of Genes and QTLs in Pigeonpea BT – The Pigeonpea Genome* (R. K. Varshney, R. K. Saxena, & S. A. Jackson (eds.); pp. 55–64). Springer International Publishing. https://doi.org/10.1007/978-3-319-63797-6_6

Mishra, R. K., Bohra, A., Kamaal, N., Kumar, K., Gandhi, K., Sujayanand, G. K., Saabale, P. R., Satheesh Naik, S. J., Sarma, B. K., Kumar, D., Mishra, M., Srivastava, D. K., & Singh, N. P. (2018). Utilization of biopesticides as sustainable solutions for management of pests in legume crops: Achievements and prospects. *Egyptian Journal of Biological Pest Control, 28*(1), 1–11. https://doi.org/10.1186/s41938-017-0004-1

Moorthy, P. N. K., Kumar, N. R. P., Mani, M., & Saroja, S. (2022). Pests and their management in leguminous vegetables: (French Beans, Sweet Peas, Cowpea, Vegetable Pigeon Pea, Dolichos Bean, Red Kidney and Cluster Beans). *Trends in Horticultural Entomology, 2022*, 1031–1049. https://doi.org/10.1007/978-981-19-0343-4_43

Moparthi, S., Burrows, M., Mgbechi-Ezeri, J., & Agindotan, B. (2021). Fusarium spp. associated with root rot of pulse crops and their cross-pathogenicity to cereal crops in montana. *Plant Disease, 105*(3), 548–557. https://doi.org/10.1094/PDIS-04-20-0800-RE

Mostofa, M. G., Rahman, M. M., Ghosh, T. K., Kabir, A. H., Abdelrahman, M., Rahman Khan, M. A., Mochida, K., & Tran, L. S. P. (2022). Potassium in plant physiological adaptation to abiotic stresses. *Plant Physiology and Biochemistry, 186*, 279–289. https://doi.org/10.1016/j.plaphy.2022.07.011

Moura, L. de O., da Silva, M. F., da Cunha, F. F., Picoli, E. A. de T., Silva, F. C. dos S., & da Silva, F. L. (2023). Water deficit as a trigger to immature soybean pod opening. *Journal of Agronomy and Crop Science*, *209*(3), 390–401. https://doi.org/10.1111/jac.12634

Mukhopadhyay, R., & Kumar, D. (2020). Trichoderma: A beneficial antifungal agent and insights into its mechanism of biocontrol potential. *Egyptian Journal of Biological Pest Control*, *30*(1). https://doi.org/10.1186/s41938-020-00333-x

Muleta, D., Ryder, M. H., & Denton, M. D. (2017). The potential for rhizobial inoculation to increase soybean grain yields on acid soils in Ethiopia. *Soil Science and Plant Nutrition*, *63*(5), 441–451. https://doi.org/10.1080/00380768.2017.1370961

Musyoka, D. M., Njeru, E. M., Nyamwange, M. M., & Maingi, J. M. (2020). Arbuscular mycorrhizal fungi and Bradyrhizobium co-inoculation enhances nitrogen fixation and growth of green grams (Vigna radiata L.) under water stress. *Journal of Plant Nutrition*, *43*(7), 1036–1047. https://doi.org/10.1080/01904167.2020.1711940

Nadeem, M., Li, J., Yahya, M., Sher, A., Ma, C., Wang, X., & Qiu, L. (2019). Research progress and perspective on drought stress in legumes: A review. *International Journal of Molecular Sciences*, *20*(10), 2541. https://doi.org/10.3390/ijms20102541

Nagaraja, A., Chethana, B. S., & Jain, A. K. (2021). Biotic stresses and their management. *Millets and Pseudo Cereals*, *2020*, 119–142. https://doi.org/10.1016/b978-0-12-820089-6.00007-0

Nair, R. M., Pandey, A. K., War, A. R., Hanumantharao, B., Shwe, T., Alam, A. K. M. M., Pratap, A., Malik, S. R., Karimi, R., Mbeyagala, E. K., Douglas, C. A., Rane, J., & Schafleitner, R. (2019). Biotic and abiotic constraints in Mungbean production—Progress in genetic improvement. *Frontiers in Plant Science*, *10*, 1340. https://doi.org/10.3389/fpls.2019.01340

Naseri, B., Gheitury, M., & Veisi, M. (2021). Infections of bean plant and field soil are linked to region, root rot pathogen and agro-ecosystem. *Hellenic Plant Protection Journal*, *14*(1), 14–23. https://doi.org/10.2478/hppj-2021-0002

Ncama, K., Aremu, O. A., & Sithole, N. J. (2021). Plant adaptation to environmental stress: Drought, chilling, heat, and salinity. *Environment and Climate-Smart Food Production*, *2021*, 151–179. https://doi.org/10.1007/978-3-030-71571-7_5

Nigam, R., & Singh, J. (2022). Integrated diseases management of lentil wilt caused by Fusarium oxysporum f. sp. lentis [L.] Medik. *Journal of Agriculture Research and Technology*, *Special*(01), 130–137. https://doi.org/10.56228/jart.2022.sp121

Nirmalkar, V. K., Singh, S., Tiwari, R. S., Said, P. P., & Kaushik, D. K. (2017). Field efficacy of richoderma harzianum and Rhizobium against wilt complex of Chickpea. *International Journal of Current Microbiology and Applied Sciences*, *6*(7), 1421–1429. https://doi.org/10.20546/ijcmas.2017.607.170

Njeru, E. M., Muthini, M., Muindi, M. M., Ombori, O., Nchore, S. B., Runo, S., & Maingi, J. M. (2020). Exploiting arbuscular mycorrhizal fungi-rhizobia-legume symbiosis to increase smallholder farmers' crop production and resilience under a changing climate. *Climate Impacts on Agricultural and Natural Resource Sustainability in Africa*, *2020*, 471–488. https://doi.org/10.1007/978-3-030-37537-9_27

Nkomo, G. V., Sedibe, M. M., & Mofokeng, M. A. (2021). Production constraints and improvement strategies of Cowpea (Vigna unguiculata L. Walp.) genotypes for drought tolerance. *International Journal of Agronomy*, *2021*, 1–9. https://doi.org/10.1155/2021/5536417

Nosheen, S., Ajmal, I., & Song, Y. (2021). Microbes as biofertilizers, a potential approach for sustainable crop production. *Sustainability (Switzerland)*, *13*(4), 1–20. https://doi.org/10.3390/su13041868

Nyaga, J. W., & Njeru, E. M. (2020). Potential of native rhizobia to improve Cowpea growth and production in semiarid regions of Kenya. *Frontiers in Agronomy*, *2*, 606293. https://doi.org/10.3389/fagro.2020.606293

Ojiewo, C., Monyo, E., Desmae, H., Boukar, O., Thudi, C. M. M., Pandey, M. K., K., R., M., S. P., K., G. S., Asnake, C., Ganga, F. N., CV, R., SameerKumar, Okori, P., Janila, P., Akpo, J. C. R. C., Essegbemon, G., Varshney, L., Williams, O. S. N. B. F. P. B. M. K. M., … K., M. R. (2018). Genomics, genetics and breeding of tropical legumes for better livelihoods of smallholder farmers. *Plant Breeding*, *138*(4), 487–499. https://doi.org/https://doi.org/10.1111/pbr.12554

Ongoma, V., Chen, H., Gao, C., Nyongesa, A. M., & Polong, F. (2018). Future changes in climate extremes over Equatorial East Africa based on CMIP5 multimodel ensemble. *Natural Hazards*, *90*(2), 901–920. https://doi.org/10.1007/S11069-017-3079-9/METRICS

Osuna-Caballero, S., Rispail, N., Barilli, E., & Rubiales, D. (2024). Management and breeding for rust resistance in legumes. *Journal of Plant Pathology*, *2024*, 1–27. https://doi.org/10.1007/s42161-024-01679-z

Otieno, H. (2023). Prevalence of potato cyst (Globodera spp.) nematode and potato root-knot (Meloidogyne spp.) nematode in Kenya and potential management strategies: A review. *Archives of Agriculture and Environmental Science*, *8*(2), 250–258. https://doi.org/10.26832/24566632.2023.0802023

Pallangyo, B., Mdily, K., Mkondo, C., & Kibola, A. (2019). Crop pests, control measures and potential impacts in Kihansi Catchment area. *Tanzania Journal of Science*, *45*(4), 650–660. www.ajol.info/index.php/tjs/

Panwar, M., Tewari, R., & Nayyar, H. (2016). Native halo-tolerant plant growth promoting rhizobacteria Enterococcus and Pantoea sp. improve seed yield of Mungbean (Vigna radiata L.) under soil salinity by reducing sodium uptake and stress injury. *Physiology and Molecular Biology of Plants*, *22*(4), 445–459. https://doi.org/10.1007/s12298-016-0376-9

Parthasarathi, T., Firdous, S., Mariya David, E., Lesharadevi, K., & Djanaguiraman, M. (2022). Effects of high temperature on crops. *Advances in Plant Defense Mechanisms*. *2022*, 1–18 https://doi.org/10.5772/intechopen.105945

Penn, C. J., & Camberato, J. J. (2019). A critical review on soil chemical processes that control how soil ph affects phosphorus availability to plants. *Agriculture*, *9*(6), 120. https://doi.org/10.3390/agriculture9060120

Qian, D., Wang, M., Niu, Y., Yang, Y., & Xiang, Y. (2025). Sexual reproduction in plants under high temperature and drought stress. *Cell Reports*, *44*(3), 1–16. https://doi.org/10.1016/j.celrep.2025.115390

Qin, D., Liu, X.-Y., Miceli, C., Zhang, Q., & Wang, P. (2019). Soybean plants expressing the Bacillus thuringiensis cry8-like gene show resistance to Holotrichia parallela. *BMC Biotechnology*, *19*(1), 66. https://doi.org/10.1186/s12896-019-0563-1

Raimi, A., Adeleke, R., & Roopnarain, A. (2017). Soil fertility challenges and Biofertiliser as a viable alternative for increasing smallholder farmer crop productivity in sub-Saharan Africa. *Cogent Food and Agriculture*, *3*(1), 1–26. https://doi.org/10.1080/23311932.2017.1400933

Rane, J., Singh, A. K., Kumar, M., Boraiah, K. M., Meena, K. K., Pradhan, A., & Vara Prasad, P. V. (2021). The adaptation and tolerance of major cereals and legumes to important abiotic stresses. *International Journal of Molecular Sciences*, *22*(23), 12970. https://doi.org/10.3390/ijms222312970

Rani, A., Devi, P., Jha, U. C., Sharma, K. D., Siddique, K. H. M., & Nayyar, H. (2020). Developing climate-resilient Chickpea involving physiological and molecular approaches with a focus on temperature and drought stresses. *Frontiers in Plant Science*, *10*, 1–29. https://doi.org/10.3389/fpls.2019.01759

Ranjan, A., Sinha, R., Sharma, T. R., Pattanayak, A., & Singh, A. K. (2021). Alleviating aluminum toxicity in plants: Implications of reactive oxygen species signaling and crosstalk with other signaling pathways. *Physiologia Plantarum*, *173*(4), 1765–1784. https://doi.org/10.1111/ppl.13382

Rao, G. P., & Reddy, M. G. (2020). Overview of yield losses due to plant viruses. In L.P. Awasthi (ed.), *Applied Plant Virology* (pp. 531–562). Elsevier. https://doi.org/10.1016/b978-0-12-818654-1.00038-4

Reen, R. A., Mumford, M. H., & Thompson, J. P. (2019). Novel sources of resistance to root-lesion nematode (Pratylenchus thornei) in a new collection of wild cicer species (C. Reticulatum and C. Echinospermum) to improve resistance in cultivated Chickpea (C. Arietinum). *Phytopathology*, *109*(7), 1270–1279. https://doi.org/10.1094/PHYTO-02-19-0047-R

Reena, & Prasad, S. (2022). Biotic stresses of major pulse crops and their management strategies. *Abiotic and Biotic Stress Management in Plants*, *2*, 31–48. https://doi.org/10.1201/9781003286134-2

Richard, B., Qi, A., & Fitt, B. D. L. (2022). Control of crop diseases through Integrated Crop Management to deliver climate-smart farming systems for low- and high-input crop production. *Plant Pathology*, *71*(1), 187–206. https://doi.org/10.1111/ppa.13493

Roriz, M., Carvalho, S. M. P., Castro, P. M. L., & Vasconcelos, M. W. (2020). Legume biofortification and the role of plant growth-promoting bacteria in a sustainable agricultural era. *Agronomy*, *10*(3), 235. https://doi.org/10.3390/agronomy10030435

S. Basu, P., Kumar Chaturvedi, S., Mall Gaur, P., Mondal, B., Kumar Meena, S., Das, K., Kumar, V., Tewari, K., & Sharma, K. (2022). Physiological mechanisms of tolerance to drought and heat in major pulses for improving yield under stress environments. *Advances in Plant Defense Mechanisms, 2022*, 357–386. https://doi.org/10.5772/intechopen.106054

Saggar, S., & Bolan, N. S. (2024). Secondary nutrients: Sulphur, calcium, and magnesium. *Handbook of Processes and Modeling in the Soil-Plant System*, *2003*, 561–588. https://doi.org/10.1201/9781003578543-18

Salman, Ö., & Boyraz, N. (2023). Efficacy of Some Trichoderma Isolates as Biocontrol Agents Against Rhizoctonia solani Kühn in Bean (Phaseolus vulgaris L.). *Selcuk Journal of Agricultural and Food Sciences*, *37*(1), 133–144. https://doi.org/10.15316/sjafs.2023.014

Sangiogo, M., Rodriguez, D. P., Moccellin, R., Bermudez, J. M. M., Corrêa, B. O., & Moura, A. B. (2018). Foliar spraying with bacterial biocontrol agents for the control of common bacterial blight of bean. *Pesquisa Agropecuaria Brasileira*, *53*(10), 1101–1108. https://doi.org/10.1590/S0100-204X201800 1000003

Santos, M. S., Nogueira, M. A., & Hungria, M. (2019). Microbial inoculants: Reviewing the past, discussing the present and previewing an outstanding future for the use of beneficial bacteria in agriculture. *AMB Express*, *9*(1), 1–22. https://doi.org/10.1186/s13568-019-0932-0

Senthamil, E., Halli, H. M., Basavaraj, P. S., Angadi, S. S., Gangana Gowdra, V. M., Harisha, C. B., Boraiah, K. M., Sandeep Adavi, B., Salakinkoppa, S. R., Mohite, G., & Reddy, K. S. (2025). Waterlogging effects on root morphology, yield, and stress tolerance in Cowpea (Vigna unguiculata L. Walp) grown on semi-arid vertisols. *Journal of Agronomy and Crop Science*, *211*(1), e70014. https://doi.org/10.1111/jac.70014

Sharma, G., Aminedi, R., Saxena, D., Gupta, A., Banerjee, P., Jain, D., & Chandran, D. (2019). Effector mining from the Erysiphe pisi haustorial transcriptome identifies novel candidates involved in pea powdery mildew pathogenesis. *Molecular Plant Pathology*, *20*(11), 1506–1522. https://doi.org/10.1111/mpp.12862

Sharma, S., Sharma, J., Soni, V., Kalaji, H. M., & Elsheery, N. I. (2021). Waterlogging tolerance: A review on regulative morpho-physiological homeostasis of crop plants. *Journal of Water and Land Development*, *49*, 16–28. https://doi.org/10.24425/jwld.2021.137092

Singh, D. P., Singh, H. B., & Prabha, R. (2016). Microbial inoculants in sustainable agricultural productivity: Vol. 2: Functional applications. *Microbial Inoculants in Sustainable Agricultural Productivity: Vol. 2: Functional Applications*, *2*, 1–308. https://doi.org/10.1007/978-81-322-2644-4

Singh, R. K., Prasad, A., Muthamilarasan, M., Parida, S. K., & Prasad, M. (2020). Breeding and biotechnological interventions for trait improvement: Status and prospects. *Planta*, *252*(4), 54. https://doi.org/10.1007/s00425-020-03465-4

Smith, M. R., Dinglasan, E., Veneklaas, E., Polania, J., Rao, I. M., Beebe, S. E., & Merchant, A. (2022). Effect of drought and low P on yield and nutritional content in common bean. *Frontiers in Plant Science*, *13*, 814325. https://doi.org/10.3389/fpls.2022.814325

Smýkal, P., von Wettberg, E. J. B., & McPhee, K. (2020). Legume genetics and biology: From mendel's pea to legume genomics. *International Journal of Molecular Sciences*, *21*(9), 1–5. https://doi.org/10.3390/ijms21093336

Sofi, P. A., Rehman, K., Gull, M., Kumari, J., Djanaguiraman, M., & Prasad, P. V. V. (2021). Integrating root architecture and physiological approaches for improving drought tolerance in common bean (Phaseolus vulgaris L.). *Plant Physiology Reports*, *26*(1), 4–22. https://doi.org/10.1007/s40502-021-00570-8

Srinivasan, R., Sevgan, S., Ekesi, S., & Tamò, M. (2019). Biopesticide based sustainable pest management for safer production of vegetable legumes and brassicas in Asia and Africa. *Pest Management Science*, *75*(9), 2446–2454. https://doi.org/10.1002/ps.5480

Srivastava, P. K., Singh, V. P., Singh, A., Tripathi, D. K., Singh, S., Prasad, S. M., & Chauhan, D. K. (2020). Pesticides in crop production: Physiological and biochemical action. *Pesticides in Crop Production: Physiological and Biochemical Action*, *2020*, 1–292. https://doi.org/10.1002/978111 9432241

Stracquadanio, C., Quiles, J. M., Meca, G., & Cacciola, S. O. (2020). Antifungal activity of bioactive metabolites produced by trichoderma asperellum and trichoderma atroviride in liquid medium. *Journal of Fungi*, *6*(4), 1–18. https://doi.org/10.3390/jof6040263

Sunyar, B., Yeşildağ, M. F., & Alma, M. H. (2024). Effectiveness of bacillus and pseudomonas strains in biological control of common bacterial blight disease in common bean (Phaseolus vulgaris L.). *Journal of Crop Health*, *76*(6), 1357–1372. https://doi.org/10.1007/s10343-024-01064-x

Swarnalakshmi, K., Yadav, V., Tyagi, D., Dhar, D. W., Kannepalli, A., & Kumar, S. (2020). Significance of plant growth promoting rhizobacteria in grain legumes: Growth promotion and crop production. *Plants*, *9*(11), 1–25. https://doi.org/10.3390/plants9111596

Tadele, M. (2020). Impacts of soil acidity on growth performance of Faba bean (Vicia faba L.) and management options. *Academic Research Journal of Agricultural Science and Research*, *8*(4), 423–431. https://doi.org/10.14662/ARJASR2020.275

Tavares, D. S., Sant'Anna-Santos, B. F., & Gomes, M. P. (2024). Unleashing the power of fungi: Utilizing the arbuscular mycorrhizal fungi rhizophagus clarus to mitigate salinity stress and boost Cowpea bean productivity for food security. *Stresses*, *4*(2), 393–410. https://doi.org/10.3390/stresses4020026

Thompson, M. E. H., & Raizada, M. N. (2018). Fungal pathogens of Maize gaining free passage along the silk road. *Pathogens*, *7*(4), 1–16. https://doi.org/10.3390/pathogens7040081

Thu, S. W., Lu, M. Z., Carter, A. M., Collier, R., Gandin, A., Sitton, C. C., & Tegeder, M. (2020). Role of ureides in source-to-sink transport of photoassimilates in non-fixing soybean. *Journal of Experimental Botany*, *71*(15), 4495–4511. https://doi.org/10.1093/jxb/eraa146

van Heerwaarden, J., Baijukya, F., Kyei-Boahen, S., Adjei-Nsiah, S., Ebanyat, P., Kamai, N., Wolde-meskel, E., Kanampiu, F., Vanlauwe, B., & Giller, K. (2018). Soyabean response to rhizobium inoculation across sub-Saharan Africa: Patterns of variation and the role of promiscuity. *Agriculture, Ecosystems and Environment*, *261*, 211–218. https://doi.org/10.1016/j.agee.2017.08.016

Varadharajan, V., Rajendran, R., Muthuramalingam, P., Runthala, A., Madhesh, V., Swaminathan, G., Murugan, P., Srinivasan, H., Park, Y., Shin, H., & Ramesh, M. (2025). Multi-omics approaches against abiotic and biotic stress—A review. *Plants*, *14*(6), 1–33. https://doi.org/10.3390/plants14060865

Vimala Devi, P. S., Duraimurugan, P., Poorna Chandrika, K. S. V., Vineela, V., & Hari, P. P. (2020). Novel formulations of Bacillus thuringiensis var. kurstaki: an eco-friendly approach for management of lepidopteran pests. *World Journal of Microbiology and Biotechnology*, *36*(5), 1–14. https://doi.org/10.1007/s11274-020-02849-8

Wang, J., Li, R., Zhang, H., Wei, G., & Li, Z. (2020). Beneficial bacteria activate nutrients and promote wheat growth under conditions of reduced fertilizer application. *BMC Microbiology*, *20*(1), 1–12. https://doi.org/10.1186/s12866-020-1708-z

Wu, C., Tang, S., Li, G., Wang, S., Fahad, S., & Ding, Y. (2019). Roles of phytohormone changes in the grain yield of rice plants exposed to heat: A review. *PeerJ*, *7*, e7792. https://doi.org/10.7717/peerj.7792

Xiong, R., Liu, S., Considine, M. J., Siddique, K. H. M., Lam, H. M., & Chen, Y. (2021). Root system architecture, physiological and transcriptional traits of soybean (Glycine max L.) in response to water deficit: A review. *Physiologia Plantarum*, *172*(2), 405–418. https://doi.org/10.1111/ppl.13201

Yadav, S., Modi, P., Dave, A., Vijapura, A., Patel, D., & Patel, M. (2020). Effect of abiotic stress on crops. *Sustainable Crop Production, 2019*, 1–21. 10.5772/intechopen.88434.

Yanni, A. E., Iakovidi, S., Vasilikopoulou, E., & Karathanos, V. T. (2024). Legumes: A Vehicle for Transition to Sustainability. *Nutrients*, *16*(1), 98. https://doi.org/10.3390/nu16010098

Zaidi, A., Khan, M. S., & Musarrat, J. (2017). Microbes for legume improvement, second edition. *Microbes for Legume Improvement, Second Edition, July 2018*, 1–412. https://doi.org/10.1007/978-3-319-59174-2

Zaman, M. S. U., Malik, A. I., Luna, L. N., Hossain, M. A., Alam, A. K. M. M., Prodhan, M. A., & Erskine, W. (2024). Differences of waterlogging tolerance in winter pulse crop between emergence and vegetative stages. *Journal of Agronomy and Crop Science*, *210*(2), e12704. https://doi.org/10.1111/jac.12704

Zhao, G., Liu, L., Wang, Z. Y., Jin, Z., & He, J. S. (2023). Grassland Science in a New Era. *Fundamental Research*, *3*(2), 149–150. https://doi.org/10.1016/j.fmre.2023.02.001

Zheng, L., Zhao, J., Liang, X., Zhan, G., Jiang, S., & Kang, Z. (2017). Identification of a novel alternaria alternata strain able to hyperparasitize Puccinia striiformis f. sp. tritici, the causal agent of wheat stripe rust. *Frontiers in Microbiology*, *8*, 71. https://doi.org/10.3389/fmicb.2017.00071

Zhu, Q., Liu, X., Hao, T., Zeng, M., Shen, J., Zhang, F., & de Vries, W. (2020). Cropland acidification increases risk of yield losses and food insecurity in China. *Environmental Pollution*, *256*. https://doi.org/10.1016/j.envpol.2019.113145

Section II

Climate Change and Its Impact on Legume Plants

6 Building Resilience in Legumes

Strategies to Mitigate Environmental Stresses for Sustainable Agriculture

Mohamed Kouighat, Ilham Dehbi, Abdelaaziz Farhaoui, Fatima-Ezzahra Soussani, Chayma Ikan, Khadija Benamar, Rachid Ezzouggari, Hajar El Hamss, Yunfei Jiang, Abelilah Meddich, and Rachid Lahlali

6.1 INTRODUCTION

Legumes play a crucial role in agricultural sustainability due to their ability to fix atmospheric nitrogen through their symbiotic relationship with rhizobial bacteria. This process improves soil quality and reduces the need for nitrogen fertilizers, thereby promoting sustainable crop production (Ayilara et al., 2022). Additionally, legumes provide high-quality protein, fiber, and essential micronutrients, enhancing food security, especially in regions where access to animal protein is limited (Lisciani et al., 2024).

However, legume productivity faces increasing threats from abiotic and biotic stresses, which are intensifying due to climate change. The major challenges include drought, extreme temperature fluctuations, and soil salinization. Water deficits hinder biological nitrogen fixation, leading to a significant decline in yields (Islam et al., 2021). Similarly, salt accumulation in soils disrupts nutrient absorption and induces oxidative stress, resulting in agricultural losses (Abiala et al., 2018). Temperature fluctuations further impede essential physiological processes such as photosynthesis and germination, contributing to reduced productivity (Gogoi et al., 2018). Legumes also face threats from various pests and diseases, including pod borers, *Fusarium* wilt, and yellow vein mosaic virus, which can lead to total yield loss depending on the severity of the infection (Lisciani et al., 2024). Given these challenges, strengthening legume resilience is becoming increasingly critical to sustaining agricultural systems and ensuring global food security (Ayilara et al., 2022). Advancements in genetic breeding and biotechnology offer promising solutions. Genomic, pangenomic, and proteomic tools facilitate the identification and integration of resistance genes, enhancing legume resilience against various stresses (Huang, 2024). In addition to biotechnology, agroecological practices contribute to crop resilience. The use of biostimulants, such as nitrogen-fixing microbes, improves legume resistance to water stress and enhances yields (Huang, 2024). Likewise, crop management strategies such as rotation and intercropping improve soil quality, promote water absorption, and reduce pest infestations (Islam et al., 2021). Furthermore, artificial intelligence and precision agriculture play a strategic role in mitigating climate-related threats and optimizing crop management through predictive modeling. Beyond technological advancements, an integrated management approach–including policymakers, farmers, seed companies, researchers,

DOI: 10.1201/9781003477235-8

and extension services–is essential for enhancing legume resilience. A collaborative effort fosters the adoption of innovations and enables a more effective response to climate variability.

This chapter examines the key physiological and molecular mechanisms that enable legumes to withstand abiotic and biotic stresses. It highlights advancements in genetics, biotechnology, artificial intelligence, and agroecological practices while emphasizing the importance of integrating management strategies to sustain legume productivity and ensure global food security.

6.2 IMPACT OF ENVIRONMENTAL STRESSES ON LEGUMES

6.2.1 Abiotic Stress

Drought is a major abiotic factor limiting legume growth, affecting various phenological stages with different levels of sensitivity. Water deficit during flowering disrupts pollination and fertilization, significantly reducing pod numbers and seed set. Similarly, water scarcity during seed filling decreases seed weight, ultimately lowering yield. Depending on intensity and duration, drought can cause a 20–50% yield reduction, with losses exceeding 70% under prolonged extreme conditions (Nadeem et al., 2019). At the physiological and molecular levels, drought induces cellular water loss, decreases water potential, and leads to turgor loss (Fang & Xiong, 2015). These changes impair nutrient transport and cell growth. In legumes, drought stress also disrupts nitrogen fixation by altering symbiosis with *Rhizobium* bacteria, further disturbing metabolism. Photosynthesis declines due to stomatal closure, limiting CO_2 uptake and increasing reactive oxygen species (ROS) accumulation, which triggers oxidative stress (Qiao et al., 2024). Among legumes, chickpea (*Cicer arietinum*) and common bean (*Phaseolus vulgaris*) are particularly sensitive to drought, while lentil (*Lens culinaris*) and fava bean (*Vicia faba*) exhibit greater tolerance. Despite their higher water-use efficiency and osmolyte accumulation, prolonged drought compromises these adaptive mechanisms. The growth and productivity of legumes are significantly affected by salt stress, particularly in soils with high salinity levels or under saline irrigation. The primary cause of this stress is the accumulation of soluble salts, with NaCl being the predominant ion, which disrupts water uptake and cellular processes. The critical periods of exposure vary among species; however, for legumes, germination, seedling establishment, and the reproductive stage are the most vulnerable phases. Excessive salinity drastically reduces germination rates, with decreases ranging from 20% to 80%, depending on the severity of stress and the plant's resistance (Farooq et al., 2017). Vegetative growth is also severely impacted, with biomass production being reduced by up to 50% under high NaCl levels (Hameed et al., 2021). Furthermore, salinity inhibits symbiotic nitrogen fixation by impairing nodulation, directly affecting crop productivity (Rane et al., 2021).

Extreme temperatures affect legume growth and reproduction. Heat stress (>30 °C) impairs photosynthetic efficiency, denatures proteins, and reduces rhizobia activity (Aranjuelo et al., 2015). Cold (<0 °C) induces intracellular ice formation and disrupts membrane fluidity (Jahed et al., 2023). Cold temperatures delay germination and seedling emergence, reducing crop establishment in various legume crops, such as chickpea, barrel medic (*Medicago truncatula*), and common bean (Smýkal et al., 2015). Temperature extremes, along with drought, exacerbate stress, compounding challenges for legume crops. Reproductive processes are more temperature-sensitive than vegetative growth. With increasing frequency and intensity of temperature extremes due to climate change, understanding the physiological mechanisms of heat and cold stress is crucial for breeding and biotechnological interventions aimed at developing temperature-tolerant legume genotype.

6.2.2 Biotic Stress

Legume crops face multiple biotic stresses that significantly reduce yields and threaten global food security. Insects, fungi, viruses, bacteria, and parasites severely impact these crops. Major economic losses occur due to pests such as the pod borer, which devastates harvests. Severe diseases, including *Fusarium* wilt, root rot, and yellow vein mosaic virus, can destroy up to 100% of yields depending on infection severity (Mahmoud, 2021). More than 20 diseases affect soybean, with root-knot nematodes responsible for 44.7% of productivity losses (Sandhu et al., 2023). Whiteflies, aphids, and leafrollers reduce soybean yields by 30–40% (Sandhu et al., 2023). Chickpea production is severely constrained by biotic stresses such as *Fusarium* wilt, pod borer, and ascochyta blight, reducing yields to less than 1 t/ha, far below the crop's potential (Sandhu et al., 2023). Kidney, cyst, and lesion nematodes further exacerbate these losses. Despite a yield potential of 2.5–3 t/ha, mungbean productivity remains stagnated at only 0.5 t/ha (Nair et al., 2019) due to viral (mungbean yellow mosaic disease, MYMD), fungal (anthracnose, downy mildew, leaf spots), and bacterial (halo blight) diseases, causing up to 85% yield losses. Groundnut crops are affected by insect pests (borers, aphids, and mites) and fungal diseases (leaf spot, rust, and aspergillosis), leading to a 50% yield loss (Mahmoud, 2021). Viral diseases, such as peanut mottle virus (PMV) and bud necrosis, account for up to 60% of losses (Sandhu et al., 2023).

Abiotic and biotic stresses interact synergistically, exacerbating damage to legume crops. Herbivore damage from caterpillars and beetles creates entry points for opportunistic pathogens, such as bacteria and fungi. Additionally, these infestations alter plant metabolism, reducing their ability to resist diseases (Balla et al., 2021). Given these challenges, developing legume crop varieties resistant to both abiotic and biotic stresses is crucial.

6.3 MECHANISMS OF STRESS RESPONSE IN LEGUMES

6.3.1 Morphological and Physiological Adaptations

Legumes exhibit diverse adaptations that enable survival in harsh environments. Some species develop deep root systems, allowing them to access water and minerals from deeper soil layers, which is particularly beneficial in arid and nutrient-poor soils (Gao et al. 2023). Legumes also establish symbiotic relationships with nitrogen-fixing microorganisms, such as *Rhizobium*, which colonizes their root nodules. This symbiosis enhances nitrogen availability–a key nutrient often limited in soils–while improving soil quality for surrounding crops (Madouh & Quoreshi, 2023). These adaptations promote legume growth and productivity even under stressful conditions such as drought and poor soil fertility.

Physiologically, legumes employ multiple strategies to withstand environmental stresses. They regulate stomatal activity to balance gas exchange and water conservation, a crucial response to drought and salinity. To maintain cellular hydration and osmotic stability, they accumulate osmoprotectants such as proline, glycine betaine, and soluble sugars, which also stabilize proteins and membranes under stress (Islam et al., 2021). Additionally, to counteract oxidative damage, legumes activate their antioxidant defense system. This involves regulating enzymatic antioxidants (e.g., superoxide dismutase [SOD], catalase [CAT], and peroxidases) and accumulating non-enzymatic compounds such as ascorbic acid, glutathione, and flavonoids (Shomali et al., 2022). These adaptations help neutralize reactive oxygen species (ROS) and mitigate oxidative damage.

Under biotic stress, legumes enhance their defense mechanisms by synthesizing secondary metabolites, including flavonoids, tannins, and phytoalexins, which inhibit pathogen growth (Akbar et al., 2023). They also reinforce their physical defenses by depositing callose and lignin in cell walls, enhancing protection against pathogen invasion. These adaptive traits not only improve legume resilience but also contribute to sustainable agriculture, making them crucial for food security in stress-prone regions.

6.3.2 Molecular Mechanisms: Stress-signaling Pathways

At the molecular level, legumes orchestrate intricate stress-signaling pathways to detect, transmit, and respond to environmental challenges. These pathways integrate a wide range of signaling molecules, transcription factors, and stress-responsive genes to regulate adaptive physiological and biochemical processes. Under abiotic stress, key messengers such as calcium ions (Ca^{2+}), ROS, and phytohormones including abscisic acid (ABA), salicylic acid (SA), jasmonic acid (JA), and ethylene mediate rapid signal transduction (Jha et al., 2024). Calcium-dependent protein kinases (CDPKs) and mitogen-activated protein kinases (MAPKs) are important regulators that activate defense genes and boost stress signals (Srivastava & Sahoo, 2021). Transcription factors such as dehydration-responsive element-binding proteins (DREBs), NAC transcription factors, and v-Myb avian myeloblastosis viral oncogene homolog (MYB) modify gene expression to enhance drought, salinity, and metabolic adaptations via regulating osmoprotection, antioxidant defense, and heat tolerance (Manna et al., 2021). Additionally, small RNAs and epigenetic modifications, including DNA methylation, histone acetylation, and chromatin remodeling, fine-tune gene expression for stress adaptation. In response to biotic stress, legumes rely on pattern recognition receptors (PRRs) to detect pathogen-associated molecular patterns (PAMPs), triggering a cascade of immune responses (Nabi et al., 2024). This activation leads to the synthesis of antimicrobial compounds, cell wall reinforcement, and the induction of systemic acquired resistance (SAR) and induced systemic resistance (ISR) through SA- and JA-mediated pathways (Mishra et al., 2024). These defense mechanisms not only enhance resistance to pathogens but also prime the plant for future attacks. By integrating molecular, hormonal, and epigenetic regulation, legumes establish a dynamic and multilayered defense system, ensuring their resilience in fluctuating and often hostile environments (Fahad et al., 2024).

6.3.3 Role of Secondary Metabolites in Stress Tolerance

Secondary metabolites play a crucial role in enhancing stress tolerance in legumes by acting as antioxidants, osmoprotectants, signaling molecules, and antimicrobial agents. These bioactive compounds, including flavonoids, alkaloids, terpenoids, phenolics, and phytoalexins, help plants mitigate both abiotic and biotic stresses (Chen et al., 2022). Under drought and salinity stress, flavonoids and phenolics scavenge ROS, reducing oxidative damage and preserving cellular integrity (Azeem et al., 2023). Some secondary metabolites, such as glycine betaine and polyamines, function as osmoprotectants, stabilizing proteins and membranes under dehydration stress (Mishra et al., 2024). In response to biotic stress, legumes synthesize antimicrobial phytoalexins and tannins, which inhibit pathogen growth and reinforce structural defenses like lignin deposition in cell walls (Saini et al., 2024). Additionally, volatile organic compounds (VOCs) and signal molecules such as jasmonates and salicylic acid derivatives play a role in activating systemic resistance against pests and pathogens (Razo-Belman & Ozuna, 2023). These diverse functions of secondary metabolites collectively enhance the resilience of legumes, enabling them to survive and adapt to challenging environmental conditions.

6.4 INNOVATIVE STRATEGIES FOR STRESS MITIGATION

6.4.1 Development of Stress-tolerant Legume Varieties

Varietal development plays a central role in improving the resilience of legumes to environmental stresses such as drought, salinity, extreme temperatures, pests, weeds, and diseases. Conventional breeding programs, combined with modern biotechnological approaches, have identified and introduced genetic traits that promote tolerance to adverse abiotic conditions (Chetto et al., 2024). Research on chickpea has revealed genotypes with drought avoidance mechanisms, such as increased root depth and more efficient water use (Yadava et al., 2022). Mafakheri et al. (2010) revealed that the drought-tolerant chickpea variety "Bivaniej" had the highest yield, while the drought-sensitive variety "Pirouz" had the lowest yield. Similarly, common bean varieties have been developed to withstand low-phosphorus soils and high temperatures through interspecific crosses and marker-assisted selection (Delfini et al., 2021). The integration of genomic tools, such as genome-wide selection and CRISPR/Cas9 gene editing, has accelerated the introduction of multiple stress resistance traits. For instance, regulatory gene editing in soybean (*Glycine max* L.) which has resulted in varieties with improved salinity tolerance by modifying the expression of ion transporters (Zhong et al., 2024). Moreover, the multiomics approach, combining transcriptomics, metabolomics, and proteomics, has allowed a better understanding of the molecular networks involved in the stress response, thereby facilitating the identification of new target genes for varietal development (Satrio et al., 2024).

6.4.2 Application of Plant Growth-promoting Rhizobacteria (PGPR)

PGPRs enhance drought tolerance and nutrient uptake while suppressing pathogens through mechanisms such as the production of phytohormones, biofilm formation, and antioxidant activity (Shah et al., 2021). PGPRs are thus a promising ecological alternative to chemical inputs for sustainable crop management. These beneficial microbes colonize the rhizosphere and can also reside with plant roots, supporting plant development through a variety of direct and indirect mechanisms (Prashar et al., 2014). More recently, their ability to help plants cope with both biotic and abiotic stresses has gained significant attention. Regarding drought tolerance, certain rhizobacteria can enhance plant resilience through multiple pathways. These include (1) producing key phytohormones like abscisic acid (ABA), gibberellic acid, cytokinins, and indole-3-acetic acid (IAA), which regulate plant growth and stress responses; (2) releasing ACC deaminase, an enzyme that lowers ethylene levels in roots to reduce stress-related damage; (3) triggering systemic tolerance through bacterial metabolites; and (4) producing exopolysaccharides, which help retain moisture around plant roots (Jha et al., 2024). PGPR application, particularly co-inoculation with *Rhizobium leguminosarum* and *Pseudomonas putida*, mitigated the adverse effects of water stress while improving seed yield and water use efficiency (WUE). This improvement is attributed to an increase in the expression of photosynthetic pigments, better regulation of the plant's water relations, favorable changes in root parameters, and the activation of enzymatic and non-enzymatic defense systems (Chieb & Gachomo, 2023).

6.4.3 Role of Nanotechnology in Stress Management

Nanotechnology has emerged as a promising tool for managing abiotic and biotic stresses in plants, offering innovative solutions to improve nutrient uptake, soil health, and plant immunity. Nanoparticles can act as carriers for pesticides, fertilizers, and plant growth regulators, ensuring targeted delivery and minimizing environmental impact (An et al., 2022). In addition, nanoparticles are recognized for their ability to modulate antioxidant activity, improve water retention in plant tissues, and enhance stress tolerance (El-Saadony et al., 2022). Experimental studies have demonstrated the effectiveness of nanoparticles in improving the resilience of legumes to various environmental stresses. Kareem et al. (2022) evaluated the positive impact of different concentrations of zinc oxide nanoparticles (nano-ZnO) on mung bean (*Vigna radiata* L.) under heat stress. Similarly, the use of titanium dioxide nanoparticles (TiO_2 NPs) has shown positive effects on chickpea resilience to water stress. Hojjat (2020) reported that applying TiO_2 NPs significantly improved germination parameters under drought stress. Under variable humidity conditions, Ghorbani et al. (2023) observed an increase in chlorophyll and carotenoid production, highlighting the potential of nanoparticles to enhance photosynthesis and the growth of legumes under water stress.

6.4.4 Integration of Omics Approaches

The genome size of legumes is considerable, and certain species are polyploid. Fundamental research on legumes is predominantly conducted using model plant systems, particularly *Lotus japonicas* and

TABLE 6.1
Use of "Omic" Technologies Against Biotic and Abiotic Stresses in Legumes

Legumes	Stress	Omics technology	References
Cicer arietinum	Drought	Transcriptomics, proteomics, and metabolomics	(Kudapa et al., 2024)
Phaseolus vulgaris L.	Drought	Transcriptomics	(Subramani et al., 2023)
Medicago falcate	Drought	Transcriptomics	(Li et al., 2022)
Pisum sativum L.	Drought	Proteomics	(Shumilina et al., 2021)
Eremosparton songoricum	Drought	Genomics	(Zhao et al., 2022)
Arachis hypogaea L.	Drought	Metabolomics	(Gundaraniya et al., 2020)
Glycine soja	Salinity	Metabolomics and transcriptomics	(Hu et al., 2024)
Vigna radiata L.	Salinity	Metabolomics and transcriptomics	(Lim et al., 2022)
Medicago sativa L.	Salinity	Metabolomics and proteomics	(Li et al., 2021)
Phaseolus vulgaris	Salinity	Transcriptomics and metabolomics	(Zhang et al., 2023)
Phaseolus vulgaris	Cold	Metabolomics and transcriptomics	(Yang et al., 2023)
Medicago sativa L.	Aphid	Metabolomics and transcriptomics	(Liu et al., 2024)

Medicago truncatula, due to their small diploid genomes, short life cycles, and high seed yield (Hu et al., 2024). The application of these models has significantly enriched the genetic and genomic databases of legumes, reinforcing the similarities between the different genomes of this plant family. Although genome sequencing data constitute a valuable resource and an essential foundation, they remain insufficient to fully elucidate gene function and the multiple metabolic and regulatory responses triggered by persistent stress conditions in legumes. To address these limitations, innovative strategies and adapted methodologies are essential. These include the examination of quantitative and qualitative traits to validate gene expression. The implementation of transcriptomic, proteomic, and metabolomic analyses is essential, as is the integration of genetically modified organisms and marker-assisted selection to refine the understanding of stress responses (Kudapa et al., 2024). Metabolomic genomics, transcriptomics, transgenomics, proteomics, phenomics, and functional genomics, collectively referred to as "omics", represent crucial tools for the study and improvement of legumes facing environmental stresses (Table 6.1).

6.5 SOIL AND AGRONOMIC INTERVENTIONS

6.5.1 Soil Amendments and Their Impact on Stress Mitigation

Soil amendments, such as organic matter, biochar, and gypsum, can improve soil structure, increase water retention, and enhance nutrient availability. These amendments also contribute to the regulation of soil pH and microbial diversity, both of which are essential for plant health (Chen et al., 2022). Research indicates that the use of soil amendments can significantly mitigate the effects of drought and salinity, improving crop growth and yield under stressful conditions (Ullah et al., 2021). The application of various soil amendments alongside fertilizers can enhance soil physical properties and fertility (Diacono & Montemurro, 2011). Biochar, a stable carbon-rich product from biomass pyrolysis, is particularly effective in improving soil water-holding capacity, nutrient availability, and crop productivity (Diatta et al., 2020), and it may also enhance crop performance during drought conditions (Nadeem et al., 2019). For instance, combining biochar with arbuscular mycorrhizal fungi (AMF) has been shown to enhance drought tolerance in chickpeas by increasing chlorophyll content, net photosynthetic activity, relative water content, nodule number, nitrate reductase enzyme activity, and nitrogen and phosphorus uptake (Hashem et al., 2019). In soybeans, adding biochar has resulted in a 3–14% increase in grain yield under drought stress due to improved water use efficiency (Zhang et al., 2023).

6.5.2 Water Management Strategies

Efficient water management is critical in areas prone to water scarcity. Techniques such as drip irrigation, mulching, and rainwater harvesting can optimize water use, reduce water loss, and improve plant water status (Nadeem et al., 2019). Additionally, the adoption of drought-tolerant varieties and soil moisture conservation practices can mitigate the impact of water stress on crop productivity (Ray et al., 2020). Legumes like *Cicer* L., *Lathyrus* L., *Lens* Mill., and *Lupinus* L. are considered relatively drought-resistant genera, whereas *Pisum* L. and *Vicia* L. are generally considered more drought-sensitive (Latef & Ahmad, 2015). While irrigation plays a significant role in global agriculture, a substantial portion of these legumes, particularly chickpea, grass pea, lentil, and lupines, are often cultivated in marginal, rainfed areas (Siddique et al., 2012). With a drip irrigation system, yield and WUE for green bean increased with increasing irrigation interval (Abuarab et al., 2020), combined with soil mulching in common bean peak values for bean yield were achieved with farmyard manure mulch (FYM). In another study, aerated water utilized for sub-surface drip irrigation enhanced chickpea yield by 10–27% in comparison to the control. Chickpea yields achieved with subsurface drip irrigation were 81–187% greater than the national average of 13 t/ha (Nadeem et al., 2019). The implementation of aerated water irrigation augmented both the gross production water use index and the irrigation water use index for chickpea in a Vertosol (Pendergast et al., 2019).

6.5.3 Crop Rotation and Intercropping Practices for Resilience

Crop rotation and intercropping are traditional yet effective agronomic practices that enhance soil fertility, reduce pest pressure, and improve overall crop resilience. By alternating crops with varying nutrient requirements and stress tolerances, farmers can minimize soil depletion and enhance biodiversity. These practices also help break pest and disease cycles, reducing the need for chemical inputs. Continuous monocropping leads to the depletion of soil nutrients and fosters a buildup of pests and diseases (Zou et al., 2024). Crop rotation manages soil fertility and disrupts pest life cycles by incorporating diverse crops with varying nutrient needs and root structures (Shah et al., 2021). Intercropping has also improved grain yield by about 22% compared to monocultures (Li et al., 2021). Benefits like these could be explained by improved soil organic matter (SOM), higher nitrogen, and better structure in the soil due to the complex interplay between plant roots, microorganisms of the soil, and properties of soil to enhance nutrient cycling (Mohammadi et al., 2011). Recently, crop rotation and diversification have received considerable attention as sustainable methods for managing pests and diseases. By disrupting the life cycles of pests and pathogens, crop rotation prevents their buildup to damaging levels. Crop rotation renders fields unsuitable for the reproduction of pests and avoids disease buildup by eliminating continuous hosts (Boeraeve & Hatt, 2024). Diversified farming disrupts the avenues of pest movement and fosters beneficial organisms. Additionally, soil microorganisms play a crucial role in suppressing soilborne pathogens, thereby reducing disease risks to economically manageable levels when crop rotation is implemented (Panth et al., 2020).

Intercropping, the simultaneous cultivation of multiple crop species in the same field, is widely practiced globally, particularly in developing countries such as China, India, Indonesia, Niger, Mali, and Ethiopia (Solanki et al., 2017). Various intercropping systems exist worldwide, including wheat/maize, wheat/soybean, and maize/faba bean in northwest China, as well as agroforestry-based systems in tropical regions like Indonesia and Africa (Wang et al., 2015). In Europe, combinations such as barley/pea (Šarunaite et al., 2013) and wheat/pea (Pankou et al., 2022) are prevalent, while North America employs systems like wheat/canola (Mbanyele et al., 2024) and maize/soybean (West & Griffith, 1992). Research highlights yield increases of 21–23% in maize/faba bean systems compared to monocultures, underscoring the efficiency of resource utilization in intercropping (Wang et al., 2015). Nutrient acquisition benefits are particularly notable in phosphorus-deficient soils, where intercropped species enhance phosphorus uptake through root exudation processes (Bargaz et al., 2017). Similarly, intercropping enhances nitrogen acquisition, as demonstrated in wheat/maize, wheat/soybean, and maize/faba bean systems (Li et al., 2020). Despite its long-term benefits, few studies have assessed intercropping's effects on soil fertility, especially regarding soil chemistry and enzyme activities under different phosphorus application levels (Mbanyele et al., 2024).

6.6 AI AND DIGITAL TOOLS IN LEGUME STRESS MANAGEMENT

6.6.1 Predictive Models

By combining data from multiple sources (genomic, phenotypic, climatic, and agronomic), predictive models allow for the management of biotic and abiotic challenges in legumes through the prediction of crop performance and varietal choice. An example of this has been combining weather and genomic data, where Random Forest and Gradient Boosting machine learning methods were used to estimate the resistance of soybeans against heat. Additionally, a study on the multi-omics of common bean has shown that genomic prediction models (Genomic Prediction, GP), such as GBLUP and BayesB, use DNA markers for the estimation of agronomic performance, such that the response against salt stress can be anticipated (Keller et al., 2020). Cowpea (*Vigna unguiculata*) growth under varying climatic conditions has been modeled using crop models such as the Decision Support System for Agrotechnology Transfer (DSSAT), which predicted a 15% yield decrease if the average temperature increases by 2 °C (Woli et al., 2022). To promote better durum wheat intercropping management, the STICS model (v. 9.2) has also been calibrated for chickpeas under varying levels of nitrogen (Kherif et al., 2022). Artificial neural networks (ANNs), being applicable for the simulation of complex systems, are also a way artificial intelligence (AI) offers complex solutions. Indeed, ANN paired different machine learning methods, such as support vector regressions (SVM), Gaussian processes, XGBoost, and Random Forest, for chickpea in vitro regeneration of shoots estimation (Kirtis et al., 2022). Also, chickpea germination under osmotic stress has been analyzed with a hydrothermal model (HTT), with emphasis on the action of cardinal temperatures on the rate of germination (Shah et al., 2021). These methods enhance legume stress management practices and strengthen agricultural resilience to climate variations through crop modeling, artificial intelligence, and genomic predictive approaches.

6.6.2 AI in Breeding Programs

Breeding stress-resilient crop cultivars requires holistic fusion of genomics, phenomics, and environment data, even with the complexity of genotype–phenotype gap challenges and multiomics data complexity. Artificial intelligence (AI) provides a solution by enabling the development of more productive and stress-tolerant varieties. However, phenotypic plasticity complicates data interpretation, necessitating extensive phenotyping across diverse environments. A strategy combining high-throughput functional phenotyping (HFPS) with the utilization of the soil–plant–atmosphere continuum (SPAC) for quantification of plant response against different environments has been recommended by Negin and Moshelion (2016). By integrating genome-wide association studies (GWASs), this strategy enhances understanding of the genetic basis of adaptation, thereby accelerating the development of novel genetic resources for legume breeding. Machine learning (ML) techniques have proven effective in identifying genetic variants associated with abiotic stress tolerance, addressing the limitations of conventional GWAS analyses. Rairdin et al. (2022) employed an ML-based imaging phenotyping approach to study iron-deficiency chlorosis in soybean, identifying a novel locus associated with iron accumulation. Naik et al. (2017) developed a high-throughput phenotyping approach to quantify stress severity in soybean, facilitating marker identification for genomic selection (GS) and greater genetic gain. Advanced ML models also underlie genomic prediction. Convolutional neural networks (CNNs) derive predictive traits from single-nucleotide polymorphism (SNP) data (Liu et al., 2024), and artificial neural networks (ANNs) recommend a resilient genotype of common bean under stressed environments. Neural network systems including deep belief networks, multilayer perceptrons, and radial basis function networks also improved heat and drought tolerance. A notable contribution of Falk et al. (2020) is a computer vision and ML root phenotyping platform, which discovered genetic root architectural variations in soybean. Collectively, these advancements highlight the value of ML as a cost-effective, non-destructive, and high-throughput tool for marker-assisted breeding and crop improvement under climate change.

6.6.3 Precision Agriculture Applications

Precision agriculture is a key strategy for optimizing legume cultivation under stressful environmental conditions. This approach relies on the use of novel technologies such as Internet of Things (IoT) sensors, remote sensing, and AI to ensure the precise application of inputs (water, pesticides, and fertilizers) where needed. By reducing costs and minimizing the environmental impact, precision agriculture enhances crop resilience. Optimized irrigation is a critical application of precision methods. For instance, variable rate irrigation (VRI), utilizing soil moisture sensors and satellite data, enabled optimized water distribution based on local needs in cowpea cultivation, leading to 25% water savings and an 18% yield increase during a drought (Ahmed et al., 2023). Similarly, in common bean cultivation, multispectral sensors identified nitrogen-deficient areas, allowing for targeted fertilizer application, which reduced fertilizer costs by 20% and improved crop health and yield (Lazarević et al., 2024). Additionally, AI and machine learning improve disease prevention. A study integrating satellite data and predictive AI models accurately forecasted soybean rust outbreaks, a fungal disease influenced by specific temperature and moisture conditions. Preemptive alerts allowed for the preventive use of fungicides, reducing crop losses by 30% (Waseem et al., 2024). Furthermore, ground sensors and remote sensing significantly enhance water resource management. In lentil cultivation (*Lens culinaris*), real-time soil moisture maps enabled optimized irrigation scheduling, leading to 20% water savings and a 10% yield increase during extreme drought conditions (Woli et al., 2022). In conclusion, precision agriculture, by integrating multiple data sources and automating decision-making, represents a powerful tool for sustainable legume cultivation. It enhances agricultural efficiency, strengthens crop resilience to climatic stress, and contributes to global food security.

6.7 INTEGRATED STRESS MANAGEMENT FRAMEWORKS

6.7.1 Policy Interventions

Agricultural policies and multidisciplinary collaborations form the cornerstone for legume integrated stress management through the adaptation of tailored practices and the adoption of innovative technologies. Regionally, the European Union has rolled out activities such as the Legume Futures, Valeur Légale, and the Farm to Fork Strategy for the improvement of legume farming and reducing chemical inputs. Programs such as Voluntary Coupled Support (VCS) and the greening requirement, where the European Union distributes 30% of the CAP Pillar 1 direct payments for nitrogen-fixation crops, have helped improve the crop's ability to withstand adversity (Balázs et al., 2021). In Africa and Asia, the Tropical Legumes III initiative has rolled out stress-tolerant varieties and expanded seed distribution networks through partnerships with ICARDA and USAID.

Nationally, the Government of India's PMFBY initiative launched in 2016 extends subsidized crop insurance for farmers for the losses caused by the incidence of floods and drought (Kumar & Shobana, 2024). In Nigeria, subsidies enhanced access to cowpea seed resistant to drought, while Canada, Australia, and Germany funded crop diversification through financial incentives and applied study projects (Balázs et al., 2021).

Additionally, private sector players such as East-West Seed, Seed Co, and Advanta Seeds contribute significantly by developing and spreading seed varieties adapted for the environment

(Schreinemachers et al., 2021). Private–public partnerships, alongside institutions such as the International Crops Research Institute for the Semi-Arid Tropics (ICRISAT) and the Consultative Group for International Agricultural Research (CGIAR), disseminate innovative practices and technologies, contributing to crop ability and food security, specifically for Africa through projects like PABRA (Buruchara et al., 2021).

6.7.2 Extension Services and Technology Transfer

Unlike traditional models where farmers are passive adopters, the participatory approach considers them as co-drivers of innovation. Farmers, researchers, and other stakeholders can collaborate to co-design solutions that are tailored to local conditions using participatory innovation platforms. This approach is exemplified by the Alliance/PABRA program, which integrates a gender perspective in bean breeding through training in Arusha, Tanzania, in 2023 (Assefa et al., 2023). In Zimbabwe, farmer participation has been shown to enhance the adoption of new cultivars (Mutari et al., 2021). In Burkina Faso, with farmer involvement in the breeding process, the level of adoption for resistant cowpea varieties rose by 40% when compared to non-participating farmers (Vom Brocke et al., 2010). In Mali, market access and management together with variety trials during the post-harvest enhanced legume adoption (Sylla et al., 2023).

6.7.3 Role of Stakeholders in Mitigating Stress Impacts

Mitigating the effects of abiotic and biotic stresses on legumes requires the synergy of a wide range of stakeholders, from farmers to policymakers, researchers, nongovernmental organizations (NGOs), and the private sector (Figure 6.1). Each actor plays a complementary role that is essential to strengthening the resilience of production systems. Grain aggregators and input suppliers like Pioneer Hi-Bred, ICRISAT, and SeedCo are vital to guaranteeing the supply of improved seeds and other necessary agricultural inputs. Their role is supplemented by that of regulatory and extension organizations, which oversee field operations, promote the adoption of locally tailored rules, and help disseminate agricultural innovations. Further along the value chain, processors, such as AGT Food and Ingredients and Guts Agro Industry, intervene by creating market opportunities for producers. They positively drive better agricultural practices by enforcing quality standards, thus encouraging the adoption of improved and sustainable cultivation techniques. Farmers, as end users of innovations, are key players in the adoption of sustainable crop management practices. Their practical knowledge is useful in pinpointing environmental limits, which aids in directing research initiatives. On the other hand, research and technology development organizations such as IFDC, ICRAF, ILRI, and others are essential to innovation. Researchers contribute toward the enhancement of legume variety genetics, the setting up of climate-resilient agricultural practices, and the conservation of natural resources sustainably. Governments, ministries, and international organizations like the European Union and the FAO also make notable inputs. The private sector, seed corporations, and agribusiness support these activities by disseminating innovative technologies and increasing the market value of legumes, encouraging farmers to adopt more resilient practices. International and local non-governmental organizations (NGOs) such as ICRISAT, HarvestPlus, and the Catholic Relief Services also provide support by developing the capacities of local stakeholders. They facilitate the transfer of innovative technologies, support community development programs, and contribute to the resilience of agricultural systems to climate shocks.

6.8 CONCLUSION AND FUTURE PROSPECTS

This chapter highlights the crucial role of legumes in ensuring food security, enhancing climate resilience, and promoting sustainable farming practices. It outlines key strategies

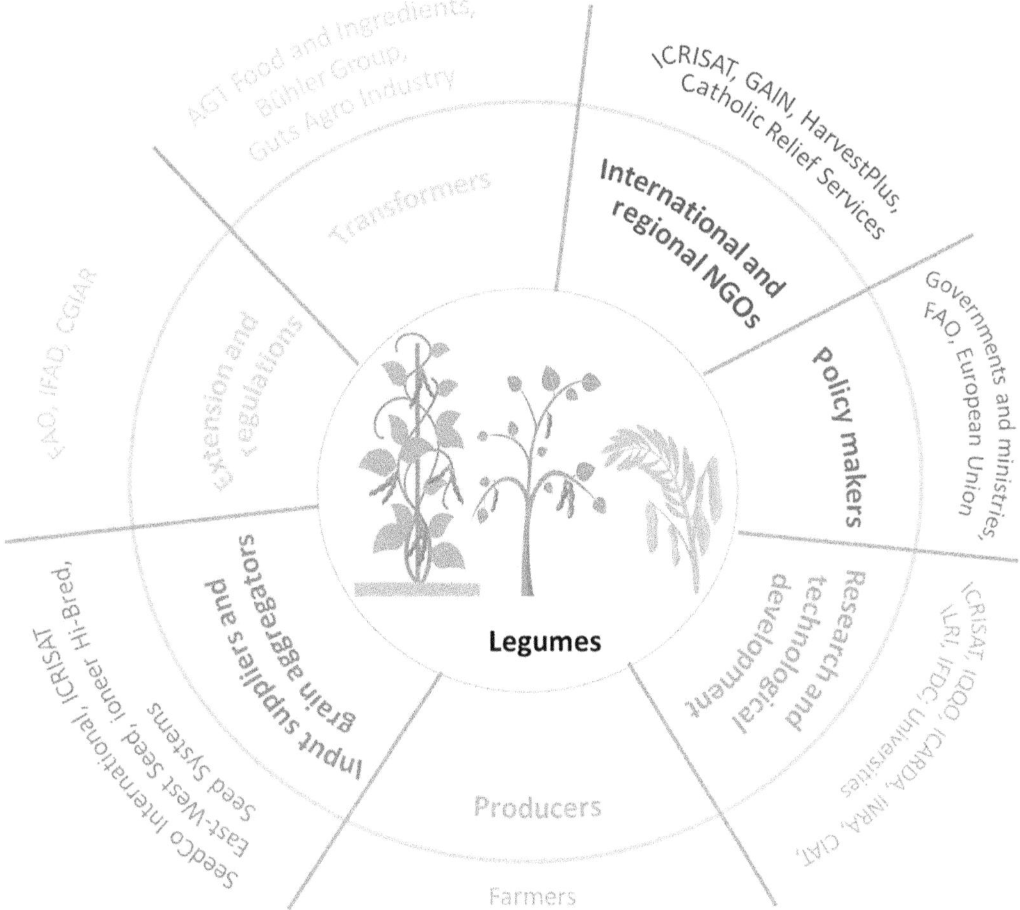

FIGURE 6.1 Interconnected role of stakeholders in legume resilience. ICRISAT (International Crops Research Institute for the Semi-Arid Tropics), GAIN (Global Alliance for Improved Nutrition), FAO (Food and Agriculture Organization of the United Nations), IFAD (International Fund for Agricultural Development), CGIAR (Consultative Group on International Agricultural Research), ILRI (International Livestock Research Institute), IFDC (International Fertilizer Development Center), ICRAF (International Centre for Research in Agroforestry), INRA (Institut National de la Recherche Agronomique), CIAT (International Center for Tropical Agriculture).

and research priorities for advancing legume development. The discussion focuses on four major areas: securing funding, developing new products, improving crop varieties, and providing training for farmers, addressing both the challenges and opportunities in these areas. The future of legume research lies in an integrated approach that combines local knowledge, technological innovation, and species diversification. The application of genomics, bioinformatics, and AI tools presents new opportunities for genetic improvement, while sustainable agriculture practices and participatory research enhance legume resilience under extreme conditions. However, achieving these advances requires close coordination between researchers, policymakers, farmers, and industry stakeholders. Strengthening collaboration, supported by adequate funding and public–private partnerships, is essential for developing solutions tailored to local contexts. Moreover, increasing the market value of legumes – through product diversification and demand stimulation – serves as a key driver for their integration into sustainable food systems. By addressing these challenges, legume research will not only enhance agricultural resilience but also contribute to global food security and climate change mitigation.

ACKNOWLEDGMENT

This research was financially supported by the Phytopathology Unit of the Department of Plant Protection and Environment at the École Nationale d'Agriculture de Meknès.

REFERENCES

Abiala, M. A., M. Abdelrahman, D. J. Burritt, and L. P. Tran. 2018. Salt Stress Tolerance Mechanisms and Potential Applications of Legumes for Sustainable Reclamation of Salt-Degraded Soils. *Land Degradation & Development* 29 (10): 3812–3822. doi:10.1002/ldr.3095.

Abuarab, M. E., S. M. Hafez, M. M. Shahein, A. M. Hassan, M. B. I. El-Sawy, M. M. El-Mogy, and E. A. Abdeldaym. 2020. Irrigation Scheduling for Green Beans Grown in Clay Loam Soil under a Drip Irrigation System. *Water SA* 46 (4): 573–582. doi:10.17159/wsa/2020.v46.i4.9070.

Ahmed, Z., D. Gui, G. Murtaza, Y. Li, and S. Ali. 2023. An Overview of Smart Irrigation Management for Improving Water Productivity under Climate Change in Drylands. *Agronomy* 13 (8): 2113.

Akbar, M. U., M. Aqeel, M. S. Shah, G. Jeelani, N. Iqbal, A. Latif, R. O. Elnour, M. Hashem, O. M. Alzoubi, T. Habeeb, M. Qasim, and A. Noman. 2023. Molecular Regulation of Antioxidants and Secondary Metabolites Act in Conjunction to Defend Plants against Pathogenic Infection. *South African Journal of Botany* 161: 247–257. doi:10.1016/j.sajb.2023.08.028.

An, C., C. Sun, N. Li, B. Huang, J. Jiang, Y. Shen, C. Wang, X. Zhao, B. Cui, C. Wang, X. Li, S. Zhan, F. Gao, Z. Zeng, H. Cui, and Y. Wang. 2022. Nanomaterials and Nanotechnology for the Delivery of Agrochemicals: Strategies Towards Sustainable Agriculture. *Journal of Nanobiotechnology* 20 (1): 1–19. doi:10.1186/s12951-021-01214-7.

Aranjuelo, I., J. Aldasoro, C. Arrese-Igor, G. Erice, and Á. Sanz-Sáez. 2015. How Does High Temperature Affect Legume Nodule Symbiotic Activity? In *Legume Nitrogen Fixation in a Changing Environment*, edited by S. Sulieman and L.-S. P. Tran, 67–87. Springer International Publishing. doi:10.1007/978-3-319-06212-9_4.

Assefa, T., S. Beebe, W. Amongi, R. Odama, E. Nchanji, N. Yao, C. Mukankusi, and J. C. Rubyogo. 2023. Capacity Building of National Agricultural Research System (NARES) Partners on Common Bean Breeding. [Blog post] Bioversity International and the International Center for Tropical Agriculture (CIAT). Published online 18 July 2023. URL: www.cgiar.org/news-events/news/capacity-building-of-national-agricultural-research-system-nares-partners-on-common-bean-breeding/.

Ayilara, M. S., M. Abberton, O. A. Oyatomi, O. Odeyemi, and O. O. Babalola. 2022. Potentials of Underutilized Legumes in Food Security. *Frontiers in Soil Science* 2: 1020193. www.frontiersin.org/articles/10.3389/fsoil.2022.1020193/full.

Azeem, M., K. Pirjan, M. Qasim, A. Mahmood, T. Javed, H. Muhammad, S. Yang, R. Dong, B. Ali, and M. Rahimi. 2023. Salinity Stress Improves Antioxidant Potential by Modulating Physio-Biochemical Responses in *Moringa oleifera* Lam. *Scientific Reports* 13 (1): 1–17. doi:10.1038/s41598-023-29954-6.

Balázs, B., E. Kelemen, T. Centofanti, M. W. Vasconcelos, and P. P. M. Iannetta. 2021. Integrated Policy Analysis to Identify Transformation Paths to More Sustainable Legume-Based Food and Feed Value-Chains in Europe. *Agroecology and Sustainable Food Systems* 45 (6): 931–953. doi:10.1080/21683565.2021.1884165.

Balla, A., A. Silini, H. Cherif-Silini, A. Chenari Bouket, W. K. Moser, J. A. Nowakowska, T. Oszako, F. Benia, and L. Belbahri. 2021. The Threat of Pests and Pathogens and the Potential for Biological Control in Forest Ecosystems. *Forests* 12 (11): 1579.

Bargaz, A., G. L. Noyce, R. Fulthorpe, G. Carlsson, J. R. Furze, E. S. Jensen, D. Dhiba, and M. E. Isaac. 2017. Species Interactions Enhance Root Allocation, Microbial Diversity and P Acquisition in Intercropped Wheat and Soybean under P Deficiency. *Applied Soil Ecology* 120: 179–188. doi:10.1016/J.APSOIL.2017.08.011.

Boeraeve, F., and S. Hatt. 2024. Integrating Agroecological Practices to Manage Pests While Combining Organic and Conservation Agriculture. In J. Wang, H. Liu, I. Menzler-Hokkanen, and H. Jiang (eds.) *The Concept of Ecostacking: Techniques and Applications*, 163–190. doi:10.1079/9781789248715.0013.

Buruchara, R. A., P. Onyango, J.-C. Rubyogo, B. Mutari, R. Zulu, S. Nkalubo, R. Chirwa, and B. Fenta. 2021. PABRA Means Partnership: Transforming Agriculture in Africa Together. https://cgspace.cgiar.org/bitstream/10568/113037/1/PABRA%20Means%20Partnership.pdf.

Chen, D., X. Wang, V. J. Carrión, S. Yin, Z. Yue, Y. Liao, Y. Dong, and X. Li. 2022. Acidic Amelioration of Soil Amendments Improves Soil Health by Impacting Rhizosphere Microbial Assemblies. *Soil Biology and Biochemistry* 167: 108599. doi:10.1016/J.SOILBIO.2022.108599.

Chetto, O., L. Belqadi, Z. E. A. Fatemi, J. Charafi, M. Kouighat, A. Najmi, M. El Fechtali, K. Houmanat, and A. Nabloussi. 2024. Assessment of Elite Faba Bean Lines for Enhanced Productivity and Resilience in Contrasting Challenging Environments Using Phenotypic and Molecular Markers. *Journal of Agriculture and Food Research* 18: 101488. doi:10.1016/j.jafr.2024.101488.

Chieb, M., and E. W. Gachomo. 2023. The Role of Plant Growth Promoting Rhizobacteria in Plant Drought Stress Responses. *BMC Plant Biology* 23 (1): 407. doi:10.1186/s12870-023-04403-8.

Delfini, J., V. Moda-Cirino, J. dos Santos Neto, D. M. Zeffa, A. F. Nogueira, L. A. B. Ribeiro, P. M. Ruas, P. Gepts, and L. S. A. Gonçalves. 2021. Genome-Wide Association Study Identifies Genomic Regions for Important Morpho-Agronomic Traits in Mesoamerican Common Bean. *Frontiers in Plant Science* 12: 748829.

Diacono, M., and F. Montemurro. 2011. Long-Term Effects of Organic Amendments on Soil Fertility. *Sustainable Agriculture* 2: 761–786. doi:10.1007/978-94-007-0394-0_34.

Diatta, A., J. Fike, M. Battaglia, J. G.-A. J. of, and undefined 2020. 2020. Effects of Biochar on Soil Fertility and Crop Productivity in Arid Regions: A Review. *Springer* 13 (14). doi:10.1007/s12517-020-05586-2.

El-Saadony, M. T., A. M. Saad, S. M. Soliman, H. M. Salem, E. S. M. Desoky, A. O. Babalghith, A. M. El-Tahan, O. M. Ibrahim, A. A. M. Ebrahim, T. A. Abd El-Mageed, A. S. Elrys, A. A. Elbadawi, K. A. El-Tarabily, and S. F. AbuQamar. 2022. Role of Nanoparticles in Enhancing Crop Tolerance to Abiotic Stress: A Comprehensive Review. *Frontiers in Plant Science* 13: 946717. doi:10.3389/FPLS.2022.946717.

Fahad, M., L. Tariq, S. Muhammad, and L. Wu. 2024. Underground Communication: Long Non-Coding RNA Signaling in the Plant Rhizosphere. *Plant Communications* 5 (7): 100927. doi:10.1016/j.xplc.2024.100927.

Falk, K. G., T. Z. Jubery, S. V. Mirnezami, K. A. Parmley, S. Sarkar, A. Singh, B. Ganapathysubramanian, and A. K. Singh. 2020. Computer Vision and Machine Learning Enabled Soybean Root Phenotyping Pipeline. *Plant Methods* 16 (1): 5. doi:10.1186/s13007-019-0550-5.

Fang, Y., and L. Xiong. 2015. General Mechanisms of Drought Response and Their Application in Drought Resistance Improvement in Plants. *Cellular and Molecular Life Sciences* 72 (4): 673–689. doi:10.1007/s00018-014-1767-0.

Farooq, M., N. Gogoi, M. Hussain, S. Barthakur, S. Paul, N. Bharadwaj, and K. H. Siddique. 2017. Effects, Tolerance Mechanisms and Management of Salt Stress in Grain Legumes. *Plant Physiology and Biochemistry* 118: 199–217.

Gao, Y., Z. Zhang, F. Zeng, and X. Ma. 2023. Root Morphological and Physiological Traits Are Committed to the Phosphorus Acquisition of the Desert Plants in Phosphorus-Deficient Soils. *BMC Plant Biology* 23 (1): 1–16. doi:10.1186/s12870-023-04178-y.

Ghorbani, R., A. Ganjeali, A. Movafeghi, and J. Nabati. 2023. "Exposure to TiO_2 Nanoparticles Improves the Physiological Characteristics of Drought-Challenged Chickpeas (*Cicer arietinum* L.)." *Legume Science* 5 (4). doi:10.1002/leg3.208.

Gogoi, N., M. Farooq, S. Barthakur, B. Baroowa, S. Paul, N. Bharadwaj, and S. Ramanjulu. 2018. Thermal Stress Impacts on Reproductive Development and Grain Yield in Grain Legumes. *Journal of Plant Biology* 61 (5): 265–291. doi:10.1007/s12374-018-0130-7.

Gundaraniya, S. A., P. S. Ambalam, and R. S. Tomar. 2020. Metabolomic Profiling of Drought-Tolerant and Susceptible Peanut (*Arachis hypogaea* L.) Genotypes in Response to Drought Stress. *ACS Omega* 5 (48): 31209–31219. doi:10.1021/acsomega.0c04601.

Hameed, A., M. Z. Ahmed, T. Hussain, I. Aziz, N. Ahmad, B. Gul, and B. L. Nielsen. 2021. Effects of Salinity Stress on Chloroplast Structure and Function. *Cells* 10 (8): 2023.

Hashem, A., A. Kumar, A. M. Al-Dbass, A. A. Alqarawi, A.-B. F. Al-Arjani, G. Singh, M. Farooq, and E. F. Abd_Allah. 2019. Arbuscular Mycorrhizal Fungi and Biochar Improves Drought Tolerance in Chickpea. *Saudi Journal of Biological Sciences* 26 (3): 614–624. doi:10.1016/j.sjbs.2018.11.005.

Hojjat, S. S. 2020. Effects of TiO_2 Nanoparticles on Germination and Growth Characteristics of Grass Pea (*Lathyrus sativus* L.) Seed under Drought Stress. *Nanotechnologies in Russia* 15 (2): 204–211. doi:10.1134/S199507802002010X.

Hu, Y., T. Luan, X. Wang, Z. Luan, Y. Hu, and M. Li. 2024. Integrated Metabolomics and Transcriptomics Analyses Reveal Resistance to Salt Stress in Wild Soybean (*Glycine soja*) During the Post-Germination Growth Period. *Journal of Agronomy and Crop Science* 210 (5). doi:10.1111/jac.12748.

Huang, D. H. D. 2024. CRISPR/Cas9 Genome Editing in Legumes: Opportunities for Functional Genomics and Breeding. *Legume Genomics and Genetics* 15. https://cropscipublisher.com/index.php/lgg/article/view/3965.

Islam, M. S., S. Fahad, A. Hossain, M. K. Chowdhury, M. A. Iqbal, A. Dubey, A. Kumar, K. Rajendran, S. Danish, M. H. U. Rahman, M. A. Raza, M. Arif, S. Saud, M. A. Hossain, E. Waraic, Z. Ahmad, S. Hussain, A. Çīg, M. Erman, ... A. E. L. Sabagh. 2021. Legumes under Drought Stress: Plant Responses, Adaptive Mechanisms, and Management Strategies in Relation to Nitrogen Fixation. *Engineering Tolerance in Crop Plants Against Abiotic Stress*, December, 179–207. doi:10.1201/9781003160717-9.

Jahed, K. R., A. K. Saini, and S. M. Sherif. 2023. Coping with the Cold: Unveiling Cryoprotectants, Molecular Signaling Pathways, and Strategies for Cold Stress Resilience." *Frontiers in Plant Science* 14: 1246093.

Jha, Y., K. A. Yadav, and H. I. Mohamed. 2024. "Plant Growth-Promoting Bacteria and Exogenous Phytohormones Alleviate the Adverse Effects of Drought Stress in Pigeon Pea Plants. *Plant and Soil* 505 (1–2): 163–183. doi:10.1007/s11104-023-06155-8.

Kareem, H. A., M. F. Saleem, S. Saleem, S. A. Rather, S. H. Wani, M. H. Siddiqui, S. Alamri, R. Kumar, N. B. Gaikwad, Z. Guo, J. Niu, and Q. Wang. 2022. Zinc Oxide Nanoparticles Interplay with Physiological and Biochemical Attributes in Terminal Heat Stress Alleviation in Mungbean (*Vigna radiata* L.). *Frontiers in Plant Science* 13 (February): 1–15. doi:10.3389/fpls.2022.842349.

Keller, B., D. Ariza-Suarez, J. De la Hoz, J. S. Aparicio, A. E. Portilla-Benavides, H. F. Buendia, B. Raatz. 2020. Genomic Prediction of Agronomic Traits in Common Bean (*Phaseolus vulgaris* L.) under Environmental Stress. *Frontiers in Plant Science* 11: 1001.

Kherif, O., M. Seghouani, E. Justes, D. Plaza-Bonilla, A. Bouhenache, B. Zemmouri, ... M. Latati. 2022. The First Calibration and Evaluation of the STICS Soil-Crop Model on Chickpea-Based Intercropping System under Mediterranean Conditions. *European Journal of Agronomy* 133: 126449.

Kirtis, A., M. Aasim, and R. Katırcı. 2022. Application of Artificial Neural Network and Machine Learning Algorithms for Modeling the *In Vitro* Regeneration of Chickpea (*Cicer arietinum* L.). *Plant Cell, Tissue and Organ Culture (PCTOC)* 150 (1): 141–152.

Kudapa, H., A. Ghatak, R. Barmukh, P. Chaturvedi, A. Khan, S. Kale, L. Fragner, A. Chitikineni, W. Weckwerth, and R. K. Varshney. 2024. Integrated Multi-Omics Analysis Reveals Drought Stress Response Mechanism in Chickpea (*Cicer arietinum* L.). *The Plant Genome* 17 (1). doi:10.1002/tpg2.20337.

Kumar, J. S., and D. Shobana. 2024. Role of Pradhan Mantri Fasal Bima Yojana (PMFBY) in Agricultural Development and Farmer Welfare in Tamil Nadu. *Shanlax International Journal of Economics* 13 (1): 33–46.

Latef, A. A. H. A., and P. Ahmad. 2015. Legumes and Breeding under Abiotic Stress. In Azooz, M. M., and P. Ahmad (eds.) *Legumes Under Environmental Stress:* Yield, Improvement and Adaptations, 1–20. Wiley. doi:10.1002/9781118917091.ch1.

Lazarević, B., J. Gunjača, T. Safner, M. Vidak, T. Javornik, and K. Carović-Stanko. 2024. Multispectral and Chlorophyll Fluorescence Imaging for Detection of Nutrient Deficiency Symptoms in Common Bean. *Journal of Central European Agriculture* 25 (2): 416–432.

Li, J., J. Essemine, C. Shang, H. Zhang, X. Zhu, J. Yu, G. Chen, M. Qu, and D. Sun. 2020. Combined Proteomics and Metabolism Analysis Unravels Prominent Roles of Antioxidant System in the Prevention of Alfalfa (*Medicago sativa* L.) against Salt Stress. *International Journal of Molecular Sciences* 21 (3): 909. doi:10.3390/ijms21030909.

Li, Q., L. Gu, J. Song, C. Li, Y. Zhang, Y. Wang, Y. Pang, and B. Zhang. 2022. Physiological and Transcriptome Analyses Highlight Multiple Pathways Involved in Drought Stress in *Medicago falcata*. *PLoS One* 17 (4): e0266542. doi:10.1371/journal.pone.0266542.

Li, X. F., Z. G. Wang, X. G. Bao, J. H. Sun, S. C. Yang, P. Wang, C. B. Wang, J. P. Wu, X. R. Liu, X. L. Tian, Y. Wang, J. P. Li, Y. Wang, H. Y. Xia, P. P. Mei, X. F. Wang, J. H. Zhao, R. P. Yu, W. P. Zhang, … L. Li. 2021. Long-Term Increased Grain Yield and Soil Fertility from Intercropping. *Nature Sustainability* 4 (11): 943–950. doi:10.1038/s41893-021-00767-7.

Lim, I., M. Kang, B. C. Kim, and J. Ha. 2022. Metabolomic and Transcriptomic Changes in Mungbean (*Vigna radiata* (L.) R. Wilczek) Sprouts under Salinity Stress. *Frontiers in Plant Science* 13: 1030677. doi:10.3389/fpls.2022.1030677.

Lisciani, S., S. Marconi, C. Le Donne, E. Camilli, A. Aguzzi, P. Gabrielli, L. Gambelli, K. Kunert, D. Marais, and B. J. Vorster. 2024. Legumes and Common Beans in Sustainable Diets: Nutritional Quality, Environmental Benefits, Spread and Use in Food Preparations. *Frontiers in Nutrition* 11: 1385232. www.frontiersin.org/articles/10.3389/fnut.2024.1385232/full.

Liu, H., M. Xu, Y. Guo, Z. Dan, X. Liu, J. Zhang, C. Li, S. Jia, L. Jia, A. Yu, and L. Cong. 2024. Combined Transcriptome and Metabolome Analysis of Alfalfa Responses to Aphid Infestation. *Agriculture* 14 (9): 1545. doi:10.3390/agriculture14091545.

Madouh, T. A., and A. M. Quoreshi. 2023. The Function of Arbuscular Mycorrhizal Fungi Associated with Drought Stress Resistance in Native Plants of Arid Desert Ecosystems: A Review. *Diversity* 15 (3). doi:10.3390/d15030391.

Mafakheri, A., A. Siosemardeh, B. Bahramnejad, P. C. Struik, and E. Sohrabi. 2010. Effect of Drought Stress on Yield, Proline and Chlorophyll Contents in Three Chickpea Cultivars. *Australian Journal of Crop Science* 4 (8): 580–585.

Mahmoud, G. A.-E. 2021. Biotic Stress to Legumes: Fungal Diseases as Major Biotic Stress Factor. In *Sustainable Agriculture Reviews 51*, edited by P. Guleria, V. Kumar, and E. Lichtfouse, 181–212. Springer International Publishing. doi:10.1007/978-3-030-68828-8_7.

Manna, M., T. Thakur, O. Chirom, R. Mandlik, R. Deshmukh, and P. Salvi. 2021. Transcription Factors as Key Molecular Target to Strengthen the Drought Stress Tolerance in Plants. *Physiologia Plantarum* 172 (2): 847–868. doi:10.1111/ppl.13268.

Mbanyele, V., R. O. Enesi, L. Shaw, and L. Y. Gorim. 2024. A Review of Intercropping Systems in Western Canada. *Agronomy Journal* 116 (5): 2089–2108. doi:10.1002/agj2.21622.

Mishra, S., R. Roychowdhury, S. Ray, A. Hada, A. Kumar, U. Sarker, T. Aftab, and R. Das. 2024. Salicylic Acid (SA)-Mediated Plant Immunity against Biotic Stresses: An Insight on Molecular Components and Signaling Mechanism. *Plant Stress* 11 (December 2023): 100427. doi:10.1016/j.stress.2024.100427.

Mohammadi, K., G. Heidari, S. Khalesro, and Y. Sohrabi. 2011. Soil Management, Microorganisms and Organic Matter Interactions: A Review. *African Journal of Biotechnology* 10 (86): 19840–19849. doi:10.5897/AJBX11.006.

Mutari, B., J. Sibiya, E. Bogweh Nchanji, K. Simango, and E. Gasura. 2021. Farmers' Perceptions of Navy Bean (*Phaseolus vulgaris* L.) Production Constraints, Preferred Traits and Farming Systems and Their Implications on Bean Breeding: A Case Study from South East Lowveld Region of Zimbabwe. *Journal of Ethnobiology and Ethnomedicine* 17 (1): 13. doi:10.1186/s13002-021-00442-3.

Nabi, Z., S. Manzoor, S. U. Nabi, T. A. Wani, H. Gulzar, M. Farooq, V. M. Arya, F. S. Baloch, C. Vlădulescu, S. M. Popescu, and S. Mansoor. 2024. Pattern-Triggered Immunity and Effector-Triggered Immunity: Crosstalk and Cooperation of PRR and NLR-Mediated Plant Defense Pathways during Host–Pathogen Interactions. *Physiology and Molecular Biology of Plants* 30 (4): 587–604. doi:10.1007/s12298-024-01452-7.

Nadeem, M., J. Li, M. Yahya, A. Sher, C. Ma, X. Wang, and L. Qiu. 2019. Research Progress and Perspective on Drought Stress in Legumes: A Review. *International Journal of Molecular Sciences* 20 (10): 2541. doi:10.3390/ijms20102541.

Naik, H. S., J. Zhang, A. Lofquist, T. Assefa, S. Sarkar, D. Ackerman, A. Singh, A. K. Singh, and B. Ganapathysubramanian. 2017. A Real-Time Phenotyping Framework Using Machine Learning for Plant Stress Severity Rating in Soybean. *Plant Methods* 13: 1–12.

Nair, R. M., A. K. Pandey, A. R. War, B. Hanumantharao, T. Shwe, A. Alam, A. Pratap, S. R. Malik, R. Karimi, and E. K. Mbeyagala. 2019. Biotic and Abiotic Constraints in Mungbean Production—Progress in Genetic Improvement. *Frontiers in Plant Science* 10: 1340.

Negin, B., and M. Moshelion. 2016. The Advantages of Functional Phenotyping in Pre-Field Screening for Drought-Tolerant Crops. *Functional Plant Biology* 44 (1): 107–118.

Pankou, C., A. Lithourgidis, G. Menexes, and C. Dordas. 2022. Importance of Selection of Cultivars in Wheat–Pea Intercropping Systems for High Productivity. *Agronomy* 12 (10): 2367. doi:10.3390/agronomy12102367.

Panth, M., S. C. Hassler, and F. Baysal-Gurel. 2020. Methods for Management of Soilborne Diseases in Crop Production. *Agriculture* 10 (1): 16. doi:10.3390/agriculture10010016.

Pendergast, L., S. P. Bhattarai, and D. J. Midmore. 2019. Evaluation of Aerated Subsurface Drip Irrigation on Yield, Dry Weight Partitioning and Water Use Efficiency of a Broad-Acre Chickpea (*Cicer arietinum* L.) in a Vertosol. *Agricultural Water Management* 217: 38–46. doi:10.1016/j.agwat.2019.02.022.

Prashar, P., N. Kapoor, and S. Sachdeva. 2014. Rhizosphere: Its Structure, Bacterial Diversity and Significance. *Reviews in Environmental Science and Bio/Technology* 13 (1): 63–77. doi:10.1007/s11157-013-9317-z.

Qiao, M., C. Hong, Y. Jiao, S. Hou, and H. Gao. 2024. Impacts of Drought on Photosynthesis in Major Food Crops and the Related Mechanisms of Plant Responses to Drought. *Plants* 13 (13): 1808.

Rairdin, A., F. Fotouhi, J. Zhang, D. S. Mueller, B. Ganapathysubramanian, A. K. Singh, S. Dutta, S. Sarkar, and A. Singh. 2022. Deep Learning-Based Phenotyping for Genome Wide Association Studies of Sudden Death Syndrome in Soybean. *Frontiers in Plant Science* 13: 966244.

Rane, J., A. K. Singh, M. Kumar, K. M. Boraiah, K. K. Meena, A. Pradhan, and P. V. Prasad. 2021. The Adaptation and Tolerance of Major Cereals and Legumes to Important Abiotic Stresses. *International Journal of Molecular Sciences* 22 (23): 12970.

Ray, R. L., P. A. Y. Ampim, and M. Gao. 2020. Crop Protection Under Drought Stress. In Jabran, K., S. Florentine, and B. S. Chauhan (eds.) *Crop Protection Under Changing Climate*, 145–170. Springer International Publishing. doi:10.1007/978-3-030-46111-9_6.

Razo-Belman, R., and C. Ozuna. 2023. Volatile Organic Compounds: A Review of Their Current Applications as Pest Biocontrol and Disease Management. *Horticulturae* 9 (4). doi:10.3390/horticulturae9040441.

Saini, N., A. Anmol, S. Kumar, A. W. Wani, M. Bakshi, and Z. Dhiman. 2024. Exploring Phenolic Compounds as Natural Stress Alleviators in Plants—A Comprehensive Review. *Physiological and Molecular Plant Pathology* 133 (August): 102383. doi:10.1016/j.pmpp.2024.102383.

Sandhu, R., S. K. Bangarwa, M. Attri, S. Tiwai, S. Kohli, and N. Chaudhary. 2023. Effects of Biotic Stresses and Their Mitigation Strategies in Legumes: A Review. *Legume Research-An International Journal* 1 (10): 193–202.

Šarunaite, L., I. Deveikyte, A. Arlauskiene, Ž. Kadžiuliene, and S. Maikšteniene. 2013. Pea and Spring Cereal Intercropping Systems: Advantages and Suppression of Broad-Leaved Weeds. *Polish Journal of Environmental Studies* 22 (2): 541–551.

Satrio, R. D., M. H. Fendiyanto, and M. Miftahudin. 2024. Tools and Techniques Used at Global Scale Through Genomics, Transcriptomics, Proteomics, and Metabolomics to Investigate Plant Stress Responses at the Molecular Level. In Shahid, M., and R. Gaur, (eds.) *Molecular Dynamics of Plant Stress and Its Management*, 555–607. Springer Link. doi:10.1007/978-981-97-1699-9_25.

Schreinemachers, P., J. Howard, M. Turner, S. N. Groot, B. Dubey, L. Mwadzingeni, T. Chagomoka, M. Ngugi, V. Afari-Sefa, P. Hanson, and M. C. S. Wopereis. 2021. Africa's Evolving Vegetable Seed Sector: Status, Policy Options and Lessons from Asia. *Food Security* 13 (3): 511–523. doi:10.1007/s12571-021-01146-y.

Shah, A., M. Nazari, M. Antar, L. A. Msimbira, J. Naamala, D. Lyu, M. Rabileh, J. Zajonc, and D. L. Smith. 2021. PGPR in Agriculture: A Sustainable Approach to Increasing Climate Change Resilience. *Frontiers in Sustainable Food Systems* 5. doi:10.3389/fsufs.2021.667546.

Shomali, A., S. Das, N. Arif, M. Sarraf, N. Zahra, V. Yadav, S. Aliniaeifard, D. K. Chauhan, and M. Hasanuzzaman. 2022. Diverse Physiological Roles of Flavonoids in Plant Environmental Stress Responses and Tolerance. *Plants* 11 (22). doi:10.3390/plants11223158.

Shumilina, J., D. Gorbach, V. Popova, A. Tsarev, A. Kusnetsova, M. Grashina, M. Dorn, E. Lukasheva, N. Osmolovskaya, E. Romanovskaya, V. Zhukov, C. Ihling, T. Grishina, T. Bilova, and A. Frolov. 2021. Protein Glycation and Drought Response of Pea (*Pisum sativum* L.) Root Nodule Proteome: A Proteomics Approach. *Biological Communications* 66 (3): 210–214. doi:10.21638/SPBU03.2021.303.

Siddique, K. H. M., C. Johansen, N. C. Turner, M. H. Jeuffroy, A. Hashem, D. Sakar, Y. Gan, and S. S. Alghamdi. 2012. Innovations in Agronomy for Food Legumes: A Review *Agronomy for Sustainable Development* 32 (1): 45–64. doi:10.1007/S13593-011-0021-5.

Smýkal, P., C. J. Coyne, M. J. Ambrose, N. Maxted, H. Schaefer, M. W. Blair, J. Berger, S. L. Greene, M. N. Nelson, N. Besharat, T. Vymyslický, C. Toker, R. K. Saxena, M. Roorkiwal, M. K. Pandey, J. Hu, Y. H. Li, L. X. Wang, Y. Guo, … R. K. Varshney. 2015. Legume Crops Phylogeny and Genetic

Diversity for Science and Breeding. *Critical Reviews in Plant Sciences* 34 (1–3): 43–104. doi:10.1080/07352689.2014.897904.
Solanki, M. K., Z. Wang, F. Y. Wang, C. N. Li, T. J. Lan, R. K. Singh, P. Singh, L. T. Yang, and Y. R. Li. 2017. Intercropping in Sugarcane Cultivation Influenced the Soil Properties and Enhanced the Diversity of Vital Diazotrophic Bacteria. *Sugar Tech* 19 (2): 136–147. doi:10.1007/S12355-016-0445-Y.
Srivastava, R., and L. Sahoo. 2021. Balancing Yield Trade-Off in Legumes during Multiple Stress Tolerance via Strategic Crosstalk by Native NAC Transcription Factors. *Journal of Plant Biochemistry and Biotechnology* 30 (4): 708–729. doi:10.1007/s13562-021-00749-y.
Subramani, M., C. A. Urrea, R. Habib, K. Bhide, J. Thimmapuram, and V. Kalavacharla. 2023. Comparative Transcriptome Analysis of Tolerant and Sensitive Genotypes of Common Bean (*Phaseolus vulgaris* L.) in Response to Terminal Drought Stress. *Plants* 12 (1): 210. doi:10.3390/plants12010210.
Sylla, A., J. O. Yila, S. Diallo, and S. Traoré. 2023. Importance of the Social Structures in Cowpea Varietal Demands for Women and Men Farmers in Segou Region, Mali. *Sustainability* 15 (4): 3433.
Ullah, N., A. Ditta, M. Imtiaz, X. Li, A. U. Jan, S. Mehmood, M. S. Rizwan, and M. Rizwan. 2021. Appraisal for Organic Amendments and Plant Growth-Promoting Rhizobacteria to Enhance Crop Productivity under Drought Stress: A Review. *Journal of Agronomy and Crop Science* 207 (5): 783–802. doi:10.1111/JAC.12502.
Vom Brocke, K., G. Trouche, E. Weltzien, C. P. Barro-Kondombo, E. Gozé, and J. Chantereau. 2010. Participatory Variety Development for Sorghum in Burkina Faso: Farmers' Selection and Farmers' Criteria. *Field Crops Research* 119 (1): 183–194.
Wang, Z. G., X. G. Bao, X. F. Li, X. Jin, J. H. Zhao, J. H. Sun, P. Christie, and L. Li. 2015. Intercropping Maintains Soil Fertility in Terms of Chemical Properties and Enzyme Activities on a Timescale of One Decade. *Plant and Soil* 391 (1–2): 265–282. doi:10.1007/S11104-015-2428-2.
Waseem, M., A. Raza, and A. Malik. 2024. AI-Driven Crop Yield Prediction and Disease Detection in Agroecosystems. In Kaunert, C., K. Vig, B. Singh, and S. Dutta (eds.) *Maintaining a Sustainable World in the Nexus of Environmental Science and AI*, 229–258. IGI Global. doi:10.4018/979-8-3693-6336-2.ch009.
West, T. D., and D. R. Griffith. 1992. Effect of Strip-Intercropping Corn and Soybean on Yield and Profit. *Journal of Production Agriculture* 5 (1): 107–110. doi:10.2134/JPA1992.0107.
Woli, P., G. R. Smith, C. Long, and F. M. Rouquette Jr. 2022. Assessing Cowpea-Wheat Double Cropping Strategies in the Southern United States Using the DSSAT Crop Model. *Agricultural Sciences* 13 (6): 758–775.
Yadava, D. K., H. K. Dikshit, G. P. Mishra, and S. Tripathi, eds. 2022. *Fundamentals of Field Crop Breeding*. Springer Nature Singapore. doi:10.1007/978-981-16-9257-4.
Yang, X., C. Liu, M. Li, Y. Li, Z. Yan, G. Feng, and D. Liu. 2023. Integrated Transcriptomics and Metabolomics Analysis Reveals Key Regulatory Network That Response to Cold Stress in Common Bean (*Phaseolus vulgaris* L.). *BMC Plant Biology* 23 (1): 85. doi:10.1186/s12870-023-04094-1.
Zhang, Q., S. Wang, B. Qin, H. Sun, X. Yuan, Q. Wang, J. Xu, Z. Yin, Y. Du, J. Du, and C. Li. 2023. Analysis of the Transcriptome and Metabolome Reveals Phenylpropanoid Mechanism in Common Bean (*Phaseolus vulgaris*) Responding to Salt Stress at Sprout Stage. *Food and Energy Security* 12 (5). doi:10.1002/fes3.481.
Zhao, M., Y. Haxim, Y. Liang, S. Qiao, B. Gao, D. Zhang, and X. Li. 2022. Genome-Wide Investigation of AP2/ERF Gene Family in the Desert Legume *Eremosparton songoricum*: Identification, Classification, Evolution, and Expression Profiling under Drought Stress. *Frontiers in Plant Science* 13. doi:10.3389/fpls.2022.885694.
Zhong, X., L. Hu, and G. Tang. 2024. The Application of Genome Editing Technologies in Soybean (*Glycine max* L.) for Abiotic Stress Tolerance. In *Plant Genome Editing Technologies*, edited by J.-T. Chen and S. Ahmar, 221–237. Springer Nature Singapore. doi:10.1007/978-981-99-9338-3_8.
Zou, Y., Z. Liu, Y. Chen, Y. Wang, and S. Feng. 2024. Crop Rotation and Diversification in China: Enhancing Sustainable Agriculture and Resilience. *Agriculture* 14 (9): 1465. doi:10.3390/agriculture14091465.

7 Physiological Responses Induced by Altered Secondary Metabolism of Legumes in Response to Climate Variability

Simon Wambui Mburu, Eric Kimani Kuria, Mourine Mutai, Gilbert Koskey, and Ezekiel Mugendi Njeru

7.1 INTRODUCTION

The growing pressure from human population growth, climate change, and land degradation are major drivers of global food insecurity. In response, legumes are increasingly fronted as potential high-yield crops for ensuring food and nutrition security (Rahman et al., 2022; Tian et al., 2021). These crops are rich in protein and essential minerals, hence they are indispensable to human health and wellbeing. Legume cultivation spans a wide spectrum including grain, forage, vegetables, oilseeds, and range legumes, making them a valuable source of dietary diversity and nutritional requirements for humans and animals (Kumari et al., 2025; Lisciani et al., 2024). Vegetable legumes, for instance, provide beneficial bioactive compounds that function as coenzymes and cofactors in key metabolic reactions (Dave et al., 2024). Beyond their nutritional value, legumes improve soil health by fixing atmospheric nitrogen, recycling nutrients, enhancing soil porosity through their robust root systems, and promoting soil aggregation (Xing et al., 2025). When cultivated in intercropping systems, legumes further boost overall plant productivity of the other crops, suppress weeds growth due to their rapid early growth (Koskey et al., 2022), and enhance water infiltration (Kumawat et al., 2022). Importantly, legumes synthesize unique secondary metabolites such as polyphenols, saponins, and isoflavones, vital for physiological processes in plants.

Despite their numerous nutritional and agronomic benefits, legumes are highly vulnerable to climate change impacts such as drought and flooding, significantly harming their productivity and quality (Dave et al., 2024). Climate change entails the gradual alteration of key environmental parameters that govern the ecosystem dynamics. Even though the precise frequency and nature of extreme climate events remain uncertain, current climatic trends have already compromised legume growth and development (Khan et al., 2025). According to Khan et al. (2025), legumes will encounter even more increasingly variable extreme climatic conditions in the coming years. The current climate-related pressures affecting legumes include intensified extreme weather events, unpredictable rainfall distribution, rising temperatures, more complex pathogen and pest interactions, and rising atmospheric carbon dioxide (CO_2) levels (Sita et al., 2017). Among these, severe drought represents the most widespread abiotic stress affecting legume yield and weakening their resistance to diseases. Additionally, key physiological processes in legumes are severely disrupted at temperatures beyond 30–35 °C, with temperatures above 40 °C often leading to complete plant

DOI: 10.1201/9781003477235-9

failure (Izaurralde et al., 2011; Sita et al., 2017). Elevated CO_2 levels and higher temperatures have also been shown to influence legumes' biochemical and physiological traits (Tariq et al., 2024). Collectively, these climatic challenges pose severe threats to legume physiology, growth, and productivity, underscoring the urgent need for breeding initiatives that enhance resilience in legumes to climate change.

Secondary metabolites are essential components in plants, playing important roles in resilience against environmental stresses. Based on their species, legumes synthesize a wide range of secondary metabolites through complex biochemical mechanisms. Some of these metabolites function as solutes, aiding in maintaining water homeostasis in drought conditions, while others serve as antifeedants, providing defense against pathogens (Shah & Smith, 2020). As noted by Srivastava et al. (2021), the production of secondary metabolites is often closely tied to specific ecological and physiological functions that facilitate plant adaptation to their environments. The biosynthesis of these metabolites involves distinct metabolic pathways and varies considerably among legume species. Major classes of secondary metabolites include flavonoids, alkaloids, saponins, phenolic acids, and terpenoids, among others (Bulut et al., 2023; Eswaran et al., 2025). Despite their recognized importance, their evidence on how the evolution of secondary metabolism underpins legume physiological responses to climate change remains limited. This chapter delves into a comprehensive overview of legume physiological adaptations to climate change and the role of altered secondary metabolism.

7.2 ROLE OF SECONDARY METABOLITES ON LEGUME PHYSIOLOGICAL STRESS RESPONSES

Secondary metabolites including phenolic, flavonoids, alkaloids, and terpenoids, are involved in inducing tolerance against climate change-associated stress including drought, heat stress, high carbon dioxide, ultraviolet-B radiation, and nutrient deficiency (Trush & Pal'ove-Balang, 2023). Secondary metabolite levels vary depending on seasons, climatic conditions, and diurnal cycles (Qaderi et al., 2023). As a result, legumes encounter various environmental stressors, hence evolving survival strategies by producing a wide range of secondary metabolites essential for defense and stress adaptation (Fazeli-Nasab et al., 2025). These metabolites are synthesized through specific biochemical pathways often regulated at transcriptional and enzymatic levels in response to external stimuli (Jan, Asaf, Numan, Lubna, et al., 2021).

Phenolics, an important class of secondary metabolites, are synthesized through the phenylpropanoid pathway which involves the deamination of phenylalanine by phenylalanine ammonia-lyase (Saltveit, 2017). This pathway results in the production of various secondary metabolites including lignin, flavonoids, tannins, and simple phenolic acids (Table 7.1). In stress events such as drought and pathogen attack, phenolics act as potent antioxidants that scavenge reactive oxygen species (ROS) generated during oxidative stress (Hajam et al., 2023). In addition, phenolics contributes to cell wall fortification through lignification, forming a barrier against pathogen penetration (Khanday et al., 2024). For instance, chickpeas grown under salinity stress accumulate more phenolics such as ferulic acid and p-coumaric acid to enhance plant tolerance to stress (Singh & Roychoudhury, 2023).

Alkaloids are derived from amino acids such as lysine and tryptophan (Table 7.1). These secondary metabolites are produced via the decarboxylation and subsequent modification of amino acids (Bhambhani et al., 2021). Indole alkaloid and phyrrolizidine alkaloids are produced by chickpea and cowpea. Given their potent and deterrent properties, they protect plants against microbial pathogens and insects (Bhambhani et al., 2021). In contrast, terpenoids including monoterpenes and sesquiterpenes are synthesized via the mevalonate (MVA) pathway in the cytosol or methylerythritol phosphate (MEP) pathway in plastids (Zandalinas et al., 2022). These secondary metabolites facilitate host defense by attracting predators of herbivores or act directly as phytoalexins as in the case

TABLE 7.1
Secondary Metabolites Involved in Legume Physiological Response to Climate-Induced Stresses

Secondary metabolite	Chemical class	Legume	Example in legume	Role in the legume	References
Flavonoids	Polyphenols	Cowpea, pea	Quercetin, kaempferol, genistein	Signaling of symbiotic relationship in legumes to facilitate nodulation. Reduces oxidative damage from environmental stress by regulating expression of stress-related genes	(Patil et al., 2024)
Isoflavonoids	Polyphenols	Common bean, soybean	Diadzein, genistein	Enhance cold tolerance and act as phytoalexins against pathogens	(Tang et al., 2025; Kim, 2022)
Saponins	Glycosides	Soybean	Soyasaponins	Enhance membrane stabilization under stress and provide antimicrobial properties against plant pathogens	(T. F. Yu et al., 2024)
Tannin	Polyphenols	Clover	Condensed tannins	Enhance drought tolerance	(Afkar, 2023)
Alkaloids	Nitrogen containing	Cowpea	Vacianine, lupanine	Provide defense against pests and contribute to drought and pathogen resistance	(Avanza et al., 2021)
Phenolic acids	Phenolics	Soybean	Ferulic acids, caffeic acid	Strengthening the plant cell walls and provide antioxidant defense	(T. F. Yu et al., 2024)
Terpenoids	Terpenes	Pea, lentils, common beans, faba	Monoterpenes, sesquiterpenes	Signaling for plant defense against oxidative stress occasioned by climate change effects (salinity)	(Zandalinas et al., 2022)
Coumarins	Phenolic lactones	Soybean	Scopoletin	Improve iron acquisition in nutrient stress and provide antimicrobial properties	(Erb & Kliebenstein, 2020; Beyer et al., 2019)
Phytoalexins	Polyphenol	Soybean	Glyceollin, pisatin	Provides direct defense against plant microbes	(Tang et al., 2025; Dong & Song, 2020)

of soybean which produces glyceollin that is highly induced when fungi attack the plant. Moreover, saponins disrupt the bacterial, fungal, and insect membranes by forming pores. Often, the production of saponins is high in salt stress and fungal infections, enhancing the plant membrane stability and resistance to pathogens.

7.3 IMPACT OF CLIMATE-INDUCED STRESSES ON LEGUME PHYSIOLOGICAL FUNCTIONING

In the face of accelerating climate change, the agricultural landscape is undergoing unprecedented transformations. Although legumes are among the crops most vulnerable to environmental upheavals, they remain a cornerstone of global food security, soil fertility, and sustainable farming systems (Khan et al., 2025). Legumes are celebrated for their high seed protein content and for their ability to fix atmospheric nitrogen, enriching the soil and reducing the need for synthetic fertilizers (Mburu et al., 2020). However, their physiological functioning is increasingly compromised by a suite of climate-induced stresses including rising temperatures, altered precipitation patterns, drought, salinity, and increased incidence of pests and diseases (Dave et al., 2024). These stresses disrupt essential physiological processes such as photosynthesis, transpiration, nutrient uptake, and reproductive development.

7.3.1 Effects of elevated atmospheric CO_2 on legumes

Li et al. (2018) state that short exposure to high carbon dioxide (CO_2) in C3 plants increases their potential for net leaf photosynthesis. This effect is primarily attributed to a high rate of carboxylation and reduced oxidation efficiencies of RuBp carboxylase oxygenase enzyme under increased atmospheric CO_2 (Erb & Zarzycki, 2018). Enhanced photosynthetic activity under these conditions often translates to greater biomass accumulation and high yields. However, this enhanced productivity may bring a nutritional cost, where concentrations of key compounds such as proteins, minerals, and vitamins in legumes tend to decline due to elevated CO_2 (Adireddy et al., 2024). However, prolonged exposure may not impact the photosynthesis rate since the plant has to acclimatize to elevated CO_2, largely due to reduced stomatal and mesophyll conductance. This acclimation can lead to increased temperature, up to 2 °C (Xu et al., 2016).

Higher CO_2 levels also influence stomatal function and development. Habermann et al. (2019) reported decreases in stomatal density, index, and conductance, which collectively result in reduced transpiration and conservation of soil moisture. Within legume leaves, higher CO_2 stimulates photosynthetic activity, boosts starch accumulation, improves water-use efficiency, and supports the functionality of photosystem II (Li et al., 2008). However, the associated rise in leaf temperature alters the PS-II photosynthetic apparatus, damages the chloroplast which is heat sensitive, and occasions redistribution of plant resources from the shoots to the roots (Doran & Zeiss, 2021). Despite these complexities, legumes respond to elevated CO_2 with increased leaf area, higher rates of leaf-level photosynthesis, improved water use efficiency, and reduced stomatal conductance (Sollenberger & Kohmann, 2024). This is in line with Habermann et al. (2019), who demonstrated anatomical and functional changes in legume stomata due to elevated CO_2 conditions.

Elevated CO_2 also influences N fixation in legumes. In alfalfa, Sollenberger and Kohmann (2024) reported that an enhanced photosynthesis rate under elevated CO_2 promoted greater sugar translocation from leaves to roots, thereby supporting nodule formation. Bellido et al. (2023) further noted that nodulation in beans offers a large carbon sink which reduces the potential acclimatization to increased CO_2 levels. This observation is supported by carbon-for-nitrogen exchange within root nodules, where an extra C sink is created which in turn minimizes the occurrence of photosynthetic acclimatization (Bellido et al., 2023). According to Madhu et al. (2015), elevated CO_2 can significantly enhance nodulation by 114% in soybeans, 22% in faba bean (Parvin et al., 2019), and 38% in pigeon peas (Sreeharsha et al., 2019). Concurrently, high CO_2 was linked to a 38% rise in leghemoglobin content in pea nodules and enhancing nitrogenase activity (Cabrerizo et al., 2001).

The response of legumes to elevated CO_2 varies across developmental stages, with early growth phases exhibiting negative effects as compared to the maturity phase (Angelotti et al., 2020). High CO_2 decreases legume dark respiration at early stages (Cavalcante et al., 2016). However, it accelerates the reproductive development and advances early initiation of flowering while decreasing pod development time (Li et al., 2018). This contrasts with Laza et al. (2021) who reported that despite increasing aboveground biomass by 58%, high CO_2 increased peanut pod production but reduced the peanut kernel yield and oil. According to Li et al. (2015), at the pod development stage, elevated CO_2 reduced the concentration of nitrogen in cowpea leaves but increased phosphorus and potassium uptake.

Apart from plant physiology, elevated CO_2 affects the nutritional components of legume seeds. In a study by Köhler et al. (2019), high CO_2 in soybean occasions a decline in seed protein levels due to an imbalance in nitrogen fixation. The low N level results from increased variation between carbohydrate and N fixation levels under high CO_2. While elevated CO_2 decreases the legume seed protein, it increases the oil quality by elevating oleic acid and reducing the linoleic acid content (Burkey et al., 2007; Li et al., 2018). Similarly, the mineral content is also affected by notable reductions in seed copper (13.93% to 26.19%) and calcium (24% to 8.20%) under elevated CO_2. Also, Mishra and Agrawal (2014) documented a significant decline in mung bean seed sugars (9.3–15.1%) and starch content (15.5%) upon exposure to high levels of CO_2.

7.3.2 Impact of Waterlogging and High Precipitation on Legumes

Soil flooding and submergence (waterlogging) due to high precipitation are detrimental abiotic factors that constrain legume production. The impact of waterlogging depends on the legume stage of growth, legume genotype, soil type, and duration of submergence (Langan et al., 2022). The ability to survive and recover following waterlogging decreases with plant age. According to Avenhaus et al. (2016), waterlogging in legumes changes the soil environment by inhibiting diffusion of vital gases such as oxygen. This causes a deficiency of oxygen in the soil and inhibits root respiration, leading to decreased shoot and root function and growth (Olorunwa et al., 2022). In addition, it occasions reduced photosynthesis, plant growth, grain yield, formation, function, and survival of nodules, biological nitrogen fixation (BNF), and causes plant death (Pucciariello et al., 2019).

Additionally, waterlogging increases the occurrence of pests and diseases. Waterlogged soils provide favorable conditions for soil-borne pathogens affecting legumes such as fungi (*Phytophtora*, *Pythium*, and *Fusarium*) and bacteria thriving in oxygen-deprived and saturated soils (Moura et al., 2022). In addition, the weakened root system due to hypoxic conditions makes them susceptible to infections, thus lowering the plant's vigor and natural defenses against bacterial, fungal, and pest-related diseases (Gopalakrishnan & Srinivas, 2019). Moreover, stagnant water around the plant promotes foliar diseases. Waterlogging conditions attract pests such as aphids, which are vectors of viral plant infection.

Hypoxic conditions that mainly occur due to waterlogging suppress the physiological processes in legumes by constraining the availability of oxygen in the legume roots, disrupting cellular respiration and reducing the production of energy (ATP) (Avenhaus et al., 2016). This not only impairs root growth but also nutrient uptake and nitrogen fixation since the symbiotic rhizobia are sensitive to low oxygen levels (Striker & Colmer, 2017). Consequently, this results in chlorosis, stunted growth, and yield loss. Given the increasing frequency of extreme weather events and poor drainage in many agricultural regions, there is a critical need to breed legume lines that can tolerate hypoxic stress. Such breeding efforts should focus on survival traits such as enhancing the formation of aerencyma. Colmer and Voesenek (2009) improved antioxidant capacity and enabled more efficient aerobic metabolism to support root function and maintain legume productivity in extreme environments such as waterlogged soils (Pucciariello et al., 2019).

7.3.3 Effect of Fluctuating Temperatures on Legumes

Temperature variations exert significant influence on legume physiological processes supporting growth and development. In soybeans, an increase in day/night temperature from 29–23 °C to 32–29 °C resulted in a shortened vegetative period by approximately seven days across different varieties (Cavalcante et al., 2016). Angelotti et al. (2020) similarly observed a contraction of the growth cycle by 11 days when temperatures rose by 5 °C. In cowpeas, Angelotti et al. (2020) and Cavalcante Junior et al. (2016) indicated that growth is reduced by 14–23 days under elevated environmental temperatures.

The severity of heat stress varies depending on the developmental stage of the legume. During the vegetative phase, heat stress is particularly damaging, leading to pollen sterility and premature flower drop (Raja et al., 2019). Extended exposure to temperatures exceeding 28 °C can be harmful to the seeds and, for instance, causes a reduction in germination and death of soybean seeds (50–88%) and beans (36–38%) (Staniak et al., 2023). Mung beans subjected to temperatures above 50 °C for just 10–30 minutes had their germination reduced significantly (Patriyawaty et al., 2018). Heat stress also adversely affects reproductive traits in other legumes, such as the reduction in flowering branches and flower number in peanuts. These physiological constraints on seed germination and reproductive success under high temperatures in legumes can translate into significantly reduced yields, as reported in peanuts (Halder et al., 2020), mung beans (Kaur et al., (2015), lentils (Seif et al., 2012), and chickpeas (Sita et al., 2017).

Temperature plays a pivotal role in legume nodulation by influencing not only the host plant but also the symbiotic rhizobia, which form the nitrogen-fixing nodules. Nodulation is a complex biological process that is initiated by molecular communication between the legume root and rhizobia, leading to nodule organogenesis and nitrogen fixation (Ghantasala & Choudhury, 2022). Temperature affects every stage of this process from initiation signaling to infection, nodule development, and nitrogenase activity. Low temperatures occasionally reduce nodulation and delay initiation, mainly due to the suppression of rhizobia growth and metabolic activity that is temperature-dependent (Bhattacharya, 2022). Production of Nod factors including lipochitooligosaccharides essential for initiating nodulation are reduced by suboptimal temperatures, leading to weaker signaling and low infection threads (Yeremko et al., 2025; Patra & Mandal, 2022). Also, low temperature causes legume root hair deformation, a prerequisite for rhizobia entry, further limiting successful colonization by the N-fixing rhizobia (Yeremko et al., 2025). Temperatures below 15 °C have been reported to lower the number and size of legume nodules (Thurston et al., 2022). Moreover, low temperatures inhibit nitrogenase enzyme activity due to decreased ATP availability and suppressed leghemoglobin function, leading to poor nitrogen fixation efficiency (Bhattacharya, 2022).

Conversely, high temperatures denature proteins involved in signaling and nitrogen fixation, reduce rhizobia viability, and limit the development of functional nodules (Domingo-Serrano et al., 2025). According to Serova et al. (2023), high temperatures reduced nodulation due to failure of infection Thread formation and premature nodule senescence. Also, they induce oxidative stress, damaging plant cells and suppressing nodulation-related gene expression (Zhang et al., 2025). Similarly, they decrease photosynthetic allocation to roots, affecting the energy supply needed for nodule formation and maintenance. High temperatures indirectly affect nodulation by altering soil microbial communities and nutrient availability, influencing competitive dynamics between the effective and ineffective rhizobia strains. In high temperatures, selection pressure may favor heat-tolerant rhizobia, which are less efficient, and leads to suboptimal nitrogen fixation (Zhang et al., 2024).

7.3.4 Effect of Drought on Legumes

In legumes, drought stress suppresses photosynthesis by reducing water availability crucial for physiological functions and disrupting the internal structure of leaf tissues (Khatun et al., 2021). Drought induces stomatal closure, which reduces carbon dioxide intake and lowers the rate of photosynthetic carbon assimilation. Consequently, the plant chlorophyll content reduces while photosystem II activity is inhibited, leading to the accumulation of reactive oxygen species (ROS), which damage chloroplast membranes and photosynthetic proteins (Pospíšil, 2016). Reduced photosynthetic capacity limits the production of carbohydrates needed for energy and biomass accumulation. Legumes heavily rely on photosynthesis to support nodule development and nitrogen fixation, and so drought-induced reductions in photosynthesis have cascading effects on overall plant productivity.

Water deficit also reduces the mobility of rhizobia in the soil, limiting their interaction with host plant root hairs. Furthermore, drought suppresses gene expression in nodulation signaling pathways, inhibiting infection thread formation and the development of root nodules (Iqbal et al., 2022). Even when nodules form, their size and function are often reduced under drought conditions due to limited nutrients and carbohydrate transport. Additionally, nitrogenase activity within the nodules declines significantly because the oxygen diffusion barrier is altered and energy supply is limited, compromising N-fixation (Saikia et al., 2024).

According to Lumactud et al. (2023), the cumulative effect of drought on photosynthesis and nodulation contribute to stunted growth and poor development in soybean. Water deficit affects the shoot and root systems by reducing cell turgor, inhibiting cell expansion and division. Consequently, vegetative growth is curtailed with a noticeable reduction in leaf area, plant height, and biomass accumulation. Although root growth may initially increase as an adaptive mechanism of exploring deeper water reserves, prolonged drought eventually leads to root shrinkage and damage. Drought alters hormonal balance such as increased abscisic acid levels which inhibit growth-promoting hormones such as auxins and cytokinins. The net effect is a decrease in the yield and quality of legume crops.

7.4 BIOCHEMICAL PATHWAY REPROGRAMMING IN LEGUMES UNDER CLIMATE VARIABILITY

Anthropogenic climate change is known to influence the production of secondary metabolites in legumes, reshaping key biochemical pathways crucial for ecological interactions, adaptation, and defense (Khatun et al., 2021). As sessile organisms, plants primarily cope with fluctuating environmental conditions by producing alternative chemical metabolites. These secondary metabolite profile modifications enhance the biological fitness of the plant, strengthening its resilience to biotic and abiotic stresses (Figure 7.1). Previous studies have shown that metabolic shifts in legumes often occur in response to environmental stimuli (Araújo et al., 2015). Even though genetic make-up is vital in the biosynthesis and accumulation of secondary metabolites, many of these compounds are induced by external factors, resulting in considerable intraspecific variation. Recent studies further reveal that these climate-driven adjustments are not merely passive responses but involve precise regulatory networks that optimize stress tolerance (Bhat et al., 2022; Kumar et al., 2021). Under adverse conditions, legumes can activate complex biochemical pathways to produce specialized metabolites tailored for defense. Climate-related factors, in particular, have a substantial impact on

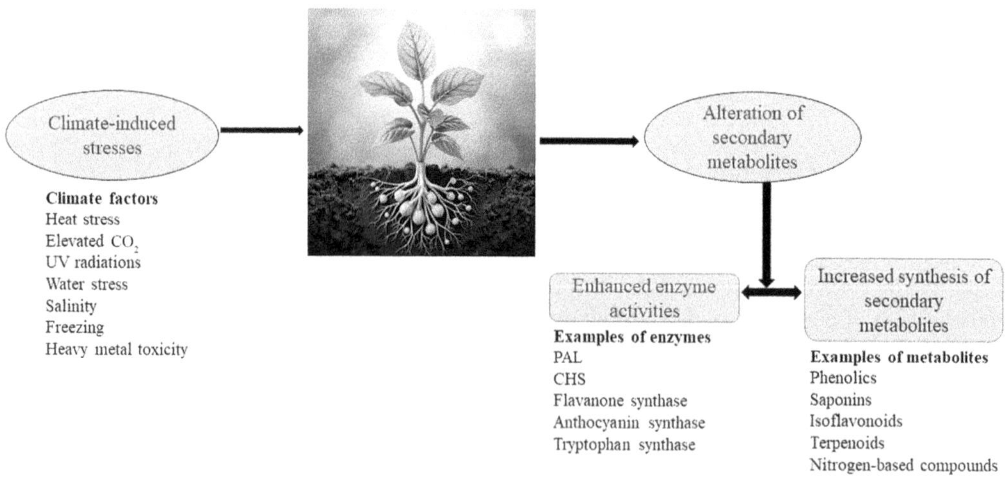

FIGURE 7.1 Alteration in legume secondary metabolites due to climate-induced stresses. PAL, Phenylalanine ammonia-lyase; CHS, chalcone synthase.

both the composition and quantity of natural products in legumes. The nature of these products is to help the plant adapt to a wide array of external stressors. It is important to recognize that the production of secondary metabolites is a highly conserved process associated with specific cues, plant developmental stages, or particular plant tissue. The amount of secondary metabolites produced varies depending on seasons, climatic conditions, and diurnal cycles (Qaderi et al., 2023). Therefore, studying physiological responses and transcriptional patterns under different abiotic stresses can provide valuable insights into the determinants of climate resilience, which can be leveraged to enhance desirable traits in plants.

7.5 CLIMATE-INDUCED TRANSCRIPTIONAL AND METABOLIC REPROGRAMMING

Legumes show complex transcriptional and physiological reprogramming in response to various abiotic stresses. Under stress conditions, these responses are modulated by intricate transcriptional factors that regulate gene expression. For instance, phytocystatins in legumes regulate protein turnover during drought stress by increasing protease activity and expression while concurrently reducing cystatin concentration (Kunert et al., 2015). Despite these insights, different cystatins for current mutants have not been explored in terms of their precise functions. In cowpea, multicystatin genes are known to be stress-inducible, and their expression is further regulated by phytocystatin-encoding genes (Diop et al., 2004). Similarly, altered cystatins and cysteine proteases in lupins are among the most responsive proteins to drought, demonstrating their critical roles in resilience and recovery. In this context, studies have shown that increased cystatin expression demonstrates that this protein enhances recovery from drought (Kunert et al., 2015).

In legumes such as soybean (*Glycine max*), drought stress activates the upregulation of key transcriptional regulators, including DREB proteins, NAC domain-containing factors, and WRKY proteins. These factors coordinate responses by regulating hormonal signaling, enhancing the biosynthesis of osmoprotectant, and promoting antioxidant defense mechanisms (Aleem et al., 2021; Kao et al., 2025; Li et al., 2022). A genome-wide survey of the DREB gene family further revealed cultivar-specific expression profiles during water-deficit conditions, suggesting a complex regulation in dehydration responses.

In contrast, exposure to high temperatures disrupts flavonoid biosynthesis by suppressing the activity of key enzymes such as chalcone synthase and chalcone isomerase, which are central to this pathway. This compromises heat stress tolerance in legumes such as common beans (*Phaseolus vulgaris*) due to the decreased flavonoid pools. Conversely, short-term heat events have been shown to stimulate isoflavonoid production in soybean, suggesting dynamic and stress-dependent modulation of secondary metabolites (Trush & Pal'ove-Balang, 2023; Veremeichik et al., 2021). Similarly, chickpeas (*Cicer arietinum*) under high temperature stress show an elevated expression of small heat shock proteins that help to maintain cellular homeostasis and protein integrity (Kalefetoğlu MacAr & Ekmekçi, 2009; Keerthi Sree et al., 2023). Further, RNA-seq profiling of six chickpea genotypes under control and heat stress conditions identified 14,544 differentially expressed genes (DEGs), including 800 transcription factors across 53 families. This highlights a broad transcriptional reprogramming that facilitates thermal resilience. Importantly, combined stressors such as drought and high temperatures synergistically enhance the production of specific secondary metabolites such as tannin and flavonoid in chickpea under dual stress conditions (Fazeli-Nasab et al., 2025). These complex, interactive stress responses emphasize the need for multifactorial experimental approaches to fully elucidate the intricate regulation of secondary metabolism under climate-driven stress scenarios.

Elevated atmospheric CO_2 concentrations enhance legumes' carbon assimilation, promoting high accumulation of carbon-based secondary metabolites. Kretzschmar et al. (2009) reported that soybeans exposed to elevated CO_2 above 700 ppm exhibited a significant rise in the concentrations of daidzein and genistein, essential isoflavonoids involved in plant defense and microbial symbiosis. Conversely, high atmospheric CO_2 concentrations can also lead to dilution effects and shifts for carbon–nitrogen balance, reducing nitrogen-containing alkaloids. Similarly, extreme drought induces specific metabolic reprograming. Based on a study on *Medicago truncatula* under water stress, the legume showed enhanced accumulation of specific saponins and proanthocyanidins which are linked to pathogen resistance and oxidative stress management (Liu et al., 2018). These findings underscore how altered secondary metabolites act as first responders to both carbon and water availability. Moreover, metabolomics profiling coupled with RNA-seq has uncovered co-expression modules linking secondary-metabolite biosynthetic genes with stress-signal transduction pathways, providing candidate targets for engineering resilience. A comprehensive understanding of the transcriptional networks and physiological adaptation is crucial in tailoring climate-resilient legumes.

7.6 ENVIRONMENTAL STRESSORS AND THEIR INFLUENCE ON SECONDARY METABOLITE PRODUCTION IN LEGUMES

Climate change affects plants through multiple factors, including increased UV-B radiation, elevated CO_2 levels, nutrient deficiencies, heavy metal accumulation, temperature fluctuations, and drought (Zhang et al., 2025). The following section discusses the impacts of some of these stressors on secondary metabolite production in legumes

7.6.1 Ultraviolet B Radiation

Ultraviolet B radiation (UV-B), with a wavelength ranging between 315–280 nm, is a natural constituent of solar energy that is experienced on Earth's surface because of stratospheric ozone (O_3) layer thinning. It acts as an environmental stressor that influences production of secondary metabolites especially in legumes (Wang et al., 2024). According to Monje-Rueda et al. (2023) exposure of *Lotus japonicas* to elevated UV-B levels induces oxidative stress, prompting plants to enhance the biosynthesis of protective compounds such as flavonoids, phenolic acids, and tannins. These secondary metabolites act as natural sunscreens, absorbing and dissipating UV radiation to safeguard plant tissues. In cowpea, elevated UV-B radiation depresses biomass yield and symbiotic nitrogen fixation, demonstrating a shift in metabolic priorities toward stress mitigation mechanisms (Ayalew et al., 2022). In soybean, UV-B regulates isoflavone biosynthesis as a complex adaptive response to UV-induced stress (Wang et al., 2024). The alteration in secondary metabolism not only protects the plants but also has implications for their nutritional properties.

The impact of UV-B radiation on secondary metabolites varies depending on the legume species. In peas, UV-B exposure significantly increases flavonoid content, enhancing the plant antioxidant capacity (Choudhary & Agrawal, 2014). However, excessive UV-B inhibits its growth and biomass reduction. In soybean, UV-B enhances antioxidant capacity in callus cultures, thus there is production of health-promoting compounds under controlled UV-B exposure (Wang et al., 2023). The species-specific response to UV-B radiation in legumes is influenced by various factors including exposure duration, intensity, and the plant development stage. Therefore, understanding these dynamics is essential for optimizing legume cultivation practices to enhance desired secondary metabolites.

7.6.2 Elevated Carbon Levels

In plants, secondary metabolite synthesis demands huge carbon commitment involving subsequent storage in trichomes or internal organs such as lacifers, resin ducts, and vacuoles (Bhatla & Lal, 2023). In transcriptomic analysis of chickpeas grown under elevated atmospheric CO_2, genes are regulated in secondary metabolite biosynthesis pathway (Kumar et al., 2019). In a study done on soybean by Kim (2022), elevated exposure to CO_2 leads to increased isoflavones, a secondary metabolite within the flavonoid class. In cowpea, high CO_2 increased the plant phenolic compound and subsequently plant antioxidant capacity and stress response. Similarly, Elateeq et al. (2023) reported that high CO_2 in chickpeas increases the plant terpenoids and alkaloids through changes in nitrogen allocation and metabolic pathways.

7.6.3 Nitrogen, Phosphorus, and Potassium Deficiency

Nutrient deficiencies in legumes, particularly nitrogen, phosphorus, and potassium, significantly influence the biosynthesis of phenolic compounds, a class of secondary metabolites involved in plant defense and stress response (Singh et al., 2017). Nitrogen, a core element in amino acids and proteins, directly affects the shikimate pathway, which is responsible for producing phenolics (Singh et al., 2017). A deficiency results in altered phenolic content, leading to a defense mechanism against nutrient stress. Legumes may increase flavonoid synthesis in roots to promote symbiosis with rhizobia (Singh & Siddiqui, 2023). However, the total pool of phenolic may decline due to reduced carbon availability for secondary metabolism. In addition, phosphorus deficiency increases the production of pigments essential for antioxidant properties such as anthocyanins, while potassium shortage can result in oxidative stress, triggering phenolic biosynthesis (Hernández-Ruiz et al., 2025). Therefore, nutrient availability regulates the quality and quantity of phenolic compounds in legume tissues.

Additionally, alkaloid production in legumes is affected by nitrogen availability. In nitrogen-deficit conditions, alkaloid synthesis significantly decreases due to competition for limited nitrogen, especially when protein synthesis is prioritized for plant survival. However, in some legumes, stress-induced pathways can lead to increased specific alkaloids as part of a defense response. Terpenoids, synthesized from the mevalonate and methylerythritol phosphate pathway, are impacted more by carbon and phosphorus availability. Phosphorus deficiency limits ATP and NADPH generation and affects energy-intensive steps of terpenoids biosynthesis. Deficits in the micronutrients magnesium and iron, which are cofactors in terpenoid production, can hinder the synthesis of essential oils, phytoalexins, and other defense-related compounds, compromising the ability of plants to withstand biotic stress.

7.6.4 Heavy Metal Toxicity

Heavy metals such as cadmium (Cd), lead (Pb), and arsenic (As) disrupt the legume cellular homeostasis and interfere with key metabolic pathways. Metal toxicity induces oxidative stress in legume crops including common beans (Tian et al., 2024). In response to this stress, secondary metabolites such as flavonoids, phenolic acids, and alkaloids are produced. These metabolites aid in scavenging the reactive oxygen species (ROS) generated by the metal stress (Stambulska & Bayliak, 2020). While some secondary metabolites increase a protective mechanism, others may decrease due to metabolic disruption (Elateeq et al., 2023). The imbalance in secondary metabolites can result in altered physiological functions, affecting plant growth, reproduction, and resilience. Therefore, metal toxicity not only disrupts primary metabolism but also reshapes the profile and concentration of secondary metabolites in legumes, impacting plant overall health and productivity (Pandey et al., 2022).

Changes in secondary metabolite production under metal toxicity are closely tied to the plant signaling pathway. However, this depends on the metal type and concentration (Berni et al., 2019). In pulses, high-protein legumes, metal stress activates various signaling molecules such as jasmonic acid, salicylic acid, and ethylene, which regulate the expression of genes involved in plant secondary metabolism (Sultana et al., 2024; Erb & Kliebenstein, 2020; Fang et al., 2019). For instance, in legume–rhizobia symbiosis, accumulation of flavonoids and anthocyanins is often upregulated to provide antioxidative protection (Stambulska & Bayliak, 2020). Additionally, some metals may affect the activity of key enzymes such as phenylalanine ammonia-lyase (PAL), initiating the synthesis of phenolic compounds (Magray et al., 2023). The disruption or stimulation of these enzymatic pathways results in quantitative and qualitative changes in secondary metabolite profiles in alfalfa and chickpea (Misra et al., 2023).

7.6.5 Temperature Variations

Temperature significantly influences the biosynthesis of secondary metabolites in legumes, affecting the production of compounds such as flavonoids, alkaloids, phytoalexins, phenolic, and terpenoids (Ku et al., 2020). In soybeans, high temperatures decrease the concentration of isoflavones, especially genistein and daidzein, which are vital for plant defense (Kim, 2022; Caldwell et al., 2005) (Table 7.2). This decline is associated with the downregulation of enzymes in the phenylpropanoid pathway, a crucial route for flavonoid biosynthesis (Bi et al., 2022). Similarly, in mung beans, higher temperatures occasion a decline in flavonoid content, demonstrating that thermal stress can limit plant antioxidant defense mechanisms (Huang et al., 2022). In common beans, temperature variations modulate the level of flavanol content, with lower temperatures favoring high flavanol accumulation, enhancing the plant antioxidant capacity. Conversely, higher (warm) temperatures reduce flavonoid levels, potentially compromising plant resilience to oxidative stress. This temperature-dependent fluctuation underscores the need for optimal thermal conditions to maintain legumes' nutritional and protective qualities. Interestingly, this pattern does not hold universally; in peas, for example, phytoalexin production appears to correlate positively with temperature increases, suggesting a species-specific response (Singh et al., 2017). On the other hand, terpenoids in legumes are sensitive to temperature changes. According to Karolkowski et al. (2021), high temperatures increase the emission of volatile minor terpenes, while major terpenes remain unaffected. These metabolic shifts underscore the complex and often unpredictable influence of thermal stress on secondary metabolite profiles.

7.6.6 Drought

Drought stress decreases the water available for plant use and antioxidant machinery. Plants acclimatize to moderate drought stress by maintaining cell turgor pressure, while in extreme drought conditions, there is an imbalance in plant osmotic potential (Zivcak et al., 2016). Secondly, legumes respond to limited water availability through the accumulation of solutes including secondary metabolites (Khatun et al., 2021). This results in lower influx of carbon dioxide and a subsequent decrease in the consumption of Calvin cycle reduction equivalents, constraining osmotic-stress-induced protein degradation (Araújo et al., 2015).

TABLE 7.2
Effect of Environmental-Induced Factors on Legume Secondary Metabolites

Environmental factor	Metabolite class affected	Effect	Legume	Reference
Elevated CO_2	Flavonoid, phenolics	Increased	Soybean, faba	Singer et al., 2020
Heat stress	Isoflavonoids, antioxidants, alkaloids, terpenoids	Increased	Soybean, common bean, mungbeans	(Mandal et al., 2025; Priya et al., 2019)
Drought	Proline, anthocyanins	Increased	Chickpea, cowpea	(Keerthi Sree et al., 2023; Carvalho et al., 2019)
UV-B radiation	Flavonoids, phenolics, tannins	Increased	Soybean, pea	(Rai et al., 2023; Choudhary & Agrawal, 2014)
Cold stress	Sugars, phenolics	Increased	Pea, lentil, cowpea	(El Haddad et al., 2023; Bhat et al., 2022)

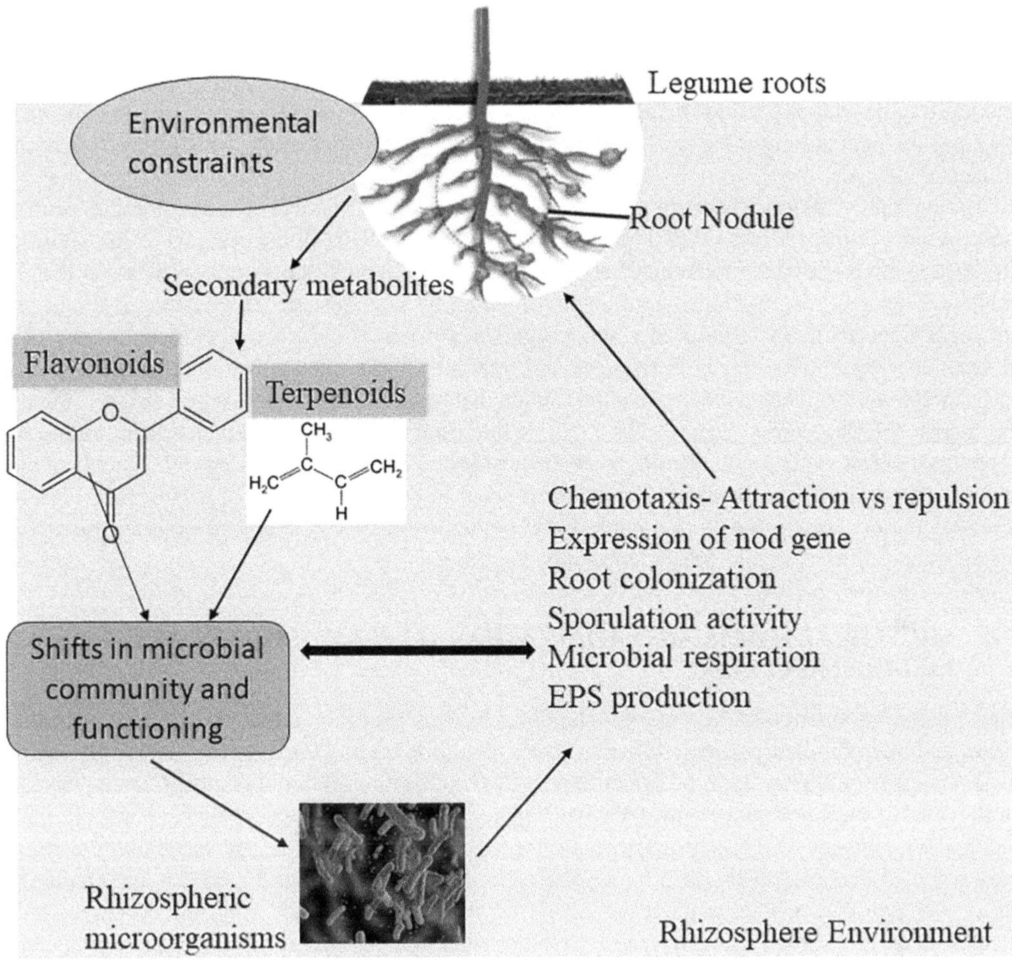

FIGURE 7.2 Below-ground effects of flavonoids and terpenoids on microbial metabolic activity and community structure (Morcillo & Manzanera, 2021). EPS, extracellular polymeric substances.

7.7 PLANT-MICROBE INTERACTIONS AND METABOLITE-MEDIATED ADAPTATION OF LEGUMES TO CLIMATE VARIABILITY

Plant growth-promoting rhizobacteria (PGPR), both symbionts and non-symbionts, are an important microbiota group in legume production and resilience against climate-induced stresses. These microbes have a rich and diverse spectrum of growth-promotion traits. The rhizospheric bacteria supply hormones and essential nutrients to plants, enhancing quality and yield while alleviating numerous biotic and abiotic stresses (Pang et al., 2021). However, secondary metabolites in legumes play important roles in stress mitigation and regulation of microbial associations and composition (Figure 7.2). Recent studies have revealed climate-induced stress's profound impact on legume–microbe symbiosis due to modification of secondary metabolites (Idris et al., 2021). Environmental factors such as high acidity, salinity, and drought can suppress microbial growth in the rhizosphere, affecting their roles in crop resilience and nutrient bioavailability. The microbial growth is further suppressed by complex secondary metabolites synthesized by the plants in response to abiotic stresses. Altered legume secondary metabolite production under climate stress can have dual effects on plant health and productivity. For instance, legume stress-related signaling molecules and plant peptides enhance plant–microbial interactions, contributing to ecological stability and fitness (Wille et al., 2019). While such alteration enhances stress tolerance under environmental challenges, they can also modify root exudate compositions, potentially disrupting microbial community dynamics and limiting beneficial symbiotic functions (Turan et al., 2017). Adaptation strategies are therefore essential for maintaining effective legume–microbe symbiosis against the backdrop of escalating climate challenges.

Among the symbiotic bacteria associated with legumes, rhizobia species are the primary contributors of nitrogen input via biological nitrogen fixation (BNF) (Mburu et al., 2022). Establishing effective symbiotic associations under abiotic stress involves intricate signaling between the rhizobia and legume host, which is significantly affected by climate-induced physiological and biochemical changes. Climate-induced stress triggers a substantial shift and alteration in the profile of secondary metabolites for both legumes and rhizobia. These metabolites include ascorbic acid and superoxide dismutase, which are antioxidants that neutralize reactive oxygen species, thereby mitigating oxidative stress (Latif et al., 2020). While establishing a symbiotic association, the rhizobia must detect signaling molecules from the host roots. The roots then release flavonoids, which triggers the bacteria's activation of nodulation genes (nod/nol) (Figure 7.2). However, elevated levels of abiotic stress can lead to notable shifts in the profile of secondary metabolites, disrupting the delicate dynamics of microbial community assembly.

7.8 MICROBIAL SHIFTS AND ADAPTATION TO PLANT SECONDARY METABOLIC CHANGES

The dynamic production of secondary metabolites in legumes plays a central role in shaping their associated microbiomes, particularly under stress conditions. The PGPR are considered part of the plant since they are believed to be synchronized by the host plant in functionality. Certain bacterial strains evolve mechanisms to withstand environmental changes and stress to maintain the interrelationship between PGPR and the legume host. Recent metagenomic and metabolomic studies have revealed that microbial consortia associated with legumes undergo functional and taxonomic shifts in response to changes in plant secondary metabolite profiles (Latif et al., 2020). On the other hand, studies have shown that enhanced biosynthesis of isoflavonoids during abiotic stress can have selective activity, favoring rhizobia growth (Table 7.3) while suppressing opportunistic pathogens (Bueno & Lopes, 2020). This bidirectional interaction may buffer plants against climate-induced stress due to the enhanced association with selected microbial communities.

Rhizosphere microbiomes adjust by modifying gene expression, including upregulation of detoxification pathways and quorum-sensing systems that help maintain compatibility with their host. For instance, rhizobia strains that synthesize hopanoids are better adapted to acid, heat, and saline stress, sustaining the symbiotic functioning even under adverse conditions (Latif et al., 2020). These adaptive mechanisms enhance resilience and maintain microbial functions even under metabolic flux. Other microorganisms produce extracellular polymeric substances (EPSs) such as polysaccharides (Table 7.3). The EPSs enhance biofilm formation, which is crucial for horizontal gene sharing, quorum sensing, resistance against biotic and abiotic stress, and effective plant colonization (Morcillo & Manzanera, 2021). Additionally, soil aggregation and water retention potential are enhanced via EPS production. Metabolic reprogramming of the host can be leveraged to selectively enrich beneficial microbial taxa or to activate latent symbiotic traits.

7.9 APPROACHES TO UNDERSTANDING SECONDARY METABOLISM IN LEGUMES UNDER CLIMATE STRESS

The biosynthesis of secondary metabolites in legumes is shaped by a complex interplay of factors, including species identity, growth stage, physiological state, and the prevailing environmental

TABLE 7.3
Non-Exhaustive List of Legume Secondary Metabolites and Their Effects on PGPR

Legume bioactive compounds	Microorganisms involved	Function	References
Phytocystatins (cysteine proteases)	Rhizobia	Stress-induced premature nodule senescence	(Kunert et al., 2015)
Flavonoids (isoflavonoids)	Rhizobia	Stimulate bacterial nod genes, inducing microbial respiration	(Pang et al., 2021)
Saponins	Rhizobia	Enhance root colonization and influence microbial community structure	(Hoagland et al., 2024)
Tannins	Rhizobia	Modulate rhizobial growth and infection efficiency	(Pang et al., 2021)
Betaines	Rhizopheric bacteria	Influence bacterial behavior in the rhizosphere	(Khatun et al., 2021)
Phenolic acids	Rhizospheric and endophytic bacteria	Act as signaling molecules for beneficial microbial recruitment	(Keerthi Sree et al., 2023)
Lectins	Rhizobia, endophytes	Mediate recognition and attachment of symbionts to roots	(Kumar et al., 2021)

factors (Afkar, 2023). These bioactive compounds are derived from primary metabolic pathways and tend to accumulate in cells after induction by various exogenous elicitors (Humbal & Pathak, 2023b). Therefore, they can be induced in vivo or in vitro through treating plants or cultured cells with elicitors in a process called elicitation (Fazili et al., 2022; Raza et al., 2023). The process has long been used in producing secondary metabolites of medicinal value (Humbal & Pathak, 2023a). Some elicitors that have been reported to influence secondary metabolism in plants include sound (Wu et al., 2023), microbes/microbial proteins and metabolites (Chamkhi et al., 2021), salinity (Raza et al., 2023), mechanical damage (Chen et al., 2023), light (Ali et al., 2023), temperature (Jamloki et al., 2021), and chemical treatment (Guru et al., 2022), among others.

There are various approaches to understanding secondary metabolism in legumes. Researchers have adopted analytical platforms such as gas chromatography-mass spectrometry (GC-MS) and liquid chromatography-mass spectrometry (LC-MS). For instance, Bulut et al. (2023) generated a detailed profile map for metabolites and lipids for key legumes in Europe, including pearl lupin, lentils, common beans, white lupin, and chickpeas. They detected over 3400 metabolites that included 224 derivatized metabolites and 2283 specific metabolites including triterpene saponins, isoflavonoid, sparteine, glycosides, lupanine, and alkaloids. These compounds are synthesized in response to environmental stresses. Because the compounds are involved in stress tolerance, their study can impact legume breeding strategies that are aimed at enhancing climate adaptability and resilience.

Through the use of infrared spectroscopy (IR), carbon nuclear magnetic resonance (13C NMR), and hydrogen nuclear magnetic resonance (1H NMR), more insights have been gained into the understanding of secondary metabolites. Nepomuceno et al. (2013) used these methods to characterize metabolites in forage legumes, identifying several potentially harmful compounds in low concentrations such as alkaloids, toxic amino acids, cyanogenic glycosides, isoflavones, and saponins that could adversely affect animals if not properly managed. They also identified non-toxic compounds that have some adverse effects on animals, such as lignins, tannins, cutins, biogenic silica, and volatile terpenoids that decrease the digestibility and palatability of the legumes. To make such crops highly suited for feeding animals such as cattle, varieties that do not produce these harmful secondary metabolites should be developed.

The use of molecular techniques is a promising approach to understanding secondary metabolism in response to various stresses in legumes. These approaches include linkage mapping, genome-wide association (GWA) mapping, genetic manipulation tools such as TALENs, ZFNs, and CRISPR (Khuram et al., 2022) for genome editing, RNA interference, and overexpression or inhibition of secondary metabolite biosynthesis pathways (Razzaq et al., 2023). For example, inhibition of the terpenoid biosynthetic pathway in lima beans was reported to make them less attractive to predatory mites and more susceptible to insect pest damage (Mumm et al., 2008).

Various genomics, proteomics, and metabolomics studies have revealed how climate change has affected secondary metabolism in legumes. A review of how genomic tools can be employed to enhance legume resilience to climate change and methods for identifying genetic variation and adaptive genes critical for improving stress tolerance in legumes was provided by Derazmahalleh et al. (2018). Since then, omics technologies have greatly improved, and various studies have demonstrated how these tools can be used to understand the role of secondary metabolism in adaptation to climate change. Such studies include one by Ku et al. (2020), highlighting the genetic basis of how domestication of legumes affects secondary metabolites and the potential implications for breeding programs aimed at enhancing beneficial secondary metabolites. Similarly, Bueno and Lopes (2020) explored the use of metabolomics in studying adaptive and signaling responses of legumes to abiotic stress. This study provided insights into the role of secondary metabolites in stress tolerance in legume crops. Additionally, through transcriptomics and metabolomics studies, Yang et al. (2023) provided insights into how common beans respond to cold stress at the molecular level, focusing on hormone regulation and metabolite accumulation. This study emphasized the role of secondary metabolites like flavonoids in stress tolerance. Another survey on lentils examined how omics technologies can be applied to improve productivity under climate change. It explored the role of secondary metabolites in stress responses and their potential use in breeding programs (Tiwari et al., 2022). The application of these technologies has greatly enhanced our understanding of the role of secondary metabolites in developing resilience in legumes in response to climate change.

7.10 BREEDING FOR RESILIENCE TO CLIMATE CHANGE BY TARGETING SECONDARY METABOLITES IN LEGUMES

Secondary metabolites play a critical role in adaptation to biotic and abiotic stress in legumes (Afkar, 2023). Therefore, they are an attractive target in the development of climate-change-resilient legumes. Many studies have described the possibility of using this approach to develop such crops (Afkar, 2023; Bueno & Lopes, 2020; Jia et al., 2020; Ku et al., 2020). Combining smart farming techniques with breeding for stress resistance in legumes has been identified as crucial for sustaining good yields under challenging climatic conditions (Dutta et al., 2022). This study recommends targeted breeding strategies for climate-resilient grain legumes, including systematic screening for terminal heat and drought tolerance across diverse germplasm, such as wild relatives. Enhancing photosynthetic efficiency and carbon assimilation under elevated CO_2 conditions, alongside biotechnological interventions like genetic engineering, are encouraged. Additionally, the conservation and utilization of genetic diversity, particularly locally adapted materials, are highlighted as key for improving resistance. Finally, a multidisciplinary approach integrating physiology, agronomy, molecular biology, and crop modeling is critical for exploitation of the genetic potential of legumes for climate change resilience.

Both genetic engineering and conventional breeding strategies have been employed to enhance stress tolerance in legumes by targeting genes involved in secondary metabolite biosynthesis. Jan et al. (2021) reviewed the complex regulation of secondary metabolites in different stress conditions.

The study had a particular focus on the role of transcription factors (TFs) utilized in modulating gene expression for important biosynthetic pathways. The study identified several TF families such as MYB, NAC, bZIP, AP2/ERF, WRKY, and bHLH as important regulators of secondary metabolism and can be manipulated for crop improvement. However, given the importance of transcription factors in plants, it is necessary to understand the specific roles and regulatory networks of TFs before manipulating their expression.

In practical applications, overexpression of sterols in soybeans has been shown to improve tolerance to abiotic stress by mitigating oxidative damage (Yu et al., 2024). This approach is particularly attractive in light of the increasing threats posed by climate variability. Additionally, overexpression of GmPI4L, a coenzyme A ligase gene, conferred enhanced resistance to *Phytophthora sojae* through elevated glyceollin accumulation (Chen et al., 2019). Similarly, genetic expression of the chalcone isomerase gene GmCHI1A in soybean hairy roots led to increased levels of daidzein, an isoflavonoid that improves resistance to the *Phytophthora sojae* P6497R strain (Zhou et al., 2018).

In addressing climate resilience, increasing the nutritional value of crops can reduce the land required for cultivation, especially for forage legumes. For example, enhancing secondary metabolite production in alfalfa by increasing genistein glucoside production through constitutive expression of isoflavone synthase gene, *MtIFS1*, from *Medicago truncatula* was reported by Deavours and Dixon (2005). The study aimed to enhance isoflavonoid levels – key polyphenolic compounds known for their health benefits and role in plant defense. However, this study reported tissue-specific expression and accumulation of the targeted isoflavonoid and influence by environmental factors such as UV radiation and pathogen infection.

Many techniques and molecular tools that can enhance the adaptation of legumes to climate change through regulation of secondary metabolites are now available. However, most have not been tested either singly or in combination. Pyramiding of genes is a strategy for developing multi-trait crops that demonstrate enhanced resilience to climate stresses.

7.11 CONCLUSIONS AND FUTURE PROSPECTS

Climate change poses major challenges in global food security through abiotic stresses such as drought, elevated temperature, and increased CO_2 levels. These stresses trigger complex alterations of secondary metabolites in legumes. The metabolic shifts include changes in biosynthesis of flavonoids, alkaloids, and phytoalexins responsible for regulating physiological processes such as stress tolerance, plant–microbe interactions, and resource allocation. This synthesis underscores the role of altered secondary metabolism in legumes as a dynamic component of adaptation strategies regulated by physiological mechanisms. The synergistic effects of plant–microbe interactions and secondary metabolites are crucial for sustainable agricultural systems under climate stress.

Despite intensive studies, the precise functioning of regulatory pathways linking climate factors to legume secondary metabolites remains unclear. Future research should focus more on utilizing multi-omics approaches to elucidate the complexity of these networks that modulate metabolic reprogramming under climate-induced stress. Bioengineering programs should have more emphasis on legume genotypes with optimized secondary metabolic profiles. This can be actualized due to the advantage of exceptional opportunities of gene-editing technologies that can fine-tune these pathways for an enhanced stress response. Further, understanding legume–microbe interactions in relation to secondary metabolism is also crucial for bioengineering climate-resilient legumes. Future research should focus on developing microbial strains with the capacity to sense, respond to, and co-evolve with changing plant secondary metabolites.

FUNDING

This study did not secure any external funding.

CONFLICTS OF INTEREST STATEMENT

The authors declare no conflicts of interest.

REFERENCES

Adireddy, R. G., Anapalli, S. S., Reddy, K. N., Mubvumba, P. and George, J. (2024). Possible impacts of elevated CO_2 and temperature on growth and development of grain legumes. *Environments, 11*(12), 273. https://doi.org/10.3390/ENVIRONMENTS11120273

Afkar, S. (2023). The role of metabolites in abiotic and biotic stress tolerance in legumes. *Recent Trends in Plant Breeding and Genetic Improvement*. https://doi.org/10.5772/INTECHOPEN.1003813

Aleem, M., Raza, M. M., Haider, M. S., Atif, R. M., Ali, Z., Bhat, J. A., & Zhao, T. (2021). Comprehensive RNA-seq analysis revealed molecular pathways and genes associated with drought tolerance in wild soybean (Glycine soja Sieb. and Zucc.). *Physiologia Plantarum, 172*(2), 707–732. https://doi.org/10.1111/PPL.13219

Ali, A., Mashwani, Z., Raja, N. I., & Kaushik, P. (2023). Phytomediated selenium nanoparticles and light regimes elicited in vitro callus cultures for biomass accumulation and secondary metabolite production in Caralluma tuberculata. *Frontiers in Plant Science, 14*(September), 1–17. https://doi.org/10.3389/fpls.2023.1253193

Angelotti, F., Barbosa, L. G., Barros, J. R. A., & Dos Santos, C. A. F. (2020). Cowpea development under different temperatures and carbon dioxide concentrations. *Pesquisa Agropecuaria Tropical, 50*, 1–7. https://doi.org/10.1590/1983-40632020V5059377

Araújo, S. S., Beebe, S., Crespi, M., Delbreil, B., González, E. M., Gruber, V., Lejeune-Henaut, I., Link, W., Monteros, M. J., Prats, E., Rao, I., Vadez, V., & Patto, M. C. V. (2015). Abiotic stress responses in legumes: Strategies used to cope with environmental challenges. *Critical Reviews in Plant Sciences, 34*, 237–280. https://doi.org/10.1080/07352689.2014.898450

Avanza, M. V., Álvarez-Rivera, G., Cifuentes, A., Mendiola, J. A., & Ibáñez, E. (2021). Phytochemical and functional characterization of phenolic compounds from Cowpea (Vigna unguiculata (L.) Walp.) obtained by green extraction technologies. *Agronomy, 11*(1), 162. https://doi.org/10.3390/AGRONOMY11010162

Avenhaus, U., Cabeza, R. A., Liese, R., Lingner, A., Dittert, K., Salinas-Riester, G., Pommerenke, C., & Schulze, J. (2016). Short-term molecular acclimation processes of legume nodules to increased external oxygen concentration. *Frontiers in Plant Science, 6*(JAN2016), 165883. https://doi.org/10.3389/FPLS.2015.01133

Ayalew, T., Yoseph, T., & Cadisch, G. (2022). Elevated UV-B radiation depressed biomass yield and symbiotic N2-fixation in Bradyrhizobium inoculated cowpea varieties. *Symbiosis, 87*(3), 201–211. https://doi.org/10.1007/S13199-022-00868-7

Bellido, E., de la Haba, P., & Agüera, E. (2023). Responses in nodulated bean (Phaseolus vulgaris L.) plants grown at elevated atmospheric CO2. *Plants, 12*(9), 1828. https://doi.org/10.3390/PLANTS12091828

Berni, R., Luyckx, M., Xu, X., Legay, S., Sergeant, K., Hausman, J. F., Lutts, S., Cai, G., & Guerriero, G. (2019). Reactive oxygen species and heavy metal stress in plants: Impact on the cell wall and secondary metabolism. *Environmental and Experimental Botany, 161*, 98–106. https://doi.org/10.1016/J.ENVEXPBOT.2018.10.017

Beyer, S. F., Beesley, A., Rohmann, P. F. W., Schultheiss, H., Conrath, U., & Langenbach, C. J. G. (2019). The Arabidopsis non-host defence-associated coumarin scopoletin protects soybean from Asian soybean rust. *Plant Journal, 99*(3), 397–413. https://doi.org/10.1111/TPJ.14426

Bhambhani, S., Kondhare, K. R., & Giri, A. P. (2021). Diversity in chemical structures and biological properties of plant alkaloids. *Molecules, 26*(11), 3374. https://doi.org/10.3390/MOLECULES26113374

Bhat, K. A., Mahajan, R., Pakhtoon, M. M., Urwat, U., Bashir, Z., Shah, A. A., Agrawal, A., Bhat, B., Sofi, P. A., Masi, A., & Zargar, S. M. (2022). Low temperature stress tolerance: An insight into the omics approaches for legume crops. *Frontiers in Plant Science, 13*, 888710. https://doi.org/10.3389/FPLS.2022.888710

Bhatla, S. C., & Lal, M. A. (2023). Secondary metabolites. In Frantisek Baluska (ed.) *Plant Physiology, Development and Metabolism* (pp. 765–808). Springer Singapore. https://doi.org/10.1007/978-981-99-5736-1_33

Bhattacharya, A. (2022). Low-temperature stress and nitrogen metabolism in plants: A review. In Amitav Bhattacharya (ed.) *Physiological Processes in Plants Under Low Temperature Stress* (pp. 299–407). Springer Singapore. https://doi.org/10.1007/978-981-16-9037-2_4

Bi, W., Zhao, G., Zhou, Y., Xia, X., Wang, J., Wang, G., Lu, S., He, W., Bi, T., & Li, J. (2022). Metabonomics analysis of flavonoids in seeds and sprouts of two Chinese soybean cultivars. *Scientific Reports*, *12*(1), 1–13. https://doi.org/10.1038/S41598-022-09408-1

Bueno, P. C. P., & Lopes, N. P. (2020). Metabolomics to characterize adaptive and signaling responses in legume crops under abiotic stresses. *ACS Omega*, *5*, 1752–1763. https://doi.org/10.1021/acsomega.9b03668

Bulut, M., Wendenburg, R., Bitocchi, E., Bellucci, E., Kroc, M., Gioia, T., Susek, K., Papa, R., Fernie, A. R., & Alseekh, S. (2023). A comprehensive metabolomics and lipidomics atlas for the legumes common bean, chickpea, lentil and lupin. *Plant Journal*, *116*(4), 1152–1171. https://doi.org/10.1111/TPJ.16329

Burkey, K. O., Booker, F. L., Pursley, W. A., & Heagle, A. S. (2007). Elevated carbon dioxide and ozone effects on peanut: II. Seed yield and quality. *Crop Science*, *47*(4), 1488–1497. https://doi.org/10.2135/CROPSCI2006.08.0538

Cabrerizo, P. M., González, E. M., Aparicio-Tejo, P. M., & Arrese-Igor, C. (2001). Continuous CO2 enrichment leads to increased nodule biomass, carbon availability to nodules and activity of carbon-metabolising enzymes but does not enhance specific nitrogen fixation in pea. *Physiologia Plantarum*, *113*(1), 33–40. https://doi.org/10.1034/J.1399-3054.2001.1130105.X\

Caldwell, C. R., Britz, S. J., & Mirecki, R. M. (2005). Effect of temperature, elevated carbon dioxide, and drought during seed development on the isoflavone content of dwarf soybean [Glycine max (L.) Merrill] grown in controlled environments. *Journal of Agricultural and Food Chemistry*, *53*(4), 1125–1129. https://doi.org/10.1021/JF0355351

Carvalho, M., Castro, I., Moutinho-Pereira, J., Correia, C., Egea-Cortines, M., Matos, M., Rosa, E., Carnide, V., & Lino-Neto, T. (2019). Evaluating stress responses in cowpea under drought stress. *Journal of Plant Physiology*, *241*(June), 153001. https://doi.org/10.1016/j.jplph.2019.153001

Cavalcante Junior, E. G., De Medeiros, J. F., Sobrinho, J. E., Figueirêdo, V. B., Da Costa, J. P. N., & Santos, W. D. O. (2016). Development and water requirements of cowpea under climate change conditions in the Brazilian semi-arid region. *Revista Brasileira de Engenharia Agricola e Ambiental*, *20*(9), 783–788. https://doi.org/10.1590/1807-1929/agriambi.v20n9p783-788

Chamkhi, I., Benali, T., Aanniz, T., El Menyiy, N., Guaouguaou, F.-E., El Omari, N., El-Shazly, M., Zengin, G., & Bouyahya, A. (2021). Plant-microbial interaction: The mechanism and the application of microbial elicitor induced secondary metabolites biosynthesis in medicinal plants. *Plant Physiology and Biochemistry*, *167*, 269–295. https://doi.org/https://doi.org/10.1016/j.plaphy.2021.08.001

Chen, Q., Jin, Y., Guo, X., Xu, M., Wei, G., & Lu, X. (2023). Metabolomic responses to the mechanical wounding of Catharanthus roseus ' upper leaves. *PeerJ*, *11*(e14539), 1–15. https://doi.org/10.7717/peerj.14539

Chen, X., Fang, X., Zhang, Y., Wang, X., Zhang, C., Yan, X., Zhao, Y., Wu, J., Xu, P., & Zhang, S. (2019). Overexpression of a soybean 4-coumaric acid: Coenzyme A ligase (*GmPI4L*) enhances resistance to *Phytophthora sojae* in soybean. *Functional Plant Biology*, *46*(4), 304–313.

Choudhary, K. K., & Agrawal, S. B. (2014). Ultraviolet-B induced changes in morphological, physiological and biochemical parameters of two cultivars of pea (Pisum sativum L.). *Ecotoxicology and Environmental Safety*, *100*(1), 178–187. https://doi.org/10.1016/j.ecoenv.2013.10.032

Colmer, T. D., & Voesenek, L. A. C. J. (2009). Flooding tolerance: Suites of plant traits in variable environments. *Functional Plant Biology*, *36*(8), 665–681. https://doi.org/10.1071/FP09144

Dave, K., Kumar, A., Dave, N., Jain, M., Dhanda, P. S., Yadav, A., & Kaushik, P. (2024). Climate change impacts on legume physiology and ecosystem dynamics: A multifaceted perspective. *Sustainability*, *16*(14), 6026. https://doi.org/10.3390/SU16146026

Deavours, B. E., & Dixon, R. A. (2005). Metabolic engineering of isoflavonoid biosynthesis in Alfalfa. *Plant Physiology*, *138*(4), 2245–2259. https://doi.org/10.1104/pp.105.062539

Derazmahalleh, M. M., Erskine, W., Hane, J. K., Nelson, M. N., Nguyen, H. T., Varshney, R. K., Papa, R., & Edwards, D. (2018). Adapting legume crops to climate change using genomic approaches. *Plant Cell Environment*, *42*, 6–19. https://doi.org/10.1111/pce.13203

Diop, N. N., Kidrič, M., Repellin, A., Gareil, M., D'Arcy-Lameta, A., Pham Thi, A. T., & Zuily-Fodil, Y. (2004). A multicystatin is induced by drought-stress in cowpea (Vigna unguiculata (L.) Walp.) leaves. *Febs Letters*, *577*(3), 545–550. https://doi.org/10.1016/J.FEBSLET.2004.10.014

Domingo-Serrano, L., Sanchis-López, C., Alejandre, C., Soldek, J., Palacios, J. M., & Albareda, M. (2025). A microaerobically induced small heat shock protein contributes to Rhizobium leguminosarum/Pisum sativum symbiosis and interacts with a wide range of bacteroid proteins. *Applied and Environmental Microbiology*, *91*(1). https://doi.org/10.1128/AEM.01385-24

Dong, W., & Song, Y. (2020). The significance of flavonoids in the process of biological nitrogen fixation. *International Journal of Molecular Sciences*, *21*(16), 5926. https://doi.org/10.3390/IJMS21165926

Doran, J. W., & Zeiss, M. R. (2021). Package price agroecology: The ecology of sustainable food systems. *Applied Soil Ecology*, *15*, 3–12.

Dutta, A., Trivedi, A., Nath, C. P., Sen, D., & Hazra, K. K. (2022). A comprehensive review on grain legumes as climate - smart crops: Challenges and prospects. *Environmental Challenges*, *7*(June 2021), 100479. https://doi.org/10.1016/j.envc.2022.100479

El Haddad, N., En-nahli, Y., Choukri, H., Aloui, K., Mentag, R., El-Baouchi, A., Hejjaoui, K., Rajendran, K., Smouni, A., Maalouf, F., & Kumar, S. (2023). Metabolic mechanisms underlying heat and drought tolerance in lentil accessions: Implications for stress tolerance breeding. *Plants*, *12*(23), 3962. https://doi.org/10.3390/PLANTS12233962

Elateeq, Z.-H., Khan, A. A., Khan, Y., Khan, J., Ali, A., Plant, S., Hanada, K., Barna, B., Salam, U., Ullah, S., Tang, Z.-H., Elateeq, A. A., Khan, Y., Khan, J., Khan, A., & Ali, S. (2023). Plant metabolomics: An overview of the role of primary and secondary metabolites against different environmental stress factors. *Life*, *13*(3), 706. https://doi.org/10.3390/LIFE13030706

Erb, M., & Kliebenstein, D. J. (2020). Plant secondary metabolites as defenses, regulators, and primary metabolites: The blurred functional trichotomy. *Plant Physiology*, *184*(1), 39–52. https://doi.org/10.1104/PP.20.00433

Erb, T. J., & Zarzycki, J. (2018). A short history of RubisCO: The rise and fall (?) of Nature's predominant CO2 fixing enzyme. *Current Opinion in Biotechnology*, *49*, 100–107. https://doi.org/10.1016/J.COPBIO.2017.07.017

Eswaran, S. P., Balasubramaniam, S., Sakthivel, A., Sampath, M., Sivanandhan, G., & Gnanajothi, K. (2025). Unlocking medicinal benefits and augmenting secondary metabolite synthesis in cultivated and wild legumes. In *Biotechnology, Multiple Omics, and Precision Breeding in Medicinal Plants* (1st ed., pp. 355–365). CRC Press. https://doi.org/10.1201/9781003475491-26

Fang, C., Fernie, A. R., & Luo, J. (2019). Exploring the diversity of plant metabolism. *Trends in Plant Science*, *24*(1), 83–98. https://doi.org/10.1016/J.TPLANTS.2018.09.006

Fazeli-Nasab, B., Vessal, S., Bagheri, A., & Malekzadeh-Shafaroudi, S. (2025). Evaluation of drought-tolerant chickpea genotypes (Cicer arietinum L.) using morphophysiological and phytochemical traits. *Frontiers in Plant Science*, *16*, 1529177. https://doi.org/10.3389/FPLS.2025.1529177

Fazili, M. A., Bashir, I., Ahmad, M., Yaqoob, U., & Geelani, S. N. (2022). In vitro strategies for the enhancement of secondary metabolite production in plants: A review. *Bulletin of the National Research Centre*, *46*(35), 1–12. https://doi.org/10.1186/s42269-022-00717-z

Ghantasala, S., & Roy Choudhury, S. (2022). Nod factor perception: An integrative view of molecular communication during legume symbiosis. *Plant Molecular Biology*, *110*(6), 485–509. https://doi.org/10.1007/S11103-022-01307-3

Gopalakrishnan, S., & Srinivas, V. (2019). Management of soil-borne diseases of grain legumes through broad-dpectrum actinomycetes having plant growth- promoting and biocontrol traits. *Plant Microbe Interface*, 129–144. https://doi.org/10.1007/978-3-030-19831-2_5

Guru, A., Dwivedi, P., Kaur, P., & Kumar, D. (2022). South African Journal of Botany Exploring the role of elicitors in enhancing medicinal values of plants under in vitro condition. *South African Journal of Botany*, *149*, 1029–1043. https://doi.org/10.1016/j.sajb.2021.10.014

Habermann, E., Dias de Oliveira, E. A., Contin, D. R., San Martin, J. A. B., Curtarelli, L., Gonzalez-Meler, M. A., & Martinez, C. A. (2019). Stomatal development and conductance of a tropical forage legume are regulated by elevated [CO2] under moderate warming. *Frontiers in Plant Science*, *10*, 450408. https://doi.org/10.3389/FPLS.2019.00609

Habermann, E., San Martin, J. A. B., Contin, D. R., Bossan, V. P., Barboza, A., Braga, M. R., Groppo, M., & Martinez, C. A. (2019). Increasing atmospheric CO_2 and canopy temperature induces anatomical and physiological changes in leaves of the C4 forage species Panicum maximum. *PLoS One*, *14*(2), e0212506. https://doi.org/10.1371/JOURNAL.PONE.0212506

Hajam, Y. A., Lone, R., & Kumar, R. (2023). Role of plant phenolics against Reactive Oxygen Species (ROS) induced oxidative stress and biochemical alterations. *Plant Phenolics in Abiotic Stress Management* (pp. 125–147). https://doi.org/10.1007/978-981-19-6426-8_7

Halder, D., Kheroar, S., Srivastava, R. K., & Panda, R. K. (2020). Assessment of future climate variability and potential adaptation strategies on yield of peanut and Kharif rice in eastern India. *Theoretical and Applied Climatology*, *140*(3–4), 823–838. https://doi.org/10.1007/S00704-020-03123-5

Hernández-Ruiz, R. G., Olivares-Ochoa, X. C., Salinas-Varela, Y., Guajardo-Espinoza, D., Roldán-Flores, L. G., Rivera-Leon, E. A., & López-Quintero, A. (2025). Phenolic compounds and anthocyanins in legumes and their impact on inflammation, oxidative stress, and metabolism: Comprehensive review. *Molecules*, *30*(1), 174. https://doi.org/10.3390/MOLECULES30010174

Hoagland, R. E., Zablotowicz, R. M., & Oleszek, W. A. (2024). Effects of Alfalfa Saponins on in vitro physiological activity of soil and Rhizosphere bacteria. In Ravinder K. Kohli, Harminder pal Singh, and Daizy R. Batish (Eds.) *Allelopathy in Agroecosystems* (1st ed., pp. 349–361). CRC Press. https://doi.org/10.1201/9781003578253-16

Huang, P. H., Cheng, Y. T., Chan, Y. J., Lu, W. C., & Li, P. H. (2022). Effect of heat treatment on nutritional and chromatic properties of Mung Bean (Vigna radiata L.). *Agronomy*, *12*(6), 1365. https://doi.org/10.3390/AGRONOMY12061365

Humbal, A., & Pathak, B. (2023a). Influence of exogenous elicitors on the production of secondary metabolite in plants: A review (" VSI: Secondary metabolites "). *Plant Stress*, *8*(100166), 1–13. https://doi.org/10.1016/j.stress.2023.100166

Humbal, A., & Pathak, B. (2023b). Plant stress influence of exogenous elicitors on the production of secondary metabolite in plants: A review (" VSI: Secondary metabolites ") ☆. *Plant Stress*, *8*(May), 100166. https://doi.org/10.1016/j.stress.2023.100166

Idris, A. O. A., Noh, E. A. G., Idris, A. O. A., & Noh, E. A. G. (2021). Legumes and nodule associated bacteria interaction as key factor for abiotic stresses impact mitigation. In J. Jimenez-Lopez & A. Clemente (Eds.), *Legumes Research* (1st ed., Vol. 1). IntechOpen. https://doi.org/10.5772/INTECHOPEN.99478

Iqbal, N., Sadras, V. O., Denison, R. F., Zhou, Y., & Denton, M. D. (2022). Clade-dependent effects of drought on nitrogen fixation and its components – Number, size, and activity of nodules in legumes. *Field Crops Research*, *284*, 108586. https://doi.org/10.1016/J.FCR.2022.108586

Izaurralde, R. C., Thomson, A. M., Morgan, J. A., Fay, P. A., Polley, H. W., & Hatfield, J. L. (2011). Climate impacts on agriculture: Implications for Forage and Rangeland Production. *Agronomy Journal*, *103*(2), 371–381. https://doi.org/10.2134/AGRONJ2010.0304

Jamloki, A., Bhattacharyya, M., Nautiyal, M. C., & Patni, B. (2021). Elucidating the relevance of high temperature and elevated CO_2 in plant secondary metabolites (PSMs) production. *Heliyon*, *7*(8), e07709. https://doi.org/10.1016/j.heliyon.2021.e07709

Jan, R., Asaf, S., Numan, M., Lubna, & Kim, K. M. (2021). Plant secondary metabolite biosynthesis and transcriptional regulation in response to biotic and abiotic stress conditions. *Agronomy*, *11*(5), 968. https://doi.org/10.3390/AGRONOMY11050968/S1

Jia, Y., Li, X., Liu, Q., Hu, X., Li, J., Dong, R., Liu, P., Liu, G., Luo, L., & Chen, Z. (2020). Physiological and transcriptomic analyses reveal the roles of secondary metabolism in the adaptive responses of Stylosanthes to manganese toxicity. *BMC Genomics*, *21*(861), 1–17.

Kalefetoğlu MacAr, T., & Ekmekçi, Y. (2009). Alterations in photochemical and physiological activities of chickpea (Cicer arietinum L.) cultivars under drought stress. *Journal of Agronomy and Crop Science*, *195*(5), 335–346. https://doi.org/10.1111/J.1439-037X.2009.00374.X

Kao, P. H., Baiya, S., Lee, C. W., Tseng, C. W., Chen, S. Y., Huang, Y. H., & Kao, C. F. (2025). Identification of key drought-tolerant genes in soybean using an integrative data-driven feature engineering pipeline. *Journal of Big Data*, *12*(1), 1–30. https://doi.org/10.1186/S40537-025-01078-W

Karolkowski, A., Guichard, E., Briand, L., & Salles, C. (2021). Volatile compounds in pulses: A review. *Foods*, *10*(12), 3140. https://doi.org/10.3390/FOODS10123140

Kaur, R., Bains, T. S., Bindumadhava, H., & Nayyar, H. (2015). Responses of mungbean (Vigna radiata L.) genotypes to heat stress: Effects on reproductive biology, leaf function and yield traits. *Scientia Horticulturae*, *197*, 527–541. https://doi.org/10.1016/J.SCIENTA.2015.10.015

Keerthi Sree, Y., Lakra, N., Manorama, K., Ahlawat, Y., Zaid, A., Elansary, H. O., Sayed, S. R. M., Rashwan, M. A., & Mahmoud, E. A. (2023). Drought-induced morpho-physiological, biochemical, metabolite responses and protein profiling of Chickpea (Cicer arietinum L.). *Agronomy*, *13*(7), 1814. https://doi.org/10.3390/AGRONOMY13071814

Khan, M. N., Wahab, S., Wahid, N., Shah, S. N., Ullah, B., Kaplan, A., Razzaq, A., Bibi, M., Suleman, F., & Ali, B. (2025). Impact of climate change on yield and quality of legumes. In Shah Fahad, Muhammad Adnan, Iqbal Munir, Rattan Lal, Taufiq Nawaz, and Shah Saud (Eds.) *Challenges and Solutions of Climate Impact on Agriculture* (pp. 85–111). Elsevier. https://doi.org/10.1016/B978-0-443-23707-2.00004-0

Khanday, A. H., Badroo, I. A., Wagay, N. A., & Rafiq, S. (2024). Role of phenolic compounds in disease resistance to plants. In *Plant Phenolics in Biotic Stress Management* (pp. 455–479). https://doi.org/10.1007/978-981-99-3334-1_19

Khatun, M., Sarkar, S., Era, F. M., Islam, A. M., Anwar, M. P., Fahad, S., Datta, R., & Islam, A. A. (2021). Drought stress in grain legumes: Effects, tolerance mechanisms and management. *Agronomy*, *11*(12), 2374. https://doi.org/10.3390/AGRONOMY11122374

Khuram, M., Muhammad, A., Ramala, A., Ahmad, M., & Latif, K. (2022). CRISPR-Cas9 based stress tolerance: New hope for abiotic stress tolerance in chickpea (Cicer arietinum). *Molecular Biology Reports*, *49*, 8977–8985.

Kim, I. S. (2022). Current perspectives on the beneficial effects of soybean isoflavones and their metabolites on plants. *Food Science and Biotechnology*, *31*(5), 515–526. https://doi.org/10.1007/S10068-022-01070-7

Köhler, I. H., Huber, S. C., Bernacchi, C. J., & Baxter, I. R. (2019). Increased temperatures may safeguard the nutritional quality of crops under future elevated CO2 concentrations. *Plant Journal*, *97*(5), 872–886. https://doi.org/10.1111/TPJ.14166

Koskey, G., Leoni, F., Carlesi, S., Avio, L., & Bàrberi, P. (2022). Exploiting plant functional diversity in durum wheat–lentil relay intercropping to stabilize crop yields under contrasting climatic conditions. *Agronomy*, *12*(1), 210. https://doi.org/10.3390/AGRONOMY12010210/S1

Kretzschmar, F. dos S., Aidar, M. P. M., Salgado, I., & Braga, M. R. (2009). Elevated CO2 atmosphere enhances production of defense-related flavonoids in soybean elicited by NO and a fungal elicitor. *Environmental and Experimental Botany*, *65*(2–3), 319–329. https://doi.org/10.1016/J.ENVEXPBOT.2008.10.001

Ku, Y.-S., Contador, C. A., Ng, M.-S., Yu, J., Chung, G., & Lam, H. (2020). The effects of domestication on secondary metabolite composition in legumes. *Frontiers in Genetics*, *11*(September), 1–20. https://doi.org/10.3389/fgene.2020.581357

Kumar, J., Mir, R. R., Shafi, S., Sen Gupta, D., Djalovic, I., Miladinovic, J., Kumar, R., Kumar, S., & Kumar, R. (2021). Genomics associated interventions for heat stress tolerance in cool season adapted grain legumes. *International Journal of Molecular Sciences*, *23*(1), 399. https://doi.org/10.3390/IJMS23010399

Kumar, M., Chauhan, A. S., Kumar, M., Yusuf, M. A., Sanyal, I., & Chauhan, P. S. (2019). Transcriptome sequencing of chickpea (Cicer arietinum L.) genotypes for identification of drought-responsive genes under drought stress condition. *Plant Molecular Biology Reporter*, *37*(3), 186–203. https://doi.org/10.1007/S11105-019-01147-4

Kumari, S., Dambale, A. S., Samantara, R., Jincy, M., & Bains, G. (2025). Introduction, history, geographical distribution, importance, and uses of soybean (*Glycine max* L.). In Krishna Pratap Singh, N. K. Singh, and Aravind T (Eds.) *Soybean Production Technology* (pp. 1–17). Springer Singapore. https://doi.org/10.1007/978-981-97-8677-0_1

Kumawat, A., Bamboriya, S. D., Meena, R. S., Yadav, D., Kumar, A., Kumar, S., Raj, A., & Pradhan, G. (2022). Legume-based inter-cropping to achieve the crop, soil, and environmental health security. In Ram Swaroop Meena and Sandeep Kumar (Eds.) *Advances in Legumes for Sustainable Intensification* (pp. 307–328). Elsevier. https://doi.org/10.1016/B978-0-323-85797-0.00005-7

Kunert, K. J., Van Wyk, S. G., Cullis, C. A., Vorster, B. J., & Foyer, C. H. (2015). Potential use of phytocystatins in crop improvement, with a particular focus on legumes. *Journal of Experimental Botany*, *66*(12), 3559–3570. https://doi.org/10.1093/JXB/ERV211

Langan, P., Bernád, V., Walsh, J., Henchy, J., Khodaeiaminjan, M., Mangina, E., & Negrão, S. (2022). Phenotyping for waterlogging tolerance in crops: Current trends and future prospects. *Journal of Experimental Botany*, *73*(15), 5149–5169. https://doi.org/10.1093/JXB/ERAC243

Latif, S., Weston, P. A., Barrow, R. A., Gurusinghe, S., Piltz, J. W., & Weston, L. A. (2020). Metabolic profiling provides unique insights to accumulation and biosynthesis of key secondary metabolites in annual pasture legumes of mediterranean origin. *Metabolites*, *10*(7), 267. https://doi.org/10.3390/METABO10070267

Laza, H. E., Baker, J. T., Yates, C., Mahan, J. R., Burow, M. D., Puppala, N., Gitz, D. C., Emendack, Y. Y., Layland, N., Ritchie, G. L., Chen, J., Rowland, D., Tissue, D. T., & Payton, P. R. (2021). Effect of elevated CO2 on peanut performance in a semi-arid production region. *Agricultural and Forest Meteorology*, *308–309*, 108599. https://doi.org/10.1016/J.AGRFORMET.2021.108599

Li, M., Li, H., Sun, A., Wang, L., Ren, C., Liu, J., & Gao, X. (2022). Transcriptome analysis reveals key drought-stress-responsive genes in soybean. *Frontiers in Genetics*, *13*, 1060529. https://doi.org/10.3389/FGENE.2022.1060529

Li, P., Han, X., Zong, Y., Li, H., Lin, E., Han, Y., & Hao, X. (2015). Effects of free-air CO2 enrichment (FACE) on the uptake and utilization of N, P and K in Vigna radiata. *Agriculture, Ecosystems and Environment*, *202*, 120–125. https://doi.org/10.1016/J.AGEE.2015.01.004

Li, Q.-M., Liu, B.-B., Wu, Y., & Zou, Z.-R. (2008). Interactive effects of drought stresses and elevated CO 2 concentration on photochemistry efficiency of cucumber seedlings. *Journal of Integrative Plant Biology*, *50*(10), 1307–1317. https://doi.org/10.1111/j.1744-7909.2008.00686.x

Li, Y., Yu, Z., Jin, J., Zhang, Q., Wang, G., Liu, C., Wu, J., Wang, C., & Liu, X. (2018). Impact of elevated CO2 on seed quality of soybean at the fresh edible and mature stages. *Frontiers in Plant Science*, *871*, 393146. https://doi.org/10.3389/FPLS.2018.01413

Lisciani, S., Marconi, S., Le Donne, C., Camilli, E., Aguzzi, A., Gabrielli, P., Gambelli, L., Kunert, K., Marais, D., Vorster, B. J., Alvarado-Ramos, K., Reboul, E., Cominelli, E., Preite, C., Sparvoli, F., Losa, A., Sala, T., Botha, A. M., & Ferrari, M. (2024). Legumes and common beans in sustainable diets: Nutritional quality, environmental benefits, spread and use in food preparations. *Frontiers in Nutrition*, *11*, 1385232. https://doi.org/10.3389/FNUT.2024.1385232

Liu, C., Ha, C. M., & Dixon, R. A. (2018). Functional genomics in the study of metabolic pathways in Medicago truncatula: An overview. *Methods in Molecular Biology*, *1822*, 315–337. https://doi.org/10.1007/978-1-4939-8633-0_20

Lumactud, R. A., Dollete, D., Liyanage, D. K., Szczyglowski, K., Hill, I Brett, Thilakarathna, M. S., Research, L., Centre, D., Lethbridge, A., & Correspondence, C. (2023). The effect of drought stress on nodulation, plant growth, and nitrogen fixation in soybean during early plant growth. *Journal of Agronomy and Crop Science*, *209*(3), 345–354. https://doi.org/10.1111/JAC.12627

Madhu, M., Hatfield, J. L., Madhu, M., & Hatfield, J. L. (2015). Elevated carbon dioxide and soil moisture on early growth response of soybean. *Agricultural Sciences*, *6*(2), 263–278. https://doi.org/10.4236/AS.2015.62027

Magray, J. A., Sharma, D. P., Deva, M. A., & Thoker, S. A. (2023). Phenolics: Accumulation and role in plants grown under heavy metal stress. In *Plant Phenolics in Abiotic Stress Management* (pp. 321–351). https://doi.org/10.1007/978-981-19-6426-8_15

Mandal, P., Lima, M. R. M., Wallingford, A. K., Warren, N. D., Brito, A. F., & Smith, R. G. (2025). Short-term exposure to elevated temperature and CO2 alters phytoestrogen production in red clover. *Scientific Reports*, *15*(1), 1–12. https://doi.org/10.1038/S41598-025-92250-Y

Mburu, S. W., Koskey, G., Njeru, E. M, Ombori, O., Maingi, J. M., & Kimiti, J. M. (2020). Differential response of promiscuous soybean to local diversity of indigenous and commercial bradyrhizobium inoculation under contrasting agroclimatic sones. *International Journal of Plant Production*, *14*(4), 1–12. https://doi.org/10.1007/s42106-020-00117-1

Mburu, S. W., Koskey, G., Njeru, E. M., Ombori, O., Maingi, J., & Kimiti, J. M. (2022). Genetic and phenotypic diversity of microsymbionts nodulating promiscuous soybeans from different agro-climatic conditions. *Journal of Genetic Engineering and Biotechnology*, *20*(1), 1–15. https://jgeb.springeropen.com/articles/10.1186/s43141-022-00386-5

Mishra, A. K., & Agrawal, S. B. (2014). Cultivar specific response of CO2 fertilization on two tropical mung bean (*Vigna radiata* L.) cultivars: ROS generation, antioxidant status, physiology, growth, yield and seed quality. *Journal of Agronomy and Crop Science*, *200*(4), 273–289. https://doi.org/10.1111/JAC.12057

Misra, D., Dutta, W., Jha, G., & Ray, P. (2023). Interactions and regulatory functions of phenolics in soil-plant-climate nexus. *Agronomy*, *13*(2), 280. https://doi.org/10.3390/AGRONOMY13020280

Monje-Rueda, M. D., Pal'ove-Balang, P., Trush, K., Márquez, A. J., Betti, M., & García-Calderón, M. (2023). Mutation of MYB36 affects isoflavonoid metabolism, growth, and stress responses in Lotus japonicus. *Physiologia Plantarum*, *175*(6). https://doi.org/10.1111/PPL.14084

Morcillo, R. J. L., & Manzanera, M. (2021). The effects of plant-associated bacterial exopolysaccharides on plant abiotic stress tolerance. *Metabolites*, *11*(6), 337. https://doi.org/10.3390/METABO11060337

Moura, A. B., Backhouse, D., Souza Júnior, I. T. de, & Gomes, C. B. (2022). Soilborne pathogens. In Teogenes Senna de Oliveira and Richard Willian Bell (Eds.) *Subsoil Constraints for Crop Production* (pp. 199–224). Springer Cham. https://doi.org/10.1007/978-3-031-00317-2_9

Mumm, R., Posthumus, M. A., & Dicke, M. (2008). Significance of terpenoids in induced indirect plant defence against herbivorous arthropods. *Plant, Cell and Environment*, *31*, 575–585. https://doi.org/10.1111/j.1365-3040.2008.01783.x

Nepomuceno, D. D. D., Carlos João Almeida, D. C., De Carvalho, M. G., Fernandes, R. D., & Júnior, F. E. A. C. (2013). Revista Brasileira de Zootecnia Classes of secondary metabolites identified in three legume species. *Revista Brasileira de Zootecnia*, *42*(January 2006), 700–705.

Olorunwa, O. J., Adhikari, B., Shi, A., & Barickman, T. C. (2022). Screening of cowpea (Vigna unguiculata (L.) Walp.) genotypes for waterlogging tolerance using morpho-physiological traits at early growth stage. *Plant Science*, *315*, 111136. https://doi.org/10.1016/J.PLANTSCI.2021.111136

Pandey, A. K., Zorić, L., Sun, T., Karanović, D., Fang, P., Borišev, M., Wu, X., Luković, J., & Xu, P. (2022). The anatomical basis of heavy metal responses in legumes and their impact on plant–rhizosphere interactions. *Plants*, *11*(19), 2554. https://doi.org/10.3390/PLANTS11192554

Pang, Z., Chen, J., Wang, T., Gao, C., Li, Z., Guo, L., Xu, J., & Cheng, Y. (2021). Linking plant secondary metabolites and plant microbiomes: A review. *Frontiers in Plant Science*, *12*, 621276. https://doi.org/10.3389/FPLS.2021.621276

Parvin, S., Uddin, S., Tausz-Posch, S., Fitzgerald, G., Armstrong, R., & Tausz, M. (2019). Elevated CO2 improves yield and N2 fixation but not grain N concentration of faba bean (Vicia faba L.) subjected to terminal drought. *Environmental and Experimental Botany*, *165*, 161–173. https://doi.org/10.1016/J.ENVEXPBOT.2019.06.003

Patil, J. R., Kuldeep, Mhatre, J., Yadav, · Kushi, Lal, ·,Yadav, S., Srivastava, S., & Nikalje, G. C. (2024). Flavonoids in plant-environment interactions and stress responses. *Discover Plants*, *1*(1), 1–19. https://doi.org/10.1007/S44372-024-00063-6

Patra, D., & Mandal, S. (2022). Nod–factors are dispensable for nodulation: A twist in bradyrhizobia-legume symbiosis. *Symbiosis*, *86*(1), 1–15. https://doi.org/10.1007/S13199-021-00826-9

Patriyawaty, N. R., Rachaputi, R. C. N., & George, D. (2018). Physiological mechanisms underpinning tolerance to high temperature stress during reproductive phase in mungbean (Vigna radiata (L.) Wilczek). *Environmental and Experimental Botany*, *150*, 188–197. https://doi.org/10.1016/J.ENVEXPBOT.2018.03.022

Pospíšil, P. (2016). Production of reactive oxygen species by photosystem II as a response to light and temperature stress. *Frontiers in Plant Science*, *7*(December 2016), 232551. https://doi.org/10.3389/FPLS.2016.01950

Priya, M., Sharma, L., Kaur, R., Bindumadhava, H., Nair, R. M., Siddique, K. H. M., & Nayyar, H. (2019). GABA (γ-aminobutyric acid), as a thermo-protectant, to improve the reproductive function of heat-stressed mungbean plants. *Scientific Reports*, *9*(1), 1–14. https://doi.org/10.1038/S41598-019-44163-W

Pucciariello, C., Boscari, A., Tagliani, A., Brouquisse, R., & Perata, P. (2019). Exploring legume-rhizobia symbiotic models for waterlogging tolerance. *Frontiers in Plant Science*, *10*, 440755. https://doi.org/10.3389/FPLS.2019.00578

Qaderi, M. M., Martel, A. B., & Strugnell, C. A. (2023). Environmental factors regulate plant secondary metabolites. *Plants*, *12*(3), 447. https://doi.org/10.3390/PLANTS12030447

Rahman, M. M., Alam, M. S., Islam, M. M., Kamal, M. Z. U., Rahman, G. K. M. M., Haque, M. M., Miah, M. G., & Biswas, J. C. (2022). Potential of legume-based cropping systems for climate change adaptation and mitigation. In Ram Swaroop Meena and Sandeep Kumar (Eds.) *Advances in Legumes for Sustainable Intensification* (pp. 381–402). Elsevier. https://doi.org/10.1016/B978-0-323-85797-0.00030-6

Rai, N., Neugart, S., Schröter, D., Lindfors, A. V., & Aphalo, P. J. (2023). Responses of flavonoids to solar UV radiation and gradual soil drying in two Medicago truncatula accessions. *Photochemical and Photobiological Sciences*, *22*(7), 1637–1654. https://doi.org/10.1007/S43630-023-00404-6

Raja, M. M., Vijayalakshmi, G., Naik, M. L., Basha, P. O., Sergeant, K., Hausman, J. F., & Khan, P. S. S. V. (2019). Pollen development and function under heat stress: From effects to responses. *Acta Physiologiae Plantarum*, *41*(4), 1–20. https://doi.org/10.1007/S11738-019-2835-8

Raza, A., Tabassum, J., Fakhar, A. Z., Sharif, R., Chen, H., Zhang, C., Ju, L., Fotopoulos, V., Siddique, K. H. M., Singh, R. K., Zhuang, W., & Varshney, R. K. (2023). Smart reprograming of plants against salinity stress using modern biotechnological tools. *Critical Reviews in Biotechnology*, *43*(7), 1035–1062. https://doi.org/10.1080/07388551.2022.2093695

Razzaq, M. K., Hina, A., Abbasi, A., Karikari, B., Ashraf, H. J., Mohiuddin, M., Maqsood, S., Maqsood, A., Haq, I. U., Xing, G., Raza, G., & Bhat, J. A. (2023). Molecular and genetic insights into secondary metabolic regulation underlying insect-pest resistance in legumes. *Functional & Integrative Genomics*, *32*(217), 1–18.

Saikia, N., Nithinkumar, K., Bagrecha, S., Asraful Ali, S., Jena, T., Kumari, M., Pooniya, V., Govindasamy, P., Raj, R., Lakhena, K. K., Barman, A., Kumar Singh, M., Babu, S., Saikia, N., Nithinkumar, K., A Ali, S. K., Jena, T., Kumari, M., Pooniya, V., Singh, M. K. (2024). Crop diversification with legume for sustainable production and environmental gain. In Subhash Babu, Raghavendra Singh, Sanjay Singh Rathore, Anup Das, and Vinod Kumar Singh (Eds.) *Agricultural Diversification for Sustainable Food Production* (pp. 215–239). Springer Singapore. https://doi.org/10.1007/978-981-97-7517-0_9

Saltveit, M. E. (2017). Synthesis and metabolism of phenolic compounds. In Elhadi M Yahia (Ed.) *Fruit and Vegetable Phytochemicals: Chemistry and Human Health: Second Edition* (Vol. 1, pp. 115–123). Wiley Online Library. https://doi.org/10.1002/9781119158042.ch5

Seif Barghi, S., Mostafaii, H., Peighami, F., & Zakaria, R. A. (2012). Path analysis of yield and its components in lentil under end season heat condition. *International Journal of Agriculture: Research and Review*, *2*, 969–974. www.cabidigitallibrary.org/doi/full/10.5555/20133031527

Serova, T. A., Kusakin, P. G., Kitaeva, A. B., Seliverstova, E. V., Gorshkov, A. P., Romanyuk, D. A., Zhukov, V. A., Tsyganova, A. V., & Tsyganov, V. E. (2023). Effects of elevated temperature on Pisum sativum nodule development: I—Detailed characteristic of uusual apical senescence. *International Journal of Molecular Sciences*, *24*(24), 17144. https://doi.org/10.3390/IJMS242417144/S1

Shah, A., & Smith, D. L. (2020). Flavonoids in agriculture: Chemistry and roles in, biotic and abiotic stress responses, and microbial associations. *Agronomy*, *10*(8), 1209. https://doi.org/10.3390/AGRONOMY10081209

Singer, S. D., Chatterton, S., Soolanayakanahally, R. Y., Subedi, U., Chen, G., & Acharya, S. N. (2020). Potential effects of a high CO2 future on leguminous species. *Plant-Environment Interactions*, *1*(2), 67. https://doi.org/10.1002/PEI3.10009

Singh, A., & Roychoudhury, A. (2023). Role of phenolic acids and flavonoids in the mitigation of environmental stress in plants. In Aryadeep Roychoudhury (Ed.) *Biology and Biotechnology of Environmental Stress Tolerance in Plants: Volume 1: Secondary Metabolites in Environmental Stress Tolerance* (pp. 227–248). Apple Academic Press. https://doi.org/10.1201/9781003346173-11

Singh, B., Singh, J. P., Kaur, A., & Singh, N. (2017). Phenolic composition and antioxidant potential of grain legume seeds: A review. *Food Research International*, *101*, 1–16. https://doi.org/10.1016/J.FOODRES.2017.09.026

Singh, V. P., & Siddiqui, M. H. (2023). *Plant Ionomics: Sensing, Signaling and Regulation*. John Wiley & Sons. https://doi.org/10.1002/9781119803041

Sita, K., Sehgal, A., Hanumantharao, B., Nair, R. M., Vara Prasad, P. V., Kumar, S., Gaur, P. M., Farooq, M., Siddique, K. H. M., Varshney, R. K., & Nayyar, H. (2017). Food legumes and rising temperatures: Effects, adaptive functional mechanisms specific to reproductive growth stage and strategies to improve heat tolerance. *Frontiers in Plant Science*, *8*, 298433. https://doi.org/10.3389/FPLS.2017.01658

Sollenberger, L. E., & Kohmann, M. M. (2024). Forage legume responses to climate change factors. *Crop Science*, *64*(5), 2419–2432. https://doi.org/10.1002/CSC2.21304

Sreeharsha, R. V., Mudalkar, S., Sengupta, D., Unnikrishnan, D. K., & Reddy, A. R. (2019). Mitigation of drought-induced oxidative damage by enhanced carbon assimilation and an efficient antioxidative metabolism under high CO 2 environment in pigeonpea (*Cajanus cajan* L.). *Photosynthesis Research*, *139*(1–3), 425–439. https://doi.org/10.1007/S11120-018-0586-9

Srivastava, A. K., Mishra, P., & Mishra, A. K. (2021). Effect of climate change on plant secondary metabolism: An ecological perspective. In Akhileshwar Kumar Srivastava, Vinod Kumar Kannaujiya, Rajesh Kumar Singh and Divya Singh (Eds.) *Evolutionary Diversity as a Source for Anticancer Molecules* (pp. 47–76). Elsevier. https://doi.org/10.1016/B978-0-12-821710-8.00003-5

Stambulska, U. Y., & Bayliak, M. M. (2020). Legume-Rhizobium symbiosis: Secondary metabolites, free radical processes, and effects of heavy metals. In Jean-Michel Mérillon and Kishan Gopal Ramawat (Eds.) *Reference Series in Phytochemistry*. Springer Cham. https://doi.org/10.1007/978-3-319-96397-6_43

Staniak, M., Szpunar-Krok, E., & Kocira, A. (2023). Responses of soybean to selected abiotic stresses—Photoperiod, temperature and water. *Agriculture*, *13*(1), 146. https://doi.org/10.3390/AGRICULTURE13010146

Striker, G. G., & Colmer, T. D. (2017). Flooding tolerance of forage legumes. *Journal of Experimental Botany*, *68*(8), 1851–1872. https://doi.org/10.1093/JXB/ERW239

Sultana, R., Imam, Z., Kumar, R. R., Banu, V. S., Nahakpam, S., Bharti, R., Bharadwaj, C., Singh, A. K., Pasala, R. K., Singh, D. R., & Siddiqui, M. W. (2024). Signaling and defence mechanism of Jasmonic and salicylic acid response in pulse crops: Role of WRKY transcription factors in stress response. *Journal of Plant Growth Regulation*, *44*(1), 5–21. https://doi.org/10.1007/S00344-023-11203-9

Tang, W., Li, Z., Xu, Z., Sui, X., Liang, L., Xiao, J., Song, X., Sun, B., Huang, Z., Lai, Y., Wang, C., Tang, Y., & Li, H. (2025). Transcriptomic and metabolomic analysis reveal the cold tolerance mechanism of common beans under cold stress. *BMC Plant Biology*, *25*(1), 1–21. https://doi.org/10.1186/S12870-025-06333-Z

Tariq, A., Graciano, C., Sardans, J., Zeng, F., Hughes, A. C., Ahmed, Z., Ullah, A., Ali, S., Gao, Y., & Peñuelas, J. (2024). Plant root mechanisms and their effects on carbon and nutrient accumulation in desert ecosystems under changes in land use and climate. *New Phytologist*, *242*(3), 916–934. https://doi.org/10.1111/NPH.19676

Thurston, C. L., Grossman, J. M., Fudge, R., Maul, J. E., Mirsky, S., & Wiering, N. (2022). Cold stress reduces nodulation and symbiotic nitrogen fixation in winter annual legume cover crops. *Plant and Soil*, *481*(1–2), 661–676. https://doi.org/10.1007/S11104-022-05667-Z

Tian, Y., Xie, L., Hao, S., & Zhou, X. (2024). Application of selenium to reduce heavy metal(loid)s in plants based on meta-analysis. *Chemosphere*, *364*, 143150. https://doi.org/10.1016/J.CHEMOSPHERE.2024.143150

Tian, Z., Wang, J. W., Li, J., & Han, B. (2021). Designing future crops: Challenges and strategies for sustainable agriculture. *The Plant Journal*, *105*(5), 1165–1178. https://doi.org/10.1111/TPJ.15107

Tiwari, M., Singh, B., Min, D., & Jagadish, S. V. K. (2022). Omics path to increasing productivity in less-studied crops under changing climate — lentil a case study. *Frontiers in Plant Science*, *13*(May), 1–22. https://doi.org/10.3389/fpls.2022.813985

Trush, K., & Pal'ove-Balang, P. (2023). Biosynthesis and role of isoflavonoids in legumes under different environmental conditions. *Plant Stress*, *8*, 100153. https://doi.org/10.1016/J.STRESS.2023.100153

Turan, M., Kitir, N., Elkoca, E., Uras, D., Ünek, C., Nikerel, E., Soğutmaz Özdemir, B., Tarhan, L., Eşitken, A., Yildirim, E., Mokhtari, N. E. P., Tüfenkçi, S., Karaman, M. R., & Günes, A. (2017). Nonsymbiotic and symbiotic bacteria efficiency for legume growth under different stress conditions. In Almas Zaidi, Mohammad Saghir Khan, and Javed Musarrat (Eds.) *Microbes for Legume Improvement* (pp. 387–404). Springer Cham. https://doi.org/10.1007/978-3-319-59174-2_16

Veremeichik, G. N., Grigorchuk, V. P., Butovets, E. S., Lukyanchuk, L. M., Brodovskaya, E. V., Bulgakov, D. V., & Bulgakov, V. P. (2021). Isoflavonoid biosynthesis in cultivated and wild soybeans grown in the field under adverse climate conditions. *Food Chemistry*, *342*, 128292. https://doi.org/10.1016/J.FOODCHEM.2020.128292

Wang, M., Liu, G., Guo, T., Xie, C., Wang, P., & Yang, R. (2023). UV-B radiation enhances isoflavone accumulation and antioxidant capacity of soybean calluses. *Frontiers in Nutrition*, *10*, 1139698. https://doi.org/10.3389/FNUT.2023.1139698

Wang, M., Wang, Y., Bilal, M., Xie, C., Wang, P., Rui, X., & Yang, R. (2024). UV-B radiation exhibited tissue-specific regulation of isoflavone biosynthesis in soybean cell suspension cultures. *Foods*, *13*(15), 2385. https://doi.org/10.3390/FOODS13152385

Wille, L., Messmer, M. M., Studer, B., & Hohmann, P. (2019). Insights to plant–microbe interactions provide opportunities to improve resistance breeding against root diseases in grain legumes. *Plant Cell and Environment*, *42*(1), 20–40. https://doi.org/10.1111/PCE.13214;

Wu, L., Yang, N., Guo, M., Zhang, D., Ghiladi, R. A., Bayram, H., & Wang, J. (2023). The role of sound stimulation in production of plant secondary metabolites. *Natural Products and Bioprospecting*, *9*, 40. https://doi.org/10.1007/s13659-023-00409-9

Xing, Y., Wang, X., & Mustafa, A. (2025). Exploring the link between soil health and crop productivity. *Ecotoxicology and Environmental Safety*, *289*, 117703. https://doi.org/10.1016/J.ECOENV.2025.117703

Xu, Z., Jiang, Y., Jia, B., & Zhou, G. (2016). Elevated-CO2 response of stomata and its dependence on environmental factors. *Frontiers in Plant Science*, *7*(May 2016), 169580. https://doi.org/10.3389/FPLS.2016.00657

Yang, X., Liu, C., Li, M., Li, Y., Yan, Z., Feng, G., & Liu, D. (2023). Integrated transcriptomics and metabolomics analysis reveals key regulatory network that response to cold stress in common Bean (Phaseolus vulgaris L .). *BMC Plant Biology*, *23*(85), 1–18. https://doi.org/10.1186/s12870-023-04094-1

Yeremko, L., Czopek, K., Staniak, M., Marenych, M., & Hanhur, V. (2025). Role of environmental factors in legume-rhizobium symbiosis: A review. *Biomolecules*, *15*(1), 118. https://doi.org/10.3390/BIOM15010118

Yu, T. F., Hou, Z. H., Wang, H. L., Chang, S. Y., Song, X. Y., Zheng, W. J., Zheng, L., Wei, J. T., Lu, Z. W., Chen, J., Zhou, Y. Bin, Chen, M., Sun, S. L., Jiang, Q. Y., Jin, L. G., Ma, Y. Z., & Xu, Z. S. (2024). Soybean steroids improve crop abiotic stress tolerance and increase yield. *Plant Biotechnology Journal*, *22*(8), 2333–2347. https://doi.org/10.1111/PBI.14349

Zandalinas, S. I., Balfagón, D., Gómez-Cadenas, A., & Mittler, R. (2022). Plant responses to climate change: Metabolic changes under combined abiotic stresses. *Journal of Experimental Botany*, *73*(11), 3339–3354. https://doi.org/10.1093/JXB/ERAC073

Zhang, H., He, L., Li, H., Tao, N., Chang, T., Wang, D., Lu, Y., Li, Z., Mai, C., Zhao, X., Niu, B., Ma, J., & Wang, L. (2025). Role of GmFRI-1 in regulating soybean nodule formation under cold stress. *International Journal of Molecular Sciences*, *26*(3), 879. https://doi.org/10.3390/IJMS26030879/S1

Zhang, H., Lang, Z., Zhu, J. K., & Wang, P. (2025). Tackling abiotic stress in plants: Recent insights and trends. *Stress Biology*, *5*(8), 1–7. https://doi.org/10.1007/s44154-025-00216-x

Zhang, Y., Ku, Y. S., Cheung, T. Y., Cheng, S. S., Xin, D., Gombeau, K., Cai, Y., Lam, H. M., & Chan, T. F. (2024). Challenges to rhizobial adaptability in a changing climate: Genetic engineering solutions for stress tolerance. *Microbiological Research*, *288*, 127886. https://doi.org/10.1016/J.MICRES.2024.127886

Zhou, Y., Huang, J., Zhang, X., Zhu, L., Wang, X., Guo, N., Zhao, J., & Xing, H. (2018). Overexpression of Chalcone Isomerase (CHI) increases resistance against phytophthora sojae in soybean. *Journal of Plant Biology*, *61*(5), 309–319. https://doi.org/10.1007/s12374-018-0017-7

Zivcak, M., Brestic, M., & Sytar, O. (2016). Osmotic adjustment and plant adaptation to drought stress. In Mohammad Anwar Hossain, Shabir Hussain Wani, Soumen Bhattacharjee, David J Burritt and Lam-Son Phan Tran (Eds.) *Drought Stress Tolerance in Plants, Vol 1: Physiology and Biochemistry* (pp. 105–143). Springer Cham. https://doi.org/10.1007/978-3-319-28899-4_5

8 Application of Genome-Editing Technology for the Development of Climate-Resilient Soybeans (*Glycine max* L. Merrill)

Pirtunia Nyadzani Mushadu, Lucy Molatudi, and Phetole Mangena

8.1 INTRODUCTION

Food security is a major global concern, and high food production levels are necessary in meeting demands due to the expected increases in the global human population estimated to reach above 9 billion by 2050 [1–3]. However, the increased food demand also elevates the demand for soybeans, resulting from the use of soy-oil for biodiesel production, human consumption, and high-protein meal-based animal feeds [4]. According to reports, issues such as climate change, rising carbon emissions [5], newly emerging global pandemics such as COVID-19 [6], stagnant agricultural technology, and a lack of modernized cost-effective rapid breeding tools subsequently lead to a decreased soybean value chain and thus increased food insecurity [5].

Employing conventional breeding approaches in fulfilling the global food demand in the 21st century remains insufficient, particularly with more than a 23% deficit in current agricultural output levels to meet global production demand [7]. Furthermore, conventional breeding that has been widely used to improve crop species also generally takes several years to progress from the initial stages of phenotypic and genotypic screening to the first successful crosses. It also relies solely on the amount of genetic variability available, thus consuming a lot of time as well as resources, causing serious limitations to crop genetic improvement, for soybean as well as other leguminous crops [2]. Various other limitations have also been observed, including limited genetic diversity in soybean varieties derived from their original progenitors (*Glycine soja*) that were developed through these conventional techniques [8].

As such, modern technological advancements in genetic breeding are highly required to overcome those limitations in genetic variability that continue to offset crop improvement, which still cannot be efficiently achieved via cross breeding, plant transformation, or the use of induced physical and chemical mutagenesis [7]. Modern tools such as genome editing serve as straightforward, precise, and rapid methods for genetic engineering [9] whereby they introduce desired and targeted DNA sequence mutations in the form of base deletions, insertions, or substitutions. According to Mushtaq et al. [7], the first successful plant gene-editing trial using tobacco protoplasts encouraged researchers all over the world to switch their attention toward the development of novel gene-editing protocols for a wide-ranging number of monocot and dicot plant species [7]. This innovation has modernized the biological world by enabling researchers to edit genes by adding desirable and removing undesirable alleles [10].

 DOI: 10.1201/9781003477235-10

Gene editing was regarded by many scientists as the greatest innovation that has ever been introduced due to its capacity to produce novel gene combinations or genetically modified crops that are considered to be "transgene-free" plants [6,11,12]. Furthermore, genome editing can potentially be used to counteract many breeding constraints and limitations encountered during conventional breeding. They include three major genome-editing technologies, namely zinc finger nucleases (ZFNs), transcription activators-like effector nucleases (TALENs), and the newly developed and preferred clustered regularly interspaced short palindromic repeats (CRISPR) [6,7,13,14]. Furthermore, they all employ a sequence-specific nuclease (SSN) called engineered nuclease that can either be induced to recognize specific DNA sequences or generate double-stranded breaks (DSBs) on the target genome [11,15]. The engineered nuclease consists of two parts: part 1, targeting the DNA while guiding the nuclease enzyme to specific targeted site; and part 2, used to cut the genome at specific sites [7,12]. Ever since its discovery, genome editing has been used to improve agricultural and agronomic traits of different commercially viable crops such as tomato, soybean, wheat, and rice [6–8,11].

In soybeans, genome editing has been successfully applied since 2011 and has highly facilitated precision breeding leading to the re-evaluation of soybean genes that are essential for high yield, yield quality, and resistance to biotic and abiotic stress factors [6, 16]. This tool has enhanced genetic variability in soybeans, which is a key source of traits for crop improvement [2,15] to confer resistance to disease, insects, and nematodes, a wide range of adaptations, tolerance to abiotic stresses such as very low/high soil pH, salinity, drought stress, and herbicide tolerance [17]. Genome editing serves as the most ideal tool for the genetic manipulation of soybean due to its gene redundancy and as a diploid, self-pollinating crop, that contains highly duplicated genes [15,18], falling under the Leguminosae family, Papilionoideae subfamily, tribe Phaseoleae, and genus *Glycine* with 40 chromosomes (2n = 40) [19,17]. It is one of the most cultivated crops worldwide [20] and is regarded as the most valued oilseed crop accounting for a total of more than 36.65 million tons of global vegetable oil production [17].

As such, soybeans are cultivated primarily as a source of protein and vegetable oil, wherein their seeds contain approximately 40% protein, 20% oil, 10% soluble carbohydrates, and more than 5% vitamins. Soybeans are also rich in isoflavone, phytate sugar, and various other nutritional mineral components [21]. Nevertheless, compared to other plant species, genomic editing in soybeans remains more challenging with fewer successes recorded, as indicated in Figure 8.1 [15,18,22–24]. Like in monocots, the application of gene-editing technology in recalcitrant crops like soybeans has the potential to accelerate their breeding beyond expectations. Therefore, this review highlights the application of genome editing in soybean, and outlines its current status and future outlook with the aim of providing a better understanding of improving this crop using genome-editing technology.

8.2 ROLE OF GENOME EDITING IN CROP IMPROVEMENT

Since their discovery, gene-editing techniques such as site-specific transcription activator-like effector nucleases (TALENs), clustered regularly interspaced short palindromic repeats-associated protein 9 (CRISPR-Cas9), and zinc-finger nuclease (ZFNs) have been employed for the genetic improvement of one or two agronomic traits in soybean. The CRISPR-Cas9 system remains the most commonly used tool among the abovementioned techniques. However, the first genetically modified soybean was successfully developed using ZFNs which targeted GMDcld4a and GMcld4b genes in hairy roots [8]. Furthermore, ZFNs are regarded as one of the oldest genome-editing methods [15,25], developed in the early 1990s by Sangamo BioSciences. ZFNs use premeditated restriction enzymes with a specific DNA sequence that binds zinc finger motifs and non-specific cleavage domains of FokI endonucleases [25].

The ZFN module is composed of a custom-engineering Cys2-His2 (C2H2) zinc finger protein (ZFP) array and a FokI nuclease recognizing specific DNA sequences as a DNA-binding domain,

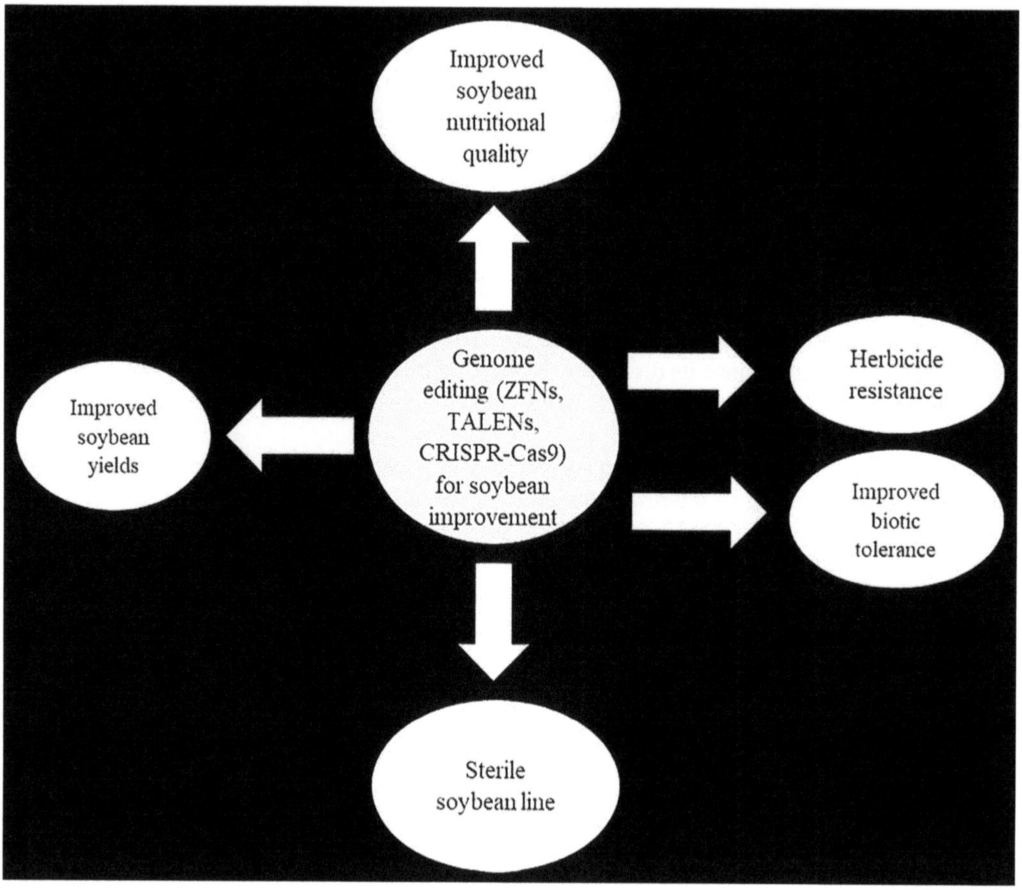

FIGURE 8.1 Application of genome editing for genetic improvement in soybeans (*Glycine max* L. Merrill.).

while FokI creates the DSBs by cutting the targeted DNA site [15,25]. Other crop species that have been manipulated include tobacco (*Nicotiana tabacum*), maize (*Zea mays*), and *Arabidopsis* (*Arabidopsis thaliana*) as well as human cells to treat or prevent diseases such as cystic fibrosis, hemophilia, sickle cell disease, cancer, heart disease, mental illness, and human immunodeficiency virus [25,26]. Haun et al. [23] reported the first genome-editing-mediated soybean with two target sites on the DNA using a TALENs system [6,23]. TALENs are traditionally regarded as the long transcription activator-like effector (TALE) sequence occurring naturally and joining the FokI domain with the carboxylic-terminal end of manipulated TALE arrays.

TALENs have been considered a ZFN substitute following their development in 2010. This tool contains a customized binding DNA domain that is joined with the non-specific FokI nuclease domain. Nevertheless, as compared to ZFNs, TALENs involve individual nucleotide repeat target site and amino acid sequences of TAL effector protein interactions. The interactions generate overhangs using the FokI nuclease domain to persuade site-specific DNA cleavages [25]. However, several drawbacks continue to limit the use of both ZFN and TALEN systems in the improvement of soybean, which can be overcome through the use of the CRISPR system. ZFN and TALEN systems are both difficult to engineer due to their highly repetitive sequences and the complex nature of their interactions with DNA [15].

CRISPR-Cas9 is regarded as the most powerful tool for gene editing and was introduced in 2012 to overcome limitations experienced in ZNF and TALEN systems as indicated above [9,27].

CRISPR is an RNA-directed DNA endonuclease adapted from the immune system of a bacterial antiviral defense system. The system consists of a molecule of CRISPR RNA (crRNA) that recognizes targettarget DNA sequences, *trans*-activating maturation of the crRNA, and lastly a Cas9 protein for cleaving the DNA [28]. This system is also programmable, site specific, more popular, robust, efficient, easy to design, and cost effective [27]. The first successful report of CRISPR genome editing was reported by Jacobs et al. [29] and regarded as a powerful tool in higher plants for site-directed mutagenesis, including the genetic modification of leguminous crops such as soybean [9,18].

8.3 SOYBEAN TRAIT IMPROVEMENT VIA GENOMIC EDITING

The most critical components that affect the yields of soybeans and other leguminous crops are the number of biotic and abiotic factors that negatively impact crop growth, productivity, and yield [30]. Additionally, these factors also reduce the accumulation of metabolites, including those serving as bioactive compounds such as flavonoids, terpenoids, and alkaloids required for human health and sustaining dynamic interactions with various stress factors (Table 8.1). However, genetic manipulation through techniques such as genomic editing has greatly circumvented these environmental effects [8]. For instance, soybean growth and productivity are affected by a range of factors such as drought, pests, and pathogens [31]. Genome editing in turn, has been applied in soybean breeding to generate novel DNA sequences that are intended to improve yields, yield quality, and produce cultivars showing resistance to various environmental stresses [8]. Some of these genetic and agronomic characteristics that have been improved using genomic editing are discussed below.

TABLE 8.1
Estimates of the Nutraceutical Composition of African and Asian Orphan Legumes in Comparison with Soybean

Component	Africa	Asia	Soybean
Carbohydrates (%)	2–61	57–60	11–25
Lipids (%)	0.8–37	5	8.1–24
Dietary fiber (%)	4.0–23	–	9.19–16.5
Proteins (%)	18–32	17–21	25–45
Vitamin A (mg/g)	0.002–0.27	–	1
β-Carotene	0.01–0.032	–	13
Vitamin B (mg/g)		–	
Thiamine	0.3–0.7	–	0.87
Riboflavin	0.06–0.2	–	0.87
Niacin	1.8–10	–	1.62
Pyridoxine	0.2–1.7	–	0.38
Folate	0.14–0.44	278	0.375
Vitamin C (mg/g)	6	–	6
Vitamin E (mg/g)	0.42–6.3	–	0.85
Calcium (mg/g)	40–300	–	277
Iron (mg/g)	1.3–33	4.6	15.7
Magnesium (mg/g)	180–380	119.6	280
Phosphorus (mg/g)	250–550	–	704
Potassium (mg/g)	750–1300	1223	1797
Sodium (mg/g)	0.2–30	–	2
Zinc (mg/g)	2–4.6	4.0	4.89
Copper (mg/g)	0.6–0.9	–	1.66

There is limited available data on nutrient composition of this group of legumes.

8.3.1 Grain and Yield Quality Enhancements

Soybean remains an important agricultural commodity grown worldwide for food, feed, and biodiesel production. The value of this crop's yield depends on the rate of flowering, pod production, and the size of seeds contained within the pods [13]. For example, to increase the yield of soybeans, CRISPR genome-editing technology was employed to edit the *GmJAG1* gene in cultivar Hauchan 6 which is a low-latitude spring soybean variety. Two RNA guides were designed to target the *GmJAG1* gene, and Cas-9-free homozygous mutants were successfully obtained in the T1 segregation population, which was then named gmjag soybean variety. As indicated, the novel gmjag soybean varieties contain a *GmJAG1* or *GmJAGGED1* gene that regulates the number of seeds per pod (NSPP) and leaflet shape. The gmjag variety produced a high number of four seeded pods, leading to an increase in the total number of seeds per plant, subsequently increasing overall yield of the variety as compared to Hauchan 6 which did not have a considerable number of plants with four-seeded pods [30].

Furthermore, soybean seeds remain an important part of the plant, and contain oils and proteins, with a considerable amount of high-quality and high-quantity storage proteins, including other major biochemical components that influence the overall quality of food and feed products derived from soy grains [32]. Two major families of storage proteins, conglycinins (7s) and glycinins (11s), account for about 70% of soy seed total proteins. Genes encoding these proteins or their mutant alleles are necessary since their subunits tend to affect the chemical composition of nutrients, amino acids, vitamins, and the overall physicochemical properties of soy food or feed meals.

Additionally, application of the CRISPR/Cas-9 system coupled with *Agrobacterium rhizogenes*-mediated hairy root transformation was effectively employed to edit the genome for high protein expressions in soybean. The sgRNA that is driven by AtU6 promoter was cloned individually into the pZG23Cos vector. These targeted genes were Glyma.20g148400, Glyma.03g163500, and Glyma.19164900. This tool was then successfully used to mutate and genetically alter the seed storage protein content in soybean, which was also reported in other leguminous crops [13]. In line with the above reports, genetic diversity studies continued to identify some yield-associated genes in both leguminous and non-leguminous crops. In comparison, oils produced from engineered soybeans were high in polysaturated fats, while soybean lines with a low content of polyunsaturated fatty acids were generated using the TALEN system.

8.4 TARGETED GENOME EDITING USING TALEN

As one of the most indispensable gene editing tool, TALEN was also efficiently used to introduce two fatty acid desaturase-2 genes (*FAD2-1A* and *FAD2-1B*) in soybean. As alluded to above, such genes are known to convert monosaturated fats such as oleic acid to polysaturated linoleic fatty acids. Since its establishment, TALEN gene-editing technology has been capable of precisely cleaving conserved double-stranded DNA sequences in a variety of host systems [13,23]. In this process, non-homologous end joining (NHEJ) DNA is responsible for repairing the introduced DSBs without using homologous templates, which often leave a small deletion or insertion at the cut site that may also cause gene function losses [23]. Moreover, both *FAD2-1A* and *FAD2-1B* genes that were successfully edited using this technology produced four times more oleic acid than the wild-type parent of genetically engineered soybeans [23].

Soybeans containing high oleic acid oil contents are more desirable because they exhibit increased shelf-life, enhanced heat stability, and improved value of nutrients [23,33]. The food products derived from such improved lines contain a combination of high nutritional content, including macronutrients, micronutrients, phytochemicals, and organoleptic attributes such as taste, flavor, aroma, storage, and stability. As such, genome-editing technology enables modifications of targeted genes to improve the crop's nutritional profile and storage [33]. Other important and major nutritional components found in these legumes include a class of water-soluble carbohydrates present in soybean seeds called raffinose family oligosaccharides (RFOs). Although generally indigestible and able to cause flatulence, these water-soluble carbohydrates have been reported to be beneficial in humans and animals that are monogastric by maintaining cell membrane stability and acting as osmoprotectants, among other properties.

Furthermore, TALEN system genome editing was also been employed to induce knockout in galactinol synthase (GOLS) gene, the *GmGOLS1A* and its homolog *GmGOLS1B* in soybeans with the aim of lowering/reducing the level of RFOs to minimize indigestibility and flatulence caused by soybeans. In the targeted site of the two genes, the construct design effectively induced various deletions resulting in single and double mutants with reduced oligosaccharide phenotypes without affecting other key agronomic traits such as the germination rate and morphology [34].

8.5 CHARACTER IMPROVEMENT FOR ABIOTIC STRESS TOLERANCE

The main objective of soybean breeding across the world remains to produce and develop cultivars that show resistance to both abiotic (salinity, soil fertility, and drought) and biotic (viruses, fungi, and bacteria) stresses, as well as tolerance to herbicides [35]. CRISPR/Cas9-gRNA genome-editing technology has been employed in soybean to successfully edit the *ALS1* gene and enable its mutagenic integration into this crop's genome. DD20 and DD43 genomic sites, which are both located in chromosome 4, were mutagenized. DD43 homolog-directed recombination was transmitted to the T1 population and the progenies of a T1 integration event showed segregation according to the Mendelian law of inheritance, and the developed soybean mutant conferred resistance to the weed herbicide chlorsulfuron [36].

CRISPR/Cas9 technology was employed to create soybeans with an early flowering germplasm. Soybean, as a short-day crop, presented sensitivity to the photoperiod due to its gradual adaptation to various latitudes and photoperiods. Photoperiod, though, serves as a crucial factor that determines the crop's adaptation ability to different ecological regions, and plays a key role in the development and differentiations of flower buds [37]. As such, flowering time also serves as a key factor for successful yield development and adaptation, demonstrating the plant's sensitivity to seasonal changes based on day length [38,39]. Presently, only 11 genes (*E1–E10* and *J*) have been identified which control soybean growth period under a wide range of latitudes [38].

The *E1* gene, for instance, was reported to have the greatest flowering inhibitory effect among all the 11 genes. However, *Agrobacterium*-mediated transformation coupled with the CRISPR/Cas9 system was used to generate targeted mutants controlling flowering using the *E1* gene. The two novel mutant types had 11bp and 40bp deletions at the E1 coding region, disinhibiting the *GmFT2a/5a* gene that regulates flowering time and then increasing its expression, resulting in early flowering of soybean mutant plants [38].

8.6 TRAIT IMPROVEMENT FOR BIOTIC STRESS RESISTANCE

Pathogens such as fungi, bacteria, and viruses cause various plant diseases that result in both growth and yield losses in soybeans. Some of the most detrimental pathogens and insect pests in legumes are summarized in Table 8.2 [31]. Globally, the loss of crops due to biotic stresses ranges between 20–40% crop damage per hectare [27]. Among such pathogens, soybean mosaic virus diseases also cause yield loss and reduce the quality of soy grains. To circumvent these effects, a CRISPR/Cas9 multiplex gene-editing system was employed to target the *GmF3H1*, *GmFH2*, and *GmFNSII-1* genes to generate homozygous triple mutants with increased iso-flavonoids content expression, thus providing resistance to soybean mosaic virus (SMV) and the basis for further breeding [40]. Iso-flavonoids are regarded as vital secondary metabolites synthesized in plants, playing a significant role in disease resistance of the plant and serving as a messenger for node gene inductions. Furthermore, soybeans are reported to contain four times higher isoflavonoids content than any other legumes [21].

TABLE 8.2
Some of the Most Common Biotic Stress Factors Negatively Affecting Leguminous Crops under Diverse Environmental Conditions [31]

Pathogen/species name	Category	Disease/common name
Pseudomonas syringae pv. phaseolicola	Bacteria	Halo blight
Pseudomonas syringae pv. syringae	Bacteria	Bacterial brown spot
Xanthomonas campestris pv. phaseoli	Bacteria	Bacterial blight
Fusarium oxysporum f.sp.ciceris	Fungi	Fusarium wilt
Fusarium solani	Fungi	Black root rot
Leveillula taurica cv. Arn	Fungi	Powdery mildew
Erysiphe spp.	Fungi	Powdery mildew
Uromyces cicer-arietini [Gorgn.]	Fungi	Rust
Rhizoctonia spp.	Fungi	Dry/wet root rots
Sclerotium rolfsii	Fungi	Collar rot
Meloidogyne spp.	Nematode	Root knot
Alfalfa mosaic virus (AMV)	Virus	
Beet Western yellow virus (BWYV)	Virus	
Broad bean mosaic virus (BBMV)	Virus	
Seed borne mosaic virus (SBMV)	Virus	
Broad bean wilt virus (BBWV)	Virus	
Bean golden mosaic virus (BGMV)	Virus	
Empoasca spp.	Insect pest	Leafhopper
Aphis craccivora	Insect pest	Aphid
Ophiomyia phaseoli syn. *Melanagromyza phaseoli*	Insect pest	Bean fly
Ootheca mutabilis	Insect pest	Beetle
Mylabris spp.	Insect pest	Bettle
Medythia guaterna	Insect pest	Beetle
Nezara spp.	Insect pest	Bug
Anoplocnemis spp.	Insect pest	Bug
Riptortus spp.	Insect pest	Bug
Acanthomia spp.	Insect pest	Bug

8.7 INDUCING MALE STERILITY IN SOYBEAN USING GENOME EDITING

Using a male-sterility maternal line is one of the most effective and efficient ways of preventing self-pollination [27]. Male sterility for hybrid vigor utilization and hybrid seed production is a very useful trait [41]. One of the challenges in agriculture is that hybrid seed production under a commercial setting remains slow because restricting self-pollination of female parent lines in crops such as soybean to eliminate the development of homozygous seeds is difficult [27]. Recently, a male sterile soybean mutant line was developed using CRISPR/Cas9 technology serving as a valuable commodity for both basic research and breeding applications to construct out-crossing populations. Two homologous recombinations were generated using CRISPR technology, where one had a single-nucleotide insertion (MUT1a), and the other had seven deletions (MUT7d). The resulting homozygous plants of both edited lines phenocopied the male sterility trait of the MS1 mutant plants [24].

8.8 CURRENT AND FUTURE OUTLOOK OF GENOMIC EDITING IN SOYBEAN

Spontaneous mutation, induced mutation, and conventional breeding have been commonly used in soybeans to obtain genetic resources with diverse characteristics for breeding programs [42].

However, uncertainty about the mutations and the difficulty in providing diverse genetic resources in soybean and other leguminous crops due to their recalcitrant nature continue to motivate scientists globally to investigate new ways of introducing precise gene recombinations with desirable traits and at targeted sites. As such, the application of genome editing to edit crops such as soybean has been rapidly intensified [15], with new soy mutants already developed using ZFNs. Current trends demonstrate that studies involving genome editing will exponentially increase and the CRISPR system will probably dominate, especially in the genetic improvement of valuable commodity crops like soybean.

This may include the improvement of soybeans for targeted traits influencing yields, growth and development, grain quality, stress tolerance, herbicide tolerance, and resistance to diseases [8,11,15,43,23,30]. Currently, soybean remains among the top and major crop commodities at the forefront of approved genetically modified crop production worldwide, in addition to maize, cotton, wheat, and rice [42]. Recent reports have also indicated that the growth and development-related genes were edited to manipulate patterns of flowering and pod development, as well as seed filling in soybean [13]. The quality of oils has also been furthermore improved by targeting genes involved in fatty acid metabolism, including generating new insights into the association between genes and their controlled traits which remain critical for the improvement of leguminous crops such as soybean [23].

It is presently clear that genome editing has revolutionized plant breeding techniques by elucidating gene functioning and improving the understanding of plant genetics [15,44]. On the other hand, some challenges in its applications for crop improvement remain, however the technology contains many advantages compared to traditional breeding. These challenges may include the inability of genome editing to deal with the high-level complexity of host genomes, which often results in observable difficulties during functional gene analysis [15]. Insufficiencies in fundamental genetics and molecular insights into the biology of crops, including soybeans, also remains a major bottleneck due to the deficit of targeted gene candidates and identification of genes responsible for the expression of desirable agronomic characteristics.

The limitations in the method of delivering the reagents [45] and lack of guarantees to precisely mutate any target site in the host cell genome form part of the major obstacles that need to be overcome. As far as CRISPR continues to be widely explored for creating novel gene combinations and breeding better varieties [15], this technology will endure all hurdles while successfully delivering edited mutant lines via Agrobacteria-mediated and in vitro regeneration protocols as part of its process for soybean genetic improvement [45–47].

8.9 CONCLUSION

In conclusion, the above review discussion clearly indicates that the discovery and routine application of genome editing has opened new wide-ranging opportunities for the genetic modification of soybean and other leguminous and non-leguminous crops. Some of these opportunities will be instrumental in complementing most of the conventional agricultural practices indicated in Figure 8.2. Further insights generated in this technology will assist in building the next generation of breeding tools for researchers to apply and transform the agricultural industry. It suffices to say that future agricultural productivity will soon depend on breeders' ability to constantly produce new varieties that are adapted to the changing environmental conditions [47]. Clearly, this depends on researchers' crafting of new breeding tools that need to be explored and optimized to exhaust the full capacity of genome-editing technology and its application in recalcitrant crops such as soybeans.

Genome editing has been applied to improve the agronomic traits in soybean [46], but the possibility of developing soybean varieties that can also be cultivated year-round, including during the winter season, has not yet been explored. As such, genome-editing technology remains a key tool in

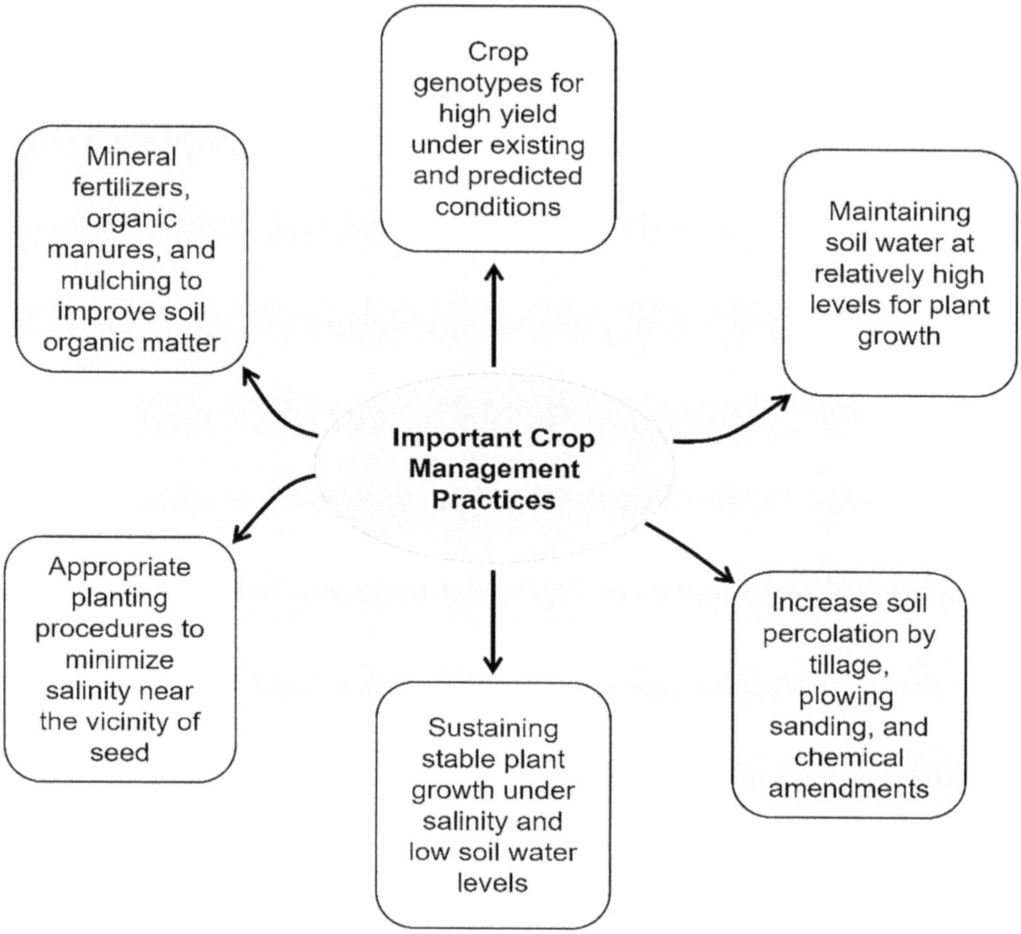

FIGURE 8.2 Basic crop management practices required for improved growth, development, and yield of crops, including soybean and other leguminous or non-leguminous plants under varied environmental conditions [48].

potentially achieving long-term breeding goals of creating "smart crops" as it contests to be a leader in this era of crop biotechnology development and finally meeting escalating future demands for soybean food and feed for human and animal consumption.

REFERENCES

[1] Sasson A. (2012). Food security for Africa: An urgent global challenge. *Agriculture & Food Security* 1(1): 1–6. DOI: 10.1186/2048-7010-1-2

[2] Zakari SA, Nuhu AA, Aliyu KT, Tadda SA. (2019). Advancement and applications of gene editing tools for crop improvements: The Contemporary. *FUDMA Journal of Sciences* 3(4): 46–60.

[3] Razzaq A, Saleem F, Kanwal M, Mustafa G, Yousaf S, Imran Arshad HM, Hameed MK, Khan MS, Joyia FA. (2019). Modern trends in plant genome editing: an inclusive review of the CRISPR/Cas9 toolbox. *International Journal of Molecular Sciences* 20(16): 4045 1–44. DOI: 10.3390/ijms20164045

[4] Orf J. (2010). Introduction. In Bilyeu K, Ratnaparkhe BM, Kole C (eds.) *Genetics, Genomics, and Breeding of Soybean*. CRC Press, Science Publisher, Enfield, New Hampshire, USA. pp. 1–19.

[5] Hanjra MA, Ferede T, Blackwell J, Jackson T, Abbas A. (2013). Global food security: facts, issues, interventions, and public policy implications. In Hanjra MA (Ed.) *Global food Security: Emerging Issues and Economic Implications*. Nova Science Publishers, New York. pp. 1–35.
[6] Xia X, Cheng X, Li R, Yao J, Li Z, Cheng Y. (2021). Advances in application of genome editing in tomato and recent development of genome editing technology. *Theoretical and Applied Genetics* 134: 2727–2747. DOI: 10.1007/s00122-021-03874-3
[7] Mushtaq M, Ahmad Dar A, Skalicky M, Tyagi A, Bhagat N, Basu U, Bhat BA, Zaid A, Ali S, Dar TU, Rai GK. (2021). CRISPR-based genome editing tools: Insights into technological breakthroughs and future challenges. *Genes*. 12(6): 797. DOI: 10.3390/genes12060797
[8] Xu H, Zhang L, Zhang K, Ran Y. (2020). Progresses, challenges, and prospects of genome editing in soybean (*Glycine max*). *Frontiers in Plant Science* 22(11): 1593. DOI: 10.3389/fpls.2020.571138
[9] Kanazashi Y, Hirose A, Takahashi I, Mikami M, Endo M, Hirose S, Toki S, Kaga A, Naito K, Ishimoto M, Abe J. (2018). Simultaneous site-directed mutagenesis of duplicated loci in soybean using a single guide RNA. *Plant Cell Reports* 37(3): 553–63. DOI: 10.1007/s00299-018-2251-3
[10] Manghwar H, Lindsey K, Zhang X, Jin S. (2019). CRISPR/Cas system: recent advances and future prospects for genome editing. *Trends in Plant Science* 24(12): 1102–1125. DOI: 10.1016/j.tplant.2019.09.006
[11] Zhang Y, Massel K, Godwin ID, Gao C. (2018). Applications, and potential of genome editing in crop improvement. *Genome Biology* 19(1): 1. DOI: 10.1186/s13059-018-1586-y
[12] Mishra R, Joshi RK, Zhao K. (2018). Genome editing in rice: Recent advances, challenges, and future implications. *Frontiers in Plant Science*s 19(9): 1361. DOI: 10.3389/fpls.2018.01361
[13] Li C, Nguyen V, Liu J, Fu W, Chen C, Yu K, Cui Y. (2019). Mutagenesis of seed storage protein genes in soybean using CRISPR/Cas9. *BMC Research Notes* 12(1): 1–7. DOI: 10.1186/s13104-019-4207-2
[14] Ahmar S, Saeed S, Khan MH, Ullah Khan S, Mora-Poblete F, Kamran M, Faheem A, Maqsood A, Rauf M, Saleem S, Hong WJ. (2020). A revolution toward gene-editing technology and its application to crop improvement. *International Journal of Molecular Sciences* 21(16): 5665. DOI: 10.3390/ijms21165665
[15] Bao A, Zhang C, Huang Y, Chen H, Zhou X, Cao D. (2020). Genome editing technology and application in soybean improvement. *Oil Crop Science* 5(1): 31–40. DOI: 10.1016/j.ocsci.2020.03.001
[16] Bhowmik P, Konkin D, Polowick P, Hodgins CL, Subedi M, Xiang D, Yu B, Patterson N, Rajagopalan N, Babic V, Ro DK. (2021). CRISPR/Cas9 gene editing in legume crops: Opportunities and challenges. *Legume Science* 3(3): e96. DOI: 10.1002/leg3.96.
[17] Kumudini S. (2010). Soybean: Soybean and Development. In Singh G. (ed.) *Botany, Production and Uses*. CAB International, London UK. pp. 48–73.
[18] Cai Y, Chen L, Liu X, Sun S, Wu C, Jiang B, Han T, Hou W. (2015). CRISPR/Cas9-mediated genome editing in soybean hairy roots. *PLoS One* 10(8): e0136064. DOI: 10.1371/journal.pone.0136064
[19] Ishaq MN, Ehirim BO. (2014). Improving soybean productivity using biotechnology approach in Nigeria. *World Journal of Agricultural Sciences* 2(2): 013–018.
[20] Oerke, EC, Dehne HW. (2004). Safeguarding production—losses in major crops and the role of crop protection. *Crop Protection* 23(4): 275–285. DOI: 10.1016/j.cropro.2003.10.001
[21] Messina MJ. (1999). Legumes and soybeans: Overview of their nutritional profiles and health effects. *The American Journal of Clinical Nutrition* 70(3): 439s–450s. DOI: 10.1093/ajcn/70.3.439s
[22] Curtin SJ, Xiong Y, Michno JM, Campbell BW, Stec AO, Čermák T, Starker C, Voytas DF, Eamens AL, Stupar RM. (2018). Crispr/cas9 and TALENs generate heritable mutations for genes involved in small RNA processing of *Glycine max* and *medicago truncatula*. *Plant Biotechnology Journal* 16(6): 1125–1137. DOI: 10.1111/pbi.12857
[23] Haun W, Coffman A, Clasen BM, Demorest ZL, Lowy A, Ray E, Retterath A, Stoddard T, Juillerat A, Cedrone F, Mathis L. (2014). Improved soybean oil quality by targeted mutagenesis of the fatty acid desaturase 2 gene family. *Plant Biotechnology Journal* 12(7): 934–940. DOI: 10.111/pbi.1220
[24] Jiang B, Chen L, Yang C, Wu T, Yuan S, Wu C, Zhang M, Gai J, Han T, Hou W, Sun S. (2021). The cloning and CRISPR/Cas9-mediated mutagenesis of a male sterility gene MS1 of soybean. *Plant Biotechnology Journal* 19(6): 1098. DOI: 10.1111/pbi.13601
[25] Arora L, Narula A. (2017). Gene editing and crop improvement using CRISPR-Cas9 system. *Frontiers in Plant Science* 8(8): 1932. DOI: 10.3389/fpls.2017.01932

[26] Sander JD, Dahlborg EJ, Goodwin MJ, Cade L, Zhang F, Cifuentes D, Curtin SJ, Blackburn JS, Thibodeau-Beganny S, Qi Y, Pierick CJ. (2011). Selection-free zinc-finger-nuclease engineering by context-dependent assembly (CoDA). *Nature Methods* 8(1): 67–69. DOI: 10.1038/nmeth.1542

[27] Ahmad S, Tang L, Shahzad R, Mawia AM, Rao GS, Jamil S, Wei C, Sheng Z, Shao G, Wei X, Hu P. (2021). CRISPR-based crop improvements: A way forward to achieve zero hunger. *Journal of Agricultural and Food Chemistry* 69(30): 8307–8323. DOI: 10.1021/acs.jafc.1c02653

[28] Mao Y, Botella JR, Liu Y, Zhu JK. (2019). Gene editing in plants: progress and challenges. *National Science Review* 6(3): 421–437. DOI: 10.1093/nsr/nwz005

[29] Jacobs TB, LaFayette PR, Schmitz RJ, Parrott WA. (2015). Targeted genome modifications in soybean with CRISPR/Cas9. *BMC Biotechnology* 15(1): 1–10. DOI: 10.1186/s12896-015-0131-2

[30] Cai Z, Xian P, Cheng Y, Ma Q, Lian T, Nian H, Ge L. (2021). CRISPR/Cas9-mediated gene editing of GmJAGGED1 increased yield in the low latitude soybean variety Huachun 6. *Plant Biotechnology Journal* 19(10): 1–3. DOI: 10.1111/pbi.13673

[31] Mangena P. (2020). *In vivo* and *in vitro* application of colchicine on germination and shoot proliferation in soybean [*Glycine max* (L.) Merr.]. *Asian Journal of Crop Science* 12: 34–42. DOI: 10.3923/ajcs.2020.34.42.

[32] Krishnan HB. (2001). Biochemistry and molecular biology of soybean seed storage proteins. *Journal of New Seeds* 2(3): 1–25. DOI: 10.1300/J153v02n03_01

[33] Miladinovic D, Antunes D, Yildirim K, Bakhsh A, Cvejić S, Kondić-Špika A, Jeromela AM, Opsahl-Sorteberg HG, Zambounis A, Hilioti Z. (2021). Targeted plant improvement through genome editing from laboratory to field. *Plant Cell Reports* 21: 1–7. DOI: 10.1007/s00299-020-02655-4

[34] Le H, Nguyen NH, Ta DT, Le TN, Bui TP, Le NT, Nguyen CX, Rolletschek H, Stacey G, Stacey MG, Pham NB. (2020). CRISPR/Cas9-mediated knockout of galactinol synthase-encoding genes reduces raffinose family oligosaccharide levels in soybean seeds. *Frontiers in Plant Science* 11: 612942. DOI: 10.3389/fpls.2020.612942

[35] Panthee DR. Varietal improvement in soybean. In Singh G. (ed.) *The Soybean Botany, Production and Uses*. CAB International, London UK. pp. 92–112.

[36] Li Z, Liu ZB, Xing A, Moon BP, Koellhoffer JP, Huang L, Ward RT, Clifton E, Falco SC, Cigan AM. (2015). Cas9-guide RNA directed genome editing in soybean. *Plant Physiology* 169(2): 960–970. DOI: 10.1104/pp.15.00783

[37] Yang W, Wu T, Zhang X, Song W, Xu C, Sun S, Hou W, Jiang B, Han T, Wu C. (2019). Critical photoperiod measurement of soybean genotypes in different maturity groups. *Crop Science* 59(5): 2055–2061. DOI: 10.2135/cropsci2019.03.0170

[38] Han J, Guo B, Guo Y, Zhang B, Wang X, Qiu LJ. (2019). Creation of early flowering germplasm of soybean by CRISPR/Cas9 technology. *Frontiers in Plant Science* 10: 1446. DOI: 10.3389/fpls.2019.01446

[39] Li Z, Cheng Q, Gan Z, Hou Z, Zhang Y, Li Y, Li H, Nan H, Yang C, Chen L, Lu S. (2021). Multiplex CRISPR/Cas9-mediated knockout of soybean LNK2 advances flowering time. *The Crop Journal* 9(4): 767–776. DOI: 10.1016/j.cj.2020.09.005

[40] Zhang P, Du H, Wang J, Pu Y, Yang C, Yan R, Yang H, Cheng H, Yu D. (2020). Multiplex CRISPR/Cas9-mediated metabolic engineering increases soya bean isoflavone content and resistance to soya bean mosaic virus. *Plant Biotechnology Journal* 18(6): 1384–1395. DOI: 10.1111/pbi.13302

[41] Wan X, Wu S, Li Z, Dong Z, An X, Ma B, Tian Y, Li J. (2019). Maize genic male-sterility genes and their applications in hybrid breeding: Progress and perspectives. *Molecular Plant* 12(3): 321–342. DOI: 10.1016/j.molp.2019.01.014

[42] Georges F, Ray H. (2017). Genome editing of crops: A renewed opportunity for food security. *Genetically Modified Crops and Food* 8(1): 1–12. DOI: 10.1080/21645698.2016.1270489

[43] Pratap A, Gupta SK, Kumar J, Solanki RK. (2012). Soybean. In S. K Gupta (Ed.) *Technological Innovations in Major World Oil Crops* (Vol. 1, pp. 293–321). Springer, New York, NY. DOI: 10.1007/978-1-4614-0356-2_12

[44] Bonawitz ND, Ainley WM, Itaya A, Chennareddy SR, Cicak T, Effinger K, Jiang K, Mall TK, Marri PR, Samuel JP, Sardesai N. (2019). Zinc finger nuclease-mediated targeting of multiple transgenes to an endogenous soybean genomic locus via non-homologous end joining. *Plant Biotechnology Journal* 17(4): 750–761. DOI: 10.1111/pbi.13012

[45] Du H, Zeng X, Zhao M, Cui X, Wang Q, Yang H, Cheng H, Yu D. (2016). Efficient targeted mutagenesis in soybean by TALENs and CRISPR/Cas9. *Journal of Biotechnology* 217: 90–97. DOI: 10.1016/j.jbiotec.2015.11.005

[46] Jaganathan D, Ramasamy K, Sellamuthu G, Jayabalan S, Venkataraman G. (2018). CRISPR for crop improvement: An update review. *Frontiers in Plant Science* 9: 985. DOI: 10.3389/fpls.2018.00985

[47] Fita A, Rodríguez-Burruezo A, Boscaiu M, Prohens J, Vicente O. (2015). Breeding and domesticating crops adapted to drought and salinity: A new paradigm for increasing food production. *Frontiers in Plant Science* 6: 978. DOI: 10.3389/fpls.2015.00978

[48] Mangena P. (2020). Introduction to legumes. In Mangena P (ed.) *Legumes: Nutritional Value, Health Benefits and Management.* Nova Science Publishers, New York. pp. 1–14. DOI: 10.52305/WOFN2369

9 Epigenetic Modifications in Legumes in Response to Climate Change

Saurabh Chaudhary

9.1 INTRODUCTION

Legumes, the second most consumed edible food source after cereals, play a crucial role in our diet (Mousavi-Derazmahalleh et al., 2019). They are rich in plant-based proteins and are an excellent alternative to meat products (Mefleh et al., 2022; Singh et al., 2022). Since legumes are also a source of many bioactive compounds and antioxidants, they also serve as therapeutics in some cases (Myrtsi et al., 2023). Many legume species serve as feed for land and aquatic animals (Yanni et al., 2024). Among their multiple uses, the natural ability to fix atmospheric nitrogen makes legumes the most eco-friendly and sustainable crop for farming (Foyer et al., 2016; Jhu & Oldroyd, 2023). However, the impact of climate change on legumes is an urgent issue that requires immediate attention. Global climate change results in several environmental stresses such as extreme cold, heat, drought, salinity, flooding, and many others that significantly threaten legume crop yield and grain quality (Farooq et al., 2018; Mousavi-Derazmahalleh et al., 2019; Rane et al., 2021). The urgency of this issue underscores the importance of our research and the need for immediate action.

The changing climate conditions intensify the global food security challenge by impacting the overall agricultural system. In general, plants respond to these variations in the climate by fine-tuning their transcriptional and translational machinery at the molecular level. The primary effect occurs at the gene regulation or expression level, further altering the plant proteome and phenotype. With the efforts of researchers in the last few decades, it is evident that the nuclear organization and chromatin architecture play a crucial role in regulating gene expression and, ultimately, protein abundance. Also, post-transcriptional changes in chromatin architecture, such as histone modification, other modifications, such as changes in DNA methylation, variation in nucleosome positioning, RNA polymerase II processivity, and non-coding RNA activities epigenetically (without altering the DNA sequence) determine the transcriptional fate. In legumes, like other crops, several epigenetic changes or modifications occur in response to variations in climatic conditions. This chapter details the significance of epigenetic modifications or changes in response to the changing climate conditions in legumes. The discussion also includes epigenetic tools and challenges in developing climate-smart legume crops, concluding with future implementations.

9.2 EPIGENETIC MECHANISMS AND REGULATIONS IN PLANTS IN GENERAL (FIGURE 9.1)

Plants develop and activate several mechanisms, including alterations in nuclear organization (epigenetic), in response to environmental stress. Numerous epigenetic mechanisms that regulate gene expression without changing DNA sequences have been discovered and studied in both model and crop species. Though the role of epigenetic mechanisms in response to changing climatic conditions

 DOI: 10.1201/9781003477235-11

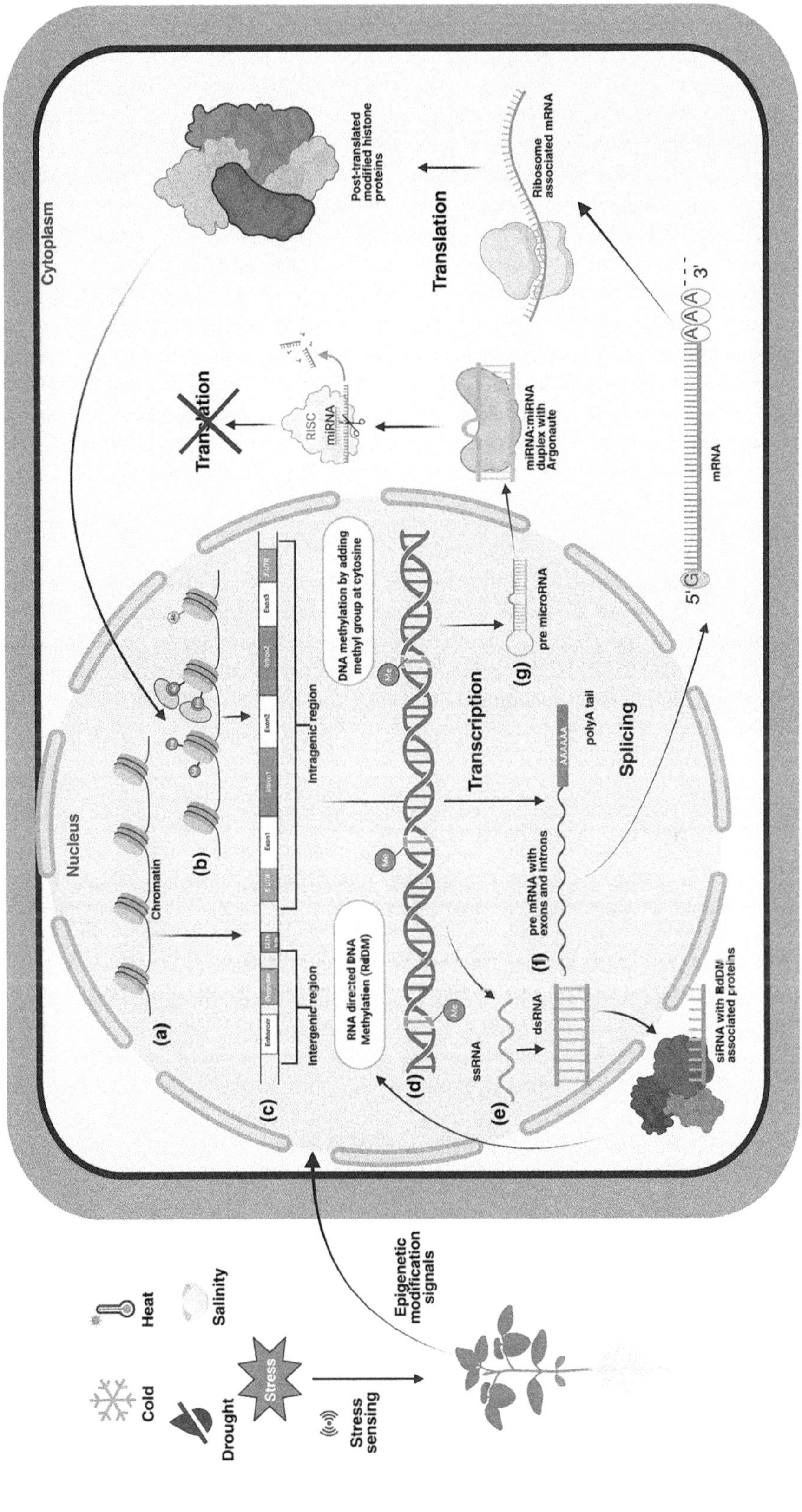
Cold
Heat
Drought
Stress
Salinity
Stress sensing
Epigenetic modification signals
Cytoplasm
Nucleus
Chromatin
(a)
(b)
(c)
Intergenic region
Intragenic region
RNA directed DNA Methylation (RdDM)
DNA methylation by adding methyl group at cytosine
(d)
(e)
ssRNA
dsRNA
(f)
pre mRNA with exons and introns
polyA tail
Splicing
Transcription
(g)
pre microRNA
siRNA with RdDM associated proteins
miRNA:miRNA duplex with Argonaute
RISC
miRNA
Translation
Post-translated modified histone proteins
Ribosome associated mRNA
mRNA
5' G
A A A 3'

FIGURE 9.1 Epigenetic mechanisms and regulations in plants in response to environmental stresses. Upon sensing environmental stresses, plants send signals for epigenetic modifications at the cellular level. The chromatin (a) in the nucleus comprises eight histones wrapped by a 147 bp-long DNA-making unit called a nucleosome at regular intervals. These histones can be modified post-translationally by adding methyl or acetyl groups or ubiquitin to the histone proteins (b). Such modifications impact the processivity of RNA polymerase II (Pol II) and regulate the expression of stress-responsive genes. Furthermore, the chromatin contains various genes whose overall structure (c) includes an enhancer, promoter, TATA box, 5' UTRs, exons, introns, and 3' UTRs. Distinct gene (d) structural units can be directly methylated or demethylated by enzymes or RNA-directed DNA methylation pathways (RdDM). Single-stranded RNA (ssRNA) (e) is transcribed from DNA, which subsequently gives rise to double-stranded RNA (dsRNA) followed by small interfering RNA (siRNA) along with RdDM-associated proteins. Generally, DNA is transcribed to produce pre-mRNA (f) with a poly-A tail, incorporating exons and introns using Pol II. This pre-mRNA is then transported to the cytoplasm after splicing out the introns, resulting in mRNA only with exons (the coding part). In the cytoplasm, these mRNAs associate with ribosomal machinery and are translated into various proteins required for distinct cellular processes. In a different epigenetic pathway, non-coding RNA genes such as microRNA (miRNA) genes give rise to pre-microRNA(pre-miRNA) (g) in the nucleus, which is transported to the cytoplasm to form mature miRNA. These mature miRNAs interact with various proteins and inhibit the translation of specific mRNAs. (Created in BioRender. https://BioRender.com/v97c176)

is still emerging, many studies have reported in model plants how different epigenetic marks, such as DNA methylation, histone modifications, chromatin remodeling, and non-coding RNAs, regulate the expression of many genes in response to climate change. However, with the advancement in next-generation sequencing (NGS) technologies and bioinformatics tools, more epigenetic data are accumulating daily in crops, specifically legumes (Varotto et al., 2020; Yung et al., 2023). The current section is dedicated to the basic epigenetic mechanism observed in plants and their regulation in response to the environment.

9.2.1 DNA Methylation and Demethylation

DNA methylation is the most prevalent epigenetic modification studied in plants. To respond to various environmental stresses, plant chromatin architecture can be altered by DNA methylation via different pathways and enzymes in three contexts: CG, CHG, and CHH (H = A, T, or C). Among these contexts, CG methylation is the most common form of methylation observed in plants, facilitated by the enzyme METHYLTRANSFERASE 1 (MET1), while the CHROMOMETHYLASE 3 (CMT3) enzyme is responsible for CHG methylation (Cao & Jacobsen, 2002; Zhang et al., 2018). The third context, CHH, is also crucial for plant development and stress responses, mediated by two distinct de novo pathways. The first RNA-directed DNA methylation (RdDM) pathway involves the biogenesis of small interfering RNAs (Si-RNAs) that target the ARGONAUTE (AGO) family, which in turn is methylated via DOMAINS REARRANGED METHYLTRANSFERASE 2 (DRM2) (Matzke & Mosher, 2014). In the second pathway, the interaction between CHROMOMETHYLASE 2 (CMT2) and DECREASE IN DNA METHYLATION 1 (DDM1) at histone H1-enriched loci occurs, which further leads to CHH methylation in that region (Zemach et al., 2005, 2013). Since DNA methylation is a reversible epigenetic modification, DNA can be actively demethylated by a family of enzymes, DNA glycosylases, and base excision repair (Li et al., 2018; Penterman et al., 2007). In response to climate changes, plants respond through DNA methylation by regulating the expression of stress-responsive genes, maintaining genome stability and regulating transposable elements and their related genes.

9.2.2 Histone Modifications: Post-Transcriptional Modifications (PTMs)

The nucleosome, the fundamental unit of all eukaryotic chromosomal DNA, consists of a 147-bp segment of genomic DNA wrapped around histone octamers. These octamers are composed of two copies of four core histone proteins: H2A, H2B, H3, and H4. This chromatin architecture influences the accessibility of coding regions and other regulatory portions of DNA, further regulating transcriptional activity and, ultimately, gene expression. In addition to these four histone proteins, an additional linker, histone protein H1, is linked to the unpacked DNA between nucleosomes, playing a crucial role in overall chromatin compaction. In plants, these histones can be readily modified in response to the environment at their N-terminal tail through reversible post-translational modifications (PTMs), particularly acetylation and deacetylation, methylation and demethylation, phosphorylation, and ubiquitination of histones (Fillingham & Greenblatt, 2008; Kouzarides, 2007). Among these, acetylation, methylation, and their reversible counterparts, deacetylation and demethylation, have been extensively studied in plants, especially in response to environmental stresses. Acetylation of histones, including acetylation on histone 3 (H3), lysine 9 (H3K9ac), lysine 14 (H3K14ac), lysine 27 (H3K27ac), and lysine 36 (H3K36ac), is known to disrupt nucleosome structure and loosen chromatin to facilitate transcription through gene activation. Conversely, the deacetylation of histones leads to transcriptional repression (Berger, 2007). The other well-studied post-translational modification (PTM) in histones is methylation, which occurs at arginine and lysine residues of histone proteins. Depending on the modification site, histone methylation can activate or repress gene expression and subsequent transcriptional activity. For example, methylation on H3, at lysine 4 (H3K4me) and lysine36 (H3K36me) activates the genes by loosening the chromatin state, whereas at lysine 9 (H3K9me) and lysine 27 (H3K27me), it causes gene repression by creating heterochromatin. Furthermore, methylation can be mono, di, or tri, depending on the number of methyl groups added. In plants, the transposable elements and repeated sequences enriched in heterochromatin are marked by H3K9me1 (mono) and H3K9me2 (di) methylation, which serves to repress transcriptional activity (Varotto et al., 2020).

9.2.3 Plant Non-Coding RNAs and Chromatin Remodeling

In plants, non-coding RNAs (except housekeeping such as transfer RNA, ribosomal RNA, and small nucleolar RNA) are also regarded as epigenetic markers that regulate the expression of stress-responsive genes and those involved in development, growth, and reproduction. Long non-coding RNAs (lncRNAs >200 nucleotides) and small non-coding RNAs (small RNAs: microRNAs and small interfering RNAs, 18–24 nucleotides long) modulate various gene expressions at both transcriptional and post-transcriptional levels (Bhogireddy et al., 2021; Wang et al., 2017). The small RNAs are generated using different pathways and enzymes, including dicer-like 1 (DCL1), HYL1, and zinc finger complex proteins using 18–24 nucleotide templates from mRNAs (Guleria et al., 2011). Meanwhile, lncRNAs are transcribed using a template more than 200 nucleotides long (Mach, 2017). Another class of small RNA called circular RNAs (ci-RNAs) regulates gene expression at the post-transcriptional level (Bhogireddy et al., 2021).

Among sRNAs, miRNAs are involved in post-transcriptional mechanisms to regulate gene activity by either silencing or activating. On the other hand, siRNAs are known for their gene-silencing function, mainly at the transcriptional and/or post-transcriptional level (Guleria et al., 2011; Zhan & Meyers, 2025). Plant small RNAs are used as a tool for RNA interference (RNAi) for post-transcriptional gene silencing to investigate the function and regulation of a gene (Hung & Slotkin, 2021). Plant siRNAs are further involved in another pathway, where they transfer from one cell to another and regulate epigenetic modifications via the RNA-dependent DNA methylation (RdDM) pathway (Xie & Yu, 2015). In addition to small RNAs, lncRNAs interact with nucleic acids and proteins, acting as scaffolds to form complexes in the nucleus and mediating chromatin remodeling (Han & Chang, 2015).

Overall, several epigenetic modifications in plants, including DNA methylation, histone modifications and variants, chromatin remodeling, and non-coding RNA-mediated epigenetic changes, play a crucial role in responding to changing climatic conditions. Researchers have started to make efforts to utilize these epigenetic modifications or marks to develop climate-smart crops, including legumes, using breeding or epigenome editing.

9.3 EPIGENETIC CHANGES IN LEGUME CROPS: CASE STUDIES (TABLE 9.1)

In response to climate change, several epigenetic modifications in major legume crops, including soybean, alfalfa or lucerne, barrel clover, chickpea, peanut, pigeon pea, and others (Table 9.1), have been reported over the last decade {reviewed by Yung et al. [2023]). The previous section discussed the key epigenetic modifications, their mechanisms, and how they are regulated in plants. This section examines primary legume crops and the epigenetic modifications or changes observed in response to various stresses (Table 9.1).

9.3.1 Soybean (*Glycine Max* L.)

Soybean (*Glycine max* L.) is a vital oil crop and food legume, constituting 53% of global oilseed production. It plays a significant role in agricultural systems across major countries such as the USA, China, Brazil, Argentina, and India. Over the past two decades, advancements in crop improvement programs using genetic and molecular breeding tools have enhanced soybean production and productivity. However, like other essential crops, soybean yield and production are also threatened by the changing climate, posing a challenge to food security. Research on epigenetic modifications in response to climate change in soybeans has accelerated progress in developing stress-resilient varieties over the past few years. Some are discussed below.

In 2020, Zhang et al. reported a decrease in the DNA methylation level at the transcription factor-encoding gene GmMYB84 promoter region in response to salinity stress (Zhang et al., 2020). The overexpression of GmMYB84 increases the salinity stress tolerance by increasing the abundance of the transcripts. Therefore, the study concluded that DNA methylation modulates the salinity stress-induced expression of the soybean transcription factor-encoding gene GmMYB84, thereby conferring salinity stress tolerance. Another report suggested a global decrease in DNA methylation and identified 3305 lncRNAs in response to prolonged salinity stress in the soybean root (Chen et al., 2019). A differential DNA methylation pattern was observed in the soybean root hairs in response to heat stress (Hossain et al., 2017). The study suggested changes in DNA methylation are directly or indirectly linked to the expression of genes and transposons in a tissue- and stress-dependent manner (Hossain et al., 2017). Another report on soybean suggested an increase in DNA methylation and a correlation between siRNA and transposable element genes modified by CHH methylation in response to phosphorus stress (Chu et al., 2020a).

Like alteration in DNA methylation, histone modifications are also observed in soybean under variable stresses. For example, a 2018 study reported 28 histone deacetylation (HDAC) genes in the soybean genome. The differential expression of GmHDAC transcripts in response to abiotic cues such as cold and heat stress suggests these epigenetic modifiers may play a role in the transcriptional response to stress in soybeans (Yang et al., 2018). Sun et al. noted differential methylation at H3K27me3 in soybeans in response to salinity stress (Sun et al., 2019). They employed high-throughput RNA-Seq and ChIP-Seq (discussed later) technologies to investigate genome-wide transcriptional and histone modifications during salt stress. The epigenetic marker H3K27me3 exhibited changes in its deposition under these conditions. Most genes are silenced during salt stress, which correlates with establishing new H3K27me3 marks in the promoter or coding regions absent in

TABLE 9.1
Epigenetic Modifications in Response to Environmental Stress Studied in the Last Decade in Important Legume Crops (adapted and updated from Yung et al., 2023)

Common name	Scientific name	Epigenetic Modifications	Environmental Cues	Major Findings	References
Alfalfa	*Medicago sativa* L.	DNA methylation	Drought stress	Methylation level decreases	(Ventouris et al., 2020)
		DNA methylation	Salt stress	Methylation level decreases at the MsMYB4	(Dong et al., 2020)
		DNA methylation	Cadmium stress	Reduction in DNA methylation level	(Chen et al., 2025)
		DNA methylation	Drought stress	Differential DNA methylation in metabolic related genes	(Bai et al., 2025)
		Histone modifications	Salt stress	H3K4me3 and H3K9ac increases	(Dong et al., 2020)
		MicroRNA (miRNA)	Drought stress	348 known and 281 novel miRNAs were reported	(Li et al., 2017)
		DNA methylation	Salt stress	Methylation level increases	(Al-Lawati et al., 2016)
Barrel clover	*Medicago truncatula* L.	DNA methylation	Salt stress	Differential methylated regions	(Yaish et al., 2018)
		Histone modifications	Desiccation stress	H3Kac and H3K27me3 decreases	(Sano et al., 2022)
		Histone modifications	Heat stress	H3Kac decreases and H3K27me3 increases	(Malabarba et al., 2022)
		MicroRNA (miRNA)	Drought stress and rewatering	Differential expression of 57 miRNA	(Shi et al., 2021)
		Long non-coding RNA (lncRNA)	Phosphorous deficiency	Identification of 10,785 putative lncRNA	(Wang et al., 2017)
		Long non-coding RNA (lncRNA)	Osmotic and salt stress	Identification of 23,324 putative lncRNA	(Wang et al., 2015)
Castor bean	*Ricinus communis* L.	Histone modifications	Salt stress	H3K4me3 increases and H3K27me3 decreases	(Han et al., 2020)
Chickpea	*Cicer arietinum* L.	DNA methylation	Cold stress	Methylated band at 5'-CCGG-3' increases	(Rakei et al., 2016)
		DNA methylation	Salt stress	Differential methylated regions in transposable element (TE) genes	(Gupta & Garg, 2023)
		Histone modifications	Drought stress	H3K9ac increases	(Sen et al., 2017)
		Histone modifications	Salt stress	H3K9ac increases	(Sen et al., 2017)
		MicroRNA (miRNA)	Wilt and salt stress	181 potential miRNAs identified	(Kohli et al., 2014)
		MicroRNA (miRNA)	Salt and drought stress	284 potential miRNAs identified	(Khandal et al., 2017)
		Long non-coding RNA (lncRNA)	Salt stress	3452 putative lncRNA identified	(Kumar et al., 2021)

(continued)

TABLE 9.1 (Continued)
Epigenetic Modifications in Response to Environmental Stress Studied in the Last Decade in Important Legume Crops (adapted and updated from Yung et al., 2023)

Common name	Scientific name	Epigenetic Modifications	Environmental Cues	Major Findings	References
		Long non-coding RNA (lncRNA)	Salt and drought stress	3457 putative lncRNA identified	(Singh et al., 2017)
Common bean	*Phaseolus vulgaris* L.	MicroRNA (miRNA)	Drought stress	169 potential miRNAs identified	(Wu et al., 2017)
Common vetch	*Vicia sativa* L.	MicroRNA (miRNA)	Drought stress	379 potential miRNA identified	(Zhu et al., 2021)
Faba bean	*Vicia faba* L.	DNA methylation	Drought stress	Decrease in global DNA methylation level	(Abid et al., 2017)
Horse gram	*Macrotyloma uniflorum* L.	MicroRNA (miRNA)	Soil moisture deficit	15 potential miRNAs identified	(Yasin et al., 2020)
Hyacinth bean	*Lablab purpureus* L.	DNA methylation	Heat and chemical stress	Differential methylated band from 5'-CCGG-3' increases	(Rai et al., 2018)
Mungbean	*Vigna radiata* L.	DNA methylation	Drought stress	Differential methylated regions	(Zhao et al., 2022)
Peanut	*Arachis hypogaea* L.	Histone modifications	Drought stress and rewatering	Acetylation at histone 3 (H3ac) decreases and increases during recovery	(Liu et al., 2020)
		Histone modifications	Drought stress	Acetylation at histone 3 (H3ac) increases	(Zhang et al., 2018)
		MicroRNA (miRNA)	Calcium deficit	161 potential miRNAs identified	(Chen et al., 2019)
Pigeon pea	*Cajanus cajan* L.	DNA methylation	Salt stress	Decrease in global DNA methylation level	(Awana et al., 2019)
		MicroRNA (miRNA)	Drought stress	116 potential miRNAs identified	(Buch et al., 2020)
Soyabean	*Glycine max* L.	DNA methylation	Salt stress	DNA methylation level increases at the *GmMYB84* transcription factor promoter	(Zhang et al., 2020)
		DNA methylation	Salt stress	Decrease in global DNA methylation level	(Chen et al., 2019)
		DNA methylation	Heat stress	Differential DNA methylation all three context CG, CHG, and CHH in roots	(Hossain et al., 2017)
		DNA methylation	Phosphorous stress	Differential DNA methylation and siRNAs regulation	(Chu et al., 2020b)
		Histone modifications	Cold stress	H3K4me2 increases whereas H3K4me3 and H3Kac decreases	(Yang et al., 2018)
		Histone modifications	Drought stress	Global histone2B (H2B) monoubiquitylation (H2Bub1) increases	(Chen et al., 2020)
		Histone modifications	Heat stress	H3Kac increases	(Yang et al., 2018)
		Histone modifications	Salt stress	Differential methylation at H3K27me3	(Sun et al., 2019)

TABLE 9.1 (Continued)
Epigenetic Modifications in Response to Environmental Stress Studied in the Last Decade in Important Legume Crops (adapted and updated from Yung et al., 2023)

Common name	Scientific name	Epigenetic Modifications	Environmental Cues	Major Findings	References
		Histone modifications	Salt priming	Differential methylation at H3K4me3 and H3K9ac in different tissues	(Yung et al., 2022)
		MicroRNA (miRNA)	Salt stress	71 potential miRNAs identified	(Sun et al., 2016)
		Long non-coding RNA (lncRNA)	Salt stress	3,305 putative lncRNA identified	(Chen et al., 2019)
		Long non-coding RNA (lncRNA)	Multiple stresses	46,974 lncRNA in G. max and 35,675 lncRNA in G. Soja identified	(Lin et al., 2020)
		Long non-coding RNA (lncRNA)	Phosphorous deficiency	4,166 putative lncRNA identified	(Zhang et al., 2021)

control plants. The study concluded by identifying soybean histone modifiers, which could facilitate de novo histone methylation and gene silencing in response to salt stress. Moreover, GmUBC9 is a gene for a ubiquitin-conjugating enzyme found in soybeans, with overexpression leading to increased drought stress tolerance. Histone monoubiquitination 2 (HUB2), an E3-like protein that participates in the ubiquitylation of histone H2B (H2Bub1), has been shown to interact with GmUBC9 in both *Arabidopsis* and soybean. In drought conditions, levels of H2Bub1 rise, and the transcription of drought-response genes is activated in GmUBC9 transgenic soybeans (Chen et al., 2020). In another case study, Yung et al. observed histone modifications in response to salt stress priming (Yung et al., 2022). The authors conducted genome-wide profiling of histone-3-lysine-4-dimethylation (H3K4me2), H3K4me3, and histone-3-lysine-9-acetylation (H3K9ac). They revealed that priming induced significant alterations in these histone marks, coordinating the stress response, ion homeostasis, and cell wall modification (Yung et al., 2022). In addition to DNA methylation and histone modifications, several stress-related lncRNA and miRNA were also identified in soybean. For example, 3305, 46,974, and 4166 putative lncRNAs were identified in response to salt stress (Chen et al., 2019), multiple stresses (Lin et al., 2020), and phosphorus deficiency (Zhang et al., 2021), respectively, in soybean. Also, 71 salt-responsive potential miRNAs were reported in soybean (Sun et al., 2016).

Though many studies, as discussed above, validate the role of epigenetic modifications in response to several environmental stresses, and many more studies are in the pipeline, it is time to accelerate the research toward developing climate-smart soybean varieties using epigenomic editing (discussed later in the chapter) or breeding at the epigenomic level.

9.3.2 Alfalfa (*Medicago sativa* L.)

Alfalfa or lucerne (*Medicago sativa* L.), is a vital legume crop from the Fabaceae family, ranking among the most significant forage crops grown in warm temperate and cool subtropical regions. It produces more digestible energy and protein per hectare than many crops and is a crucial source of vitamin A and other vitamins (Sullivan et al., 2004). This makes alfalfa a key provider of livestock protein and minerals. It grows vigorously, regenerating multiple times after cutting and can last three or more years with proper management. Alfalfa is vital in cropping systems for its deep roots and ability to fix atmospheric nitrogen into an accessible form for plants. When included in crop rotations, alfalfa enhances the productivity of subsequent crops. Several epigenetic modifications have been observed in alfalfa under various environmental stresses. For example, drought (Bai et al., 2025; Ventouris et al., 2020), salt (Al-Lawati et al., 2016; Dong et al., 2020), and cadmium (Chen et al., 2025) stresses have been reported to affect DNA methylation at multiple levels in alfalfa. In the study by Ventouris et al., DNA methylation dynamics were examined under recurrent drought stress in two subspecies of alfalfa (Lamia and Chaironia). Lamia shows a decrease in DNA methylation, which is a result of the reoccurrence of drought stress. In contrast, changes in Chaironia were similar to the control, even after the recurrence of stress (Ventouris et al., 2020). Bai et al. showed drought stress memory and recovery response with DNA methylation changes in another study. Differential DNA methylation was observed during drought training, facilitating selective expression of energy metabolism-related genes (Bai et al., 2025). Moreover, the DNA methylation level at the promoter regions regulates the expression of various stress-responsive genes. Under salt stress, a decrease in methylation level at the promoter region of the MsMYB4 transcription factor is crucial for its activation and salinity tolerance in alfalfa (Dong et al., 2020). Similarly, activation of MsMYB4 also correlated with increased histone H3K4 trimethylation and H3K9 acetylation in certain promoter regions in response to salt stress (Dong et al., 2020). Al-Lawati et al. in 2016 and Chen et al. in 2025 also reported a decrease in global DNA methylation level under salt (Al-Lawati et al., 2016) and cadmium stress (Chen et al., 2025), respectively, in alfalfa. In addition to the above-mentioned epigenetic changes, a study on the differential expression of miRNAs was reported in response to drought stress in alfalfa (Li et al., 2017). In total, 348 known and 281 novel miRNAs in response to drought were reported, whereas 12 miRNAs were observed as drought-responsive differentially expressed (Li et al., 2017).

Although research on epigenetic modifications related to climate change is limited due to its small yet well-annotated genome, alfalfa holds potential as a model legume crop for epigenomic editing aimed at developing climate-smart varieties for other crops, including legumes.

9.3.3 Chickpea (*Cicer Arietinum* L.)

Chickpea (*Cicer arietinum* L.) is a vital legume crop worldwide, providing an excellent plant-based protein source. Like other legume crops, environmental stresses also affect its yield and productivity. Therefore, techniques such as transcriptomics, proteomics, and metabolomics were extensively used to identify genes and proteins associated with abiotic stress tolerance in chickpeas (Arriagada et al., 2022). However, epigenomic data on chickpea under several environmental stresses are still scarce but are beginning to emerge.

Under cold stress, two chickpea genotypes (Sel96Th11439, a cold-tolerant genotype, and ILC533, a cold-susceptible genotype) were examined for DNA methylation changes (Rakei et al., 2016). The tolerant genotype showed notable changes in DNA methylation/demethylation patterns, often practical factors in protecting cells against cold-induced oxidative stress (Rakei et al., 2016). Moreover, Gupta and Garg, in 2023, analyzed global DNA methylation and its impact on gene expression in chickpea under salinity stress (Gupta & Garg, 2023). The authors observed increased CG methylation in the gene bodies and CHH methylation in transposable element (TE) regions. Moreover, differential methylation analysis revealed that the stress-tolerant genotype showed more hypermethylation regions than the sensitive genotype across the genome in response to salt stress (Gupta & Garg, 2023). In addition, the histone modifications study under drought and salt stresses suggested increased H3K9ac in chickpea (Sen et al., 2017).

In chickpeas, 181 potentials, including 122 conserved and 59 novel miRNAs, were reported using high-throughput sequencing in response to wilt and salt stress (Kohli et al., 2014). Moreover, another study reported 284 potential miRNAs in response to salt and drought stress (Khandal et al., 2017). Similar to miRNAs, 3452 salt stress-responsive (Kumar et al., 2021) and 3457 salt and drought stress-responsive (Singh et al., 2017) lncRNAs were reported in chickpea studies.

9.3.4 Pigeon Pea (*Cajanus cajan* L.) and Peanut (*Arachis hypogaea* L.)

Pigeon pea (*Cajanus cajan* L.) and peanut (*Arachis hypogaea* L.) are two other main legume crops utilized worldwide in diets as a source of plant-based proteins, carbohydrates, vitamins, minerals, and fiber. In pigeon pea, a decrease in global DNA methylation level was observed under salt stress (Awana et al., 2019). The study selected two contrasting genotypes, ICP1071 (sensitive) and ICP7 (tolerant), for bisulfite sequencing under salt stress. Gene expression data suggested the CcCYP (Cytochrome P450) gene was upregulated in roots, and CcCDR (cold and drought regulatory protein-encoding gene) was upregulated in shoots of the salt-tolerant genotype to enhance stress protection. DNA methylation level decreases in the genotypes correlated with salt tolerance. Salt stress caused a 2.6-fold increase in gene expression due to methylation changes in all three cytosine residues in ICP7 shoots (Awana et al., 2019). Another study identified 116 potential miRNAs in pigeon pea in response to drought stress (Buch et al., 2020). Meanwhile, in peanuts, 161 potential miRNAs were reported to be in response to calcium deficiency (Chen et al., 2019).

In peanut transcription factor, AhGLK1 activates AhPORA gene expression, stimulates chlorophyll biosynthesis, and interacts with the histone deacetylase gene AhHDA1. Further, AhHDA1 binds to the AhGLK1 promoter, altering histone deacetylation to inhibit AhGLK1 expression. In 2020, Liu et al. showed that using a ChIP-seq experiment, AhCAB (chlorophyll a/b encoding gene) was identified as a target of AhHDA1 and AhGLK1. ChIP-qPCR further confirmed this by revealing that AhHDA1 and AhGLK1 bind to the promoters of AhCAB and AhPORA, regulating their expression during water stress and recovery. Overall, the study proposed that AhHDA1 and AhGLK1 act as an ON/OFF switch for AhCAB and AhPORA during drought stress and rewatering conditions, with AhGLK1 activating and AhHDA1 suppressing their expression (Liu et al., 2020). In another study under drought stress, an increase in acetylation at histone 3 (H3ac) was reported in peanut (Zhang et al., 2018).

9.3.5 Other Legume Crops

The previous sections discuss various epigenetic changes in response to environmental stresses, using examples from key legume crops. The current section provides a few examples from other legume crops, including castor bean (*Ricinus communis* L.), common bean (*Phaseolus vulgaris* L.), common vetch (*Vicia sativa* L.), faba bean (*Vicia faba* L.), horse gram (*Macrotyloma uniflorum* L.), hyacinth bean (*Lablab purpureus* L.), and mungbean (*Vigna radiata* L.). Under salt stress, castor bean, a crucial nonedible oil legume, showed an increase in trimethylation at histone 3 lysine 4 (H3K4me3), whereas there was a decrease in trimethylation at histone 3 lysine 27 (H3K27me3) (Han et al., 2020). The integration of Chip-Seq and RNA-Seq data in the study suggested various salt-responsive genes and histone methylation sites, notably the transcription of RADIALIS-LIKE SANT (RSM1), a MYB-related factor linked to ABA-mediated salt signaling, which is possibly regulated by these H3K4me3-H3K27me3 modifications (Han et al., 2020). For DNA methylation changes in response to drought stress, faba bean (Abid et al., 2017) and mungbean (Zhao et al., 2022), showed a decrease in global methylation level and differentially methylated regions, respectively. The latter study suggested that wild mungbean is more resistant to drought stress, with more hypermethylated DMRs and less hypomethylated DMRs after drought stress, corresponding to more downregulated DEGs than upregulated DEGs. Among the three DNA methylation contexts, CG, CHG, and CHH, asymmetric CHH contexts are more dynamic and prone to be altered by drought stress and genotypic variations (Zhao et al., 2022). Similarly, chemical stresses in hyacinth bean (*Lablab purpureus* L.) elevate the temperature and increase the differential methylated band from 5'-CCGG-3' (Rai et al., 2018).

Like histone modifications and methylation, the drought stress-responsive 169 and 379 potential miRNAs were reported in common beans (Wu et al., 2017) and common vetch (Zhu et al., 2021), respectively. Meanwhile, in horse gram under soil moisture deficiency, 15 potential miRNAs were reported (Yasin et al., 2020).

9.4 GOLD STANDARD TECHNIQUES TO MEASURE EPIGENETIC MODIFICATIONS IN LEGUMES (FIGURE 9.2)

The measurement and analysis of epigenetic changes in response to climate changes in legumes or any other crop or plant species rely on various sophisticated techniques. The choice of technique depends on the specific epigenetic modification being studied, the scale of the analysis, and the desired resolution. In the past few decades, the improvement in next-generation sequencing (NGS) has revolutionized epigenetics research, enabling high-resolution genome-wide analysis of epigenetic changes in tissue- and condition-dependent manners for any organism, including legume crops. The current section explains various NGS-based techniques, which have been and can be utilized in studying epigenetic modifications (DNA methylation, histone modifications, and chromatin accessibility) in legume crops.

9.4.1 Bisulfite Sequencing for DNA Methylation Analysis (Figure 9.2A[a])

Despite being labor- and computation-intensive, bisulfite-based sequencing methods are the gold standard for measuring DNA methylation due to their single-nucleotide resolution, flexibility across organisms, and low input requirements (10–100 pg of DNA) (Singer, 2019). The genomic DNA extracted from the desired tissue samples under the desired conditions is sonicated to smaller fragments and treated with sodium bisulfite (commercial kits are available). This treatment transforms the epigenetic information of genomic DNA into genetic information. The bisulfite conversion rapidly turns unmethylated cytosines into uracil while protecting 5-methylcytosine. For precise methylation analysis, high cytosine conversion rates to uracil (over 99%) are crucial, which is validated against unmethylated λ-bacteriophage DNA. The harsh bisulfite treatment fragments DNA, creating single-stranded components that lose CG interstrand pairing. PCR amplification replaces uracil with thymine, allowing comparison with a reference sequence to identify methylation status. Whole-genome bisulfite sequencing (WGBS) is the current standard, employing random PCR for DNA amplification and providing a genome-wide detailed methylation landscape (Yong et al., 2016). However, a comprehensive assessment through WGBS is often unnecessary. The other bisulfite-based technique, reduced representation bisulfite sequencing (RRBS), efficiently targets CpG-rich regions. Modified RRBS measures 10–20% of genomic CpGs, lowering costs, enabling multiplexing, and increasing statistical power.

The bioinformatics analysis of WGBS and RRBS data involves processing bisulfite-converted sequencing reads to quantify and interpret DNA methylation patterns accurately. Several valuable tools were developed for this purpose. However, the typical steps involved:

1. Quality control entails evaluating the quality of raw sequencing data with tools like FastQC and trimming adapters and low-quality sequences using tools such as Trimgalore, Trimmomatic, or Fastp, resulting in high-quality reads.
2. Alignment of high-quality reads to the reference genome. For this purpose, specialized and sophisticated tools such as Bismark or BSeeker2 were developed to convert cytosine to thymine.
3. After aligning to establish the methylation status, methylation can be performed using R-based tools such as methylKit. This will also calculate the percentage of methylation levels for all three CpG, CHH, and CHG contexts.

4. For differential methylation analysis, the methylation levels between two samples (control and treated) can be compared, revealing differentially methylated regions (DMRs). This can be accomplished using R-based statistical tools such as methylKit, limma, or edgeR.
5. In the final step, the identified DMRs can be annotated for genomic features such as gene bodies, promoter, or intergenic regions. Additionally, pathway enrichment analysis can be performed using tools such as DAVID.

Although WGBS or RRBS are gold standard techniques for measuring genome-wide or targeted DNA methylation, they can sometimes be challenging. Bisulfite-based approaches can be labor-intensive, expensive, and require sophisticated bioinformatics analysis. Additionally, expertise is needed to handle the input DNA due to the DNA template's degradation risk after harsh bisulfite treatment.

Besides bisulfite sequencing for analyzing DNA methylation, techniques like methylated DNA immunoprecipitation (MeDIP), methylation arrays, and long-read genome sequencing using either Oxford Nanopore or PacBio Sciences are also available (Gouil & Keniry, 2019; Tost, 2009).

9.4.2 Chromatin Immunoprecipitation (ChIP) for Histone Modifications (Figure 9.2A[b])

The conventional method for studying histone modifications and variants consists of chromatin immunoprecipitation with antibodies that detect specific modifications, followed by high-throughput analysis of the immunoprecipitated fractions (ChIP-seq or ChIP-on-chip) (Brabencová et al., 2017). The analysis starts with cross-linking using formaldehyde, followed by chromatin extraction, DNA sonication, and immunoprecipitation (Song et al., 2014). The immunoprecipitation requires a specific antibody, which is typically designed to recognize histone modifications of interest and extract the DNA with those modifications. The isolated DNA can then be sequenced using high-throughput NGS (Chip-Seq), and quantified using PCR (ChIP-PCR) or microarray (Chip-on-chip). Here is a summary of the steps in a ChIP-Seq experiment adopted from Song et al. (Song et al., 2014).

Harvest legume crop materials after treatments, collecting 1–2 g in a 50-mL Falcon tube. Rinse twice with ddH_2O, then immerse in 37 mL of formaldehyde cross-linking solution. Pack the tube with nylon mesh to submerge the plants, then perform vacuum infiltration at room temperature three times for 5 minutes each, releasing the vacuum in between to allow buffer penetration. Gently shake the desiccator to eliminate air bubbles; seedlings should appear "water-soaked." Stop cross-linking by adding glycine quenching solution to reach a final concentration of 0.125 M (mix 2.5 mL of 2 M stock with 37 mL of buffer) and apply vacuum for 5 minutes. Rinse formaldehyde with 40 mL of ddH_2O at least three times. Remove excess water by placing seedlings on paper before freezing in liquid nitrogen. Cross-linked material can be stored at −80 °C for months or further processed for chromatin extraction. Grind cross-linked material in liquid nitrogen to a fine powder and resuspend in extraction buffer. The material will be mixed by vortexing and incubated for 5 minutes until homogeneous. Ensure buffer volume exceeds powder by five times. Filter through double-layer Miracloth into an ice-cold falcon tube until transparent. Spin for 20 min at 2880 × g at 4 °C. Gently remove supernatant and resuspend pellet in 1 mL of extraction buffer. Transfer to a 1.5-mL Eppendorf tube and centrifuge at 12,000 × g for 10 min at 4 °C. Remove supernatant and resuspend the pellet in 300 µL of extraction buffer. Overlay onto 900 µL of extraction buffer in a fresh 1.5-mL Eppendorf tube and spin for 45 min to 1 h at 16,000 × g.

Remove the supernatant and resuspend the chromatin pellet in a DNA LoBind tube with 320 µL nuclei lysis buffer by pipetting. Save a 10-µL aliquot of "unsheared" chromatin for later analysis. Sonicate the chromatin for three 5-minute sessions at "low" using the sonicator and cooling the water bath as needed. This fragments DNA to 200–800 bp, centered at 500 bp. The sonicated chromatin

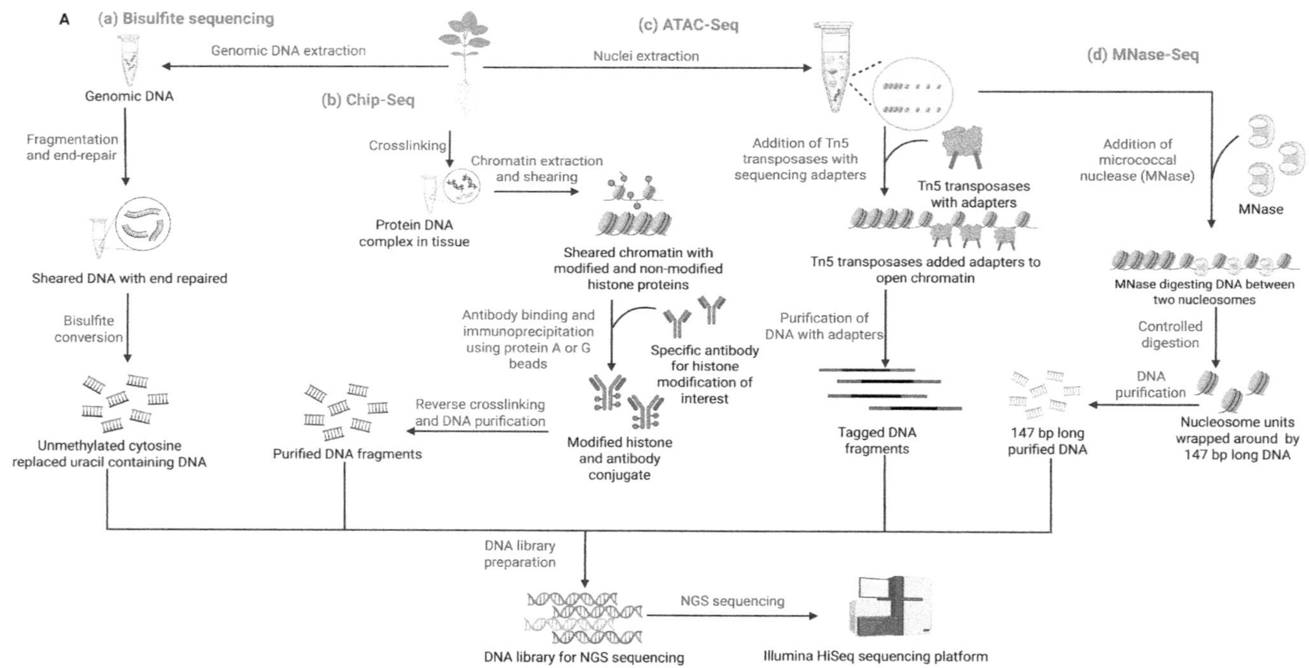

FIGURE 9.2 Techniques to detect epigenetic modifications in legumes. (A) Experimental designs: (a) Bisulfite sequencing detects genome-wide DNA methylation changes from extracted genomic DNA, which is sheared. Bisulfite treatment converts unmethylated cytosines into uracils, and the final DNA is used for next-generation sequencing (NGS) library preparation. (b) Chromatin immunoprecipitation sequencing (ChIP-Seq) involves cross-linking DNA and proteins with formaldehyde in plant tissue, followed by chromatin extraction and shearing. The chromatin incubates with a specific antibody for histone modifications of interest and pulls out the antibody-bound chromatin with protein A or G beads. Reverse cross-linking releases DNA, which is used for NGS library preparation. (c) Assay for transposase-accessible chromatin using sequencing (ATAC-seq) starts with isolated nuclei containing chromatin. Tn5 transposases add sequencing adapters for NGS library preparation to accessible chromatin, allowing direct use following purification. (d) Micrococcal nuclease sequencing (MNase-Seq) uses MNase enzymes to digest accessible chromatin, leaving nucleosomes wrapped in 147 bp DNA, which is then purified for NGS. (B) Bioinformatic analysis. This part illustrates tools for analyzing sequencing data from (a) bisulfite, (b) ChIP, (c) ATAC, and (d) MNase sequencing, with essential tools shown in green. Data can be visualized on free genome browsers like IGV, IGB, or Jbrowse. (Created in https://BioRender.com.)

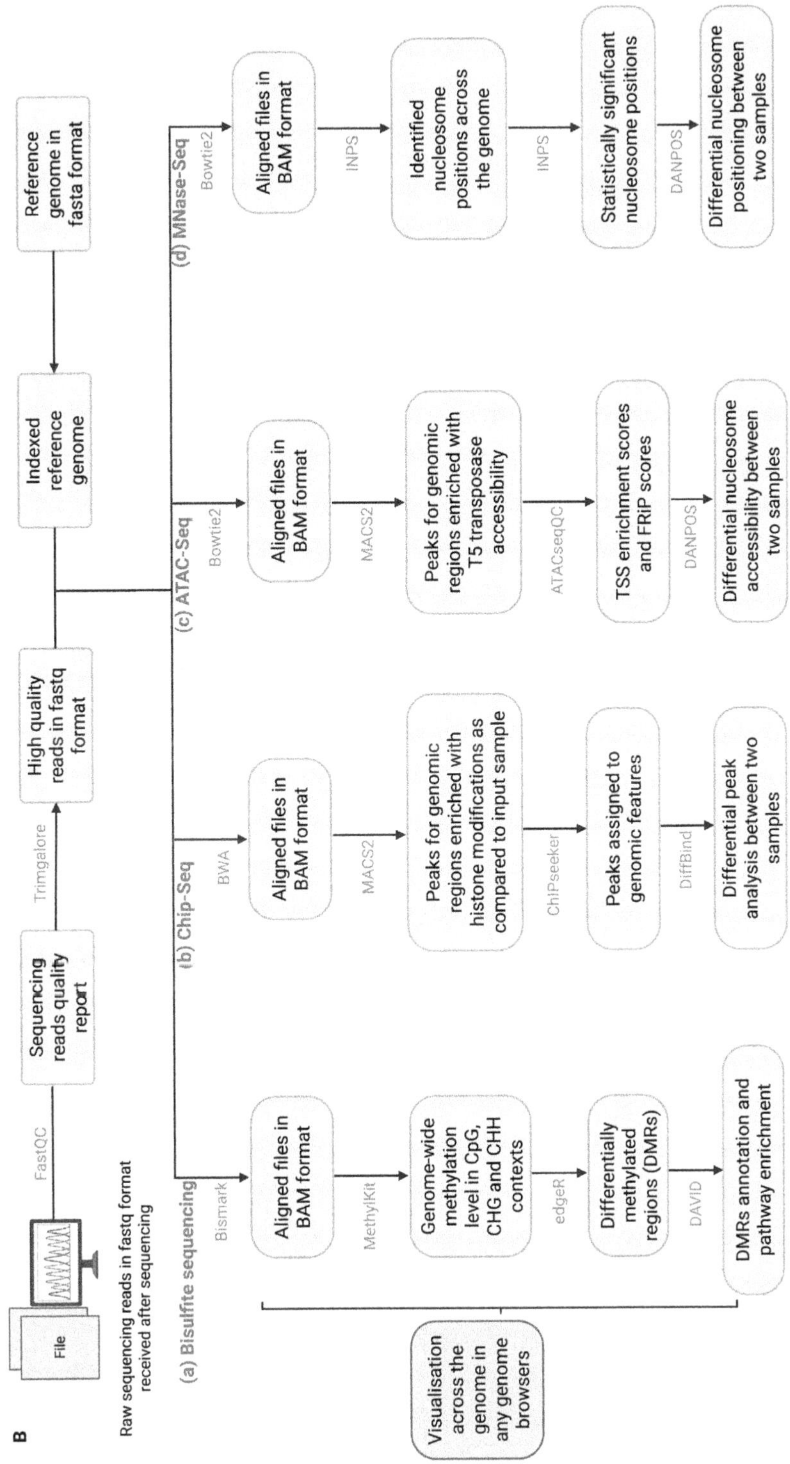

FIGURE 9.2 (Continued)

can be frozen in liquid nitrogen at −80 °C or processed for immunoprecipitation, which can be performed with the desired antibody for specific histone modification. After immunoprecipitation, the final steps involve the recovery of DNA with modified histones. Finally, the chipped DNA can be used for high-throughput NGS or ChIP-PCR for quantification. The bioinformatics analysis for the ChIP-Seq data involves the basic steps:

1. Quality control involves assessing the quality of raw sequencing data using tools like FastQC, trimming adapters, and low-quality sequences with tools such as Trimgalore, Trimmomatic, or Fastp, resulting in high-quality reads.
2. Align high-quality reads with the reference genome using Bowtie2 or BWA. This step identifies the genomic locations of the sequencing reads. Tools such as samtools can evaluate mapping statistics to ensure successful alignment.
3. Aligned reads are further used for peak calling, leveraging algorithms like MACS2. This process identifies genomic regions exhibiting significant read enrichment compared to the input control or background. Histone modification ChIP-seq typically yields "broad peaks," in contrast to transcription factor ChIP-seq. Consequently, peak callers should be applied with parameters that consider these differences.
4. Peaks are marked to determine their positions concerning genomic elements, including genes, promoters, and enhancers. Tools such as ChIPseeker and HOMER assist in this process. Gene ontology determines the biological functions or pathways enriched about the genes near the identified peaks.
5. Detect enriched DNA sequence motifs in peak regions, indicating the binding sites of transcription factors or other DNA-binding proteins. Commonly utilized tools include MEME and HOMER. Compare the identified motifs through motif enrichment analysis with existing motif databases.
6. Analyze ChIP-seq data across different experimental conditions to detect variations in protein binding. You can use tools like DiffBind or DESeq2 for this purpose. Present the findings with visualization tools like the Integrative Genomics Viewer (IGV) or R packages like ggplot2.

ChIP-Seq is a labor-intensive experiment; however, it is regarded as the gold standard for histone modifications and other epigenetic marks, such as RNA polymerase II processivity. Other methods, such as mass spectrometry (LC-mass-spec), can also be utilized to study PTMs, including histone modifications.

9.4.3 Chromatin Accessibility Analysis Using ATAC-Seq (Figure 9.2A[c]) and MNase-Seq (Figure 9.2A[d])

The arrangement of chromatin architecture is essential for numerous biological functions, particularly in controlling transcription, as nucleosomes (the fundamental unit of chromatin) and transcription factors facilitate access to DNA sequences and play a crucial role in transcriptional regulation. Additionally, nucleosomes with varying packing configurations have unique roles in regulating gene expression in response to environmental stresses. Thus, globally mapping nucleosome positions in legume crops offers valuable insights into how these positioning and occupancy patterns relate to transcriptional regulation. Overall, nucleosome positioning assesses chromatin accessibility, and two primary approaches (ATAC-Seq and MNase-Seq) can be utilized, as discussed below.

The Assay for Transposase-Accessible Chromatin using Sequencing (ATAC-Seq) utilizes a hyperactive Tn5 transposase to cut and tag open chromatin regions with sequencing adapters simultaneously (Grandi et al., 2022). It primarily detects open chromatin areas, such as nucleosome-depleted regions and transcription factor binding sites. As the latest method for studying open chromatin, ATAC-Seq exploits the hyperactive Tn5 transposase to fragment DNA and insert adaptor sequences into active regulatory regions in vivo. In the procedure, gentle detergents isolate nuclei, which are then labeled in vitro with sequencing adaptors using purified Tn5 transposase. Due to steric hindrance, most adaptors incorporate into accessible chromatin regions, are PCR amplified for library preparation, and then followed by paired-end sequencing. This method generates reads from small amounts of plant material, reflecting accessible regions related to nucleosome positioning and transcription factor-binding sites. Consequently, ATAC-Seq offers numerous benefits, including a quick and robust approach to predicting nucleosome positions with minimal input material. It also directly assesses chromatin accessibility and is relatively straightforward, requiring fewer steps in the protocol (Bajic et al., 2018).

Another essential technique uses micrococcal nuclease (MNase) to digest linker DNA between nucleosomes, effectively pinpointing nucleosome locations and, by extension, areas of closed chromatin. MNase-digested DNA, paired with high-throughput sequencing (MNase-Seq), has been extensively utilized for globally mapping nucleosome positions in eukaryotic organisms with known genomes. This method primarily involves isolating plant nuclei, treating the purified nuclei with MNase, recovering MNase-trimmed mononucleosomal DNA of about 150 bp through gel electrophoresis, preparing the MNase-seq library, and ultimately performing high-throughput NGS and data analysis (Zhang & Jiang, 2018). Like other techniques, MNase-Seq has both advantages and disadvantages. It accurately maps high-resolution nucleosome positions and provides detailed information on nucleosome occupancy. However, MNase-Seq requires a significant amount of starting plant material, relies on an indirect measure of chromatin, and involves a labor-intensive protocol.

The bioinformatics analysis for the ATAC-Seq and MNase-Seq involves similar basic steps:

1. Quality control involves assessing the quality of raw sequencing data using tools like FastQC and trimming adapters and low-quality sequences with tools such as Trimgalore, Trimmomatic, or Fastp, resulting in high-quality reads.
2. Align the high-quality reads with the reference genome using Bowtie2. This step identifies the genomic locations of the sequencing reads. Tools like Samtools can evaluate mapping statistics to ensure successful alignment.
3. In ATAC-Seq, aligned reads are utilized for peak calling by leveraging algorithms such as MACS2 or Genrich. This process identifies genomic regions exhibiting significant read representation, indicating areas of increased transposase accessibility. Consider adjusting reads to account for Tn5 insertion offsets. In MNase-Seq, aligned reads are employed to determine nucleosome positions by analyzing read density and distribution. Tools such as DANPOS and INPS are designed to locate the dyad positions of nucleosomes. MNase-Seq data are used to infer open chromatin, with areas of low read density regarded as open.
4. In ATAC-Seq, the next step involves conducting quality control of peaks, which provides detailed TSS enrichment. This includes evaluating the strength of the signal at transcription start sites and analyzing fragment size distribution and FRiP scores (Fraction of Reads in Peaks). In contrast, MNase-Seq utilizes the identified nucleosome positions to quantify nucleosome occupancy, spacing, and regularity across the genome.
5. Lastly, both peaks from ATAC-Seq and nucleosome positions from MNase-Seq can utilize DiffBind and DNAPOS, respectively, to perform differential accessibility and differential nucleosome positioning between two samples.
6. The various genome browser tools, including IGV, IGB, and JBrowse, can visualize genome-wide chromatin peaks and nucleosome positions.

Despite well-established techniques, protocols, and the availability of bioinformatics tools in model plant species, ATAC-Seq and MNase-Seq remain less popular in plant science research, particularly in legume crops.

9.5 FUTURE IMPLEMENTATION OF EPIGENETIC MEMORY AND EPIGENOME EDITING IN LEGUMES

It has been established that epigenetic modifications regulate the expression of numerous stress-responsive genes and are essential for legumes' and other crops' adaptation to climate change. However, many vital applications of epigenetics and their implementation in developing varieties with stress tolerance are confined to model plants or specific crop species. Consideration of these mechanisms, including epigenetic memory-based transgenerational adaptation and epigenome editing, is necessary to develop climate-smart legume crops in the future.

9.5.1 Epigenetic Memory-Based Transgenerational Adaptation

The concept of epigenetic transgenerational adaptation in plants is a captivating area of research that explores how plants can convey adaptations to environmental stressors to their offspring without changing their fundamental DNA sequence (Hauser et al., 2011). While stress-related epigenetic modifications usually return to baseline after stress, some remain stable and are passed down as epigenetic stress memory through mitotic and meiotic divisions. Others can reactivate during reproduction or embryonic development. This epigenetic memory, which does not alter nucleotide sequences, relies on chromatin adjustments that influence gene activation under varying environmental pressures. Three types of epigenetic memory, cellular, transcriptional, and transgenerational, can be inherited after stress exposure. Cellular memory refers to heritable transcriptional states established during initial development using tissue-specific transcription factors. Transcriptional memory describes mitotically heritable changes lasting several generations due to previous stress. Transgenerational memory involves meiotically hereditary changes affecting the behavior of the future generation (Baulcombe & Dean, 2014; Hauser et al., 2011; Ramakrishnan et al., 2022).

Epigenetic stress memory is well studied and validated in the model plant *Arabidopsis thaliana* (Baulcombe & Dean, 2014; Ramakrishnan et al., 2022) and the crop rice (Ganie et al., 2024); however, only a few recent reports have suggested epigenetic transgenerational inheritance in legumes (Yung et al., 2023). For instance, Kumari et al. (2018) found that the seeds generated by F1-generation endosymbiotic plants in a controlled environment were utilized for a second-generation study in the greenhouse. Specific fungal and bacterial endosymbionts promoted seed germination and facilitated root and shoot growth in second-generation seeds subjected to drought stress without endophytes (Kumari et al., 2018). Furthermore, silencing of the plant MutS HOMOLOG1 (MSH1) gene in various model plant species, including *Arabidopsis thaliana* (Virdi et al., 2015), resulted in developmental changes, altered defense, phytohormone, and abiotic stress response pathways, and modified DNA methylation patterns. This msh1-induced developmental reprogramming endures irrespective of transgene segregation, resulting in transgene-null "memory" effects. Crossing an isogenic memory line with wild-type generates progeny with more significant variation in adaptive traits (Raju et al., 2018). Therefore, Raju et al. (2018) developed the MSH1 system in soybeans, which can induce valuable epigenetic variation (Raju et al., 2018). The authors generated MSH1 epi-populations by crossing msh1-acquired soybean memory lines. In both glasshouse and field trials, these epi-lines exhibited increased variance in yield-related traits such as pods per plant, seed weight, and maturity time (Raju et al., 2018). The findings in the study highlighted the potential of MSH1-derived epigenetic variation for enhancing yield and stability in soybean and other legume crop breeding.

Since epigenetic memory-based transgenerational adaptation to develop epi-populations for climate-smart crops has emerged in recent years, future efforts must be made to utilize this phenomenon in legume crops.

9.5.2 Future Implementations of Epigenome Editing in Legumes

Gene or genome editing refers to intentionally modifying the genetic material, which includes removing, introducing, or altering the DNA sequence in a cell or organism. This process removes a specific gene or regulatory element in the genome. Gene- or genome-editing technology has been revolutionized in the last two decades with the discovery and utilization of CRISPR/Cas9 (Clustered Regularly Interspaced Short Palindromic Repeats/Cas endonuclease) in eukaryotic cells (reviewed by Gostimskaya [2022]). The CRISPR/Cas system is extensively used for editing plant genomes and generating mutants because of its straightforwardness and practicality. It is increasingly utilized to alter traits in various plants, including essential crops, and to establish new germplasm resources (Gaillochet et al., 2021). Manipulating each gene sequence in plants is always challenging due to their large and complex genomes. However, genome-editing tools such as CRISPR/Cas9 can induce mutations in an endogenous gene, altering its sequence and function by disrupting, inserting, deleting, and replacing bases at a locus within a genome. It can also regulate gene expression to manage single-nucleotide polymorphisms and create reporter fusions while maintaining gene regulation. A typical gene comprises a transcription factor binding site, a promoter region, exons (coding sequences), introns (non-coding sequences), and an untranslated region. Genome editing is skillfully executed using engineered nucleases that create double-stranded cuts or single-stranded nicks in plant DNA. These sites are modified and repaired through homologous recombination or non-homologous end joining. This form of genetic engineering is beneficial for various applications, including targeted gene mutations, chromosomal rearrangements, and the creation of transgenic plants (Khan et al., 2018).

Several goals in crop improvement were achieved through genome editing using CRISPR/Cas9 in plant genetic engineering (reviewed by Khan et al. [2018]), but its implementation for epigenetic engineering remains elusive. Since it is evident now that epigenetic modifications or changes in response to climate change alter the expression of many stress-responsive genes, epigenome editing through CRISPR/Cas9 has the potential to change gene expression and result in phenotype quicker than the genetic engineering approach. The prospects of the CRISPR/Cas9 system necessitate studying epigenetic changes in legumes. This system may facilitate plant epigenomic modifications through DNA methylation and other epigenetic factors, as most genes reside in DNA wrapped around histone proteins for compaction and gene regulation. This DNA is often inaccessible, rendering the genes contained within it typically inactive. The CRISPR/Cas9 system is associated with DNA methyltransferase or acetyltransferase at the histone "tails" to induce plant epigenetic modifications. These epigenetic factors bind to histone "tails," altering the extent to which DNA is wrapped around histones and the accessibility of genes within the DNA for activation. Changes in epigenetics through the CRISPR/Cas9 system in plants also enhance the growth potential of various crops, including legumes, in response to climate change, which presents an intriguing goal. This could increase global crop yields and, consequently, the production of legume crops, thus contributing to the food security challenge.

9.6 CONCLUSIONS

Due to their versatile nature and capacity to address global food security challenges, legume crops are a focal point for developing climate-smart agricultural products. Furthermore, advancements in NGS technologies and the development of bioinformatics tools over the past two decades have enhanced our understanding of epigenetic modifications and their regulatory roles in genes related to plant development, growth, and defense mechanisms. Several epigenetic modifications, including DNA methylation, histone modifications, and non-coding RNAs, have been discovered and shown to regulate various stress-responsive genes in numerous legume crops. However, although most of these modifications have been studied as individual elements or under single stress conditions, simultaneous genome-wide modifications and the effects of multiple stressors must be considered.

Moreover, the identified epigenetic modifications in legume crops in response to climate change must be applied to validate the concept of epigenetic memory-based transgenerational adaptations and epigenome editing in developing epi-populations of legume crops with enhanced stress tolerance. In summary, the implications of global climate change on legume crops must be addressed urgently through epigenome editing or breeding tools.

ACKNOWLEDGMENT

I thank the Leverhulme Trust, United Kingdom, and the University of Sheffield for funding. I am also grateful to Dr Shrinivas Dighe from Cardiff University for assisting in providing literature that helped to compile the current book chapter.

REFERENCES

Abid, G., Mingeot, D., Muhovski, Y., Mergeai, G., Aouida, M., Abdelkarim, S., Aroua, I., El Ayed, M., M'hamdi, M., Sassi, K., & Jebara, M. (2017). Analysis of DNA methylation patterns associated with drought stress response in faba bean (Vicia faba L.) using methylation-sensitive amplification polymorphism (MSAP). *Environmental and Experimental Botany*, *142*, 34–44. https://doi.org/10.1016/j.envexpbot.2017.08.004

Al-Lawati, A., Al-Bahry, S., Victor, R., Al-Lawati, A. H., & Yaish, M. W. (2016). Salt stress alters DNA methylation levels in alfalfa (Medicago spp). *Genetics and Molecular Research*, *15*(1), 1-16. https://doi.org/10.4238/gmr.15018299

Arriagada, O., Cacciuttolo, F., Cabeza, R. A., Carrasco, B., & Schwember, A. R. (2022). A Comprehensive Review on Chickpea (Cicer arietinum L.) Breeding for Abiotic Stress Tolerance and Climate Change Resilience. International Journal of Molecular Sciences, 23(12), 6794. https://doi.org/10.3390/ijms23126794

Awana, M., Yadav, K., Rani, K., Gaikwad, K., Praveen, S., Kumar, S., & Singh, A. (2019). Insights into salt stress-induced biochemical, molecular and epigenetic regulation of spatial responses in Pigeonpea (Cajanus cajan L.). *Journal of Plant Growth Regulation*, *38*(4), 1545–1561. https://doi.org/10.1007/s00344-019-09955-4

Bai, Z., Yang, X., Zi, N., Ren, W., Yin, J., Yuan, T., Wang, M., Yuan, F., & Liu, Y. (2025). Drought stress memory enhances the tolerance of alfalfa Medicago sativa L. in response to a subsequent drought: A physiological and omics perspective. *Environmental and Experimental Botany*, *230,* 1–11. https://doi.org/10.1016/j.envexpbot.2025.106088

Bajic, M., Maher, K. A., & Deal, R. B. (2018). Identification of open chromatin regions in plant genomes using ATAC-Seq. In John M. Walker (ed.). *Methods in Molecular Biology* (Vol. 1675, pp. 183–201). Humana Press. https://doi.org/10.1007/978-1-4939-7318-7_12

Baulcombe, D. C., & Dean, C. (2014). Epigenetic regulation in plant responses to the environment. *Cold Spring Harbor Perspectives in Biology*, *6*(9), 1–19. https://doi.org/10.1101/cshperspect.a019471

Berger, S. L. (2007). The complex language of chromatin regulation during transcription. *Nature*, *447*(7143), 407–412. https://doi.org/10.1038/nature05915

Bhogireddy, S., Mangrauthia, S. K., Kumar, R., Pandey, A. K., Singh, S., Jain, A., Budak, H., Varshney, R. K., & Kudapa, H. (2021). Regulatory non-coding RNAs: A new frontier in regulation of plant biology. *Functional and Integrative Genomics*, *21*(3), 313–330. Springer Science and Business Media Deutschland GmbH. https://doi.org/10.1007/s10142-021-00787-8

Brabencová, S., Ihnatová, I., Potěšil, D., Fojtová, M., Fajkus, J., Zdráhal, Z., & Lochmanová, G. (2017). Variations of histone modification patterns: Contributions of inter-plant variability and technical factors. *Frontiers in Plant Science*, *8*, 2084. https://doi.org/10.3389/fpls.2017.02084

Buch, D. U., Sharma, O. A., Pable, A. A., & Barvkar, V. T. (2020). Characterization of microRNA genes from Pigeonpea (Cajanus cajan L.) and understanding their involvement in drought stress. *Journal of Biotechnology*, *321*, 23–34. https://doi.org/10.1016/j.jbiotec.2020.06.019

Cao, X., & Jacobsen, S. E. (2002). *Locus-Specific Control of Asymmetric and CpNpG Methylation by the DRM and CMT3 Methyltransferase Genes*. Proceedings of the National Academy of Sciences, *99*(suppl_4), 16491–16498. https://doi.org/10.1073/pnas.162371599

Chen, H., Yang, Q., Chen, K., Zhao, S., Zhang, C., Pan, R., Cai, T., Deng, Y., Wang, X., Chen, Y., Chu, W., Xie, W., & Zhuang, W. (2019). Integrated microRNA and transcriptome profiling reveals a miRNA-mediated regulatory network of embryo abortion under calcium deficiency in peanut (Arachis hypogaea L.). *BMC Genomics*, *20*(1), 392. https://doi.org/10.1186/s12864-019-5770-6

Chen, K., Tang, W. S., Zhou, Y. Bin, Xu, Z. S., Chen, J., Ma, Y. Z., Chen, M., & Li, H. Y. (2020). Overexpression of GmUBC9 gene enhances plant drought resistance and affects flowering time via histone H2B monoubiquitination. *Frontiers in Plant Science*, *11*, 555794. https://doi.org/10.3389/fpls.2020.555794

Chen, L., Li, X., Liu, H., He, F., Li, M., Long, R., Wang, X., Kang, J., & Yang, Q. (2025). Comprehensive analysis of epigenetic modifications in alfalfa under cadmium stress. *Journal of Hazardous Materials*, *482*, 136545. https://doi.org/10.1016/j.jhazmat.2024.136545

Chen, R., Li, M., Zhang, H., Duan, L., Sun, X., Jiang, Q., Zhang, H., & Hu, Z. (2019). Continuous salt stress-induced long non-coding RNAs and DNA methylation patterns in soybean roots. *BMC Genomics*, *20*(1), 730. https://doi.org/10.1186/s12864-019-6101-7

Chu, S., Zhang, X., Yu, K., Lv, L., Sun, C., Liu, X., Zhang, J., Jiao, Y., & Zhang, D. (2020a). Genome-wide analysis reveals dynamic epigenomic differences in soybean response to low-phosphorus stress. *International Journal of Molecular Sciences*, *21*(18), 1–21. https://doi.org/10.3390/ijms21186817

Chu, S., Zhang, X., Yu, K., Lv, L., Sun, C., Liu, X., Zhang, J., Jiao, Y., & Zhang, D. (2020b). Genome-wide analysis reveals dynamic epigenomic differences in soybean response to low-phosphorus stress. *International Journal of Molecular Sciences*, *21*(18), 1–21. https://doi.org/10.3390/ijms21186817

Dong, W., Gao, T., Wang, Q., Chen, J., Lv, J., & Song, Y. (2020). Salinity stress induces epigenetic alterations to the promoter of MsMYB4 encoding a salt-induced MYB transcription factor. *Plant Physiology and Biochemistry*, *155*, 709–715. https://doi.org/10.1016/j.plaphy.2020.08.015

Farooq, M., Hussain, M., Usman, M., Farooq, S., Alghamdi, S. S., & Siddique, K. H. M. (2018). Impact of abiotic stresses on grain composition and quality in food legumes. *Journal of Agricultural and Food Chemistry*, *66*(34), 8887–8897. https://doi.org/10.1021/acs.jafc.8b02924

Fillingham, J., & Greenblatt, J. F. (2008). A histone code for chromatin assembly. In *Cell*, *134*(2), 206–208. Elsevier B.V. https://doi.org/10.1016/j.cell.2008.07.007

Foyer, C. H., Lam, H. M., Nguyen, H. T., Siddique, K. H. M., Varshney, R. K., Colmer, T. D., Cowling, W., Bramley, H., Mori, T. A., Hodgson, J. M., Cooper, J. W., Miller, A. J., Kunert, K., Vorster, J., Cullis, C., Ozga, J. A., Wahlqvist, M. L., Liang, Y., Shou, H., … Considine, M. J. (2016). Neglecting legumes has compromised human health and sustainable food production. *Nature Plants*, *2*(8), 16112. https://doi.org/10.1038/NPLANTS.2016.112

Gaillochet, C., Develtere, W., & Jacobs, T. B. (2021). CRISPR screens in plants: Approaches, guidelines, and future prospects. In *Plant Cell,* 33(4), 794–813. American Society of Plant Biologists. https://doi.org/10.1093/plcell/koab099

Ganie, S. A., McMulkin, N., & Devoto, A. (2024). The role of priming and memory in rice environmental stress adaptation: Current knowledge and perspectives. In *Plant Cell and Environment,* 47(5), 1895–1915. John Wiley. https://doi.org/10.1111/pce.14855

Gostimskaya, I. (2022). CRISPR–Cas9: A history of its discovery and ethical considerations of its use in genome editing. *Biochemistry (Moscow),* 87(8), 777–788. https://doi.org/10.1134/S0006297922080090

Gouil, Q., & Keniry, A. (2019). Latest techniques to study DNA methylation. *Essays in Biochemistry,* 63(6), pp. 639–648. Portland Press. https://doi.org/10.1042/EBC20190027

Grandi, F. C., Modi, H., Kampman, L., & Corces, M. R. (2022). Chromatin accessibility profiling by ATAC-seq. *Nature Protocols,* 17(6), 1518–1552. Nature Research. https://doi.org/10.1038/s41596-022-00692-9

Guleria, P., Mahajan, M., Bhardwaj, J., & Yadav, S. K. (2011). Plant small RNAs: Biogenesis, mode of action and their roles in abiotic stresses. *Genomics, Proteomics and Bioinformatics,* 9(6), 183–199. https://doi.org/10.1016/S1672-0229(11)60022-3

Gupta, K., & Garg, R. (2023). Unravelling differential DNA methylation patterns in genotype dependent manner under salinity stress response in Chickpea. *International Journal of Molecular Sciences*, *24*(3), 1863. https://doi.org/10.3390/ijms24031863

Han, B., Xu, W., Ahmed, N., Yu, A., Wang, Z., & Liu, A. (2020). Changes and associations of genomic transcription and histone methylation with salt stress in castor bean. *Plant and Cell Physiology*, *61*(6), 1120–1133. https://doi.org/10.1093/pcp/pcaa037

Han, P., & Chang, C. P. (2015). Long non-coding RNA and chromatin remodeling. *RNA Biology*, *12*(10), 1094–1098. https://doi.org/10.1080/15476286.2015.1063770

Hauser, M. T., Aufsatz, W., Jonak, C., & Luschnig, C. (2011). Transgenerational epigenetic inheritance in plants. *Biochimica et Biophysica Acta – Gene Regulatory Mechanisms, 1809*(8), 459–468. https://doi.org/10.1016/j.bbagrm.2011.03.007

Hossain, M. S., Kawakatsu, T., Kim, K. Do, Zhang, N., Nguyen, C. T., Khan, S. M., Batek, J. M., Joshi, T., Schmutz, J., Grimwood, J., Schmitz, R. J., Xu, D., Jackson, S. A., Ecker, J. R., & Stacey, G. (2017). Divergent cytosine DNA methylation patterns in single-cell, soybean root hairs. *New Phytologist*, *214*(2), 808–819. https://doi.org/10.1111/nph.14421

Hung, Y. H., & Slotkin, R. K. (2021). The initiation of RNA interference (RNAi) in plants. In *Current Opinion in Plant Biology,* 61, 102014. Elsevier. https://doi.org/10.1016/j.pbi.2021.102014

Jhu, M. Y., & Oldroyd, G. E. D. (2023). Dancing to a different tune, can we switch from chemical to biological nitrogen fixation for sustainable food security? *PLoS Biology*, *21*(3), e3001982. https://doi.org/10.1371/journal.pbio.3001982

Khan, M. H. U., Khan, S. U., Muhammad, A., Hu, L., Yang, Y., & Fan, C. (2018). Induced mutation and epigenetics modification in plants for crop improvement by targeting CRISPR/Cas9 technology. *Journal of Cellular Physiology,* 233(6), 4578–4594. Wiley-Liss. https://doi.org/10.1002/jcp.26299

Khandal, H., Parween, S., Roy, R., Meena, M. K., & Chattopadhyay, D. (2017). MicroRNA profiling provides insights into post-transcriptional regulation of gene expression in chickpea root apex under salinity and water deficiency. *Scientific Reports*, *7*(1), 4632. https://doi.org/10.1038/s41598-017-04906-z

Kohli, D., Joshi, G., Deokar, A. A., Bhardwaj, A. R., Agarwal, M., Katiyar-Agarwal, S., Srinivasan, R., & Jain, P. K. (2014). Identification and characterization of wilt and salt stress-responsive MicroRNAs in chickpea through high-throughput sequencing. *PLoS ONE*, *9*(10), e108851. https://doi.org/10.1371/journal.pone.0108851

Kouzarides, T. (2007). Chromatin modifications and their function. *Cell,* 128(4), 693–705. https://doi.org/10.1016/j.cell.2007.02.005

Kumar, N., Bharadwaj, C., Sahu, S., Shiv, A., Shrivastava, A. K., Reddy, S. P. P., Soren, K. R., Patil, B. S., Pal, M., Soni, A., Roorkiwal, M., & Varshney, R. K. (2021). Genome-wide identification and functional prediction of salt- stress related long non-coding RNAs (lncRNAs) in chickpea (Cicer arietinum L.). *Physiology and Molecular Biology of Plants*, *27*(11), 2605–2619. https://doi.org/10.1007/s12298-021-01093-0

Kumari, V., Germida, J., & Vujanovic, V. (2018). Legume endosymbionts: Drought stress tolerance in second-generation chickpea (Cicer arietinum) seeds. *Journal of Agronomy and Crop Science*, *204*(6), 529–540. https://doi.org/10.1111/jac.12283

Li, Y., Kumar, S., & Qian, W. (2018). Active DNA demethylation: Mechanism and role in plant development. *Plant Cell Reports,* 37(1), 77–85. Springer Verlag. https://doi.org/10.1007/s00299-017-2215-z

Li, Y., Wan, L., Bi, S., Wan, X., Li, Z., Cao, J., Tong, Z., Xu, H., He, F., & Li, X. (2017). Identification of drought-responsive micrornas from roots and leaves of alfalfa by high-throughput sequencing. *Genes*, *8*(4), 119. https://doi.org/10.3390/genes8040119

Lin, X., Lin, W., Ku, Y. S., Wong, F. L., Li, M. W., Lam, H. M., Ngai, S. M., & Chan, T. F. (2020). Analysis of soybean long non-coding RNAs reveals a subset of small peptide-coding transcripts1[OPEN]. *Plant Physiology*, *182*(3), 1359–1374. https://doi.org/10.1104/PP.19.01324

Liu, X., Li, L., Zhang, B., Zeng, L., & Li, L. (2020). AhHDA1-mediated AhGLK1 promoted chlorophyll synthesis and photosynthesis regulates recovery growth of peanut leaves after water stress. *Plant Science*, *294*, 110461. https://doi.org/10.1016/j.plantsci.2020.110461

Mach, J. (2017). The long-noncoding RNA ELENA1 functions in plant immunity. In *Plant Cell,* 29, 916. American Society of Plant Biologists. https://doi.org/10.1105/tpc.17.00343

Malabarba, J., Chen, Z., Windels, D., & Verdier, J. (2022). Chromatin Immunoprecipitation dataset of H3ac and H3K27me3 histone marks followed by DNA sequencing of Medicago truncatula embryos during control and heat stress conditions to decipher epigenetic regulation of desiccation tolerance acquisition. *Data in Brief, 40*. https://doi.org/10.1016/j.dib.2022.107793

Matzke, M. A., & Mosher, R. A. (2014). RNA-directed DNA methylation: An epigenetic pathway of increasing complexity. *Nature Reviews Genetics, 15*(6), 394–408. Nature Publishing Group. https://doi.org/10.1038/nrg3683

Mefleh, M., Pasqualone, A., Caponio, F., & Faccia, M. (2022). Legumes as basic ingredients in the production of dairy-free cheese alternatives: A review. *Journal of the Science of Food and Agriculture, 102*(1), 8–18. John Wiley. https://doi.org/10.1002/jsfa.11502

Mousavi-Derazmahalleh, M., Bayer, P. E., Hane, J. K., Valliyodan, B., Nguyen, H. T., Nelson, M. N., Erskine, W., Varshney, R. K., Papa, R., & Edwards, D. (2019). Adapting legume crops to climate change using genomic approaches. *Plant Cell and Environment, 42*(1), 6–19). Blackwell. https://doi.org/10.1111/pce.13203

Myrtsi, E. D., Evergetis, E., Koulocheri, S. D., & Haroutounian, S. A. (2023). Bioactivity of wild and cultivated legumes: Phytochemical content and antioxidant properties. *Antioxidants, 12*(4), 852. https://doi.org/10.3390/antiox12040852

Penterman, J., Zilberman, D., Hoe Huh, J., Ballinger, T., Henikoff, S., & Fischer, R. L. (2007). *DNA Demethylation in the Arabidopsis Genome*. Proceedings of the National Academy of Sciences, *104*(16), 6752–6757. www.pnas.org/cgi/content/full/

Rai, K. K., Rai, N., & Rai, S. P. (2018). Salicylic acid and nitric oxide alleviate high temperature induced oxidative damage in Lablab purpureus L plants by regulating bio-physical processes and DNA methylation. *Plant Physiology and Biochemistry, 128*, 72–88. https://doi.org/10.1016/j.plaphy.2018.04.023

Raju, S. K. K., Shao, M. R., Sanchez, R., Xu, Y. Z., Sandhu, A., Graef, G., & Mackenzie, S. (2018). An epigenetic breeding system in soybean for increased yield and stability. *Plant Biotechnology Journal, 16*(11), 1836–1847. https://doi.org/10.1111/pbi.12919

Rakei, A., Maali-Amiri, R., Zeinali, H., & Ranjbar, M. (2016). DNA methylation and physio-biochemical analysis of chickpea in response to cold stress. *Protoplasma, 253*(1), 61–76. https://doi.org/10.1007/s00709-015-0788-3

Ramakrishnan, M., Zhang, Z., Mullasseri, S., Kalendar, R., Ahmad, Z., Sharma, A., Liu, G., Zhou, M., & Wei, Q. (2022). Epigenetic stress memory: A new approach to study cold and heat stress responses in plants. *Frontiers in Plant Science,* 13, 1075279. Frontiers Media S.A. https://doi.org/10.3389/fpls.2022.1075279

Rane, J., Singh, A. K., Kumar, M., Boraiah, K. M., Meena, K. K., Pradhan, A., & Vara Prasad, P. V. (2021). The adaptation and tolerance of major cereals and legumes to important abiotic stresses. *International Journal of Molecular Sciences*, 22(23), 12970. MDPI. https://doi.org/10.3390/ijms222312970

Sano, N., Malabarba, J., Chen, Z., Gaillard, S., Windels, D., & Verdier, J. (2022). Chromatin dynamics associated with seed desiccation tolerance/sensitivity at early germination in Medicago truncatula. *Frontiers in Plant Science, 13*, 1059493. https://doi.org/10.3389/fpls.2022.1059493

Sen, S., Chakraborty, J., Ghosh, P., Basu, D., & Das, S. (2017). Chickpea WRKY70 regulates the expression of a homeodomain-leucine zipper (HD-Zip) i transcription factor CaHDZ12, which confers abiotic stress tolerance in transgenic tobacco and Chickpea. *Plant and Cell Physiology, 58*(11), 1934–1952. https://doi.org/10.1093/pcp/pcx126

Shi, R., Jiao, W., Yun, L., Zhang, Z., Zhang, X., Wang, Q., Li, Y., & Mi, F. (2021). Utilization of transcriptome, small RNA, and degradome sequencing to provide insights nto drought stress and rewatering treatment in Medicago ruthenica. *Frontiers in Plant Science, 12*, 675903. https://doi.org/10.3389/fpls.2021.675903

Singer, B. D. (2019). A practical guide to the measurement and analysis of DNA methylation. *American Journal of Respiratory Cell and Molecular Biology, 61*(4), 417–428. American Thoracic Society. https://doi.org/10.1165/rcmb.2019-0150TR

Singh, N., Jain, P., Ujinwal, M., & Langyan, S. (2022). Escalate protein plates from legumes for sustainable human nutrition. *Frontiers in Nutrition, 9*, 977986. Frontiers Media S.A. https://doi.org/10.3389/fnut.2022.977986

Singh, U., Khemka, N., Rajkumar, M. S., Garg, R., & Jain, M. (2017). PLncPRO for prediction of long non-coding RNAs (lncRNAs) in plants and its application for discovery of abiotic stress-responsive lncRNAs in rice and chickpea. *Nucleic Acids Research, 45*(22), e183. https://doi.org/10.1093/nar/gkx866

Song, J., Rutjens, B., & Dean, C. (2014). Detecting histone modifications in plants. *Methods in Molecular Biology, 1112*, 165–175. https://doi.org/10.1007/978-1-62703-773-0_11

Sullivan, M. L., Hatfield, R. D., Thoma, S. L., & Samac, D. A. (2004). Cloning and characterization of red clover polyphenol oxidase cDNAs and expression of active protein in Escherichia coli and transgenic alfalfa. *Plant Physiology, 136*(2), 3234–3244. https://doi.org/10.1104/pp.104.047449

Sun, L., Song, G., Guo, W., Wang, W., Zhao, H., Gao, T., Lv, Q., Yang, X., Xu, F., Dong, Y., & Pu, L. (2019). Dynamic changes in genome-wide Histone3 Lysine27 trimethylation and gene expression of

Soybean roots in response to salt stress. *Frontiers in Plant Science*, *10*, 1031. https://doi.org/10.3389/fpls.2019.01031

Sun, Z., Wang, Y., Mou, F., Tian, Y., Chen, L., Zhang, S., Jiang, Q., & Li, X. (2016). Genome-wide small RNA analysis of soybean reveals auxin-responsive microRNAs that are differentially expressed in response to salt stress in root apex. *Frontiers in Plant Science*, *6*(JAN2016), 1273. https://doi.org/10.3389/fpls.2015.01273

Tost, J. (Ed.). (2009). *DNA Methylation* (Vol. 507). Humana Press. https://doi.org/10.1007/978-1-59745-522-0

Varotto, S., Tani, E., Abraham, E., Krugman, T., Kapazoglou, A., Melzer, R., Radanović, A., & Miladinoviæ, D. (2020). Epigenetics: Possible applications in climate-smart crop breeding. *Journal of Experimental Botany, 71*(17), 5223–5236. Oxford University Press. https://doi.org/10.1093/jxb/eraa188

Ventouris, Y. E., Tani, E., Avramidou, E. V., Abraham, E. M., Chorianopoulou, S. N., Vlachostergios, D. N., Papadopoulos, G., & Kapazoglou, A. (2020). Recurrent water deficit and epigenetic memory in medicago sativa L. Varieties. *Applied Sciences (Switzerland)*, *10*(9), 3110. https://doi.org/10.3390/app10093110

Virdi, K. S., Laurie, J. D., Xu, Y. Z., Yu, J., Shao, M. R., Sanchez, R., Kundariya, H., Wang, D., Riethoven, J. J. M., Wamboldt, Y., Arrieta-Montiel, M. P., Shedge, V., & Mackenzie, S. A. (2015). Arabidopsis MSH1 mutation alters the epigenome and produces heritable changes in plant growth. *Nature Communications*, *6*, 6386. https://doi.org/10.1038/ncomms7386

Wang, J., Meng, X., Dobrovolskaya, O. B., Orlov, Y. L., & Chen, M. (2017). Non-coding RNAs and their roles in stress response in plants. *Genomics, Proteomics and Bioinformatics, 15*(5), 301–312. Beijing Genomics Institute. https://doi.org/10.1016/j.gpb.2017.01.007

Wang, T. Z., Liu, M., Zhao, M. G., Chen, R., & Zhang, W. H. (2015). Identification and characterization of long non-coding RNAs involved in osmotic and salt stress in Medicago truncatula using genome-wide high-throughput sequencing. *BMC Plant Biology*, *15*(1), 78–86. https://doi.org/10.1186/s12870-015-0530-5

Wu, J., Wang, L., & Wang, S. (2017). MicroRNAs associated with drought response in the pulse crop common bean (Phaseolus vulgaris L.). *Gene*, *628*, 78–86. https://doi.org/10.1016/j.gene.2017.07.038

Xie, M., & Yu, B. (2015). Send orders for reprints to reprints@benthamscience.ae siRNA-Directed DNA methylation in plants. *Current Genomics*, *16*(1), 23–31.

Yaish, M. W., Al-Lawati, A., Al-Harrasi, I., & Patankar, H. V. (2018). Genome-wide DNA Methylation analysis in response to salinity in the model plant caliph medic (Medicago truncatula). *BMC Genomics*, *19*(1), 78. https://doi.org/10.1186/s12864-018-4484-5

Yang, C., Shen, W., Chen, H., Chu, L., Xu, Y., Zhou, X., Liu, C., Chen, C., Zeng, J., Liu, J., Li, Q., Gao, C., Charron, J. B., & Luo, M. (2018a). Characterization and subcellular localization of histone deacetylases and their roles in response to abiotic stresses in soybean. *BMC Plant Biology*, *18*(1), 226. https://doi.org/10.1186/s12870-018-1454-7

Yanni, A. E., Iakovidi, S., Vasilikopoulou, E., & Karathanos, V. T. (2024). Legumes: A vehicle for transition to sustainability. *Nutrients, 16*(1), 98. Multidisciplinary Digital Publishing Institute (MDPI). https://doi.org/10.3390/nu16010098

Yasin, J. K., Mishra, B. K., Pillai, M. A., Verma, N., Wani, S. H., Elansary, H. O., El-Ansary, D. O., Pandey, P. S., & Chinnusamy, V. (2020). Genome wide in-silico miRNA and target network prediction from stress responsive Horsegram (Macrotyloma uniflorum) accessions. *Scientific Reports*, *10*(1), 17203. https://doi.org/10.1038/s41598-020-73140-x

Yong, W. S., Hsu, F. M., & Chen, P. Y. (2016). Profiling genome-wide DNA methylation. *Epigenetics and Chromatin, 9*(1), 26. BioMed Central Ltd. https://doi.org/10.1186/s13072-016-0075-3

Yung, W. S., Huang, C., Li, M. W., & Lam, H. M. (2023). Changes in epigenetic features in legumes under abiotic stresses. *Plant Genome*, *16*(4), e20237. https://doi.org/10.1002/tpg2.20237

Yung, W. S., Wang, Q., Huang, M., Wong, F. L., Liu, A., Ng, M. S., Li, K. P., Sze, C. C., Li, M. W., & Lam, H. M. (2022). Priming-induced alterations in histone modifications modulate transcriptional responses in soybean under salt stress. *Plant Journal*, *109*(6), 1575–1590. https://doi.org/10.1111/tpj.15652

Zemach, A., Kim, M. Y., Hsieh, P. H., Coleman-Derr, D., Eshed-Williams, L., Thao, K., Harmer, S. L., & Zilberman, D. (2013). The arabidopsis nucleosome remodeler DDM1 allows DNA methyltransferases to access H1-containing heterochromatin. *Cell*, *153*(1), 193–205. https://doi.org/10.1016/j.cell.2013.02.033

Zemach, A., Li, Y., Wayburn, B., Ben-Meir, H., Kiss, V., Avivi, Y., Kalchenko, V., Jacobsen, S. E., & Grafi, G. (2005). DDM1 binds Arabidopsis methyl-CpG binding domain proteins and affects their subnuclear localization. *Plant Cell, 17*(5), 1549–1558. https://doi.org/10.1105/tpc.105.031567

Zhan, J., & Meyers, B. C. (2025). Plant small RNAs: Their biogenesis, regulatory roles, and functions. *Annual Review of Plant Biology, 12*, 34. https://doi.org/10.1146/annurev-arplant-070122

Zhang, B., Su, L., Hu, B., & Li, L. (2018). Expression of AhDREB1, an AP2/ERF transcription factor gene from peanut, is affected by histone acetylation and increases abscisic acid sensitivity and tolerance to osmotic stress in Arabidopsis. *International Journal of Molecular Sciences, 19*(5), 1441. https://doi.org/10.3390/ijms19051441

Zhang, H., Lang, Z., & Zhu, J. K. (2018). Dynamics and function of DNA methylation in plants. In *Nature Reviews Molecular Cell Biology, 19*(8), 489–506. Nature Publishing Group. https://doi.org/10.1038/s41580-018-0016-z

Zhang, J., Xu, H., Yang, Y., Zhang, X., Huang, Z., & Zhang, D. (2021). Genome-wide analysis of long non-coding RNAs (lncRNAs) in two contrasting soybean genotypes subjected to phosphate starvation. *BMC Genomics, 22*(1), 433. https://doi.org/10.1186/s12864-021-07750-8

Zhang, W., & Jiang, J. (2018). Application of MNase-seq in the global mapping of nucleosome positioning in plants. In John M. Walker (ed.). *Methods in Molecular Biology,* 1830, 353–366. Humana Press. https://doi.org/10.1007/978-1-4939-8657-6_21

Zhang, W., Wang, N., Yang, J., Guo, H., Liu, Z., Zheng, X., Li, S., & Xiang, F. (2020). The salt-induced transcription factor GmMYB84 confers salinity tolerance in soybean. *Plant Science, 291,* 110326. https://doi.org/10.1016/j.plantsci.2019.110326

Zhao, P., Ma, B., Cai, C., & Xu, J. (2022). Transcriptome and methylome changes in two contrasting mungbean genotypes in response to drought stress. *BMC Genomics, 23*(1), 80. https://doi.org/10.1186/s12864-022-08315-z

Zhu, Y., Liu, Q., Xu, W., Yao, L., Wang, X., Wang, H., Xu, Y., Li, L., Duan, C., Yi, Z., & Lin, C. (2021). Identification of novel drought-responsive miRNA regulatory network of drought stress response in common vetch (Vicia sativa). *Open Life Sciences, 16*(1), 1111–1121. https://doi.org/10.1515/biol-2021-0109

10 Legume Microbiome and Mitigation of Global Climate Change

Shiwani Guleria Sharma, Rahul Khanna, and Nishu Sharma

10.1 INTRODUCTION

Alterations in the climate are an alarming complication that the world is facing. Rising temperatures, changes in rainfall patterns, and increased carbon dioxide levels are affecting ecosystems, food production, and human life. To address this challenge, scientists are exploring natural ways to combat the impacts of fluctuations in climate (Andrews et al., 2017). One promising area of research is the study of the legume microbiome and its role in reducing greenhouse gases and improving soil health.

Legumes are a group of plants that include beans, peas, lentils, and soybeans. These plants are special because they can form symbiotic relationship with bacteria in the soil, particularly a group called rhizobia (Barea et al., 2020). These bacteria live in small structures called nodules on the roots of legumes. Together, they utilize inorganic nitrogen and convert it into absorbable form (Brockwell et al., 2020). This process, called biological nitrogen fixation, reduces the need for chemical fertilizers, which contribute to greenhouse gas emissions. The legume microbiome is more than just rhizobia. It encompasses a range of organisms including bacteria and fungi, that live on and around the plant (Lobell et al., 2023). These microbes help legumes grow better by improving nutrient absorption, protecting them from diseases, and even making them more resistant to drought and heat stress (Delgado-Baquerizo et al., 2018). This means legumes and their microbiomes can play an important role in sustainable agriculture and reducing the effects of climate change. Their contribution is when they grow, in addition to soil conditioning they trap carbon dioxide from the air and retain it (Farooq et al., 2019). This helps lower the amount of carbon dioxide present in the atmosphere. Additionally, planting legumes as part of crop rotations can improve soil health and increase crop yields without relying heavily on synthetic inputs, making farming more eco-friendly (Ahmed et al., 2022).

By understanding and using the legume microbiome, we can develop agricultural practices that are better for the environment. This can help reduce the impact of climate change while ensuring food security for a growing global population. Scientists are studying how to enhance the relationship between legumes and their microbes to make the system even more effective (Geurts & Bisseling, 2022).

This chapter will explore the potential of the legume microbiome in addressing climate change. It will focus on the processes by which these microbes help reduce greenhouse gases, improve soil health, and support sustainable farming practices. The findings from this research could lead to practical solutions for building a more eco- friendly and sustainable agriculture.

DOI: 10.1201/9781003477235-12

10.2 SYNOPSIS OF GLOBAL CLIMATE CHANGE

Global climate instability is one of the foremost concerns that is gathering attention. It signifies a prolonged shift in temperature, weather patterns, and other environmental conditions across the planet (Lobell et al., 2023). These alterations are caused by thedual effect of natural elements, such as volcanic eruptions and solar energy, and human activities, specifically the burning of fossil fuels, deforestation, and industrial processes. Over the past century, human actions have significantly accelerated the rate of climate change, creating problems that affect all living things on Earth.

The leading factor contributing to climate instability is the rise in greenhouse gases in the atmosphere. These gases capture heat from the sun, resulting in a "greenhouse effect" that elevates the planet's temperature. Although the greenhouse effect is a natural phenomenon that maintains the Earth's warmth sufficient to sustain life, human activities have exacerbated it by emitting substantial quantities of these gases (Nelson et al., 2023). Burning coal, oil, and natural gas for energy, cutting down forests, and using certain farming practices are major contributors to this increase (Alfonso et al., 2021).

The impacts of climate change are already being experienced globally. Temperatures are increasing, resulting in the melting of glaciers. This contributes to rising sea levels, which endanger coastal regions and small islands. Extreme weather phenomena, such as hurricanes, heatwaves, droughts, and heavy rainfall, are becoming more common and intense. These alterations disrupt ecosystems, harm wildlife, and complicate food production and access to clean water (Zandalinas et al., 2021).

Climate change also has serious economic and social consequences. Flooding and extreme weather events damage homes, infrastructure, and farmland, leading to financial losses and displacement of communities. Poor and vulnerable populations are most affected because they have fewer resources to cope with these changes. Additionally, shifts in climate patterns can increase the spread of diseases, as warmer temperatures create favorable conditions for certain pests and pathogens (Nelson et al., 2023).

Efforts to address climate change focus on two main strategies: mitigation and adaptation. Abiotic mitigation involves reducing the causes of climate change by limiting greenhouse gas emissions and increasing carbon storage. This includes using sustainable and renewable abiotic sources of energy, planting more trees, and improving energy efficiency. Adaptation involves preparing for and adjusting to the effects of climate alteration. Examples include building flood defenses, developing drought-resistant genetically modified crops, and building advanced infrastructure to combat extreme weather (Bacelar et al., 2024). However, biological mitigation involves modifying the microbial communities involved in agriculture (Verma et al., 2020)

International cooperation is essential to effectively tackle climate change. Pacts like the Paris Agreement strive to cap the global temperature increase at under 2°C compared to pre-industrial levels (Nelson et al., 2023). Countries are encouraged to set targets for reducing emissions and to support sustainable development practices. However, meeting these goals requires significant commitment and collaboration among governments, businesses, and individuals.

At the individual level, people can contribute by adopting environmentally friendly habits. Using energy-efficient appliances, reducing waste, recycling, and conserving water are simple ways to help. Shifting to a plant-based diet, using public transport, and supporting renewable energy projects also make a difference. Educating others about climate change and advocating for policies that protect the environment are important steps toward creating a sustainable future (Nelson et al., 2023).

Climate change is not just an environmental hazard but a financial burden too. Therefore, addressing it requires immediate and sustained action from everyone, including governments, industries, and individuals. By working together and using science, technology, and innovation, we can find solutions to reduce its impacts and build a more sustainable world for future generations. While the challenge is great, the opportunities to create a healthier, more resilient planet are within our reach.

10.3 IMPACT OF WORLD CLIMATE FLUCTUATIONS ON AGRICULTURE

Global climate disruption is having a noteworthy impact on agriculture around the world. Agriculture is closely tied to weather and climate, as crops, livestock, and farming systems depend on stable environmental conditions. Changes in temperature, rainfall patterns, and the frequency of extreme weather events are making it harder for farmers to produce food. This is creating challenges for food security, livelihoods, and the global economy.

10.3.1 Effects of Rising Temperatures

One of the biggest impacts of climate change on agriculture is rising temperatures. The global average temperature is rising at the rate of 0.17 °F per year (Melillo et al., 2017). Many crops grow best within a specific temperature range, and higher temperatures can reduce yields. For example, crops like wheat, rice, and maize (corn) may produce less grain when exposed to heat stress during critical growth stages. (Wheeler & von Braun, 2023). Livestock also suffer in hotter climates, as heat stress can lower milk and meat production and increase the risk of diseases.

It has been experimentally found that, with an increase in temperature, microbial respiration is anticipated to undergo consecutive steps of adaptive changes that unfold in four key stages. Initially, there is a rapid loss of carbon through microbial respiration, triggered by the increased global warmth. Furthermore, this is followed by a reconstruction of the microbial community as it acclimatizes to the new environmental conditions. As a result, this restructuring gives rise to a more varied and less nutrient-demanding microbial community, which demonstrates increased rates of soil respiration in heated plots compared to control sites. Ultimately, there is a decline in more resistant carbon reserves, which is expected to further modify the composition of the microbial community (Romero et al., 2017).

10.3.2 Change in Rainfall Patterns

Changing rainfall patterns are another major challenge. Some regions are experiencing more intense droughts, while others are facing heavier and more unpredictable rainfall. Droughts reduce water availability for crops and livestock, leading to lower yields and increased costs for irrigation. In contrast, excessive rainfall causes flooding, which damages crops, washes away topsoil, and spreads plant diseases. These changes make it difficult for farmers to plan their planting and harvesting schedules (Chowdhury et al., 2019).

As soil moisture levels rise, soil pores become filled with water, creating oxygen-deficit conditions that facilitate processes like methanogenesis and denitrification, leading to credible emissions of methane and nitrous oxide. Variations in moisture can result in differing responses from microbial communities (Sorensen et al., 2016). Furthermore, another notable concern is the impact of saltwater intrusion in coastal areas, driven by rising sea levels, which are currently increasing at an average rate of 3.2 mm/year. This disturbance infuses salt and sulfates into coastal soils, where sulfate acts as a terminal electron acceptor, altering redox dynamics. Consequently, this leads to increased microbial biomass and enhanced mineralization of soil organic carbon, resulting in elevated carbon dioxide production (Moomaw et al., 2018).

10.3.3 Drought Conditions

It has been anticipated that increasing drought conditions will lead to a decline in microbial functions essential for ecosystem sustainability (Cook et al., 2015). As soil moisture diminishes, soil pores become deficient in water, building isolated resource islands (McHugh et al., 2017). This results in reduced decomposition and respiration of soil organic carbon (SOC) into CO_2. Consequently, these changes contribute to a range of responses, including decreased productivity in drier environments and a reduction in carbon loss due to suppressed microbial respiration (Woodcroft et al., 2018).

10.3.4 Elevated CO_2

With the increase in greenhouse gases, because of excessive fuel combustion and industries, there is an increase in atmospheric CO_2. This elevated level of CO_2 (eCO_2) significantly influences soil microbiomes and carbon cycling, as shown by long-term field studies across various ecosystems. A decade-long cross-biome study revealed that different ecosystems respond uniquely to eCO_2, with an increase in Acidobacteria observed in some cases (Dunbar et al., 2012), while Australian grasslands exhibited shifts in archaea, fungi, and specific bacterial groups (Hayden et al., 2012).

Functionally, eCO_2 enhances microbial genes related to decomposition, nitrogen fixation, and carbon metabolism. However, it also triggers a priming effect, where increased carbon inputs from plant roots accelerate the decomposition of existing soil organic matter. The impacts of eCO_2 are further influenced by plant types (such as C3 vs. C4 plants) and environmental factors like water and nutrient availability, all of which are complicated by warming and other climate variables (Jansson et al., 2018).

10.3.5 Increased Agricultural Infections

Another important impact of climate change on agriculture is the spread of pests and diseases. Fluctuations in temperature and rainfall result in a warm environment for parasites, such as insects, to flourish. This increases the risk of crop damage and can lead to higher use of pesticides, which may harm the environment and human health. Similarly, livestock diseases can spread more easily in warmer and wetter climates, threatening animal health and food production (Wheeler & von Braun, 2023)

To reduce the impact of climate change on agriculture, farmers and governments are adopting new strategies (Wheeler & von Braun, 2023). These include developing drought-resistant and heat-tolerant crops, improving irrigation systems, and using techniques like crop rotation and agroforestry to protect soil health. Precision farming, which uses technology to monitor and manage crops more efficiently, is also helping farmers cope with changing conditions. Another innovative solution is introduction of biotic techniques like modification of legume holobiomes which are known to help crops to deal with climate change (Monteoliva et al, 2023)

10.4 LEGUME HOLOBIOME

Plant microbiomes consist of a complicated network of microbial populations that communicate with host plant species or may impact the fitness of the plant by modifying or altering the immediate environment of the soil (Leach et al., 2017). A legume and its microbiome that act as a single unit are called a legume holobiome. It is a remarkable implementation to mitigate the effects of climate change along with the provision of sustainable and nutritious food for the ever-increasing global population. There exists a mutualistic relationship between legume crops with either bacterial or fungal partners, particularly as bilateral engagements. Collectively, their input toward boosting plant growth and functioning is affected by the genotype of the crop, and the interaction between the microbial population and other environmental factors. However, these interactions takes place subsequently (Tsiknia et al., 2021). These interactions with the microbiome endow the plant with several advantages such as enhanced biotic and abiotic stress tolerance, pathogen control, and increased acquisition of nutrients (Monteoliva et al., 2022). Thus, adaptability of plants toward changing environments is a factor of their phenotypic plasticity and ability to recruit, maintain, or alter their microbiome (Uroz et al., 2019). Diverse types of microflora conferring a variety of benefits have been reported to be found in the root zone of leguminous crops. For example, Hartman et al. (2017) reported that the root microbiome of *Trifolium pratense* was predominately inhabited by rhizobia that provide a nutrient supply along with bacterial genera providing protection from diseases.

10.5 TYPES OF LEGUME BACTERIA

Nodules of leguminous plants host two distinct types of bacterial populations: (1) rhizobia, which are primarily responsible for nodule development, and (2) endophytic bacteria, whose roles in the legume root nodule remain largely unclear and make up the nodule microbiome (Figure 10.1). While rhizobia have been extensively studied for many years, research on nodule endophytes has only gained attention in recent years (Velázquez et al., 2017).

10.5.1 Legume-Associated Rhizobacteria

Bacteria can enhance plant growth through various mechanisms, such as endophytic colonization, symbiotic associations, or free-living activities. Rhizobia, a specific group of bacteria, form symbiotic relationships with the roots of legume plants. These bacteria play a crucial role in biological nitrogen fixation (BNF), converting atmospheric nitrogen into a usable form for plants. This process provides a vital macronutrient, nitrogen, which is essential for plant growth. The initiation of BNF involves a complex molecular dialog between the bacterial symbiont and the legume host. The bacteria secrete Nod factors, while the plant releases flavonoids. This reciprocal signaling mechanism allows the partners to recognize each other and initiate nodule formation (Oldroyd, 2013). Once nodules are formed, the rhizobia convert atmospheric nitrogen into nitrogenous compounds, which are then utilized by the plant. In return, the plant provides the bacteria with carbon-rich photosynthates (Laranjo et al., 2014).

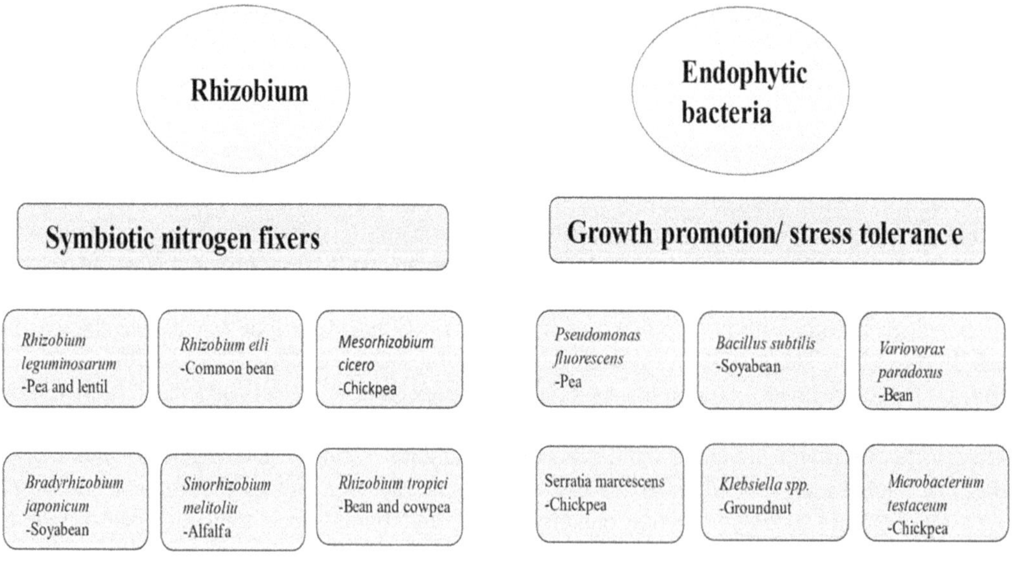

FIGURE 10.1 Rhizobial and endophytic bacteria and their beneficial associations with crops (Hinini et al., 2024).

10.5.2 Diversity of *Rhizobium* Species

Rhizobium species include 11 genera of alpha-proteobacteria and three genera of beta-proteobacteria. Alpha-rhizobia consists of 113 described species distributed among the following genera: *Bradyrhizobium* (15), *Mesorhizobium* (29), *Rhizobium* (43), *Sinorhizobium/Ensifer* (13), *Azorhizobium* (3), *Microvirga* (3), *Ochrobactrum* (2), *Burkholderia* (6), *Cupriavidus* (2), *Herbaspirillum* (1), *Aminobacter (1), Devosia* (1), *Methylobacterium* (1), *Phylobacterium* (1), and *Shinella* (1) (Laranjo et al., 2014)

10.5.3 Legume Nodule Microbiome

Along with rhizobacteria, non-rhizobial endophytes (NREs) also colonize legume root nodules. Recent advancements in sequencing technology have revealed that rhizobia are not the sole inhabitants of these nodules (Muresu et al., 2008). Previous culture-based methods often overlooked non-culturable or morphologically distinct bacteria (Howieson & Dilworth, 2016).

While NREs have not been definitively shown to fix nitrogen within legume nodules, the presence of nifH and narG genes in some isolates suggests potential nitrogen-fixing capabilities (Moore et al., 2020). However, not all NREs possess these genes. For example, *Bacillus* species, which constitute a significant portion of nodule endophytes in wild legumes, lack nitrogen-fixing genes (De Meyer et al., 2015).

NREs offer various benefits to plants. They can enhance plant growth by promoting phosphate solubilization, particularly in nutrient-deficient soils (Muresu et al., 2019). Additionally, NREs can suppress plant pathogens. Some NREs isolated from soybean nodules exhibit antifungal and antibacterial properties (Tokgöz et al., 2020). For instance, *Proteus* spp. have been shown to inhibit the growth of various pathogens both in vitro and in plant growth chamber experiments. As nodules senesce, NREs may be released into the soil, potentially benefiting subsequent crops (Hansen et al., 2020; Müller et al., 2001).

In conclusion, climate change poses serious challenges to agriculture, affecting crops, livestock, and the environment. However, with innovative solutions, better farming practices, and global cooperation, we can work toward a more resilient agricultural system that supports food security and sustainable development for future generations.

10.6 ROLE OF THE LEGUME HOLOBIOME IN MITIGATING GLOBAL CLIMATE CHANGE

Global climate change is adversely and drastically affecting agriculture. With the increasing temperature, uneven rainfall pattern, and global warming leading to changes in biodiversity, there is a decrease in production in the agricultural sector. In response to these challenges, interest is growing in biological-based solutions. Among all these solutions, the legume holobiome that comprises of bacteria and fungi is a revolutionary resolution to the problem (Lopez et al., 2023).

Recent research has demonstrated that microbial communities accompanied by legumes can enhance carbon organic levels of soil, improve soil texture, and increase nutrient availability, which increases their abiotic stress resistance (Zhang et al., 2023).

10.6.1 Enhanced Nitrogen Cycle Efficiency

Legumes are primarily cultivated for their nitrogen-fixing abilities, which reduce the reliance on synthetic fertilizers and consequently lower nitrous oxide (N_2O) emissions, a potent greenhouse gas. Recent studies highlight the significant role of the legume microbiome in enhancing nitrogen fixation efficiency (Mathesius et al., 2021). Nitrogen is a vital nutrient for all organisms on the planet. Its primary entry into biological systems occurs through nitrogen fixation, a process that converts atmospheric nitrogen into absorbable forms like ammonia, nitrate, or nitrite (Kebede et al., 2021). Diverse organisms, including bacteria and archaea, are capable of nitrogen fixation. Key nitrogen-fixing groups encompass photosynthetic cyanobacteria and heterotrophic bacteria from various phyla, such as Proteobacteria, Bacteroidetes, Actinobacteria, Firmicutes, Chlorobi, Chloroflexi, and Aquificae, as well as methanogenic archaea (Boyd & Peters, 2013). Moreover, nitrogen fixation is significantly more efficient in symbiotic relationships compared to free-living bacteria. For instance, marine symbiotic systems can exhibit nitrogen fixation rates up to 420 times higher than terrestrial microbial communities (Foster et al., 2011).

10.6.2 Soil Carbon Sequestration

The legume holobiome in agriculture plays a significant role in long-term carbon retention in soil. Root exudates from legumes, combined with microbial interactions in the rhizosphere, facilitate the development of stable organic matter. This process not only enhances soil health but also contributes to the overall carbon storage capacity of the ecosystem (Amelung et al., 2020). According to research conducted on grain legumes including the common pea, groundnut, soyabean, finger millet, and lentil by Kuyah et al. (2023), it was concluded that cropping systems with legumes increased soil carbon sequestration by up to 50%. These discoveries have potential applications in policies and incentives aimed at smallholder farmers.

10.6.3 Heat and Drought Resistance

Drought and elevated temperature have been noted for significantly diminishing crop productivity and yields, posing a threat to food security. With the ongoing climate changes, it is anticipated that the intensity of these abiotic stresses will increase, potentially jeopardizing global food security (Kumari et al., 2021). Even a slight increase in temperature can be regarded as a stressor for crops, with the effects largely influenced by the intensity of exposure. Moreover, it also hinders metabolic pathways and various physiological processes in plants, including photosynthetic reserves, assimilation of nitrogen, protein breakdown, and the buildup of byproducts from lipid peroxidation (Basu et al., 2022). It has been shown that legume holobiomes like rhizobial and mycorrhizal fungi can enhance water intake and stress tolerance in various types of crops, making them resistant to unfavorable dry conditions. Furthermore, plant growth-promoting rhizobacteria (PGPR) play a significant role in alleviating drought strain in legumes (Khan et al., 2019). Their mechanisms for mitigating drought include phosphorus solubilization, production of organic acids, nitrogen fixation, siderophore synthesis, and the enzyme ACC deaminase (Niu et al., 2018). It has been discovered that hormonal signaling is involved in enhancing water use efficincy, with phytohormones such as gibberellins, auxins, cytokinins, abscisic acid, and ethylene being regulated by these beneficial microbes (Sehgal et al., 2018).

10.6.4 Reduction of Greenhouse Gases

Agriculture is accountable for a nearly 10% hike in greenhouse gases according to the European Union (FAOSTAT, 2020). These gases, like methane and carbon dioxide, are known to trap heat in the Earth's atmosphere, and their increased concentration is causing global temperatures to rise. It has been confirmed that legumes contribute to a reduction in the release of these gases in rice paddies and flooded agriculture systems (Ruisi et al., 2017). A possible reason for this is the role of legume microbial communities in promoting methanotrophy (Sans-Cobena et al., 2017). Crop growing using legume systems is mushrooming as a sustainable alternative to harsh chemical fertilizers. Research conducted by Oliveira et al. (2021) determined that incorporating legumes in crops reduced nitrous oxide and methane gas in a wet year in various varieties of pea crop, which would have plummetted if synthetic fertilizers had been used.

10.7 CONCLUSION

Climate change casts a long shadow over global agriculture, jeopardizing crop yields, livestock health, and ecosystem balance. The escalating temperatures, erratic rainfall, and frequent extreme weather events pose a significant threat to food security, particularly in regions already grappling with vulnerability. However, innovative solutions, such as harnessing the power of the legume holobiome, offer a glimmer of hope in mitigating these adverse impacts. Furthermore, the legume holobiome, a complex network of symbiotic relationships between legumes and their associated microorganisms, including rhizobia, plays a fundamental role in enhancing soil fertility, boosting nitrogen fixation efficiency, and promoting carbon sequestration. These microbial interactions empower legumes to withstand abiotic stresses like drought and heat, thereby enhancing crop resilience. Moreover, legumes contribute to reducing greenhouse gas emissions, making them a valuable asset in sustainable agricultural systems. Therefore, integrating legume-based cropping systems into global agricultural practices presents a multi-faceted approach to combatting the climate crisis, simultaneously promoting food security and environmental sustainability. As research delves deeper into the untapped potential of legume microbiomes, it becomes evident that these natural solutions will be instrumental in developing resilient and sustainable farming systems in the face of climate change.

REFERENCES

Ahemad, M., and Kibret, M. (2022). Mechanisms and applications of plant growth-promoting rhizobacteria: Current perspective. *J. King Saud Univ. Sci.* 26(1), 1–20. https://doi.org/10.1016/j.jksus.2013.05.001

Alfonso, S., Gesto, M., and Sadoul, B. (2021). Temperature increase and its effects on fish stress physiology in the context of global warming. *J. Fish Biol.* 98(6), 1496–1508. https://doi.org/10.1111/jfb.14674

Amelung, W., Bossio, D., de Vries, W., Kögel-Knabner, I., Lehmann, J., Amundson, R., et al. (2020). Towards a global-scale soil climate mitigation strategy. *Nat. Commun.* 11, 1–10. https://doi.org/10.1038/s41467-020-18887-7

Andrews, M., and Andrews, M. E. (2017). Specificity in legume-rhizobia symbioses. *Int. J. Mol. Sci.* 18(4), 705. https://doi.org/10.3390/ijms18040705

Bacelar, E., Pinto, T., Anjos, R., Morais, M. C., Oliveira, I., Vilela, A., and Cosme, F. (2024). Impacts of climate change and mitigation strategies for some abiotic and biotic constraints influencing fruit growth and quality. *Plants* 13(14), 1942. https://doi.org/10.3390/plants13141942

Barea, J. M., Pozo, M. J., Azcón, R., and Azcón-Aguilar, C. (2020). Microbial co-operation in the rhizosphere. *J. Exp. Bot.* 56(417), 1761–1778. https://doi.org/10.1093/jxb/eri197

Basu, P. S., Chaturvedi, S. K., Gaur, P. M., Mondal, B., Meena, S. K., Das, K., et al. (2022). Physiological mechanisms of tolerance to drought and heat in major pulses for improving yield under stress environments. In Josphert N. Kimatu (ed.). *Advances in Plant Defense Mechanisms*. IntechOpen. https://doi.org/10.5772/intechopen.100000

Boyd, E. S., and Peters, J. W. (2013). New insights into the evolutionary history of biological nitrogen fixation. *Front. Microbiol.* 4, 201. https://doi.org/10.3389/fmicb.2013.00201

Brockwell, J., and Bottomley, P. J. (2020). Recent advances in inoculant technology and application. In Evans, H. J., Bottomley, P. J., and Newton, W. E. (Eds.), *Nitrogen Fixation: Fundamentals and Applications* (pp. 211–223). Springer. https://doi.org/10.1007/978-94-011-0051-9_22

Cook, B. I., Ault, T. R., and Smerdon, J. E. (2015). Unprecedented 21st century drought risk in the American Southwest and Central Plains. *Sci. Adv.* 1, e1400082. https://doi.org/10.1126/sciadv.1400082

De Meyer, S. E., De Beuf, K., Vekeman, B., and Willems, A. (2015). A large diversity of non-rhizobial endophytes found in legume root nodules in Flanders (Belgium). *Soil Biol. Biochem.* 83, 1–11. https://doi.org/10.1016/j.soilbio.2015.01.002

Delgado-Baquerizo, M., Oliverio, A. M., Brewer, T. E., Benavent-González, A., Eldridge, D. J., Bardgett, R. D., et al. (2018). A global atlas of the dominant bacteria found in soil. *Science* 359(6373), 320–325. https://doi.org/10.1126/science.aap9516

Dunbar, J., et al. (2012). Common bacterial responses in six ecosystems exposed to 10 years of elevated atmospheric carbon dioxide. *Environ. Microbiol.* 14, 1145–1158. https://doi.org/10.1111/j.1462-2920.2012.02735.x

FAOSTAT. (2020). Statistical Database. Food and Agriculture Organization.

Farooq, M., Wahid, A., Lee, D. J., Ito, O., and Siddique, K. H. M. (2019). Advances in drought resistance of cereals. *Crit. Rev. Plant Sci.* 28(4), 199–217. https://doi.org/10.1080/07352680903024283

Foster, R. A., Kuypers, M. M. M., Vagner, T., Paerl, R. W., Musat, N., and Zehr, J. P. (2011). Nitrogen fixation and transfer in open ocean diatom–cyanobacterial symbioses. *ISME J.* 5, 1484–1493. https://doi.org/10.1038/ismej.2011.26

Geurts, R., and Bisseling, T. (2022). Rhizobium nod factor perception and signalling. *Plant Cell* 14(Suppl 1), S239–S249. https://doi.org/10.1105/tpc.000554

Hansen, B. L., Pessotti, R. D. C., Fischer, M. S., Collins, A., El-Hifnawi, L., Liu, M. D., et al. (2020). Cooperation, competition, and specialized metabolism in a simplified root nodule microbiome. *MBio* 11, e01917-20. https://doi.org/10.1128/mBio.01917-20

Hartman, K., van der Heijden, M. G. A., Roussely-Provent, V., and Schlaeppi, J. C. W. K. (2017). Deciphering composition and function of the root microbiome of a legume plant. *Microbiome* 5, 2. https://doi.org/10.1186/s40168-016-0220-z

Hayden, H. L., Mele, P. M., Bougoure, D. S., Allan, C. Y., Norng, S., Piceno, Y. M., Brodie, E. L., DeSantis, T. Z., Andersen, G. L., Williams, A. L., and Hovenden, M. J. (2012). Changes in the microbial community structure of bacteria, archaea, and fungi in response to elevated CO2 and warming in an Australian native grassland soil. *Environ. Microbiol.* 14, 3081–3096. https://doi.org/10.1111/j.1462-2920.2012.02824.x

Hinini, M., and Aurag, J. (2024). Prevalence, diversity and applications potential of nodules endophytic bacteria: A systematic review. *Front. Microbiol.* 15, 1386742.

Howieson, J. G., and Dilworth, M. J. (2016). *Working with Rhizobia.* Australian Centre for International Agricultural Research. Available at: http://aciar.gov.au/files/aciar_mn_173_web-updated_31_may_2016.pdf

Jansson, J. K., and Hofmockel, K. S. (2018). The soil microbiome—from metagenomics to metaphenomics. *Curr. Opin. Microbiol.* 43, 162–168. https://doi.org/10.1016/j.mib.2018.01.013

Kebede, E. (2021). Contribution, utilization, and improvement of legumes-driven biological nitrogen fixation in agricultural systems. *Front. Sustain. Food Syst.* 5, 767998. https://doi.org/10.3389/fsufs.2021.767998

Khan, N., Bano, A., and Babar, M. A. (2019). Metabolic and physiological changes induced by plant growth regulators and plant growth-promoting rhizobacteria and their impact on drought tolerance in *Cicer arietinum* L. *PLoS ONE* 14, e0213559. https://doi.org/10.1371/journal.pone.0213559

Kumari, V. V., Roy, A., Vijayan, R., Banerjee, P., Verma, V. C., Nalia, A., Pramanik, M., Mukherjee, B., Ghosh, A., Reja, M. H., Chandran, M. A. S., Nath, R., Skalicky, M., Brestic, M., & Hossain, A. (2021). Drought and Heat Stress in Cool-Season Food Legumes in Sub-Tropical Regions: Consequences, Adaptation, and Mitigation Strategies. *Plants*, *10*(6), 1038. https://doi.org/10.3390/plants10061038

Kuyah, S., Muoni, T., Bayala, J., Chopin, P., Dahlin, AS, Hughes, K., Jonsson, M., Kumar, S., Sileshi, GW, Dimobe, K., & Öborn, I. (2023). Grain legumes and dryland cereals contribute to carbon sequestration in the drylands of Africa and South Asia. *Agriculture, Ecosystems & Environment*, 355, 1–11.

Laranjo, M., Alexandre, A., and Oliveira, S. (2014). Legume growth-promoting rhizobia: An overview on the *Mesorhizobium* genus. *Microbiological Res.* 169, 2–17. https://doi.org/10.1016/j.micres.2013.09.012

Leach, J. E., Triplett, L. R., Argueso, C. T., and Trivedi, P. (2017). Communication in the phytobiome. *Cell* 169, 587–596. https://doi.org/10.1016/j.cell.2017.04.025

Lobell, D. B., Schlenker, W., and Costa-Roberts, J. (2023). Climate trends and global crop production since 1980. *Science* 333(6042), 616–620. https://doi.org/10.1126/science.1204531

López, J. L., Fourie, A., Poppeliers, S. W. M., Pappas, N., Sánchez-Gil, J. J., de Jonge, R., and Dutilh, B. E. (2023). Growth rate is a dominant factor predicting the rhizosphere effect. *ISME J.* 17, 1396–1405. https://doi.org/10.1038/s41396-022-01240-5

Mathesius, U., Jin, J., Li, Y., and Watt, M. (2021). Understanding plant-root interactions with rhizobacteria to improve biological nitrogen fixation in crops. In pp. 163–194. 10.19103/AS.2020.0075.09.

McHugh, T. A., et al. (2017). Climate controls prokaryotic community composition in desert soils of the southwestern United States. *FEMS Microbiol. Ecol.* 93, fix116. https://doi.org/10.1093/femsec/fix116

Melillo, J. M., et al. (2017). Long-term pattern and magnitude of soil carbon feedback to the climate system in a warming world. *Science* 358, 101–104. https://doi.org/10.1126/science.aan2874

Monteoliva, M., Valetti, L., Taurian, T., Crociara, C. S., and Guzzo, M. C. (2022). Synthetic communities of bacterial endophytes to improve the quality and yield of legume crops. *IntechOpen* 1, 1–36. https://doi.org/10.5772/intechopen.102519

Moomaw, W. R., et al. (2018). Wetlands in a changing climate: Science, policy and management. *Wetlands* 38, 183–205. https://doi.org/10.1007/s13157-018-1023-8

Monteoliva M. I., Ruiz O. A., and Li F. (2023). Editorial: Legumes and their microbiome in climate change mitigation. *Front Plant Sci.* 2023 Jun 12;14, 1220535. doi: 10.3389/fpls.2023.1220535. PMID: 37377800; PMCID: PMC10291733.

Moore, V., Davis, B., Poskaitis, M., Maul, J. E., Kucek, L. K., and Mirsky, S. (2020). Phenotypic and nodule microbial diversity among crimson clover (*Trifolium incarnatum* L.) accessions. *Agronomy* 10, 1434. https://doi.org/10.3390/agronomy10091434

Müller, J., Wiemken, A., and Boller, T. (2001). Redifferentiation of bacteria isolated from *Lotus japonicus* root nodules colonized by *Rhizobium sp.* NGR234. *J. Exp. Bot.* 52, 2181–2186. https://doi.org/10.1093/jexbot/52.364.2181

Muresu, R., Polone, E., Sulas, L., Baldan, B., Tondello, A., Delogu, G., et al. (2008). Coexistence of predominantly nonculturable rhizobia with diverse, endophytic bacterial taxa within nodules of wild legumes. *FEMS Microbiol. Ecol.* 63, 383–400. https://doi.org/10.1111/j.1574-6941.2007.00424.x

Muresu, R., Porceddu, A., Sulas, L., and Squartini, A. (2019). Nodule-associated microbiome diversity in wild populations of *Sulla coronaria* reveals clues on the relative importance of culturable rhizobial symbionts and coinfecting endophytes. *Microbiol. Res.* 221, 10–14. https://doi.org/10.1016/j.micres.2019.01.004

Nelson, G. C., Rosegrant, M. W., Koo, J., Robertson, R., Sulser, T., Zhu, T., et al. (2023). *The Costs of Agricultural Adaptation to Climate Change.* International Food Policy Research Institute (IFPRI). Available at: www.ifpri.org

Niu, X., Song, L., Xiao, Y., and Ge, W. (2018). Drought-tolerant plant growth-promoting rhizobacteria associated with foxtail millet in a semi-arid agro-ecosystem and their potential in alleviating drought stress. *Front. Microbiol.* 8, 2580. https://doi.org/10.3389/fmicb.2017.02580

Oldroyd G. E. (2013). Speak, friend, and enter: signalling systems that promote beneficial symbiotic associations in plants. *Nat. Rev. Microbiol.* 11, 252–264. 10.1038/nrmicro2990

Oliveira, M., Castro, C., Coutinho, J., and Trindade, H. (2021). Grain legume-based cropping systems can mitigate greenhouse gas emissions from cereal under Mediterranean conditions. *Agric. Ecosyst. Environ.* 313, 107406. https://doi.org/10.1016/j.agee.2020.107406

Romero-Olivares, A. L., Allison, S. D., and Treseder, K. K. (2017). Soil microbes and their response to experimental warming over time: A meta-analysis of field studies. *Soil Biol. Biochem.* 107, 32–40. https://doi.org/10.1016/j.soilbio.2016.12.026

Roy Chowdhury, T., Lee, J. Y., Bottos, E. M., Brislawn, C. J., White III, R. A., et al. (2019). Metaphenomic responses of a native prairie soil microbiome to moisture perturbations. *mSystems* 4, e00061–19. https://doi.org/10.1128/mSystems.00061-19

Ruisi, P., Amato, G., Badagliacca, G., Frenda, A. S., Giambalvo, D., and Di Miceli, G. (2017). Agro-ecological benefits of faba bean for rainfed Mediterranean cropping. *Ital. J. Agron.* 12, 233–245. https://doi.org/10.4081/ija.2017.865

Sanz-Cobena, A., Lassaletta, L., Aguilera, E., Prado, A., Garnier, J., and Billen, G. (2017). Strategies for greenhouse gas emissions mitigation in Mediterranean agriculture: A review. *Agric. Ecosyst. Environ.* 238, 5–24. https://doi.org/10.1016/j.agee.2016.09.038

Sehgal, A., Sita, K., Siddique, K. H. M., Kumar, R., Bhogireddy, S., Varshney, R. K., et al. (2018). Drought or/and heat-stress effects on seed filling in food crops: Impacts on functional biochemistry, seed yields, and nutritional quality. *Front. Plant Sci.* 9, 1705. https://doi.org/10.3389/fpls.2018.01705

Sorensen, P. O., Templer, P. H., and Finzi, A. C. (2016). Contrasting effects of winter snowpack and soil frost on growing season microbial biomass and enzyme activity in two mixed-hardwood forests. *Biogeochemistry* 128, 141–154. https://doi.org/10.1007/s10533-016-0203-8

Tokgöz, S., Lakshman, D. K., Ghozlan, M. H., Pinar, H., Roberts, D. P., and Mitra, A. (2020). Soybean nodule-associated non-rhizobial bacteria inhibit plant pathogens and induce growth promotion in tomato. *Plants* 9, 1494. https://doi.org/10.3390/plants9111494

Tsiknia, M., Tsikou, D., Papadopoulou, K. K., and Ehaliotis, C. (2021). Multi-species relationships in legume roots: From pairwise legume-symbiont interactions to the plant–microbiome–soil continuum. *FEMS Microbiol. Ecol.* 97, fiaa222. https://doi.org/10.1093/femsec/fiaa222

Uroz, S., Courty, P. E., and Oger, P. (2019). Plant symbionts are engineers of the plant-associated microbiome. *Trends Plant Sci.* 24, 905–916. https://doi.org/10.1016/j.tplants.2019.06.008

Velázquez, E., Carro, L., Flores-Félix, J. D., Martínez-Hidalgo, P., Menéndez, E., Ramírez-Bahena, M. H., et al. (2017). The legume nodule microbiome: A source of plant growth-promoting bacteria. In: Kumar, V., Kumar, M., Sharma, S., and Prasad, R. (Eds.) *Probiotics and Plant Health.* Springer. https://doi.org/10.1007/978-981-10-3473-2_3

Verma, K. K., Song, X. P., Li, D. M., Singh, M., Rajput, V. D., Malviya, M. K., et al. (2020). Interactive role of silicon and plant–rhizobacteria mitigating abiotic stresses: A new approach for sustainable agriculture and climate change. *Plants* 9(9), 1055. https://doi.org/10.3390/plants9091055

Wheeler, T., and von Braun, J. (2023). Climate change impacts on global food security. *Science* 341(6145), 508–513. https://doi.org/10.1126/science.1239402

Woodcroft, B. J., Singleton, C. M., Boyd, J. A., Evans, P. N., Emerson, J. B., Zayed, A. A., et al. (2018). Genome-centric view of carbon processing in thawing permafrost. *Nature* 560, 49–54. https://doi.org/10.1038/s41586-018-0338-1

Zandalinas, S. I., Fritschi, F. B., and Mittler, R. (2021). Global warming, climate change, and environmental pollution: Recipe for a multifactorial stress combination disaster. *Trends Plant Sci.* 26(6), 588–599. https://doi.org/10.1016/j.tplants.2021.02.011

Zhang, C.; Zhao, X.; Liang, A.J.; Li, Y.Y.; Li, X.Y.; Li, D.P.; Hou, N. (2023). Insight into the soil aggregate-mediated restoration mechanism of degraded black soil via biochar addition: Emphasizing the driving role of core microbial communities and nutrient cycling. *Environ. Res.* 2023, 228, 115895.

11 Role of Legume miRNAs in Stress Adaptations

Small Size but Big Player

Indra Adhikari, Ashmita Khanal*, Avinash Shrestha*, Bandana Osti, Sopnil Jahin, and Pranav Dawar#*

11.1 INTRODUCTION

Legumes (Leguminosae or Fabaceae) are composed of over 19,000 species and are the second most economically important crop group after grains worldwide (Polhill et al., 1981). They serve as a vital source of dietary proteins, oil, minerals, and vitamins for humans. Additionally, legumes are crucial for forage and rotational cropping practices due to their exceptional capacity to fix nitrogen in the soil (Bessada et al., 2019; Calles et al., 2019). Understanding the gene structure and function in legumes is essential for preserving their unique characteristics and unlocking their full potential. Traditional breeding methods, combined with advanced biochemical and molecular techniques, have proven effective in uncovering the roles of protein-coding genes that respond to biotic and abiotic stresses and in facilitating legume developmental processes (Koornneef et al., 1997). With the rapid advancement of functional genomics, it has become possible to investigate the non-coding genes (ncRNAs), and their involvement in legume adaptation and development (Song et al., 2019). Different types of ncRNAs, including small-RNAs (sRNAs) (like microRNA [miRNA], small interfering RNA [siRNA]), Piwi-interacting RNA (piRNA), long non-coding RNA (lncRNA), as well as circular RNA (circRNA), play a major role in gene regulation in both plants and animals (Dawar et al., 2024). Additionally, the regulatory and biogenesis pathways involving such ncRNAs have been extensively studied in plants and animals, often revealing high conservation within the same family and, in some cases, across entire kingdoms, as seen with microRNAs (Dawar et al., 2024). Efforts have been made to elucidate the function of these ncRNAs in legume growth and development and stress adaptability especially biased toward sRNAs ranging from 21 to 24 nucleotides in length (reviewed in Chand Jha et al., 2021). Some miRNAs have been extensively studied for their function in regulating development and stress adaptation in legumes: miR156 (Wang et al., 2015; Feyissa et al., 2019; Min et al., 2022), miR166 (Li et al., 2017), miR167 (Liu et al., 2021), miR169 (Yu et al., 2019; Xu et al., 2021), miR171 (Lauressergues et al., 2012; Hofferek et al., 2014; Ma et al., 2014; Devers, 2019; Pei et al., 2023), miR172 (Yan et al., 2013; Holt et al., 2015; Lian et al., 2021; Ucar et al., 2024), miR390 (Xia et al., 2013; Xia et al., 2017; Wen et al., 2020), and miR393 (Yan et al., 2015; Jiang et al., 2022). Beyond their role in adaptation, miRNAs also regulate key developmental processes in legumes, including germination (Das et al., 2019; DeBoer et al., 2019), nodule development (Hoang et al., 2020; Tiwari et al., 2021), and flower and seed development (Bohra et al., 2021). The overarching regulatory mechanism of miRNA, including the direct and indirect silencing of their targets in plants via gene knockdown at post-transcriptional (Vaucheret et al., 2001) and translational (Beauclair et al., 2010; Li et al., 2013) levels, provide valuable insights into how legumes navigate the complex interactions involving symbiosis and pathogenesis as well as how miRNA regulates these processes. Legumes, widely cultivated in hot and cold temperate

DOI: 10.1201/9781003477235-13

regions, face various abiotic stresses, including heat, drought, salinity, and nutrient deficiencies, as well as biotic stresses from pests and pathogens (Singh et al., 2007). miRNAs play a vital role in facilitating adaptation during these challenging growth periods (Mantri et al., 2013). In this chapter, we focus on the nitty-gritty of miRNA regulation of stress adaptation and nodule development.

11.2 MIRNAS AND ABIOTIC STRESS RESPONSE IN LEGUMES

11.2.1 Role of miRNAs in Plant Adaptations to Drought Stress

In many regions of the world, drought is regarded as the most harmful environmental situation since it lowers agricultural productivity more than any other stress (Lambers et al., 2008). Stages of cell division, elongation, and differentiation are included in growth, which is characterized as an irreversible increase in volume, size, and weight. Drought inhibits cell reproduction and development via lowering turgor pressure, enzyme activity, and the quantity of energy available to the cells (Taiz & Zeiger, 2010). Plants like legumes have developed several coping strategies to survive drought stress; morphological, biochemical, and physiological changes (Chaves et al., 2003; Farooq et al., 2009; Nadeem et al., 2019). Numerous recent studies have demonstrated the significance of small RNAs, such as small-interfering RNAs (siRNAs), and micro-RNAs (miRNAs), during the post-transcriptional stage of gene regulation. On one hand, plants suffer due to the downregulation of transcripts targeted by drought-induced miRNAs during stress . On the other hand, target transcripts that function as the positive regulators of stress adaptation accumulate because of drought-mediated repression of respective miRNAs (Nadeem et al., 2019).

Among various first-generation miRNAs, miR156 has received the most attention as a molecular marker for improving agronomic traits in legumes. Members of the *SPL* gene family are subjected to post-transcriptional regulation by miR156. Various studies have proved that overexpression of miR156 followed by the suppression of *SPL* family members resulted in increased resistance to various abiotic stresses including heat, drought, and salt stress in alfalfa (Arshad et al., 2017; Arshad et al., 2017; Feyissa et al., 2019; Matthews et al., 2019). Not limited to alfalfa, many different legume studies have provided evidence of the miR156-*SPL* regulation module in drought stress (Barrera-Figueroa et al., 2011; Nageshbabu et al., 2013; Bhat et al., 2020; Dasmandal et al., 2020; Singh et al., 2021; Lu et al., 2023; Mittal et al., 2023). Upregulation of miR156 results in a prolonged juvenile state, providing a flight strategy for the stressed plant to bloom once the stressful conditions are over. Furthermore, the miR156-*SPL* module has been shown to interact with *DFR*, resulting in increased anthocyanin biosynthesis and enhanced ROS scavenging activity, which can be interpreted as a fight mechanism used to induce stress tolerance in *A. thaliana* (Cui et al., 2014). Similar evidence of miR156/*SPL-DFR* module-related drought stress adaptation in alfalfa has also been observed (Feyissa et al., 2019; Puri et al., 2023). miR156-mediated alfalfa's ability to withstand drought has been linked to its ability to regulate stomatal behavior, thus it can be linked to miR156's ability to regulate photosynthesis (Feyissa et al., 2019). miR156-mediated accumulation of proteins involved in secondary metabolite and antioxidant production has also been demonstrated in alfalfa in response to drought stress (Puri et al., 2023). As a result, miR156 overexpression lines were found to be able to withstand drought stress and perform essential photosynthesis and carbohydrate metabolism activities more efficiently than control plants when exposed to drought stress (Puri et al., 2023).

The miR169-*NF-Y* module is also an important player in regulating drought stress (Li et al., 2008) and in fact was reported as one of the most important miRNAs in regulating drought stress (Asefpour Vakilian, 2020). *NF-Y* regulates drought stress through both abscisic acid (ABA)-dependent and independent pathways. It increases photosynthetic efficiency along with reducing active hydrogen peroxide and increasing antioxidant enzyme activities (Zhang et al., 2023). *NF-Y*-mediated negative regulation of drought tolerance has also been reported in *B. napus*. *NF-Y* homolog, *BnaA9.NF-YA7*, decreases drought resistance through increased stomatal conductivity and HISTONE-LYSINE N-METHYLTRANSFERASE *Bna.ASHH4s*-mediated reduced transcription of *Bna.ABF3/4s* (Wang et al., 2024). Similar mixed regulations of drought stress through miR169 have been reported in legumes. In *M. truncatula*, *M. ruthentica*, *L. sativus*, and *A. hypogea* under drought stress, miR169 downregulation has been reported (Trindade et al., 2010; Wang et al., 2011; Bhat et al., 2020; Shi et al., 2021; Mittal et al., 2023). Overexpression of miR169c from soybean has also conferred increased drought sensitivity in *Arabidopsis* (Yu et al., 2019). In contrast, upregulation of miR169 was found in *P. vulgaris* and *C. cajan* under drought stress (Arenas-Huertero et al., 2009; Buch et al., 2020). Additionally, the miR171/*SCL* module is also a major director of drought resistance in many different legumes by regulating the physiological characteristics such as secondary branch production and leaf morphology (Barrera-Figueroa et al., 2011; Bhat et al., 2020; Shi et al., 2021; Singh et al., 2021; Mittal et al., 2023). It is not limited to *SCL*, but miR171 has been found to enhance drought resistance by regulating the flavonoid biosynthesis gene in rice (*O. sativa*) (Um et al., 2022). This finding provides a new horizon for researchers to look at in legumes, as secondary metabolites are key in scavenging ROS during stress period.

The restriction in horizontal proliferation of roots in response to drought stress is primarily mediated through ABA and auxin regulation (Chen et al., 2012; Jiang et al., 2022; Verma et al., 2016). miR393 targets auxin receptors, like *TIR1* and *AFB2*, and has been shown to regulate lateral root development in *A. thaliana* as well as various other drought-stressed crops. Under drought stress conditions, the increased expression of miR393 inhibits these auxin receptors, resulting in reduced lateral root proliferation. This decrease in lateral roots represents an adaptive response that aids in conserving resources, thereby enhancing drought tolerance by redirecting energy toward primary root growth to access deeper water sources (Chen et al., 2012; Si-Ammour et al., 2011; Jiang et al., 2022). In-line with the work in *Arabidopsis*, miR393f and miR393k were found to be upregulated in *P. vulgaris* under drought stress (Wu et al., 2017). Moreover, a weak induction of miR393 was also reported in drought-stressed peanut and soybean roots (Mittal et al., 2023; Xing et al., 2023). On the contrary, no differential expression in miR393 during drought stress in *M. trunculata* was reported (Li et al., 2017), similar to the results presented by a research group studying the effects of drought induced by nitric oxide application on *M. sativa* (Zhao et al., 2019; Ruan et al., 2024). This diverse miR393-mediated drought response among members of the same family may suggest additional genetic and epigenetic variations are involved in the regulation of miR393 expression in response to various environmental cues. Additionally, discrepancies in drought stress application, such as short-term versus long-term drought as well as the intensity, might trigger different miRNA responses. Other first-generation miRNAs, like miR398 and miR408, were also found to be upregulated in water-stressed *M. trunculata*, along with their respective targets *CSD* and *PCY* being downregulated in both roots and shoots. Copper homeostasis by miR398 and miR408 occurs through the inhibition of genes that encode copper-binding proteins, such as *CSD*, *PCY*, and *LACCASES*. Downregulation of these target genes facilitates copper redistribution and ROS balance maintenance under drought stress, which not only protects cells from oxidative damage but also enhances drought tolerance. Based on the sequence complementarity, miR398 has also been predicted to target genes like *COX5b* and *Histone Deacetylase HDT3*, which are both involved in stress responses. By suppressing *COX5b*, miR398 alters mitochondrial respiration, which leads to reduced energy expenditure and minimized ROS production during drought conditions. Moreover, the repression of *HDT3* results in chromatin remodeling which facilitates adaptive changes in downstream gene expression that improve drought tolerance (Trindade et al., 2010; Guan et al., 2013; De la Rosa et al., 2019). These results pointed toward the significance of copper homeostasis during drought conditions. The miR169-*NF-YA* module may indirectly influence copper homeostasis by modulating the expression of miR398 and miR408 under drought stress. Functional conservation of miR398 module has also been established in other members of legumes. For example, in *C. cajan*, under drought stress, the researchers have found convincing evidence of miR398a and miR408 upregulation (Buch et al., 2020). Similar results were observed in the case of 17 days of drought-stressed chickpea, where they found miR408 overexpression had improved drought tolerance and the plants also showed a lower expression level of *PCY* (Hajyzadeh et al., 2015). Mishra et al. (2022) also reported enhanced drought tolerance with miR408 upregulation in chickpea (Mishra et al., 2022). However, in *P. sativum* both miR398 and miR408 were found to be downregulated in response to drought stress. Additionally, only one of the members of the miR398 family, miR398c, has been reported to be downregulated in the case of drought-stressed *G. max* (soybean) and *A. hypogea* (peanut) (Zhou et al., 2020; Mittal et al., 2023). Interestingly, research groups have found bicistronic transcripts containing both miR398 and miR2119 in common beans (De la Rosa et al., 2019). miR398, along with a second-generation miRNA, miR2119, co-regulate the drought stress response by targeting *CSD1* and *ADH1*, exhibiting non-concordant expression patterns (De la Rosa et al., 2019).

Apart from miR2119, many studies have also shown the importance of second-generation miRNAs in the drought stress adaptation process. Notably, miR1514a targets Phvul.010g120700, triggering the production of phased secondary small interfering RNAs (phasiRNAs) to regulate members of the *NAC* (*NAM*, *ATAF*, and *CUC*) transcription factor family in *P. vulgaris* (Sosa-Valencia et al., 2017). NAC transcription factors are critical regulators of drought-responsive gene networks involved in osmotic stress tolerance, ROS detoxification, and stomatal behavior under drought conditions. miR1514a-mediated modulation of *NAC* TFs via phasiRNAs may enhance plant drought tolerance (Sosa-Valencia et al., 2017).

11.2.2 Role of miRNAs in Plant Adaptations to Heat Stress

Climate change is already wreaking havoc around the globe. It is projected that both natural and artificial influences will cause the Earth's temperature to increase by an average of 2–3 °C by the end of the 21st century (Eitzinger et al., 2010). According to Parker et al. (2020), heatwaves and sporadic rises in temperature have a variety of detrimental consequences on human civilization and the environment, with the potential to have a major influence on future agricultural demand. One such effect is the worldwide production of crops, which could be considerably damaged (Zampieri et al., 2017). Heat stress inhibits the growth of both roots and shoots, reducing the overall development of plants primarily by impairing nutrient uptake. This decline is mainly driven by reduced formation of roots and consequent low nutrient uptake of proteins, leading to diminished root mass and lower shoot nutritional content (Mishra et al., 2023). Several heat shock transcription factor (HSF) and heat shock protein (HSP) genes are stimulated to express when plants are exposed to heat stress (HS) which can reduce the yield of legumes (Zhang et al., 2013). HSFs quickly cause HSPs to express which act as chaperones that ensure proper protein folding, prevent protein aggregation, and assist protein turnover. Both HSFs and HSPs thus protect cells from heat stress and are essential for the HS response in plants and the development of thermotolerance (Ohama et al., 2017; Ren et al., 2019). However, thermotolerance is not significantly affected by overexpression of a single HSF or HSP gene, indicating that a complex network of HSFs, HSPs, and other regulatory elements work together to provide HS resistance. Additionally, heat stress and drought stress are often interconnected, as heat stress is frequently accompanied by low soil water availability, which may also lead to low nutrient uptake (Zandalinas et al., 2018; Sehgal et al., 2017). Plant non-coding RNAs (ncRNAs), especially miRNAs and tasiRNAs, are significant players in the response to HS because they regulate the activity of transcription factors (TFs) and/or genes primarily involved in the HS response (Matsui et al., 2013; Li et al., 2017; Zhao et al., 2020). HSPs are expressed several times more during high-temperature stress, and this expression is primarily regulated by explicit HSFs (Juneja et al., 2023). Several HSFs, HSPs, and small heat shock proteins (sHSPs) have been predicted targets of miRNAs such as miR156, miR166, miR319, miR171, and miR5213. Reports indicate that miR156 may target *CaHsp17.6A* and *CaHsp21* in chickpea (*C. arietinum*). Computational analyses have also indicated that miR319 may target *HsfA1* in chickpea, potentially linking this interaction to the heat stress response, although experimental validation is still needed. Likewise, the anticipated interactions of miR166 with *CaHsp70.2*, *CaHsp70.4*, and *CaHsp70-BIP1* have not been experimentally confirmed and require further examination (Juneja et al., 2023). Several other miRNAs that have been predicted to target HSPs in response to HS include vun-miR1525, miR2657, and miR9748 in cowpea (*V. unguiculata*) (Gul et al., 2017), thus establishing a direct link between miRNA and heat stress response in legumes. Several other studies have also reported miRNAs like miR398 and miR408 targeting stress-responsive genes in response to heat stress, establishing that miRNA-mediated heat stress regulation is a conserved mechanism across diverse plant families, including legumes, cereals, and model plants (Jagadeeswaran et al., 2009; Guan et al., 2013; Juneja et al., 2023). Similar to its role in the drought stress response, the miR398 module also regulates heat stress adaptation by targeting *CSD*, which in turn modulates cellular antioxidant mechanisms and enhances heat tolerance (Guan et al., 2013; Jagadeeswaran et al., 2009). Likewise, miR408 orchestrates responses to heat and oxidative stress in multiple crops, including wheat (*T. aestivum*) and rice, by regulating the expression of copper-binding proteins such as *PCY* and *LACCACSE* (Kantar et al., 2010; Zhang et al., 2017).

For faba beans' heat sensitivity, both miR400 and miR5658 are efficacious (Koptekin & Aktaş, 2016). Plants overexpressing miR400 exhibit increased sensitivity to heat stress, highlighting its role in thermotolerance mechanisms (Yan et al., 2012). According to Li et al. (2014) and Park et al. (2014), miR400 targets *PPR*, which plays a role in abiotic stress responses and plant development, and its expression is also induced by heat stress in *Arabidopsis*. *PRRs* (*PPR1* and *PPR2*) are mostly found in mitochondria, and the miR400-*PPR* module controls the metabolism of ROS in the mitochondria, which regulates the accumulation of H_2O_2 in mitochondria during heat stress. Additionally, miR1509 targeting *serine/threonine-protein phosphatase7*, miRc20, and ahy-miR3514 targeting *PPR*s found in peanuts (*Arachis hypogaea*) fluctuate in reaction to heat (Mittal et al., 2023). Similarly, in French beans (*P. vulgaris*), high temperatures significantly altered the expression levels of novel miRNAs. Specifically, PvuN1, PvuN3, and PvuN5 were downregulated, whereas PvuN6, PvuN7, PvuN8, and PvuN9 were upregulated in response to heat stress. The promoter regions of these novel miRNAs were shown to possess transcription factor binding sites for TFs, such as MYB and WRKY, which are known to be involved in important stress-response mechanisms. Thus, these novel miRNAs were hypothesized to play crucial roles in regulating gene expression networks essential for heat tolerance and defense against heat-induced oxidative stress (Jyothi et al., 2015).

Small RNA sequencing of flower buds from high-temperature-tolerant and high-temperature-sensitive cytoplasmic male sterility (CMS)-based F1 hybrids revealed that both conserved miRNAs (such as miR156, miR160, miR397, miR398, and miR399) and soybean-specific miRNAs (like gma-miR4413a) responded to high-temperature stress (Ding et al., 2021). miR398 is a specific and well-researched example of miRNA implicated in several abiotic stressors, including HS in legumes (Kumar et al., 2021). Through the reduced expression of its target genes, *CSD1*, *CSD2*, and *CCS* (a gene producing a copper chaperone for both *CSD1* and *CSD2*), heat-inducible miR398 regulates thermotolerance in legumes as well as other crops via ROS detoxification pathways (Jagadeeswaran et al., 2009; Trindade et al., 2010; Jovanović et al., 2014; Kumar et al., 2021). Furthermore, plants expressing the miR398-resistant variants of *CSD1*, *CSD2*, or *CCS* show lower expression levels of the genes encoding HSPs and heat-induced HSFs, making them more susceptible to HS (Guan et al., 2013). In addition to miR398, studies have revealed that drought stress induces miR169 and miR408, indicating their participation. Complementing this observation, downregulation of the respective putative targets, like *NF-YA* and *UCL*, respectively, demonstrated the negative correlation between these miRNAs and their targets in pigeon pea (*C. cajan* L.), confirming the post-transcriptional regulatory function of miRNAs in response to HS (Buch et al., 2020). The miR169-NF-YA module plays a role in adjusting heat-responsive transcription by suppressing *NF-YA*, which negatively regulates heat stress transcription factors, thus enhancing thermotolerance (Rao et al., 2022). miR408 assists in the regulation of copper homeostasis and redox balance by targeting *UCL*, which is essential for safeguarding plant cells against oxidative stress induced by heat (Pan et al., 2018). To sum up, miRNAs are critical mediators of plant adaptations and responses to heat and drought stress. The relevance of miRNAs in adjusting plant adaptation mechanisms is shown by their dynamic expression patterns. However, future studies, including functional assays, may yield important molecular bases for understanding such novel miRNA-mediated pathways, which can then potentially be used for creating crops with better performance under heat and drought stress.

11.2.3 Role of miRNAs in Plant Adaptations to Cold Stress

Cold stress includes both chilling (0–15°C) and freezing (<0°C) stress and has a substantial impact on the growth and output of plants that require relatively higher temperatures for survival and development (Liu et al., 2018). Small RNAs, including siRNAs and miRNAs, have been critical in regulating cold stress. miR160-*ARF*, miR482-*WD Repeat-containing protein 26-Like* (*WDRL*), miR2118-*Disease Resistance* (*DR*), miR396-*GRF*, miR162-*Dicer-like* (*DCL*), miR1511-*Serum Response Factor* (*SRF*), and miR1511-*SPIRAL1* were found to play a significant role in regulating cold stress response. It was demonstrated that expression of miR160, miR482, and miR2118 was upregulated in response to cold stress, leading to suppression of members of *ARF*, *WDRL*, and *DR* gene families, respectively, which positively impacts the cold stress response in peanut. The miR160-*ARF* module regulates various abiotic stress responses, including cold stress, via IAA signaling pathways. Recently, additional factors like *GNOM*, a *SEC7* containing *ARF-GEF*, have also been found to regulate auxin homeostasis in *Arabidopsis* roots, pointing to direct or indirect involvement of additional factors in the cold stress response (Ashraf et al., 2019). *WDRL*, a predicted target of miR482, modulates target gene expression via histone acetylation/deacetylation pathways during adaptation to cold temperatures, ultimately improving cold stress tolerance in peanuts (Zhang et al., 2022). On the contrary, miR396, miR162, and miR1511 negatively regulate cold stress by silencing *GRF*, *DCL*, *SRF*, and *SPIRAL1* genes (Zhang et al., 2022). The *GRFs* bind to the promoter region of *CBF* (C-repeat binding transcription factor) gene to inhibit its transcription and are important for cold signal transduction (Kim et al., 2012). Likewise, *SPIRAL1* is known to maintain microtubule (MT) stability and *SRF* helps in regulation of cold-related (COR) genes (Wang et al., 2011; Wei et al., 2021). As such miR1511-*SRF* and miR1511-*SPIRAL1* may regulate cold stress by regulating the COR pathway and mediating MT stability. On the other hand, the miR162-*DCL* module attenuates cold stress tolerance by regulating small RNA biogenesis as has been observed in rice (Vyse et al., 2020). Comparative analysis of two soybean cultivars (Augusta and Fiskeby V) revealed a negative correlation between miRNAs and their targets in seedling roots for three miRNAs (miR169 targeting *NF-YA*, miR319 targeting *GAMYB* transcription factor ,and miR398 targeting *CCS*) (Kuczyński et al., 2020). The *NF-YA* is a part of *NF-Y* which helps in plant development and abiotic stress response (Sorin at al., 2014). Similarly, *GAMYB* is important for gibberellin (GA) signaling and is crucial for flower development (Tsuji et al., 2006). *CCS*, on the other hand, helps in ROS accumulation under stress by assisting *SOD*s to process superoxide radicals (Chu et al., 2005). Thus, miR169, miR319, and miR398 regulate cold stress in soybean by negatively regulating the expression of their target genes, which are vital for abiotic stress responses. In *M. sativa*, 35 differentially expressed miRNAs were reported in response to cold stress, potentially targeting 105 genes (Shu et al., 2016). Among these targets were members of key transcription factor families, including *ARF*s, *SPL*s, *NAC*s, *AP2/ERF*s, and *GRF*s, that are essential for cold stress adaptation. Gene ontology analysis of targeted genes revealed an enrichment of biological, cellular, and metabolic GO terms associated with cold as well as general abiotic stress response (Shu et al., 2016). Using miRNA expression profiling, degradome sequencing, and functional characterization of cold-sensitive (WQL20) and cold-tolerant (WQL30) varieties of peanut, seven miRNA-target modules in addition to the preceding cultivars and discovered 162 differentially expressed miRNAs, including 18 novel miRNAs, in response to cold stress. Among them, 93 were downregulated and 137 were upregulated in at least one of the soybean cultivars. GO and KEGG enrichment of potential targets of differentially expressed miRNAs were found to be related to reactive oxygen species scavenging, flavonoid biosynthesis, and regulation of osmotic potential, suggesting a valuable role of miRNA-mediated gene regulation in adaptation to cold stress (Kuczyński et al., 2022). Similarly, a comparative RNA-seq study between two pea (*P. sativum*) genotypes i.e., frost-tolerant Champagne and frost-sensitive Terese, revealed 40 differentially expressed miRNAs, 16 of which showed similar expression patterns in both the tolerant and susceptible genotypes (Mazurier et al., 2022). This study underscored the role of miR156, miR166, miR159, and miR397 in regulating cold stress. *Psat0s740g0280*, annotated as a Fip1 motif-containing gene and predicted to be a target of miR156, regulates the expression of downstream genes through adenylation pathways. Another miR156 predicted target, *Psat4g012280*, annotated as UDP-glucuronosyl/UDP-glucosyl transferase, plays an essential role in anthocyanin production, which aids in ROS scavenging during cold stress (Mazurier et al., 2022). miR166 is known to target members of *HD-ZIP III*, important for plant growth and development. Increased expression of miR166 reduces the expression of *HD-ZIP III* and as such inhibits plant growth and development to reserve energy during cold stress conditions (Mazurier et al., 2022; Valiollahi et al., 2014). Although the miR159-*MYB* regulatory module is well known for its involvement in various abiotic and biotic stress responses, miR159 has also been predicted to play a role in cold stress adaptation in legumes such as soybean by targeting *CRK6*, a gene associated with reactive oxygen species (ROS) metabolism (Kuczynski et al., 2022). Another key miRNA, miR397, also contributes to cold stress regulation both directly by targeting *LACCASES* involved in lignin biosynthesis and indirectly, by modulating the *C-repeat Binding Factor* (CBF) regulon, a central component of the plant cold response pathway (Mazurier et al., 2022). The miR397-*LAC* module increases cell wall elasticity by increased permeability, primarily due to reduced cell wall lignin content which provides a mechanical advantage under low-temperature stress. On the other hand, the miR397-*LAC/casein kinase β subunit 3* module has been hypothesized to affect the expression of *CBF* genes, thereby affecting the expression of *CBF*-regulated cold-responsive genes and ultimately defining the cold stress response of a plant.

Overall, the number of studies studying cold-stress-related miRNAs and how miRNAs play an important role in mitigating cold stress response in legumes is limited. Future studies focusing on different legumes and their responses to cold stressors at different developmental stages will provide more insights into cold-stress-responsive miRNAs, their potential targets, and the molecular mechanism of stress regulation and adaptation in legumes.

11.2.4 Role of miRNAs in Plant Adaptations to Salinity Stress

Salinity stress is defined as a high salt content that causes osmotic and ionic stress in plants, which reduces root and leaf growth. Over time toxic accumulation of Na^+ and Ca^{2+} causes premature chlorosis and leaf senescence (Munns & Tester, 2008). In addition to heat, drought, and cold stress, miRNAs are crucial in regulating salinity responses in legumes. Identifying miRNA-target modules provides valuable insights into the dynamic regulation of salt stress adaptation in legumes as well as other flowering plants. For example, miR156 was found to positively mediate salt stress by affecting the expression patterns of its target *SPL* and non-targets *ERF/AP2* and *WRKY* transcription factors, thus regulating the expression of downstream effector salt-stress-responsive genes dictating salt stress tolerance in alfalfa (Arshad et al., 2017). The miR156-mediated salinity stress response helps in increasing the acquisition of beneficial ions like Na^+ and Zn^{2+} and decreasing the acquisition of toxic ions like Ca^{2+} and Mg^{2+}. A transcriptome study between multifoliate and trifoliate alfalfa identified 3746 differentially expressed genes (DEGs), of which 1633 were significantly upregulated and 2113 were significantly downregulated. A total of 227 DEGs were identified as TFs, including the members of the *SPL* gene family (Min et al., 2022). Sequence alignment analysis of 30 members of the *SPL* gene family in alfalfa revealed a conserved miR156 binding site for 16 of them, and their expression analysis showed a high expression of *SPL* gene family members in response to drought and salt stresses. This hints that the miR156-*SPL* module positively regulates the salt stress response by fine-tuning the expression of downstream salt stress-related genes with *SPL* binding sites, as shown in two different studies (Arshad et al., 2017; Min et al., 2022). In addition to the miR156-*SPL* module, various other miRNA-target modules have been discovered to play a significant role in the salinity stress response. A study identified 21 and 129 miRNAs expressed differentially in *M. truncatula* in response to salt and alkali treatment, respectively, with 35 miRNAs upregulated and eight miRNAs downregulated in both groups (Cao et al., 2018). The detected conserved miRNAs include, but are not limited to, the mtr-miR156, miR159, miR160, miR171, miR319, miR393, miR395, miR398, and miR408 family members. In addition, a few legume-specific miRNAs, including miR1510, miR2086, and miR2628, were also found to be differentially expressed in response to salt treatment. The abovementioned miRNA-target modules have been reported to be involved in the salt/alkali stress response, sulfur metabolism, and various carbohydrate metabolic pathways (Cao et al., 2018). Moreover, seven miRNAs in *V. unguiculata*, namely vun-miR156a, miR159b, miR160a, miR162a, miR168a, miR169b, and miR408, were found to be upregulated in root tissue in response to salt stress leading to the downregulation of their conserved targets (Paul et al., 2011). Similarly, in chickpea, upregulation of miR397, miR398, and miR164 in response to salinity and water stress, as well as downregulation of miR399 in response to salt stress, was detected (Khandal et al., 2017). The conserved targets, like *LAC4*, *Cu-SOD*, *NAC1*, and *PHO2/UBC24*, were found to show anti-concordant expression patterns with respect to respective miRNAs, suggesting miRNA-mediated suppression of lignin metabolism and root development in water-deficient and high-salinity conditions (Khandal et al., 2017). Deep small RNA sequencing experiments investigating the regulatory role of miRNA in functioning nodules under salt stress revealed 104 differentially expressed miRNAs in soybean (*G. max*), out of which 14 miRNAs were upregulated at least by a two-fold change and 14 were downregulated at least by a two-fold change (Dong et al., 2013). Among them gma-miR159, gma-miR169, and gma-miR319 exhibited the most significant downregulation and gma-miR166 showed significant upregulation by three-fold change under salt stress (Dong et al., 2013). Relevant conserved miRNA modules such as miR159-*MYB*, miR169-*NF-YA*, and miR319-*TCP* have been linked to regulating salt stress of varying degrees (Zhang et al., 2020; Zhang et al., 2022). Under drought and salt stress, *MYB* increases *CAT*, *POD*, and *SOD* activities, and as such increases plant tolerance to abiotic stress by regulating ROS homeostasis. Similarly, *NF-Y* gene family members were suggested to mediate salt and drought stress via ABA- or jasmonate-signaling pathways or both (Zhang et al., 2020).

Interestingly, miR170/171 and miR172 have shown differential expression in response to salinity stress in various legumes including *M. truncatula*, soybean, and chickpea (Dong et al., 2013; Sun et al., 2016; Khandal et al., 2017; Cao et al., 2018), suggesting its conserved role in salt stress regulation. The miR172 family has been established to target members of *AP2* which crosstalks with the ABA-signaling pathway, a key hormone involved in the salt stress response (Li et al., 2016). The miR171 module, consisting of members of the *GRAS* TF family, functions predominantly in the root system and plays an essential role in the development of both primary and adventitious roots under salt stress conditions, highlighting its importance in adaptive root architecture remodeling in response to salt stress (Pei et al., 2023). Sahito et al. (2017) also reported an elevated expression of miR172c in roots in response to salt stress. Transgenic plants overexpressing miR172c had increased root hairs, root length, and number of lateral roots compared to a control in salt stress. As a result, the root system of the transgenics overexpressing miR172 had increased salt tolerance suggesting that miR172 positively regulates salt tolerance in soybean (Sahito et al., 2017). In soybean (*G. max*), 71, 50, and 45 miRNAs were expressed differentially in response to drought, salinity, and alkalinity stress, respectively (Li et al., 2011). While some miRNAs, such as miR1520, miR4374, and miR4396, had conserved expression patterns under all stress conditions, indicating their role in regulating overlapping abiotic-stress tolerance pathways, other miRNAs, like miR394 and miR4361, had contrasting expression patterns under different stress conditions, suggesting their functional divergence at the post-transcriptional level. In addition to the differential regulation of the above-mentioned miRNAs in response to salt stress, the upregulation of miR395 in response to salt treatment (0.4M NaCl) and the upregulation of miR396 in response to both salt and drought stress has been reported in *P. vulgaris* (Nageshbabu & Jyothi, 2013). miR396 is important for promoting cell proliferation, and leaf and root growth and is known to target members of the *GRF* gene family which play an important role in regulating these aspects of plant development. Overexpression of miR396 leads to the suppression of *GRF*s, which in turn reduces plant tolerance to salt stress in *Agrostis stolonifera* (Yuan et al., 2019). Additionally, the miR396-*GRF*s module have been shown to be differentially expressed in response to salinity stress in other crops as well, such as *A. thaliana*, *O. sativa*, and *Gossypium hirsutum*, suggesting its conserved role in regulating salinity stress across different crops (Liu et al., 2008; Gao et al., 2010; Wang et al., 2013). miR395, which is known to target *APS* and *SULTR* gene family members, also exhibited increased expression under sulfate starvation in *Arabidopsis* with decreased expression of its target genes (Jones-Rhoades & Bartel, 2004; Kim et al., 2010). Thereby modulating sulfur metabolism and supporting ROS detoxification, miR395 may enhance a plant's ability to tolerate salt stress (Kim et al., 2010).

Various investigations aimed at capturing the response of *P. vulgaris* to multiple abiotic stresses found increased expression of novel miRS1, miR159, miR319, and miR393 in response to salt stress. The role of miR393 in regulating cold, drought, and salt stress in *Arabidopsis* is well established (Sunkar & Zhu, 2004). Furthermore, miR393, which targets *TIR1/AFB* auxin receptors, has been reported to be upregulated in response to salinity stress in other crops like soybean, rice, and tobacco (Xing et al., 2023). miR159, which primarily targets the *MYB* transcription factor gene family important for seed germination, has been shown to be differentially expressed under salt stress conditions in several other plants/crops like *Arabidopsis*, soybean, sugarcane, and potato (Patade & Suprasanna, 2010; Kitazumi et al., 2015; Li et al., 2023). This indicates a strong functional conservation of miRNAs throughout the flowering plants. In contrast, Li et al. (2023) proposed a novel regulatory network integrating ABA signaling and miR159, revealing that the tissue-specific expression of miR159 family members is influenced by low phosphorus, high salt, and ABA treatments (Li et al., 2023). Additionally, a comparative genomics study of *TCP* TFs in five legumes, namely *G. max*, *C. arietinum*, *P. vulgaris*, *M. truncatula*, and *L. japonicus*, found differentially expressed *TCP* gene family members with contrasting patterns across different tissues and species in response to salt and drought stress (Ling et al., 2020). Among 149 identified *TCP* genes, eight contained binding sites for miR319, while 13 harbored binding sites for miR5658, suggesting the existence of a complex *TCP*-miR319/miR5658 regulatory module involved in the salt stress response (Ling et al., 2020). TCP proteins are known to regulate salt stress by binding to the promoter region of gene *OsNHX1* which assists in cation exchange across membranes (Almeida et al., 2017).

In addition to conserved and well-characterized miRNAs, several studies have investigated the role of novel miRNAs in the salt stress response. Notably, differential expressions of novel miRNAs, including Gly04, Gly03, Gly13, Gly16a/b, and Gly20, have been reported in soybean under salt stress (Sun et al., 2016). Other reports of novel specific miRNAs in chickpeas under salt stress include miR008 (targeting chalcone synthase gene), miR011 (targeting Serpin family protein), and miR015 (targeting complex-1 and ribosomal family protein) (Kohli et al., 2014). Furthermore, various legume-specific miRNAs like gma-miR5036, miR862, miR4413, miR4416, and miR5225 have also been found to be differentially expressed in response to salt stress in soybean (Ning et al., 2019; Li et al., 2022). Though such novel findings require thorough validation, as many are based on computational predictions that often yield more false positives than true targets, the intricate role of miRNAs in regulating plant stress responses cannot be overlooked.

11.2.5 Role of miRNAs in Plant Adaptations to Heavy Metal Stress

Any metal with a density greater than 5 g/cm^3 is classified as a heavy metal (Gasic & Korban, 2006). While some of these metals like copper and zinc are essential micronutrients, excess amounts of heavy metals are damaging to plant growth and development (Gasic & Korban, 2006). To date, the roles of various miRNAs toward heavy metal stress and their responses have been investigated in several legume crops, including aluminum (Al), lead (Pb), cadmium (Cd), mercury (Hg), boron (B), copper (Cu), manganese (Mn), and zinc (Zn) (Gasic & Korban, 2006; Zhou et al., 2012; Fang et al., 2013; El-Sappah et al., 2021; Lu et al., 2023; Suresh Kumar et al., 2023; Ucar et al., 2024). High-throughput sequencing in *M. truncatula* identified 23 differentially expressed miRNAs in response to aluminum treatment (Chen et al., 2012). Four miRNA-target modules, including miR159-*MYBs*, miR160-*ARFs*, miR319-*TCPs*, and miR396-*GRFs*, were identified to be differentially regulated in response to Al treatment. These modules have been characterized for their roles in regulating key developmental events in seed germination, and root, leaf, and flower development (Chen et al., 2012). These findings suggest that Al stress can trigger widespread regulatory changes, potentially resulting in pleiotropic effects on plant growth and development. Lu et al. (2023) also demonstrated that Al stress resulted in 55 differentially expressed miRNAs, including miR2119 and miR5213, potentially targeting 2502 genes in the genome of *M. truncate* (Lu et al., 2023). They also reported that miR156g-3p (star) was significantly upregulated at 4, 24, and 48 h of Al stress and was predicted to target six genes, namely *Medtr6g004260*, *Medtr4g056140*, *Medtr7g007440*, *Medtr5g077840*, *Medtr3g078260*, and *Medtr2g090610*, regulating root tip formation. In the same study, three novel miRNAs, i.e., novel_miR_135, novel_miR_182, and novel_miR_36, were found to be upregulated at 4 h post-treatment. These novel miRNAs target members of *GRAS*, *WRKY*, and *bHlH* TFs which are known to promote root development under stress and have also been hypothesized to promote resistance against Al stress (Lu et al., 2023). In wild soybean (*G. soja*), 30 miRNAs were differentially expressed, with 12 downregulated and 18 upregulated in response to Al stress (Zeng et al., 2012). Consequently, expression of genes, including *ARFs*, *NB-ARC*, *NAM*, *MYBs*, *LRR-TIR*, and *Cation transporting ATPase* were regulated in response to Al stress. Notably, *NB-ARC* and *LRR-TIR* binding domains are characteristic of plant resistance (R) proteins involved in activating immune responses (Swiderski et al., 2009; Tang et al., 2010). miRNA-mediated cleavage of genes with these domains may weaken the immune system and make plants more prone to Al toxicity (Zeng et al., 2012). Similarly, miR160 and miR1514 targeting *ARF* and *NAM* TFs, respectively, may impair root growth and elongation, exacerbating the effects of Al toxicity (Zheng et al., 2012). Comparative analysis of two soybean genotypes under Al stress, Al-tolerant BX10 and Al-sensitive BD2, 32 miRNAs were differentially expressed (19 in tolerant, seven in sensitive, and six in both genotypes). The authors further demonstrated that conserved miRNAs related to root elongation (gma-miR166k/o, gma-miR390g, and gma-miR396c/k) and oxidative stress (gma-miR169r) were downregulated in tolerant genotypes, possibly explaining the mechanism for Al tolerance in BX10 (Huang et al., 2018). Similarly, in *P. vulgaris*, 28 miRNAs were found to be differently expressed in roots and nodules in response to Al stress. Of these, members of miR164, miR170, miR393, miR396, and miR1511 were found to be upregulated, whereas members of miR157, miR169, and miR398 were found to be downregulated (Mendoza-Soto et al., 2015). In response to Al stress in common bean roots, the reduced expression of miR1511 resulted in antagonistic increased expression of its target *ALS3* (Mendoza-Soto et al., 2015; Ángel Martín-Rodríguez et al., 2021). *ALS3* is widely known to regulate root growth inhibition by suppressing *STOP1-ALMT1* and *LPR1* pathways, which indirectly affects ROS accumulation in roots by mediating iron accumulation (Wang et al., 2019; Angel Martin-Rodriguez et al., 2021).

In addition to Al, responses to other heavy metals, like Hg and Cd, have also been studied. Zhou et al. (2008) identified the differential regulation of various miRNA families under heavy metal stress, with miR171, miR319, miR393, and miR529 being upregulated, while miR166 and miR398 were downregulated. The study also suggested that upregulated miRNAs positively regulate the heavy metal stress response, whereas downregulated miRNAs negatively regulate the heavy metal-induced stress response in *M. truncate* (Zhou et al., 2008). Furthermore, comparison between the sequencing libraries made from Hg-treated and Hg-free seedlings identified 201 miRNAs, 28 of which were significantly differentially expressed in response to Hg (Zhou et al., 2012). Six miRNAs (miR2111, miR2592, miR2593, miR2595, miR2633, and miR2661) were upregulated, whereas the others were downregulated. The study also identified three novel *M. truncatula*-specific miRNA modules, miR2681-*TIR-NBS-LRR*, miR2708-*TC114805* (encoding salinity tolerance protein), and miR2687-*XTH* (encoding xyloglucan endotransglucosylase/hydrolase), to be functional in the response of Hg stress (Zhou et al., 2012). The *TIR-NBS-LRR* gene family, known for its role in disease resistance, suggests that genes activated in response to biotic stress are also regulated under Hg stress (Zhou et al., 2012). XTH, an enzyme known to modify the cell wall, contributes toward the development of the cell wall and has been known to facilitate drought and salt tolerance in plants through mesophyll cell protection in response to drought stress (Cho et al., 2006). While its response to Hg stress remains unclear, it can be hypothesized that XTH may confer tolerance through a mechanism similar to its role in drought stress. These findings suggest that miRNA-mediated regulation of stress-responsive genes may play a crucial role in plant adaptation to Hg stress.

The response to Cd toxicity was investigated by studying cadmium-tolerant (HX3) and cadmium-sensitive (ZX24) soybean genotypes. Twenty-six miRNAs were found to be significantly differentially expressed in response to Cd stress, where the expression profiles of most of these miRNAs were similar across both genotypes, except for gma-miR1535b, which was upregulated in HX3 but downregulated in ZX24 (Fang et al., 2013). A similar approach identified organ-specific differential expression of miRNAs in roots, nodules, and leaves in response to manganese toxicity (Mn) in *P. vulgaris* (Valdés-López et al., 2010) and demonstrated that the miRNA-mediated stress response to Mn toxicity varies depending on the organ/tissue type. For example, northern blot analysis for miR398 revealed an antagonistic expression pattern in the leaves as compared to roots during Mn-mediated oxidative stress in common beans. The miR398 family has also been shown to actively participate in mitigating cellular Cu levels. In response to Cu toxicity, downregulation of miR398b results in the induction of its targets, *NOD19* (predicted target) and *CSD1*, which help maintain resistance of plants against different kinds of stress through ROS detoxification (Naya et al., 2014). Under Cu-deficient conditions, reduced expression of *CSD1* facilitated the efficient allocation of Cu to its target proteins, highlighting distinct regulatory mechanisms of miR398 in Cu stress vs. Cu toxicity conditions (Naya et al., 2014). Additionally, in response to boron toxicity in *P. vulgaris* seven miRNA-target modules, including miR812-*BTB/POZ*, miR2593-*Lipoyl transferase*, miR2608-α/β hydrolase, were reported (Jyothi et al., 2018). *BTB/POZ* function in plant signaling pathways and are members of calmodulin-binding proteins which are known to help in gametophyte formation, and the stress and hormonal response in *Arabidopsis* (Robert et al., 2009; Mandadi et al., 2009). Lipoyl transferase and α/β hydrolases are metabolic enzymes involved in cellular metabolism. Under boron toxicity, the downregulation of these enzymes disrupts cellular metabolism and chlorosis, and impaired growth leading to reduced plant biomass (Jyothi et al., 2018).

11.2.6 Role of miRNAs in Nutrient Acquisition and Homeostasis

Plants acquire nutrients from the soil via a complex network of physiological and mechanical processes. Modern farming practices, coupled with increased food demands, have heightened the necessity for soil to maintain a balanced nutrient profile and ensure the availability of essential nutrients. Legumes are widely grown worldwide to improve soil nitrogen content and soil health. Studies have shown improved nitrate (NO_3^-), ammonium (NH_4^+), phosphorus (P) and soil organic carbon availability in a legume-introduced cropping system (Phiri & Njira, 2023). Because of their unique ability to fix nitrogen, it is advised to grow legume crops in nutrient-poor soils. Many studies have attempted to illuminate the intricate mechanism of miRNA-regulated nutrient stress management; however, a complete understanding of the process is still elusive.

The regulation of nitrogen homeostasis in legumes involves conserved miRNA-target modules, providing strong evidence for the need for miRNA-specific studies to elucidate their functional roles in N stress adaptation and management. Wang et al. (2013) investigated the miRNA regulation of nitrogen (N) deficiency stress in soybean, comparing two genotypes (low N-sensitive and low N-tolerant) for low N-responsive miRNAs. Wang et al. identified 452 miRNAs, of which 150 previously known and two novel (novel-soy0043-5p, novel-soy0006-3p) miRNAs were differentially regulated in response to low N availability. Additionally, it was reported that the differential regulation of miRNAs was organ specific as well as treatment specific. For instance, members of the miR169 family were found to be repressed in both roots and shoots under low nitrogen (N) availability, whereas miR398 family members showed similar expression patterns only in shoots, pointing toward tissue-specific expression regulation of miRNAs (Wang et al., 2013). In addition, a laundry list of conserved and predicted targets of DE miRNAs in response to N limitation were reported to have roles in a wide range of biological processes including protein turnover, cellular transport, phytohormone signaling, carbohydrate metabolism, and nucleic acid metabolism. Contrary to Wang et al. (2013), Li et al. (2021) demonstrated that miR160, miR167, miR393, and miR396 were upregulated in *A. hypogaea L.* under N-deficit conditions. They proposed that the miRNA mis-regulation may have resulted in reduced root length and reduced lateral root number as modules involving miR393 targeting *AFBs* and miR160/miR167 targeting *ARF*s, being the key regulators of auxin perception and signaling, predominantly functions in maintaining the root architecture and development (Li et al., 2021). *AFB*s, in the presence of auxin, interact with *Aux/IAA*, the transcriptional repressor, promoting their ubiquitination. Increased miR393 would reduce the proteolysis of *Aux/IAA*, thus disrupting overall root architecture development. This miR393-*AFB* module was also established in *Arabidopsis* under N-deficit condition. Similarly, *ARF*s have been characterized as positive regulators of lateral root development. They play a crucial role in nuclear repositioning and asymmetric cell division, both of which are essential for lateral root formation and overall organogenesis. miR160/167 overexpression negatively impacts the lateral root development by repressing *ARF* gene family members (Jiang et al., 2022). Reports of mature miR398 upregulation in roots and shoots in *M. sativa* during N-stress have also been put forward (Javed et al., 2019), which is in line with the ROS scavenging function of miR398-*SOD* modules under a wide range of oxidative stress conditions.

In addition to N, identification of P-responsive miRNAs has also paved the way for understanding miRNA-mediated P homeostasis in legumes. Zhu et al., (2010) identified 167 miRNAs belonging to 35 miRNA families in white lupin (*Lupinus albus L.*) (Zhu et al., 2010). Under P-deficit conditions, 17, 9, and 10 families were found to be upregulated in roots, stem, and leaves, respectively, whereas 7, 6, and 12 families were downregulated in roots, stem, and leaves respectively. They proposed that the overexpression of miR164 and the resultant negative regulation of *NAC1* may impact the root development, hinting toward the involvement of miR164 in maintaining root architecture during P starvation (Zhu et al., 2010). Similarly, in *G. max*, 57 miRNAs from 27 families were induced under phosphorus (P) deficiency. Among them, seven families were upregulated and 17 were downregulated in leaves, while eight families were upregulated and six were downregulated in roots, highlighting a tissue-specific miRNA-mediated response to P deficiency (Zeng et al., 2010). In addition, miR399 family members were found to be upregulated, whereas miR169 family members were downregulated in both roots and leaves. Another study on P-starvation of *G. max* conducted by Xu et al. (2013) revealed 25 miRNAs that were induced and 11 miRNAs that were repressed (Xu et al., 2013). Various miRNA-target modules like miR399-*PHO2*, miR2111-*Kelch domain containing F-box protein*, and miR159-*MYB* transcription factor were identified to be P-responsive (Xu et al., 2013). Though the function of Kelch domain-containing F-box protein is still elusive in the case of legumes, the miR399-*PHO2* module works in phosphate homeostasis in various plants, including soybean, common bean, *Trifolium pratense* (red clover), as well as in nodules (Franco-Zorrilla et al., 2007; Valdés-López et al., 2008; Zeng et al., 2010; Zhu et al., 2010; Kim et al., 2011; Liu et al., 2020; Pramanik et al., 2020; Demirkol, 2022). In *M. sativa*, miR399 and miR398 were induced, whereas miR156, miR159, miR160, miR171, miR2111, and miR2643 were inhibited under P-deficiency (Li et al., 2018). In a follow-up study, Li et al., (2022) found various conserved and *Medicago*-specific miRNA-target modules, including miR156e-*SPL13*, miR160c-*ARF18*, and miR399l-*SAUR72*, an auxin-responsive protein, miR2587a-*Phosphoenolpyruvate carboxykinase* (*PCKA*), Nov-miR27-*SPX* domain–containing protein, and Nov-miR3-*Sulfoquinobosyl transferase-2 (SQD2)* as well as *glutathione S-transferase*. Systems analysis of miRNA targets involved in the phosphate starvation response revealed their involvement in carbohydrate metabolism, sulfolipid metabolism, glutathione metabolism, and hormone signal transduction pathways (Li et al., 2022).

Few studies have attempted to investigate the combinatorial effects of stress on miRNA regulation. Ning et al. (2019) reported that various miRNAs, including miR169, miR5036, miR862, and miR398, respond to Pi starvation in combination with salt stress in soybean (Ning et al., 2019). Seven miRNA families, including miR160, miR159, miR166, miR168, miR169, miR482, and miR858, were found to play positive regulatory roles, whereas miR393 family members were found to be negative regulators in response to Pi-deficit treatment. Li et al. (2023) examined the role of miR159 in soybean and revealed that miR159e is critical in regulating P absorption and transport in soybean (Li et al., 2023). They further hypothesized Leafy (*LFY*), as one of the indirect targets, is involved in maintaining the vegetative state and delaying flowering under low P availability. Post-transcriptional regulation of long non-coding RNAs through miRNAs to regulate phosphate starvation tolerance has also been explored (Zhang et al., 2021). They found shared targets for 10 miRNAs-lncRNA pairs, hinting at a complex combinatorial transcriptional/post-transcriptional gene regulation by the dual action of lncRNA and miRNAs. However, further studies on lncRNA-miRNA-target modules regulating nutrient homeostasis in legumes are required in understanding the complex nature of related gene regulatory mechanisms.

Briefly, legumes rely on a complex network of miRNA-target interactions to perceive stress signals and trigger adaptive responses against a wide range of abiotic stresses, including drought, salinity, extreme temperatures, and nutrient imbalances. Because miRNAs act as nodes in the gene regulatory networks, understanding these regulatory modules not only deepens our knowledge of legume stress physiology but also provides valuable insights into the pleiotropic effects of miRNA misregulation. A summary of key miRNA-target modules participating in the above-mentioned abiotic stresses is provided in Figure 11.1.

11.3 MIRNAS AND BIOTIC STRESS RESPONSE

The immune system of plants is a wonder of biological intricacy, featuring complex defense systems like effector-triggered immunity (ETI) and pattern-triggered immunity (PTI). When pathogen-associated molecular patterns (PAMPs) are recognized by pattern recognition receptors (PRRs), PTI triggers a broad-spectrum defense response (Postel & Kemmerling, 2009). In contrast, ETI involves the specific recognition of pathogen effectors by resistance proteins, most notably nucleotide-binding site-leucine-rich repeat (*NB-LRR*) receptors, which results in a potent defense reaction (Noman et al., 2019). In this context, miRNAs which take part in regulation of these defense mechanisms become an important conductor, controlling post-transcriptional gene expression and adjusting plant defense mechanisms against a wide range of biotic stimuli. As already discussed, *MIRNA*s, highly conserved non-coding genes, play a multifaceted role in stress regulation, also play an important role in orchestrating the complex molecular responses in sessile plants to biotic stressors, including viruses, bacteria, nematodes, fungi, and insects (Figure 11.2). Extensively studying and understanding these regulatory modules will open the door to the development of strategies to improve biotic resilience in crops through molecular breeding or genetic modification techniques.

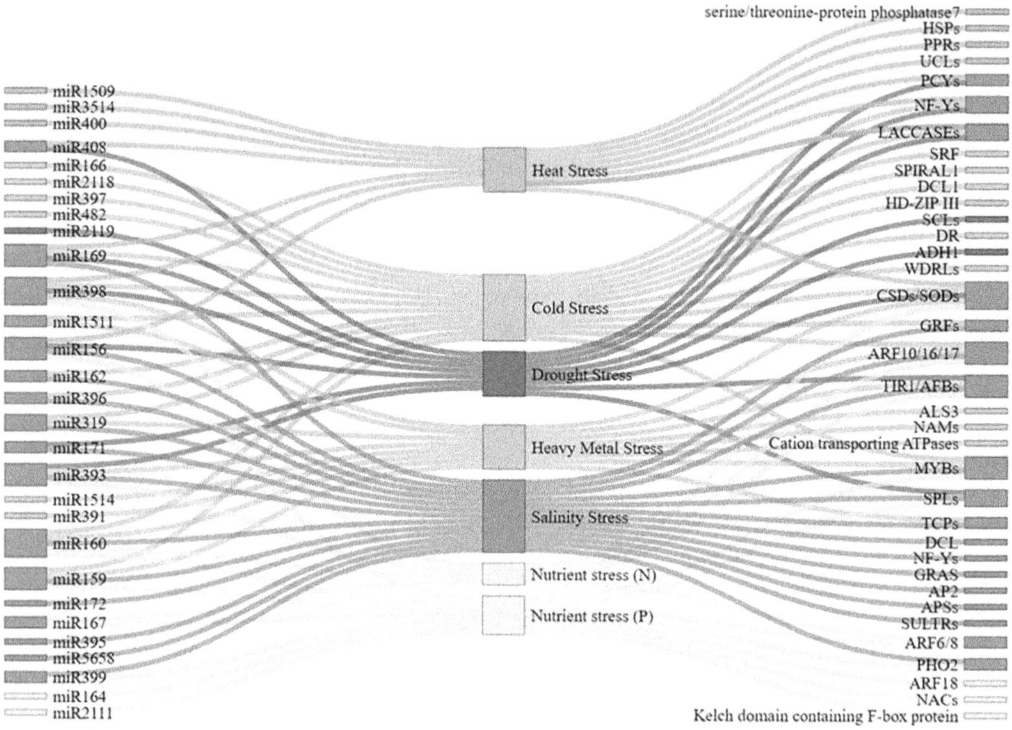

FIGURE 11.1 miRNA-target modules in abiotic stress regulation across legumes. This figure illustrates key miRNA-target modules involved in regulating a broad range of major abiotic stress responses across different legume species. The color scheme used to represent different stress types is as follows: dark turquoise, cold stress; saddle brown, heat stress; coral, heat stress; plum, heavy metal stress; pale green, nutrient stress (N) and nutrient stress (P); salinity stress, olive.

In legumes, miRNA expression regulation is a crucial mechanism for establishing and maintaining defense strategies against diverse pathogens. miRNA's ability to target mRNAs associated with immune responses, pathogen recognition, and stress signaling, influences a plant's capacity to withstand and adapt to biotic challenges. For example, miR482/2118 targets *NB-LRR* genes which are essential for pathogen detection and defense activation against fungal, bacterial, as well as viral pathogens (Zhu et al., 2013). Similarly, miR393-mediated control of auxin signaling affects bacterial pathogen defense by focusing on auxin receptors and altering the plant's perception and reaction to invaders (Jiang et al., 2022). Various other miRNAs, like miR156, miR159, miR160, and miR167, play a fundamental role in coordinating a plant's immune responses while also regulating the key developmental aspects, thereby equipping plants with the ability to fight or flight (discussed below). The complex relationship between miRNAs and their target genes emphasizes the intricacy of interactions between plants and pathogens. It also emphasizes the critical role that miRNAs play in determining how leguminous crops respond to biotic stressors, which in turn affects the crops' resilience and productivity in the face of an array of challenges.

11.3.1 Role of miRNAs in Plant Adaptations to Viral Infections

Several studies relating to the role of miRNAs against viral pathogens have identified miRNAs conserved across several commercially important legumes. A comprehensive regulatory landscape was demonstrated through the identification of 422 miRNAs exhibiting differential regulation in *P. vulgaris* upon mungbean yellow mosaic India virus (MYMIV) infection (Patwa et al., 2019). Among these, miRNAs such as miR156, miR160, miR390d, and miR319 were upregulated, and miR159, miR164c, miR166, miR398, miR172, and miR5368 were downregulated. Among those miR156, miR160, and miR395 emerged as major players in regulating defense responses targeting a spectrum of genes crucial for plant defense, including transcription factors like *SPLs*, *ARFs*, and *NB-ARC domain-containing disease resistance protein,* respectively (Mallory et al., 2005; Van Ooijen et al., 2008; Xu et al., 2016). Directly or indirectly, miRNAs also modulate the expression of genes involved in stress tolerance, such as those associated with brassinosteroid signaling, *NPR1*-mediated defense responses, and pathogenesis-related proteins (Cao et al., 1997; Ishfaqe et al., 2024; Rasheed et al., 2025). Similarly, in *Vigna mungo*, miRNAs have been demonstrated to play a pivotal role against MYMIV. A high-throughput sequencing analysis revealed a total of 53 previously identified miRNAs and 14 novel miRNA candidates specific to *V. mungo* (Kundu et al., 2017). As found in *P. vulgaris*, miRNAs, including miR156, miR159, miR160, miR166, miR398, miR1511, and miR1514, were identified as the key regulators of gene expression in response to MYMIV infection, suggesting their conserved role to virus stress response across the leguminous species (Kundu et al., 2017). In *V. mungo*, miRNA-target modules including miR164c-*NAC*, miR169-CCAAT-box TF, miR398-*SOD*, and miR1511-*Glutamine amidotransferase* were identified as some of the key components of the plant's immune response (Kundu et al., 2017). Additionally, novel miRNAs unique to *V. mungo*, vmu-miRn11 and vmu-miRn13, target genes involved in photosynthesis, which could interfere with the process and hinder the transmission of the virus (Kundu et al., 2013; Kundu et al., 2015). Conserved miRNAs such as miR168, miR169, miR396, and miR398 have been linked to resistance against viral infections in a wide range of crop species, including *G. max* (Ramesh et al., 2017). These miRNAs enhance plant defense against viral infections by specifically targeting genes involved in growth regulation, oxidative stress responses, and RNA silencing pathways (Shivaprasad et al., 2005). In particular, RNA silencing pathways essential for antiviral defense are modulated by miR164 and miR168, which respectively regulate *NACs* and *AGO1* (Várallyay et al., 2010; Luo et al., 2024). AGO1 is essential for producing 21–24 nt siRNAs which are essential for stress adaptation, and for repressing translation of downstream targets. By binding to viral short RNAs and inhibiting their action, these siRNAs support the plant's defense against viruses (Zaheer et al., 2024). Conversely, *NAC* TFs control gene expression by interacting with phytohormones and other signaling molecules such as reactive oxygen species (ROS) to trigger a hypersensitive response (HR) often resulting in programmed cell death (PCD), which is essential for localized resistance to pathogens (Nuruzzaman et al., 2013). Likewise, target genes of miR396 and miR398 are involved in developmental processes and oxidative stress responses, which strengthen physical barriers, modulate hormone signaling pathways, and trigger programmed cell death to stop the spread of viruses, enhancing resistance to viruses (Kulcheski et al., 2011; Song et al., 2011). miR160, miR167, miR390, and miR393 play significant roles in antiviral defense, primarily by modulating auxin signaling and indirectly interacting with salicylic acid (SA) signaling pathways (Luo et al., 2024). Additionally, miRNAs specific to *G. max*, the putative antiviral miRNA gma-miR5785, directly target viral transcripts, impeding viral replication and propagation (Simón-Mateo & García, 2006).

11.3.2 Role of miRNAs in Plant Adaptations to Bacterial Infections

miRNAs also orchestrate complex responses to bacterial invasion and symbiotic associations and hence play a crucial role as regulators in the complex molecular dance between leguminous plants and bacterial pathogens, as demonstrated in the case of viral attack (Singh et al., 2017). miRNAs target particular genes involved in defense (*receptor kinases*, *MYB*, *AP2/ERF*), symbiotic signaling (*TIR1/AFB*, *TML*), and developmental processes (vegetative phase to reproductive phase transition regulated by *SPLs*) in response to pathogenic bacteria, and they demonstrate dynamic regulation across a variety of leguminous crop species (Tiwari et al., 2021). In *L. japonicus*, a comprehensive study revealed the regulatory function of miRNAs during *Mesorhizobium loti* infection (Holt et al., 2015). Notably, miR172a emerges as an important player, targeting the transcription factor *AP2/ERF* family members, which are associated with bacterial resistance and stress responses (Middleton et al., 2007). The crucial role that miR172a plays in modulating the plant's defense mechanisms upon bacterial colonization is demonstrated by its upregulation, which may allow for symbiotic associations while preserving efficient defense responses. Furthermore, miRNAs such as miR482 and miR2111 along with their isoforms exhibit a response to rhizobial inoculation suggesting their involvement in regulating how plants respond to bacterial colonization (Middleton et al., 2007). Across leguminous crops, miRNAs, especially miR159, miR393, and miR156, show dynamic PTGS-mediated gene regulation in response to rhizobial infection. The miR159-*MYB* module regulates the expression of the defense-related genes such as those involved in jasmonate and salicylate signaling pathways, PAL, PR proteins; thereby enhancing the defense response of legumes against bacterial pathogens (Fei et al., 2013; Yu et al., 2023). On the other hand, miR393 and miR482 target *TIR1/AFB* auxin receptors and *NBS-LRR* genes, respectively. Both these miRNA-target modules, respectively, modulate the auxin signaling pathways and defense responses, influencing the plant's ability to recognize and respond to invading pathogens (Fei et al., 2013). Furthermore, miR156, which targets *SPB/SPL* TFs, modulates the course of plant development in response to bacterial invasion. By influencing growth phase transitions, miR156 provides the plant with an opportunity to adjust its growth cycle, thereby enhancing its ability to mitigate biotic stress and improve survival under pathogen attack (Wu et al., 2009).

11.3.3 Role of miRNAs in Plant Adaptations to Fungal Infections

The regulatory role of miRNAs against pathogens extends beyond the simple prokaryotes extending to eukaryotic fungi and other higher organisms. A comprehensive study was conducted to investigate miRNAs associated with fungal resistance in chickpea (*C. arietinum*) and to decipher the intricate molecular pathways that may underpin defense responses against fungal diseases like *Ascochyta* blight (AB) (Garg et al., 2019). A total of 651 miRNAs, including 173 novel miRNAs, were identified through high-throughput small-RNA sequencing, highlighting the complex nature of miRNA-mediated regulatory networks in chickpea. Several conserved miRNAs were identified among the differentially expressed ones, emphasizing their crucial roles in biotic stress responses, and associated developmental processes. Significant changes were observed in response to AB stress, such that increased accumulation of miR156, miR159, miR171, and miR482 resulted in significant downregulation of their respective targets, i.e., *SBPs*, *osmotin-like protein* and *PR*s, *PPR*s, and *NBS-LRRs* gene family members (Garg et al., 2019). Interestingly, a dynamic temporal response of various miRNAs, like miR171, miR319, and miR167, was also observed, indicating the transient nature of gene regulation at post-transcriptional levels. Yet again, the miR482-*NBS-LRR* module surfaced as a key regulatory component essential for resistance against biotic stressors, including AB (Zhang et al., 2022). *Aspergillus flavus*, a fungal pathogen that causes pod rot, is defended against by peanuts (*A. hypogaea*) through the action of miRNAs like miR482/2118 (Xu et al., 2022). These miRNAs target *NBS-LRR* genes which are crucial in plant defense, highlighting their significance in combating fungal infection (Shivaprasad et al., 2012). Furthermore, miRNAs that target *SPL* genes, such as miR156/157, influence secondary metabolism pathways essential for defense (Jerome Jeyakumar et al., 2020). Similarly, miRNAs such as miR166 and miR393 have been identified as critical factors in *G. max* against the fungal pathogen *Phytophthora sojae* (Wong et al., 2014). Following infection, these miRNAs exhibited rapid induction, which may have been caused by pathogen-associated molecular patterns (PAMPs). While miR393 controls auxin signaling-related genes essential for defense against *P. sojae*, miR166 targets *HD-ZIPIII* genes involved in developmental processes (Hawker & Bowman, 2004; Si-Ammour et al., 2011). Additionally, *NB-LRR* and *PPR* genes are targeted by miR482/2118 and miR482/2118-triggered phasiRNAs induced by the pathogen effectors, further fine-tune defense responses via effector-triggered immunity pathways (Subramanian et al., 2008; Zhu et al., 2013).

Across different leguminous crops, conserved miRNAs such as miR159, miR160, miR166, miR169, miR393, and miR482/2118 emerge as key regulators of fungal resistance. By targeting genes involved in stress responses, defense activation, and immune signaling pathways, these miRNAs modulate plant defense mechanisms. The regulatory networks orchestrated by miRNAs fine-tune defense responses, ensuring effective resistance against fungal pathogens.

11.3.4 Role of miRNAs in Plant Adaptations to Herbivory

Herbivore stress is also a major biotic challenge for legumes, with effects ranging from reductions in plant fitness to subtle alterations in leaf development, impacting the overall growth and development of the legume crop. Limited studies have investigated and identified miRNAs and respective targets associated with herbivore-induced stress across leguminous species like soybean and pigeon pea (Zeng et al., 2019; Malhotra et al., 2022). A multitude of miRNAs have been identified in soybean genotypes resistant to bean pyralid larvae (Zeng et al., 2019). This study has highlighted miRNA expressions specific to tolerant and sensitive cultivars, emphasizing the complexity of miRNA-mediated biotic stress tolerance. Comparative analysis of resistant genotype (Gantai-2-2) and susceptible genotype (Wan82-178) identified 70 miRNAs which exhibited differential expression in response to 48 hours of larvae feeding. Notably, soybean-specific *Gma-miR4352b* and *Gma-miR5040* were exclusively expressed in the insect-resistant Gantai-2-2, suggesting their unique presence in this genotype compared to Wan82-178. These miRNAs play a pivotal role in enhancing biotic stress tolerance in soybean by regulating genes encoding various pathways including plant hormone signal transduction, RNA transport, protein processing, zeatin biosynthesis, ubiquinone, terpenoid-quinone, and isoquinoline alkaloid biosynthesis. These pathways are essential for controlling how a plant reacts to herbivory because they coordinate the synthesis and distribution of defensive chemicals, improve the plant's capacity for damage repair, and maximize the use of resources to fortify the defense system. Plant hormone signal transduction pathways facilitate the quick release of defense hormones such as jasmonic acid, which triggers the expression of genes that oppose herbivory, while alkaloid biosynthesis pathways aid in the synthesis of secondary metabolites that are toxic and discourage herbivores.

Malhotra et al. (2022a) identified a significant number of differentially expressed miRNAs responding to *H. armigera* (pod borer) infestation in pigeon pea. Comparative analysis of a resistant wild variety, *C. scarabaeoides*, and susceptible variety, *C. cajun*, revealed differential expression patterns of miRNAs in response to pod borer infestation, indicating their potential role in biotic stress tolerance (Malhotra et al., 2022b). A total of 16 miRNA families were differentially regulated in *H. armigera* attacked *C. cajan*, nine upregulated and six downregulated, while only four miRNAs were differentially regulated in all three treatments for *C. scarabaeoides* (2, 8, and 18 h post infestation). Among the differentially expressed upregulated miRNAs, members of the miR156, miR171, and miR398 families have been previously associated with stress responses in plant species such as *Arabidopsis* and *S. habrochaites*, against *Pseudomonas syringae* (Jagadeeswaran et al., 2009) infection and whitefly infestation (Ketao et al., 2018). Another study performed by the same group compared insect-fed leaves of *C. scarabaeoides* and *C. cajan* to the adjacent unfed leaves, which revealed that of 210 identified miRNAs, 98 were differentially expressed in infected leaves compared to control leaves. miR1510b and miR482d were downregulated in systemic leaves across all three libraries, which is associated with the regulation of the nucleotide-binding leucine-rich repeat gene family, predominantly comprised of plant disease resistance genes. Downregulation of miR164 and miR167 was observed in pod-borer-infested leaves, with miR164 species known to regulate

NAC genes linked to abiotic stress resistance, and miR167 regulating members of *ARF* gene family (Malhotra et al., 2022). There exists extensive research on plants' systemic responses to pathogen attacks, however limited literature addresses how they coordinate defense signals against herbivory stress in pigeon pea.

11.3.5 Role of miRNAs in Plant Adaptations to Nematode Stress

miRNAs in legumes play a crucial role, not only against chewing pests but also against nematodes, however, little is known about miRNA involvement during nematode infestation. Li et al. (2012) investigated the effect of cyst nematode infestation on miRNA regulation in SCN (Soybean Cyst Nematode)-sensitive and -resistant cultivars of soybean and identified a total of 101 miRNAs belonging to 40 distinct families that were SCN-responsive (Li et al., 2012). Notably, miR319 was the only upregulated conserved miRNA in both cultivars. It was later found that the miR319-*TCP* module may play a crucial role in sensing and conferring resistance to nematode stress through the jasmonate signaling pathway (Zhao et al., 2015). Additionally, miR160 and miR390 were also found to be involved in modulating SCN infection by directly/indirectly targeting *ARF* genes, influencing root development as well as SCN tolerance. Another study investigating the potential role of miRNAs in the SCN response by comparing SCN-susceptible and -tolerant soybean genotypes identified 60 differentially expressed miRNAs representing 25 families. These miRNAs are involved in the defense response through the activation of PR genes via the ubiquitination pathway, ROS accumulation, and phytohormone (ethylene, JA, SA) signaling pathways (Tian et al., 2017). miR396 family members targeting *GRF* TF family members are also thought to affect the effectiveness of *Heterodera glycine*, a SCN, infection. Syncytium, a region in the vascular cylinder where *Heterodera glycine* penetrates and forms a feeding site, development phase is regulated through a downregulation of miR396. This activates *GRF*-mediated cell division and syncytium formation in root meristematic tissue. After the initial phase of development, *GRFs* are post-transcriptionally suppressed to maintain syncytium. The authors found that *GRF* suppression results in an increase in the defense response against *H. glycine*, aligning with previous research highlighting *GRFs*' dual role in regulating both active plant development and defense mechanisms (Liu et al., 2014).

Hints of different families of novel and conserved miRNA that might regulate different aspects of nematode infection can be found in the literature. Target prediction of the differentially expressed miRNAs expands the horizon of miRNA-mediated PTGS activity. Various new miRNA-target modules, including miR3522-genes encoding polyphenol oxidase, miR5037c-*GMC oxidoreductase*, miR5225-*Protein Tyrosine Kinase*, miR408a-*Multicopper Oxidase*, novel_miR_178- *AP2/EREBP*, and novel_miR_106-*LRR*, were found to be either directly or indirectly working in nematode infection and response management (Lei et al., 2019). The authors postulated these miRNA-gene modules work in plant defense, signal transduction, immune response, and nutrient metabolism under SCN invasion. Similar results were reported in Lei et al. (2022), where they found *NB-ARC domain, LRRs*, and *Hs1pro-1* homologs were targeted by miRNA and siRNA differentially expressed due to SCN infection. They reported reduced expression of 26 novel siRNAs including 17 PhasiRNAs and nine nat-siRNAs, resulting in increased expression of defense-related genes. In particular, gma-miR1514 triggered phasiRNAs were highlighted as miR1514 has previously been described to work in drought resistance through phasiRNA production regulating *NAC* expression (Sosa-Valencia et al., 2017).

By regulating defense-related genes, transcription factors, and genes involved in hormone signaling, miRNAs finetune the gene expression network during nematode infection as well as herbivory. Though a similar pattern of regulatory networks has been observed for conserved and well-established miRNAs, a plethora of novel miRNAs has also been identified to play significant roles in response to these biotic stressors. These evolutionary adaptations may have arisen from the need to distinguish beneficial and harmful microbes, given the unique ability of legume crops for symbiotic nitrogen fixation. To optimize gene expression for microbial interactions while defending against pathogens, novel miRNAs may have evolved. However, further research is necessary to fully elucidate and validate their functional roles in these gene-regulatory networks.

11.3.6 Role of miRNAs in Nodule Development in Legumes

One of the key physiological traits that make legumes highly attractive to farmers and researchers worldwide is their ability to fix nitrogen, which significantly enhances soil fertility and nutritive value. Legumes develop nodules through interactions with endosymbiotic nitrogen-fixing bacteria. Even though this interaction opens plants to pathogenic attacks, it also provides sustained nitrogen availability to the plants. Identifying symbionts and distinguishing them from soil pathogens is crucial for regulating and supporting nodule development while ensuring overall plant health. Recent advancements have provided clear insights into distinguishing key signaling pathways between symbionts and pathogens, enabling better regulation of plant–microbe interactions. After the identification of a suitable symbiont, the plant orchestrates the sequential growth of two specialized structures in the roots. The first structure, known as the infection thread, mediates the infection process. The second structure, which emerges later, functions as a complex surface for internal nutrient exchange. From recognition of Nod Factor to nitrogen fixation, more than 200 genes have been identified in maintaining, sustaining, and prolonging the *Rhizobium* symbiosis in nodules of legumes (Roy et al., 2020). Extensive research on model legumes such as *G. max*, *M. truncatula*, and *L. japonicus* has revealed the complex overarching regulation of nodules via miRNAs (Figure 11.2). Root nodule activation begins with bacterial Nod factor recognition by plant Nod factor receptor complex, *NFR1/LYK3* and *NFP*, which work alongside *symbiosis receptor like kinases SYMRK/DMI2* to initiate and regulate nodule formation (reviewed in Roy et al., 2020).

Insights into rhizobia infection and nodulation regulation through miR172 in soybean were reported by Wang et al. (2014). *NFR* was found to induce miR172c expression through a complex signal cascade, which downregulates *NNC-1*. The downregulation of *ENOD40* gene suppressor, *NNC-1*, leads to nodule organogenesis. *NNC1* encodes an *AP2* transcriptional repressor that binds to the promoter region of *ENOD40-1/2*, reducing its expression (Wang et al., 2014). Furthermore, Wang et al. (2019) speculated a formation of closed circular feedback loop of *NIN*-miR172-*NNC1*, which actively regulates nodulation in *G. max* (Wang et al., 2019). Several studies have reported the upregulation of miR172 in roots compared to leaves in model legumes, as a response to nodulation (Lelandais-Brière et al., 2009; Valdés-López et al., 2010; De Luis et al., 2012; Yan et al., 2013). Many independent studies in legumes have suggested miR172 regulation of *AP2* transcription factors (Yan et al., 2013; Wang et al., 2014; Holt et al., 2015; Tiwari et al., 2021), which in turn influence the expression of nonsymbiotic hemoglobin, playing a crucial role in nodulation control. Additionally, miR172a expression in roots and nodules has been found to be rhizobia infection-dependent (Holt et al., 2015). Moreover, miR156-mediated miR172 regulation has also emerged as a key factor in modulating vegetative and reproductive growth of *G. max*, thus affecting nodule development and maintenance (Yan et al., 2013). Wang et al., (2015) also reported miR156 overexpression resulted in reduced nodulation, pointing toward an interplay between miR156 and miR172 modules (Wang et al., 2015). Ectopic expression of gma-miR156 and gma-miR172 in tobacco has also confirmed their significance in stress adaptation and morpho-physiological trait determination (Lee et al., 2023). These observations bring up new possibilities for how miR156 and miR172 modules interact to regulate several vegetative developmental stages, including nodulation.

In *M. truncatula*, the miR166-*HD-Zip III* module has been reported to negatively impact symbiotic root nodule formation and lateral root development, a relationship functionally established using miR166 overexpression lines (Boualem et al., 2008). In addition to *Arabidopsis*, miR166-mediated regulation of *HD-Zip* gene family members has been found to be conserved across other leguminous plants including chickpea (Yadav et al., 2023), soybean (Subramanian et al., 2008),

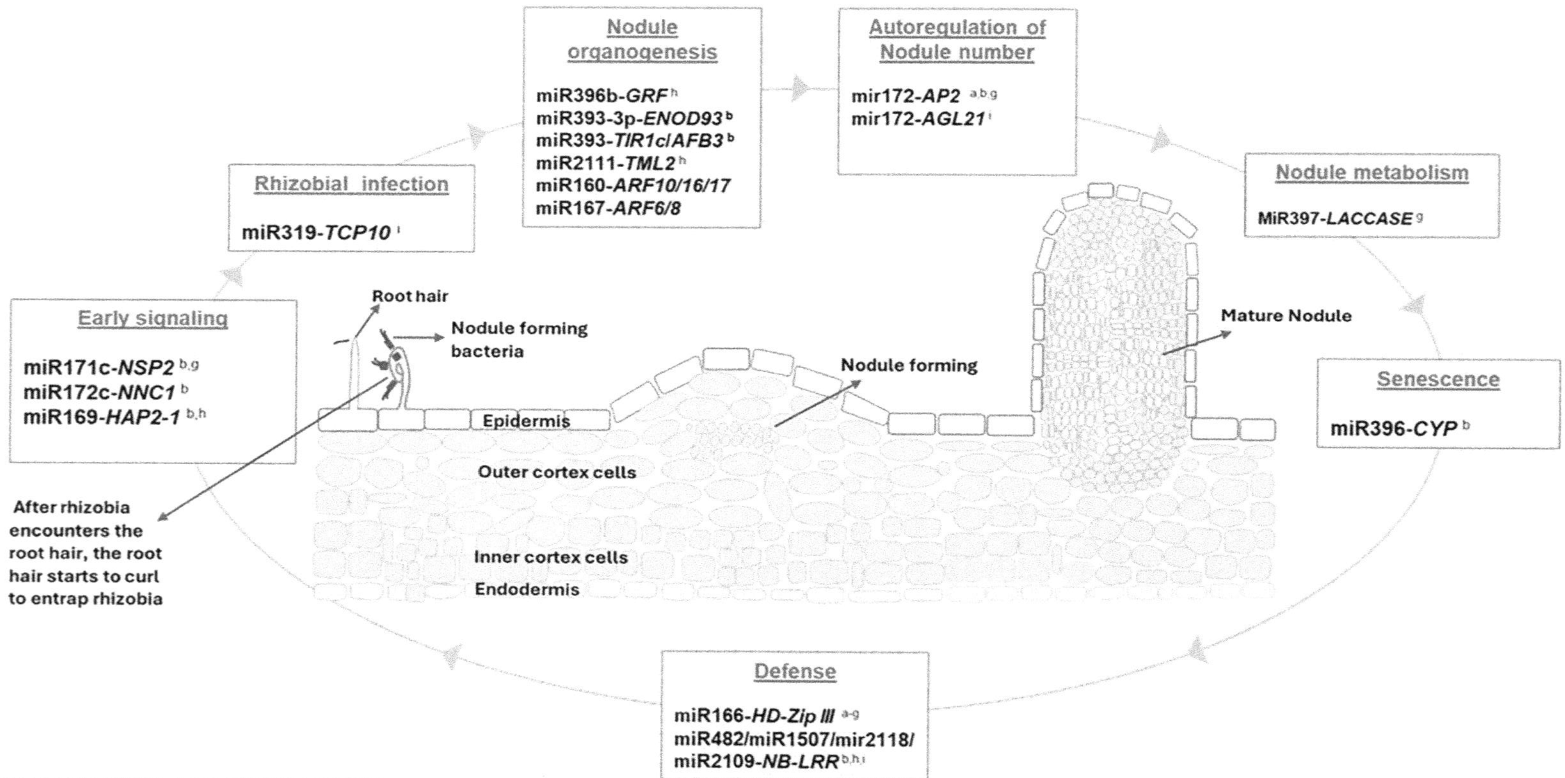

FIGURE 11.2 miRNA-mediated root nodule regulation in different genera of legumes. This figure demonstrates the roles of various miRNAs in regulating important stages of root nodule development in different legume species. The central regulatory miRNAs influence processes from early signaling and rhizobial infection to organogenesis, autoregulation, metabolism, defense, and senescence. Each stage highlights experimentally validated miRNA–mRNA interactions across various legumes, including (a) *C. arietinum*, (b) *G. max*, (c) *M. sativa*, (d) *A. hypogaea*, (e) *V. mungo*, (f) *L. albus*, (g) *L. japonicus*, (h) *M. truncatula*, and (i) *P. vulgaris*.

narrow-leafed lupin (DeBoer et al., 2019), *M. sativa* (Pokoo et al., 2018), peanut (*A. hypogea*) (Ram et al., 2019), and *V. mungo* (Kundu et al., 2017). In addition to miR166, miR396 has also been implicated in regulating nodulation in soybean by targeting *CYP*, which is an essential inducer of nodule senescence (Subramanian et al., 2008). Bazin et al. (2013) observed that miR396b plays an active role in root growth and mycorrhizal colonization by targeting *GRF1-6* and two *bHLH79-like* genes, resulting in reduced root growth and nodule formation. The miR396-*GRF* module modulates the transition between stem cells and transit-amplifying cells in *Arabidopsis* roots, thereby helping to determine cell fate (Rodriguez et al., 2015). As *PLT* induces miR396 in stem cells to repress *GRF* activity, it is plausible that *GRF* regulation in growing nodules can be hindered through *PLT*-induced miR396, thus resulting in reduced root growth and nodule formation in miR396 overexpression lines in *M. truncula*. Tiwari and Bhatia (2020) showed a different angle of miR396 regulation in chickpea as they reported an auxin-cytokinin-dependent activation of miR396. Could the ability of *PLT* and *GRF*s to sequester auxin and cytokinin be a key mechanism in regulating cell fate throughout legume plant development, with miR396 acting as a central mediator of this process?

The miR170/171 family, an ancient miRNA family conserved in the plant kingdom, is known to play a significant role in plant growth and development, hormone signaling, and stress response. In addition to regulating the expression of the *GRAS* gene family, miR171-mediated regulation extends to *NSP* and lncRNAs, as well as a few other genes (reviewed in Pei et al., 2023). miR171 also plays a critical role in nodule formation and maintenance in legumes, as miR171c targets *NSP2* and exhibits symbiosis-dependent expression in determinate nodules and shoot tissues in *L. japonicus* (De luis et al., 2012; Pei et al., 2023). Regulation of *NSP2* expression is an essential component of nod factor signaling. The downregulation of *GmSCL-6* and *GmNSP2*, which are involved in the regulation of *NIN*, *ENOD40*, and *ERF* genes, was observed because of the ectopic expression of gma-miR171o and gma-miR171q. This downregulation led to a decrease in the number of nodules in *G. max* (Hossain et al., 2019). The miR171-*NSP2* module has also been demonstrated to be functional in *M. truncula* (Devers et al., 2011; Lauressergues et al., 2012). This establishes the conserved nature of miR171-mediated nodule regulation throughout the legume species. miR393-3p was also discovered to work downstream of the miR171-*NSP2* module by targeting *Early nodulin 93* (*ENOD93*), which significantly decreases nodule formation in soybean (Yan et al., 2015).

Plant *NLR* genes play a crucial role in regulating effector-triggered immunity by promoting programmed cell death (PCD) in the host plant. In the absence of a pathogen, these genes are kept in check through miRNA-mediated regulation, aiming to decrease auto-activation and prevent undesirable PCD. During infection, the levels of miR482/2118 (and subsequently phasiRNAs) may drop to release the expression of *NLR* genes and enhance effector triggered immunity (Fei, Q., Zhang, Y., Xia, R., & Meyers, B. C., 2016. To keep plant resistance genes, i.e., members of the *Nucleotide binding and Lucine rich repeat* (*NB-LRR*) gene family, in check during nodule initiation as well as in the growing nodules, the expression of the members of miR482 superfamily, including miR1507, miR2118, and miR2109, is upregulated, as shown in *M. truncula* and soybean (Hui et al., 2010; Sós-Hegedűs et al., 2020). In addition, siRNAs produced from *NB-LRR* PHAS loci also play a critical role in regulating nodule development in different legumes, namely *M. truncula* (Zhai et al., 2011), *G. max* (Subramanian, 2012), and *P. vulgaris* (Formey et al., 2015). miRNAs, miRNA-triggered tasiRNAs, and tasiRNA-triggered phasiRNAs constitute a complicated network of regulating the expression of big *NB-LRR* gene family members in legume and non-legume crops, as reported by Zhai et al. (2011). Expression of miRNAs is not only dependent on the specific stage of rhizobial infection and nodule growth but is also dependent on the nitrogen availability to the plant, a phenomenon that has been very-well studied (Gautrat et al., 2020). Reports demonstrate that miR2111, which targets *TML1* and *TML2*, is produced in shoot and delivered to roots (Tsikou et al., 2018; Gautrat et al., 2019). In roots, miR2111 acts as a shoot-to-root signaling effector to promote root nodulation, a process that is negatively regulated by SUNN receptor in *M. truncula* (Gautrat et al., 2020) and its ortholog HARF1 receptor in *L. japonicum* (Okuma et al., 2020).

Other miRNAs, such as miR169 and miR319, have also been reported to play a significant role in nodule formation. Combier et al. (2006) demonstrated that miR169 plays an important role in regulating HAP2-1 proteins, which in turn affects nodulation in *M. truncatula*. Similar results were also shown in *G. max*, where nodule-specific expression of miR169 was significantly lower compared to roots. However, the expression of *Glyma10g10240* and *Glyma17g05920*, HAP protein coding loci, was significantly higher in nodules (Turner et al., 2012). The miR319-*TCP* module has also been demonstrated to play a significant role in regulating nodule development via jasmonic acid (JA) signaling. *TCP10* regulates *LOX* gene expression in roots, which is important for JA signaling. Overexpression of miR319d in *P. vulgaris* resulted in downregulation of *TCP*, which resulted in reduced JA production in roots. Reduced JA in roots increased the infection thread formation by bacteria, which should increase the nodule formation, but to their surprise, they found less nodulation in these overexpressed lines. They proposed that this may be due to other signaling pathway participation in this complex nodule regulation mechanism (Martín-Rodríguez et al., 2018).

An intricate hormonal balance is essential for the development of nodules, and miRNAs play a significant role in modulating the hormonal balance through a complex regulatory mechanism (reviewed in Tiwari et al., 2021). miR160 targets *ARF*, which in turn represses the activity of *GH3s* contributing to reduced auxin in root tissues of *M. trunculata*. Overexpressed lines of miR160 revealed its direct involvement in regulating *ARF* gene family members to modulate nodule organogenesis in *Medicago* (Bustos-Sanmamed et al., 2013). The detailed characteristics of miR160 expression have been emphasized, with it being functionally upstream of ENOD genes and influenced by the stage of nodule organogenesis, indicating its reliance on Nod factor signaling (Tiwari et al., 2021). The variations in the expression of mtr-miR160c and mtr-miR160d play a significant role in the spatiotemporal regulation of nodule organogenesis. Specifically, mtr-miR160c is mostly localized in the periphery region, whereas mtr-miR160d is predominantly expressed in the primordium of nodules. miR167, which specifically targets *ARF6/8*, is also linked to the regulation of auxins necessary for the growth and development of a plant. It has been found that the miR167-*ARF8* module plays a crucial role in establishing cellular responses to nitrogen availability, and thus regulating the plasticity of lateral roots (Gifford et al., 2008). miR167-mediated regulation of nodulation can be speculated through repeated reports of its abundance in developing nodules (Wang et al., 2009; De Luis et al., 2012). In addition to miR160 and miR167, the regulation of auxin accumulation in lateral roots, which determines the physiological characteristics of the lateral roots such as root length and number, has been linked to miR172-mediated *Agamous-like protein* (*AGL21*) regulation (Íñiguez et al., 2015). This regulation plays a crucial role in shaping the rhizosphere, which can alter nodule development while also having pleotropic effects on plant growth and development. Furthermore, Turner et al. (2013) demonstrated the targets of miR393, i.e., *GmTIR1c* and *GmAFB3*, exhibit diverse expression patterns during nodulation, and that overexpression of *TIR1c* and *AFB3* resulted in a higher number of infection foci and nodules (Turner et al., 2013). Mao et al. (2013) also reported that the miR393-*AFB* module regulates indeterminate but not determinate nodule development in *M. trunculata* (Mao et al., 2013), demonstrating the conserved role of the miR393-*AFB* module in nodule development across legumes.

In addition to highlighting the role of miRNA-mediated gene regulation toward nodule development, various studies have elucidated the links between miRNAs and gene networks related to nodule function and senescence. Along with a significant differential expression of members of the miR156, miR160, miR162, miR164, miR167, miR172, miR390, miR396, miR398, miR408, and miR1511 families, Figueredo et al. (2020) have identified strong upregulation of miR159, miR399, and miR3508 in the developed nodules compared to roots in *A. hypogea*. Moreover, it was discovered that the miR397-*LAC* module was active within nitrogen-fixing nodules and that it contributed to copper homeostasis related to nitrogen fixation in *L. japonicus* (De Luis et al., 2012). Like *A. hypogea*, upregulation of miR399j (an isoform of miR399) was also reported in nodules in *P. vulgaris* (Formey et al., 2015). Even though the family of miR399 was known to work on phosphate homeostasis, miR399j was proposed to regulate the expression of NB-ARC domain-containing protein(s), therefore regulating the plant's disease resistance response during formation of bacterial infection thread (Formey et al., 2015). They further suggested that a novel miRNA, miRNov494, and its predicted target, *aldehyde dehydrogenase* (phvul.004G162200.1), may also be an essential module regulating nodule development. Furthermore, miRNov153, another novel miRNA encoded in an intron of F-box protein gene known to work in root symbiosis, was predicted to target *uridine kinase* (phvul.003g180800). However, these claims require further validation.

The evidence highlights a complex interplay of phytohormones and enzymes in early flavonoid signaling, NOD factor recognition, differentiation between beneficial and harmful interactions, and the stages of nodule development, from organogenesis to maturation and senescence. MicroRNAs play a crucial role in orchestrating these processes – some well characterized, others yet to be uncovered. Advances in software and functional genomics tools promise to shed more light on miRNA functions, further enriching our understanding of nodulation. This underscores not only the conserved regulatory roles of miRNAs but also their remarkable evolutionary adaptation to the unique needs of different plant families.

11.4 CONCLUSION

Despite their small size, miRNAs finetune the expression of their targets to ensure legumes have optimal means to respond to adverse conditions, and thus ensure their growth and survival. The expanding knowledge on legume miRNA repertoires and their functional roles in stress signaling pathways allow us to understand their potential as targets for developing stress-tolerant crops. Additionally, understanding the pleiotropic impacts of miRNA-target modules would allow researchers to visualize a clearer, although not comprehensive, path toward utilizing miRNAs as targets for crop improvement platforms. So far, numerous conserved and novel targets have been identified for both conserved and novel miRNAs across the plant kingdom, including legumes. Because mature miRNAs are only 21–24 nucleotides in length and miRNA-target interactions are based on sequence complementarity, relying solely on predictions for miRNA discovery or miRNA-target interaction discovery can result in high noise, leading to a significant number of false positives over true positives. Therefore, researchers must approach prediction-based results cautiously and ensure they generate experimental proof to complement computational results before exploring a novel miRNA–target interaction for downstream crop improvement technologies. Moreover, continued exploration of tissue-specific and developmental time-specific miRNAs, particularly within stress-specific contexts, will lead the way for innovative approaches in functional omics and precision breeding. Ultimately, integrating miRNA biology into crop improvement programs could significantly contribute to sustainable agriculture and global food security amid escalating environmental challenges.

ABBREVIATIONS

ADH1 – ALCOHOL DEHYDROGENASE 1
AFB – AUXIN SIGNALING F-BOX
AGO1 – ARGONAUTE1
ALS3 – ALUMINUM SENSITIVE 3
AP2 – APETALA 2
APS – ATP SULFURYLASE
ARF – AUXIN RESPONSE FACTOR
ABF3/4 – AUXIN-SIGNALING F-BOX 3/4
BTB – BR-C, TTK AND BAB
CAT – CATALASE

COX5B – *CYTOCHROME C OXIDASE SUBUNIT 5B*
CRK6 – *CDPK-RELATED KINASE 6*
CSD – *COPPER/ZINC SUPEROXIDE DISMUTASE*
CYP – *CYSTEINE PROTEASE GENE*
DFR – *DIHYDROFLAVONOL-4-REDUCTASE*
ENOD40 – *EARLY NODULATION-40*
ENOD93 – *EARLY NODULIN 93*
EREBP – *ETHYL RESPONSIVE ELEMENT BINDING PROTEIN*
ERF – *ETHYLENE RESPONSE FACTOR*
GH3 – *GRETCHEN HAGEN3*
GRF – *GROWTH REGULATING FACTOR*
HARF1 – *HYPERNODULATION ABERRANT ROOT FORMATION1 RECEPTOR*
HD-ZIP III – *CLASS-III HOMEODOMAIN-LEUCINE ZIPPER (HD-ZIP III)*
LAC – *LACCASE*
LPR1 – *LOW PHOSPHATE ROOT 1*
LRR-TIR – *LEUCINE-RICH REPEATS–TOLL/INTERLEUKIN-1 RECEPTOR*
LYK3 – *LYSM RECEPTOR KINASE*
MYB – *MYELOBLASTOSIS*
NAM – *N-ACETYLMURAMIC ACID*
NB-ARC – *NUCLEOTIDE-BINDING ADAPTOR SHARED BY APAF-1, R PROTEINS, AND CED-4*
NFP – *NFR5/NOD FACTOR PERCEPTION*
NFR1 – *NOD FACTOR RECEPTOR 1*
NF-Y – *NUCLEAR FACTOR-Y*
NIN – *NODULE INCEPTION*
NNC-1 – *NODULE NUMBER CONTROL-1*
NSP – *NODULATION SIGNALING PATHWAY*
PCY – *PLANTACYANIN*
PHO2 – *PHOSPHATE2*
PLT – *PLETHORA*
POD – *PEROXIDASE*
POZ – *POX VIRUS AND ZINC FINGER*
PPR – *PENTATRICOPEPTIDE REPEAT*
PR – *PATHOGENESIS-RELATED*
SCL – *SCARECROW-LIKE*
SPL/SPB – *SQUAMOSA PROMOTER BINDING PROTEIN-LIKE/SQUAMOSA PROMOTER BINDING LIKE PROTEIN*
STOP1-ALMT1 – *SENSITIVE TO PROTON RHIZITOXICITY1-ALUMINUM ACTIVATED MALATE TRANSPORTER1*
SULTR – *SULPHATE TRANSPORTER*
SUNN – *SUPER NUMERIC NODULE RECEPTOR*
TCP – *TEOSINTE BRANCHED1/CYCLOIDEA/PROLIFERATING CELL FACTOR*
TIR1 – *TRANSPORT INHIBITOR RESPONSE 1*
TML1 – *TOO MUCH LOVE 1*
UBC24 – *UBIQUITIN CONJUGATING ENZYME 24*
UCL – *UCLACYANIN*
A. hypogea – *Arachis hypogea*
A. thaliana – *Arabidopsis thaliana*
B. napus – *Brassica napus*
C. arietinum – *Cicer arietinum*
C. cajan L. – *Cajanus cajan* L.

C. cajun – Cajanus cajun
C. scarabaeoides – Cajanus scarabaeoides
G. max – Glycine max
G. soja – Glycine soja
H. armigera – Helicoverpa armigera
L. japonicus – Lotus japonicus
L. sativus – Lathyrus sativus
M. ruthentica – Medicago ruthentica
M. sativa – Medicago sativa
M. trunculata – Medicago trunculata
O. sativa – Oryza sativa
P. sativum – Pisum sativum
P. vulgaris – Phaseolus vulgaris
S. habrochaites – Solanum habrochaites
T. aestivum – Triticum aestivum
V. unguiculata – Vigna unguiculata

AUTHOR CONTRIBUTIONS

I.A., A.K., A.S., P.D. – Investigation, Visualization, Writing – original draft, review, and editing. B.O., S.J. – Investigation, Writing – original draft, review, and editing. All authors have read and agreed to the published version of the manuscript.

FUNDING

This research received no external funding.

CONFLICTS OF INTEREST

The authors declare no conflicts of interest.

NOTES

* Authors contributed equally to this work.
# Corresponding author of the manuscript.

REFERENCES

Almeida DM, Gregorio GB, Oliveira MM, Saibo NJ (2017) Five novel transcription factors as potential regulators of OsNHX1 gene expression in a salt tolerant rice genotype. *Plant Molecular Biology* 93: 61–77.

Ángel Martín-Rodríguez J, Ariani A, Leija A, Elizondo A, Fuentes SI, Ramirez M, Gepts P, Hernández G, Formey D (2021) Phaseolus vulgaris MIR1511 genotypic variations differentially regulate plant tolerance to aluminum toxicity. *The Plant Journal* 105: 1521–1533.

Arenas-Huertero C, Pérez B, Rabanal F, Blanco-Melo D, De la Rosa C, Estrada-Navarrete G, Sanchez F, Covarrubias AA, Reyes JL (2009) Conserved and novel miRNAs in the legume Phaseolus vulgaris in response to stress. *Plant Molecular Biology* 70: 385–401.

Arshad M, Feyissa BA, Amyot L, Aung B, Hannoufa A (2017) MicroRNA156 improves drought stress tolerance in alfalfa (Medicago sativa) by silencing SPL13. *Plant Science* 258: 122–136.

Arshad M, Gruber MY, Wall K, Hannoufa A (2017) An insight into microRNA156 role in salinity stress responses of alfalfa. *Frontiers in Plant Science* 8: 356.

Asefpour Vakilian K (2020) Machine learning improves our knowledge about miRNA functions towards plant abiotic stresses. *Scientific Reports* 10: 3041.

Ashraf MA, Rahman A (2019) Cold stress response in Arabidopsis thaliana is mediated by GNOM ARF-GEF. *The Plant Journal* 97(3): 500–516.

Barrera-Figueroa BE, Gao L, Diop NN, Wu Z, Ehlers JD, Roberts PA, Close TJ, Zhu J-K, Liu R (2011) Identification and comparative analysis of drought-associated microRNAs in two cowpea genotypes. *BMC Plant Biology* 11: 1–11.

Bazin J, Khan GA, Combier JP, Bustos-Sanmamed P, Debernardi JM, Rodriguez R, Sorin C, Palatnik J, Hartmann C, Crespi M (2013) miR396 affects mycorrhization and root meristem activity in the legume Medicago truncatula. *The Plant Journal* 74: 920–934.

Beauclair L, Yu A, Bouché N (2010) microRNA-directed cleavage and translational repression of the copper chaperone for superoxide dismutase mRNA in Arabidopsis. *The Plant Journal* 62: 454–462.

Bessada SM, Barreira JC, Oliveira MBP (2019) Pulses and food security: Dietary protein, digestibility, bioactive and functional properties. *Trends in Food Science & Technology* 93: 53–68.

Bhat KV, Mondal TK, Gaikwad AB, Kole PR, Chandel G, Mohapatra T (2020) Genome-wide identification of drought-responsive miRNAs in grass pea (Lathyrus sativus L.). *Plant Gene* 21: 100210.

Bohra A, Gandham P, Rathore A, Thakur V, Saxena RK, Naik SS, Varshney RK, Singh NP (2021) Identification of microRNAs and their gene targets in cytoplasmic male sterile and fertile maintainer lines of pigeonpea. *Planta* 253: 1–15.

Boualem A, Laporte P, Jovanovic M, Laffont C, Plet J, Combier JP, Niebel A, Crespi M, Frugier F (2008) MicroRNA166 controls root and nodule development in Medicago truncatula. *The Plant Journal* 54: 876–887.

Buch DU, Sharma OA, Pable AA, Barvkar VT (2020) Characterization of microRNA genes from Pigeonpea (Cajanus cajan L.) and understanding their involvement in drought stress. *Journal of Biotechnology* 321: 23–34.

Bustos-Sanmamed P, Mao G, Deng Y, Elouet M, Khan GA, Bazin J, Turner M, Subramanian S, Yu O, Crespi M (2013) Overexpression of miR160 affects root growth and nitrogen-fixing nodule number in Medicago truncatula. *Functional Plant Biology* 40: 1208–1220.

Calles T, Del Castello R, Baratelli M, Xipsiti M, Navarro DK (2019) The international year of pulses-final report.

Cao C, Long R, Zhang T, Kang J, Wang Z, Wang P, Sun H, Yu J, Yang Q (2018) Genome-wide identification of microRNAs in response to salt/alkali stress in Medicago truncatula through high-throughput sequencing. *International Journal of Molecular Sciences* 19: 4076.

Cao H, Glazebrook J, Clarke JD, Volko S, Dong X (1997) The Arabidopsis NPR1 gene that controls systemic acquired resistance encodes a novel protein containing ankyrin repeats. *Cell* 88: 57–63.

Chand Jha U, Nayyar H, Mantri N, Siddique KH (2021) Non-coding RNAs in legumes: Their emerging roles in regulating biotic/abiotic stress responses and plant growth and development. *Cells* 10: 1674.

Chaves MM, Maroco JP, Pereira JS (2003) Understanding plant responses to drought—from genes to the whole plant. *Functional Plant Biology* 30: 239–264.

Chen H, Li Z, Xiong L (2012) A plant microRNA regulates the adaptation of roots to drought stress. *FEBS Letters* 586: 1742–1747.

Chen L, Wang T, Zhao M, Tian Q, Zhang W-H (2012) Identification of aluminum-responsive microRNAs in Medicago truncatula by genome-wide high-throughput sequencing. *Planta* 235: 375–386.

Cho SK, Kim JE, Park JA, Eom TJ, Kim WT (2006) Constitutive expression of abiotic stress-inducible hot pepper CaXTH3, which encodes a xyloglucan endotransglucosylase/hydrolase homolog, improves drought and salt tolerance in transgenic Arabidopsis plants. *FEBS Letters*, 580(13), 3136–3144.

Chu CC, Lee WC, Guo WY, Pan SM, Chen LJ, Li HM, Jinn TL (2005) A copper chaperone for superoxide dismutase that confers three types of copper/zinc superoxide dismutase activity in Arabidopsis. *Plant Physiology*, 139(1), 425–436.

Combier J-P, Frugier F, De Billy F, Boualem A, El-Yahyaoui F, Moreau S, Vernié T, Ott T, Gamas P, Crespi M (2006) MtHAP2-1 is a key transcriptional regulator of symbiotic nodule development regulated by microRNA169 in Medicago truncatula. *Genes & Development* 20: 3084–3088.

Cui LG, Shan JX, Shi M, Gao JP, Lin HX (2014) The miR156-SPL 9-DFR pathway coordinates the relationship between development and abiotic stress tolerance in plants. *The Plant Journal* 80: 1108–1117.

Das A, Nigam D, Junaid A, Tribhuvan KU, Kumar K, Durgesh K, Singh N, Gaikwad K (2019) Expressivity of the key genes associated with seed and pod development is highly regulated via lncRNAs and miRNAs in Pigeonpea. *Scientific Reports* 9: 18191.

Dasmandal T, Rao A, Sahu S (2020) Identification and characterization of circular RNAs regulating genes responsible for drought stress tolerance in chickpea and soybean. *Indian Journal of Genetics and Plant Breeding* 80: 1–8.

Dawar P, Adhikari I, Mandal SN, Jayee B (2024) RNA metabolism and the role of small RNAs in regulating multiple aspects of RNA metabolism. *Non-Coding RNA* 11: 1.

De la Rosa C, Covarrubias AA, Reyes JL (2019) A dicistronic precursor encoding miR398 and the legume-specific miR2119 coregulates CSD1 and ADH1 mRNAs in response to water deficit. *Plant, Cell & Environment* 42: 133–144.

De Luis A, Markmann K, Cognat V, Holt DB, Charpentier M, Parniske M, Stougaard J, Voinnet O (2012) Two microRNAs linked to nodule infection and nitrogen-fixing ability in the legume Lotus japonicus. *Plant Physiology* 160: 2137–2154.

DeBoer K, Melser S, Sperschneider J, Kamphuis LG, Garg G, Gao L-L, Frick K, Singh KB (2019) Identification and profiling of narrow-leafed lupin (Lupinus angustifolius) microRNAs during seed development. *BMC Genomics* 20: 1–16.

Demirkol G (2022) MicroRNA expression under phosphate deficiency stress in red clover (Trifolium pratense L.): A three-year field experiment. *New Zealand Journal of Agricultural Research* 65: 290–301.

Devers EA (2019) MiR 171 h restricts root symbioses and shows, like its target NSP 2, a complex transcriptional regulation in Medicago truncatula. In Frans de Bruijn (ed.). *The Model Legume Medicago Truncatula*: 975–986. https://doi.org/10.1002/9781119409144.ch125. First published:13 December 2019. Print ISBN:9781119409168 |Online ISBN:9781119409144 |DOI:10.1002/9781119409144. Copyright © 2020 John Wiley & Sons, Inc. All rights reserved. The Model Legume Medicago truncatula | Wiley Online Books

Devers EA, Branscheid A, May P, Krajinski F (2011) Stars and symbiosis: microRNA-and microRNA*-mediated transcript cleavage involved in arbuscular mycorrhizal symbiosis. *Plant Physiology* 156: 1990–2010.

Ding X, Guo J, Zhang Q, Yu L, Zhao T, Yang S (2021) Heat-responsive miRNAs participate in the regulation of male fertility stability in soybean CMS-based F1 under high temperature stress. *International Journal of Molecular Sciences* 22: 2446.

Dong Z, Shi L, Wang Y, Chen L, Cai Z, Wang Y, Jin J, Li X (2013) Identification and dynamic regulation of microRNAs involved in salt stress responses in functional soybean nodules by high-throughput sequencing. *International Journal of Molecular Sciences* 14: 2717–2738.

Eitzinger J, Orlandini S, Stefanski R, Naylor R (2010) Climate change and agriculture: Introductory editorial. *The Journal of Agricultural Science* 148: 499–500.

El-Sappah AH, Elbaiomy RG, Elrys AS, Wang Y, Zhu Y, Huang Q, Yan K, Xianming Z, Abbas M, El-Tarabily KA (2021) Genome-wide identification and expression analysis of metal tolerance protein gene family in *Medicago truncatula* under a broad range of heavy metal stress. *Frontiers in Genetics* 12. Frontiers | Genome-Wide Identification and Expression Analysis of Metal Tolerance Protein Gene Family in Medicago truncatula Under a Broad Range of Heavy Metal Stress.

Fang X, Zhao Y, Ma Q, Huang Y, Wang P, Zhang J, Nian H, Yang C (2013) Identification and comparative analysis of cadmium tolerance-associated miRNAs and their targets in two soybean genotypes. *PLoS One* 8: e81471

Farooq M, Wahid A, Kobayashi N, Fujita D, Basra SMA (2009) Plant drought stress: Effects, mechanisms, and management. *Agronomy for Sustainable Development* 29: 185–212.

Fei Q, Xia R, Meyers BC (2013) Phased, secondary, small interfering RNAs in posttranscriptional regulatory networks. *The Plant Cell* 25: 2400–2415.

Fei, Q, Zhang, Y, Xia, R, Meyers, BC (2016). Small RNAs add zing to the zig-zag-zig model of plant defenses. *Molecular Plant-Microbe Interactions*, *29*(3), 165–169.

Feyissa BA, Arshad M, Gruber MY, Kohalmi SE, Hannoufa A (2019) The interplay between miR156/SPL13 and DFR/WD40–1 regulate drought tolerance in alfalfa. *BMC Plant Biology* 19: 1–19.

Figueredo MS, Formey D, Rodríguez J, Ibáñez F, Hernández G, Fabra A (2020) Identification of miRNAs linked to peanut nodule functional processes. *Journal of Biosciences* 45: 1–7.

Formey D, Iñiguez LP, Peláez P, Li Y-F, Sunkar R, Sánchez F, Reyes JL, Hernández G (2015) Genome-wide identification of the Phaseolus vulgaris sRNAome using small RNA and degradome sequencing. *BMC Genomics* 16: 1–17.

Franco-Zorrilla JM, Valli A, Todesco M, Mateos I, Puga MI, Rubio-Somoza I, Leyva A, Weigel D, García JA, Paz-Ares J (2007) Target mimicry provides a new mechanism for regulation of microRNA activity. *Nature Genetics* 39: 1033–1037.

Gao P, Bai X, Yang L, Lv D, Li Y, Cai H, Ji W, Guo D, Zhu Y (2010) Over-expression of osa-MIR396c decreases salt and alkali stress tolerance. *Planta* 231: 991–1001.

Garg V, Khan AW, Kudapa H, Kale SM, Chitikineni A, Qiwei S, Sharma M, Li C, Zhang B, Xin L (2019) Integrated transcriptome, small RNA and degradome sequencing approaches provide insights into Ascochyta blight resistance in chickpea. *Plant Biotechnology Journal* 17: 914–931.

Gasic K, Korban S (2006) Heavy metal stress. In Rao, K, Raghavendra, A. and Reddy, K, (eds). *Physiology and Molecular Biology of Stress Tolerance* (pp. 219–254). Dordrecht: Springer.

Gautrat P, Laffont C, Frugier F (2020) Compact root architecture 2 promotes root competence for nodulation through the miR2111 systemic effector. *Current Biology* 30: 1339–1345. e1333.

Gautrat P, Mortier V, Laffont C, De Keyser A, Fromentin J, Frugier F, Goormachtig S (2019) Unraveling new molecular players involved in the autoregulation of nodulation in Medicago truncatula. *Journal of Experimental Botany* 70: 1407–1417.

Gifford ML, Dean A, Gutierrez RA, Coruzzi GM, Birnbaum KD (2008) Cell-specific nitrogen responses mediate developmental plasticity. *Proceedings of the National Academy of Sciences* 105: 803–808.

Guan Q, Lu X, Zeng H, Zhang Y, Zhu J (2013) Heat stress induction of mi R 398 triggers a regulatory loop that is critical for thermotolerance in *Arabidopsis*. *The Plant Journal* 74: 840–851.

Gul Z, Barozai MYK, Din M (2017) In-silico based identification and functional analyses of miRNAs and their targets in Cowpea (Vigna unguiculata L.). *Aims Genetics* 4: 138–165.

Hajyzadeh M, Turktas M, Khawar KM, Unver T (2015) miR408 overexpression causes increased drought tolerance in chickpea. *Gene* 555: 186–193.

Hawker NP, Bowman JL (2004) Roles for class III HD-Zip and KANADI genes in Arabidopsis root development. *Plant Physiology* 135: 2261–2270.

Hoang NT, Tóth K, Stacey G (2020) The role of microRNAs in the legume–Rhizobium nitrogen-fixing symbiosis. *Journal of Experimental Botany* 71: 1668–1680.

Hofferek V, Mendrinna A, Gaude N, Krajinski F, Devers EA (2014) MiR171h restricts root symbioses and shows like its target NSP2 a complex transcriptional regulation in Medicago truncatula. *BMC Plant Biology* 14: 1–16.

Holt DB, Gupta V, Meyer D, Abel NB, Andersen SU, Stougaard J, Markmann K (2015) micro RNA 172 (miR172) signals epidermal infection and is expressed in cells primed for bacterial invasion in Lotus japonicus roots and nodules. *New Phytologist* 208: 241–256.

Hossain MS, Hoang NT, Yan Z, Tóth K, Meyers BC, Stacey G (2019) Characterization of the spatial and temporal expression of two soybean miRNAs identifies SCL6 as a novel regulator of soybean nodulation. *Frontiers in Plant Science* 10: 452292.

Huang S, Lu G, Tang C, Ji Y, Tan G, Hu D, Cheng J, Wang G, Qi J, Yang Y (2018) Identification and comparative analysis of aluminum-induced microRNAs conferring plant tolerance to aluminum stress in soybean. *Biologia Plantarum* 62: 97–108.

Hui L, Ying D, Tianlong W, Senthil S, Oliver Y (2010, August) Misexpression of miR482, miR1512, and miR1515 Increases Soybean Nodulation, *Plant Physiology* 153(4): 1759–1770. https://doi.org/10.1104/pp.110.156950

Íñiguez LP, Nova-Franco B, Hernández G (2015) Novel players in the AP2-miR172 regulatory network for common bean nodulation. *Plant Signaling & Behavior* 10: e1062957.

Ishfaqe Q, Sami A, Zeshan Haider M, Ahmad A, Shafiq M, Ali Q, Batool A, Haider MS, Ali D, Alarifi S, Islam MS 2024. Genome wide identification of the NPR1 gene family in plant defense mechanisms against biotic stress in chili (*Capsicum annuum L.*). *Frontiers in Microbiology* 15: 1437553.

Jagadeeswaran G, Saini A, Sunkar R (2009) Biotic and abiotic stress down-regulate miR398 expression in Arabidopsis. *Planta* 229: 1009–1014.

Javed M, Sinha A, Shukla LI (2019) Evaluation of mature miR398 family, expression analysis and the post-transcriptional regulation evidence in gamma-irradiated and nitrogen-stressed *Medicago sativa* seedlings. *International Journal of Radiation Biology* 95: 585–596.

Jerome Jeyakumar JM, Ali A, Wang W-M, Thiruvengadam M (2020) Characterizing the role of the miR156-SPL network in plant development and stress response. *Plants* 9: 1206.

Jiang J, Zhu H, Li N, Batley J, Wang Y (2022) The miR393-target module regulates plant development and responses to biotic and abiotic stresses. *International Journal of Molecular Sciences* 23: 9477.

Jones-Rhoades MW, Bartel DP (2004) Computational identification of plant microRNAs and their targets, including a stress-induced miRNA. *Molecular Cell* 14(6): 787–799.

Jovanović Ž, Stanisavljević N, Mikić A, Radović S, Maksimović V (2014) Water deficit down-regulates miR398 and miR408 in pea (*Pisum sativum L.*). *Plant Physiology and Biochemistry* 83: 26–31.

Juneja S, Saini R, Adhikary A, Yadav R, Khan SA, Nayyar H, Kumar S (2023) Drought priming evokes essential regulation of Hsp gene families, Hsfs and their related miRNAs and induces heat stress tolerance in chickpea. *Plant Stress* 10: 100189.

Jyothi M, Rai D, Nagesh Babu R (2015) Identification and characterization of high temperature stress responsive novel miRNAs in French bean (Phaseolus vulgaris). *Applied Biochemistry and Biotechnology* 176: 835–849.

Jyothi M, Usha S, Suchithra B, Ulfath TS, Devaraj V, Babu RN (2018) Boron toxicity induces altered expression of miRNAs in French bean (Phaseolus vulgaris L.). *Journal of Applied Biology and Biotechnology* 6: 1–10.

Kantar M, Lucas SJ, Budak H (2010) miRNA expression patterns of Triticum dicoccoides in response to shock drought stress. *Planta* 233: 471–484.

Ketao W, Xiaomei S, Xia C, Yongchen D, Zhang S, Jianchang G (2018) Identification and characterization of microRNA during Bemisia tabaci Infestations in Solanum lycopersicum and Solanum habrochaites. *Horticultural Plant Journal* 4: 62–72.

Khandal H, Parween S, Roy R, Meena MK, Chattopadhyay D (2017) MicroRNA profiling provides insights into post-transcriptional regulation of gene expression in chickpea root apex under salinity and water deficiency. *Scientific Reports* 7: 4632.

Kim W, Ahn HJ, Chiou T-J, Ahn JH (2011) The role of the miR399-*PHO2* module in the regulation of flowering time in response to different ambient temperatures in *Arabidopsis thaliana. Molecules and Cells* 32: 83–88.

Kim JS, Mizoi J, Kidokoro S, Maruyama K, Nakajima J, Nakashima K, et al. (2012) Arabidopsis growth regulating factor7 functions as a transcriptional repressor of abscisic acid- and osmotic stress-responsive genes, including DREB2A. *Plant Cell* 24: 3393–3405. doi: 10.1105/tpc.112.100933

Kim JY, Lee HJ, Jung HJ, Maruyama K, Suzuki N, Kang H (2010) Overexpression of microRNA395c or 395e affects differently the seed germination of Arabidopsis thaliana under stress conditions. *Planta* 232: 1447–1454.

Kitazumi A, Kawahara Y, Onda TS, De Koeyer D, de los Reyes BG (2015) Implications of miR166 and miR159 induction to the basal response mechanisms of an andigena potato (Solanum tuberosum subsp. andigena) to salinity stress, predicted from network models in Arabidopsis. *Genome* 58: 13–24.

Kohli D, Joshi G, Deokar AA, Bhardwaj AR, Agarwal M, Katiyar-Agarwal S, Srinivasan R, Jain PK (2014) Identification and characterization of wilt and salt stress-responsive microRNAs in chickpea through high-throughput sequencing. *PloS One* 9: e108851.

Koornneef M, Alonso-Blanco C, Peeters A (1997) Genetic approaches in plant physiology. *The New Phytologist* 137: 1–8.

Koptekin D, Aktaş LY (2016) Identification of conserved miRNAs and their target genes in faba bean by EST based homology analysis. *Avrupa Bilim ve Teknoloji Dergisi* 5: 1–6.

Kuczyński J, Gracz-Bernaciak J, Twardowski T, Karłowski WM, Tyczewska A (2022) Cold stress-induced miRNA and degradome changes in four soybean varieties differing in chilling resistance. *Journal of Agronomy and Crop Science* 208: 777–794.

Kuczyński J, Twardowski T, Nawracała J, Gracz-Bernaciak J, Tyczewska A (2020) Chilling stress tolerance of two soya bean cultivars: Phenotypic and molecular responses. *Journal of Agronomy and Crop Science* 206: 759–772.

Kulcheski FR, de Oliveira LF, Molina LG, Almerão MP, Rodrigues FA, Marcolino J, Barbosa JF, Stolf-Moreira R, Nepomuceno AL, Marcelino-Guimarães FC (2011) Identification of novel soybean microRNAs involved in abiotic and biotic stresses. *BMC Genomics* 12: 1–17.

Kumar J, Mir RR, Shafi S, Sen Gupta D, Djalovic I, Miladinovic J, Kumar R, Kumar S, Kumar R (2021) Genomics associated interventions for heat stress tolerance in cool season adapted grain legumes. *International Journal of Molecular Sciences* 23: 399.

Kundu A, Patel A, Paul S, Pal A (2015) Transcript dynamics at early stages of molecular interactions of MYMIV with resistant and susceptible genotypes of the leguminous host, Vigna mungo. *PLoS One* 10: e0124687.

Kundu A, Paul S, Dey A, Pal A (2017) High throughput sequencing reveals modulation of microRNAs in Vigna mungo upon Mungbean Yellow Mosaic India Virus inoculation highlighting stress regulation. *Plant Science* 257: 96–105.

Kundu S, Chakraborty D, Kundu A, Pal A (2013) Proteomics approach combined with biochemical attributes to elucidate compatible and incompatible plant-virus interactions between Vigna mungo and Mungbean Yellow Mosaic India Virus. *Proteome Science* 11: 1–14.

Lambers H, Chapin F, Pons T (2008) *Plant Physiological Ecology.* 2nd ed. New York: Springer, 101: 149.

Lauressergues D, Delaux PM, Formey D, Lelandais-Brière C, Fort S, Cottaz S, Bécard G, Niebel A, Roux C, Combier JP (2012) The microRNA miR171h modulates arbuscular mycorrhizal colonization of Medicago truncatula by targeting NSP2. *The Plant Journal* 72: 512–522.

Lee S, Singh MB, Bhalla PL (2023) Functional analysis of soybean miR156 and miR172 in tobacco highlights their role in plant morphology and floral transition. *Plant Physiology and Biochemistry* 196: 393–401.

Lei P, Han B, Wang Y, Zhu X, Xuan Y, Liu X, Fan H, Chen L, Duan Y (2019) Identification of microRNAs that respond to soybean cyst nematode infection in early stages in resistant and susceptible soybean cultivars. *International Journal of Molecular Sciences* 20: 5634.

Lei P, Qi N, Yan J, Zhu X, Liu X, Xuan Y, Fan H, Chen L, Duan Y, Wang Y (2022) Genome-wide identification of small interfering RNAs from sRNA libraries constructed from soybean cyst nematode resistant and susceptible cultivars. *Gene* 832: 146557.

Lelandais-Brière C, Naya L, Sallet E, Calenge F, Frugier F, Hartmann C, Gouzy J, Crespi M (2009) Genome-wide Medicago truncatula small RNA analysis revealed novel microRNAs and isoforms differentially regulated in roots and nodules. *The Plant Cell* 21: 2780–2796.

Li B, Tao P, Xu F, He P, Wang J (2023) Function of soybean miR159 family members in plant responses to low phosphorus, high salinity, and abscisic acid treatment. *Agronomy* 13: 1798.

Li C, Nong W, Zhao S, Lin X, Xie Y, Cheung M-Y, Xiao Z, Wong AY, Chan TF, Hui JH (2022) Differential microRNA expression, microRNA arm switching, and microRNA: Long noncoding RNA interaction in response to salinity stress in soybean. *BMC Genomics* 23: 65.

Li H, Dong Y, Yin H, Wang N, Yang J, Liu X, Wang Y, Wu J, Li X (2011) Characterization of the stress associated microRNAs in Glycine maxby deep sequencing. *BMC Plant Biology* 11: 1–12.

Li L, Li Q, Davis KE, Patterson C, Oo S, Liu W, Liu J, Wang G, Fontana JE, Thornburg TE (2021) Response of root growth and development to nitrogen and potassium deficiency as well as microRNA-mediated mechanism in peanut (Arachis hypogaea L.). *Frontiers in Plant Science* 12: 695234.

Li M-Y, Wang F, Xu Z-S, Jiang Q, Ma J, Tan G-F, Xiong A-S (2014) High throughput sequencing of two celery varieties small RNAs identifies microRNAs involved in temperature stress response. *BMC Genomics* 15: 1–9.

Li S, Castillo-González C, Yu B, Zhang X (2017) The functions of plant small RNA s in development and in stress responses. *The Plant Journal* 90: 654–670.

Li S, Liu L, Zhuang X, Yu Y, Liu X, Cui X, Ji L, Pan Z, Cao X, Mo B (2013) MicroRNAs inhibit the translation of target mRNAs on the endoplasmic reticulum in Arabidopsis. *Cell* 153: 562–574.

Liu, H. H., Tian, X., Li, Y. J., Wu, C. A., & Zheng, C. C. (2008). Microarray-based analysis of stress-regulated microRNAs in Arabidopsis thaliana. *Rna*, *14*(5), 836-843.

Li W, Wang T, Zhang Y, Li Y (2016) Overexpression of soybean miR172c confers tolerance to water deficit and salt stress, but increases ABA sensitivity in transgenic Arabidopsis thaliana. *Journal of Experimental Botany*, 67(1), 175–194.

Li W-X, Oono Y, Zhu J, He X-J, Wu J-M, Iida K, Lu X-Y, Cui X, Jin H, Zhu J-K (2008) The Arabidopsis NFYA5 transcription factor is regulated transcriptionally and posttranscriptionally to promote drought resistance. *The Plant Cell* 20: 2238–2251.

Li X, Wang X, Zhang S, Liu D, Duan Y, Dong W (2012) Identification of soybean microRNAs involved in soybean cyst nematode infection by deep sequencing. *PLoS One* 7: e39650.

Li X, Xie X, Li J, Cui Y, Hou Y, Zhai L, Wang X, Fu Y, Liu R, Bian S (2017) Conservation and diversification of the miR166 family in soybean and potential roles of newly identified miR166s. *BMC Plant Biology* 17: 1–18.

Li Y, Wan L, Bi S, Wan X, Li Z, Cao J, Tong Z, Xu H, He F, Li X (2017) Identification of drought-responsive microRNAs from roots and leaves of alfalfa by high-throughput sequencing. *Genes* 8: 119.

Li Z, Tong Z, He F, Li X, Sun J (2022) Integrated mRNA and microRNA expression analysis of root response to phosphate deficiency in medicago sativa. *Frontiers in Plant Science* 13: 989048.

Li Z, Xu H, Li Y, Wan X, Ma Z, Cao J, Li Z, He F, Wang Y, Wan L (2018) Analysis of physiological and miRNA responses to Pi deficiency in alfalfa (Medicago sativa L.). *Plant Molecular Biology* 96: 473–492.

Lian H, Wang L, Ma N, Zhou C-M, Han L, Zhang T-Q, Wang J-W (2021) Redundant and specific roles of individual *MIR172* genes in plant development. *PLoS Biology* 19(2), p.e3001044.

Ling L, Zhang W, An Y, Du B, Wang D, Guo C (2020) Genome-wide analysis of the TCP transcription factor genes in five legume genomes and their response to salt and drought stresses. *Functional & Integrative Genomics* 20: 537–550.

Liu J, Rice JH, Chen N, Baum TJ, Hewezi T (2014) Synchronization of developmental processes and defense signaling by growth regulating transcription factors. *PLoS One* 9: e98477.

Liu J, Shi Y, Yang S (2018) Insights into the regulation of C-repeat binding factors in plant cold signaling. *Journal of Integrative Plant Biology* 60: 780–795.

Liu X, Chu S, Sun C, Xu H, Zhang J, Jiao Y, Zhang D (2020) Genome-wide identification of low phosphorus responsive microRNAs in two soybean genotypes by high-throughput sequencing. *Functional & Integrative Genomics* 20: 825–838.

Liu X, Huang S, Xie H (2021) Advances in the regulation of plant development and stress response by miR167. *Frontiers in Bioscience (Landmark)* 26: 655–665.

Lu G, Tian Z, Hao Y, Xu M, Lin Y, Wei J, Zhao Y (2023) Overexpression of soybean microRNA156b enhanced tolerance to phosphorus deficiency and seed yield in Arabidopsis. *Scientific Reports* 13: 652.

Lu Z, Yang Z, Tian Z, Gui Q, Dong R, Chen C (2023) Genome-wide analysis and identification of microRNAs in Medicago truncatula under aluminum stress. *Frontiers in Plant Science* 14: 1137764.

Luo C, Bashir NH, Li Z, Liu C, Shi Y, Chu H (2024) Plant microRNAs regulate the defense response against pathogens. *Frontiers in Microbiology* 15: 1434798.

Ma Z, Hu X, Cai W, Huang W, Zhou X, Luo Q, Yang H, Wang J, Huang J (2014) Arabidopsis miR171-targeted Scarecrow-Like Proteins bind to GT cis-elements and mediate gibberellin regulated chlorophyll biosynthesis under light conditions. *PLoS Genetics* 10: e1004519.

Malhotra EV, Jain R, Tyagi S, Venkat Raman K, Bansal S, Pattanayak D (2022a) Identification of dynamic microRNA associated with systemic defence against Helicoverpa armigera infestation in Cajanus scarabaeoides. *Pest Management Science* 78(7): 3144–3154.

Malhotra EV, Jain R, Tyagi S, Raman KV, Bansal S, Aminedi R, Pattanayak D (2022b) Comparative analysis of herbivory responsive miRNAs to delineate pod borer (Helicoverpa armigera) resistance mechanisms in Cajanus cajan and its wild relative Cajanus scarabaeoides. *Plant Cell Reports* 41(4): 1147–1161.

Mallory AC, Bartel DP, Bartel B (2005) MicroRNA-directed regulation of Arabidopsis AUXIN RESPONSE FACTOR17 is essential for proper development and modulates expression of early auxin response genes. *The Plant Cell* 17: 1360–1375.

Mandadi KK, Misra A, Ren S, McKnight TD (2009) BT2, a BTB protein, mediates multiple responses to nutrients, stresses, and hormones in Arabidopsis. *Plant Physiology* 150(4): 1930–1939.

Mantri N, Basker N, Ford R, Pang E, Pardeshi V (2013) The role of micro-ribonucleic acids in legumes with a focus on abiotic stress response. *The Plant Genome* 6(3): plantgenome2013–05.

Mao G, Turner M, Yu O, Subramanian S (2013) miR393 and miR164 influence indeterminate but not determinate nodule development. *Plant Signaling & Behavior* 8: e26753.

Martín-Rodríguez JÁ, Leija A, Formey D, Hernández G (2018) The MicroRNA319d/TCP10 node regulates the common bean–rhizobia nitrogen-fixing symbiosis. *Frontiers in Plant Science* 9: 396921.

Matsui A, Nguyen AH, Nakaminami K, Seki M (2013) Arabidopsis non-coding RNA regulation in abiotic stress responses. *International Journal of Molecular Sciences* 14: 22642–22654.

Matthews C, Arshad M, Hannoufa A (2019) Alfalfa response to heat stress is modulated by microRNA156. *Physiologia Plantarum* 165: 830–842.

Mazurier M, Drouaud J, Bahrman N, Rau A, Lejeune-Hénaut I, Delbreil B, Legrand S (2022) Integrated sRNA-seq and RNA-seq analyses reveal a microRNA regulation network involved in cold response in Pisum sativum L. *Genes* 13: 1119.

Mendoza-Soto AB, Naya L, Leija A, Hernández G (2015) Responses of symbiotic nitrogen-fixing common bean to aluminum toxicity and delineation of nodule responsive microRNAs. *Frontiers in Plant Science* 6: 587.

Middleton PH, Jakab J, Penmetsa RV, Starker CG, Doll J, Kalo P, Prabhu R, Marsh JF, Mitra RM, Kereszt A (2007) An ERF transcription factor in Medicago truncatula that is essential for Nod factor signal transduction. *The Plant Cell* 19: 1221–1234.

Min X, Luo K, Liu W, Zhou K, Li J, Wei Z (2022) Molecular characterization of the miR156/MsSPL model in regulating the compound leaf development and abiotic stress response in Alfalfa. *Genes* 13: 331.

Mishra S, Sahu G, Shaw BP (2022) Integrative small RNA and transcriptome analysis provides insight into key role of miR408 towards drought tolerance response in cowpea. *Plant Cell Reports*, *41*(1), 75-94.

Mishra S, Spaccarotella K, Gido J, Samanta I, Chowdhary G (2023) Effects of heat stress on plant-nutrient relations: An update on nutrient uptake, transport, and assimilation. *International Journal of Molecular Sciences* 24: 15670.

Mittal M, Dhingra A, Dawar P, Payton P, Rock CD (2023) The role of microRNAs in responses to drought and heat stress in peanut (Arachis hypogaea). *The Plant Genome* 16: e20350.

Munns R, Tester M (2008) Mechanisms of salinity tolerance. *Annual Review of Plant Biology* 59: 651–681.

Nadeem M, Li J, Yahya M, Sher A, Ma C, Wang X, Qiu L (2019) Research progress and perspective on drought stress in legumes: A review. *International Journal of Molecular Sciences* 20: 2541.

Nageshbabu R, Jyothi M (2013) Profile of small interfering RNAs from French bean Phaseolus vulgaris under abiotic stress conditions. *International Journal of Pharmacy Biological Science* 4: 176–185.

Nageshbabu R, Jyothi M, Sharadamma N, Sarika S, Rai D, Devaraj V (2013) Expression of miRNAs regulates growth and development of French bean (Phaseolus vulgaris) under salt and drought stress conditions. *International Research Jounal of Biological Sciences* 2: 52–56.

Naya L, Paul S, Valdes-Lopez O, Mendoza-Soto AB, Nova-Franco B, Sosa-Valencia G, Reyes JL, Hernandez G (2014) Regulation of copper homeostasis and biotic interactions by microRNA 398b in common bean. *PLoS One* 9: e84416.

Ning L-H, Du W-k, Song H-N, Shao H-B, Qi W-C, Sheteiwy MSA, Yu D-y (2019) Identification of responsive miRNAs involved in combination stresses of phosphate starvation and salt stress in soybean root. *Environmental and Experimental Botany* 167: 103823.

Noman A, Aqeel M, Lou Y (2019) PRRs and NB-LRRs: From signal perception to activation of plant innate immunity. *International Journal of Molecular Sciences* 20: 1882.

Nuruzzaman M, Sharoni AM, Kikuchi S (2013) Roles of NAC transcription factors in the regulation of biotic and abiotic stress responses in plants. *Frontiers in Microbiology* 4: 248.

Ohama N, Sato H, Shinozaki K, Yamaguchi-Shinozaki K (2017) Transcriptional regulatory network of plant heat stress response. *Trends in Plant Science* 22: 53–65.

Okuma N, Soyano T, Suzaki T, Kawaguchi M (2020) MIR2111-5 locus and shoot-accumulated mature miR2111 systemically enhance nodulation depending on HAR1 in Lotus japonicus. *Nature Communications* 11: 5192.

Pan J, Huang D, Guo Z, Kuang Z, Zhang H, Xie X, Ma Z, Gao S, Lerdau MT, Chu C (2018) Overexpression of microRNA408 enhances photosynthesis, growth, and seed yield in diverse plants. *Journal of Integrative Plant Biology* 60: 323–340.

Park YJ, Lee HJ, Kwak KJ, Lee K, Hong SW, Kang H (2014) MicroRNA400-guided cleavage of pentatricopeptide repeat protein mRNAs renders Arabidopsis thaliana more susceptible to pathogenic bacteria and fungi. *Plant and Cell Physiology* 55: 1660–1668.

Parker LE, McElrone AJ, Ostoja SM, Forrestel EJ (2020) Extreme heat effects on perennial crops and strategies for sustaining future production. *Plant Science* 295: 110397.

Patade VY, Suprasanna P (2010) Short-term salt and PEG stresses regulate expression of MicroRNA, miR159 in sugarcane leaves. *Journal of Crop Science and Biotechnology* 13: 177–182.

Patwa N, Nithin C, Bahadur RP, Basak J (2019) Identification and characterization of differentially expressed Phaseolus vulgaris miRNAs and their targets during mungbean yellow mosaic India virus infection reveals new insight into Phaseolus-MYMIV interaction. *Genomics* 111: 1333–1342.

Paul S, Kundu A, Pal A (2011) Identification and validation of conserved microRNAs along with their differential expression in roots of *Vigna unguiculata* grown under salt stress. *Plant Cell Tiss Organ Cult* 105: 233–242. https://doi.org/10.1007/s11240-010-9857-7

Pei LL, Zhang LL, Liu X, Jiang J (2023) Role of microRNA miR171 in plant development. *PeerJ* 11: e15632.

Phiri AT, Njira KOW (2023) Grain legume-based cropping systems' effects on soil organic carbon and nutrient dynamics. *Agricultural Research* 12: 45–52.

Pokoo R, Ren S, Wang Q, Motes CM, Hernandez TD, Ahmadi S, Monteros MJ, Zheng Y, Sunkar R (2018) Genotype-and tissue-specific miRNA profiles and their targets in three alfalfa (Medicago sativa L) genotypes. *BMC Genomics* 19: 115–131.

Polhill, RM, Raven, PH, Stirton CH (1981) Evolution and systematics of the Leguminosae. Miscellaneous: Advances in legume systematics. Part 1 [Polhill, R.M.; Raven, P.H. (Editors)], 1981, 1–26 ref. 23. Authors: R. M. Polhill, P. H. Raven, C. H. Stirton, Affiliation: Royal Botanic Gardens, Kew, Richmond, Surrey TW9 3AE, UK cabidigitallibrary.org/doi/full/10.5555/19820736582.

Postel S, Kemmerling B (2009) Plant systems for recognition of pathogen-associated molecular patterns. In *Seminars in Cell & Developmental Biology,* 20(9), 1025–1031. Elsevier. doi: 10.1016/j.semcdb.2009.06.002. Epub 2009 Jun 18. PMID: 19540353.

Pramanik K, Das A, Banerjee J, Das A, Chatterjee S, Sharma R, Kumar S, Gupta S (2020) Metagenomic insights into rhizospheric microbiome profiling in lentil cultivars unveils differential microbial nitrogen and phosphorus metabolism under rice-fallow ecology. *International Journal of Molecular Sciences* 21: 8895.

Puri A, Pajak A, Hannoufa A, Marsolais F (2023) Quantitative proteomic analysis of the role of miRNA156 in alfalfa under drought stress. *Environmental and Experimental Botany* 214: 105449.

Ram MK, Mukherjee K, Pandey DM (2019) Identification of miRNA, their targets and miPEPs in peanut (Arachis hypogaea L.). *Computational Biology and Chemistry* 83: 107100.

Ramesh SV, Chouhan BS, Kumar G, Praveen S, Chand S (2017) Expression dynamics of Glycine max (L.) Merrill microRNAs (miRNAs) and their targets during Mungbean yellow mosaic India virus (MYMIV) infection. *Physiological and Molecular Plant Pathology* 100: 13–22.

Rao, S, Gupta, A, Bansal, C, Sorin, C, Crespi, M, Mathur S (2022) A conserved HSF:miR169:NF-YA loop involved in tomato and Arabidopsis heat stress tolerance. *Plant Journal* 112: 7–26.

Rasheed MU, Malik A, Haider MZ, Sami A, Shafiq M, Ali Q, Javed MA, Ali A (2025) NPR1-like genes in Theobroma cacao: Evolutionary insights and potential in enhancing resistance to Phytophthora megakarya. *PLoS One* 20(2): e0318506.

Ren S, Ma K, Lu Z, Chen G, Cui J, Tong P, Wang L, Teng N, Jin B (2019) Transcriptomic and metabolomic analysis of the heat-stress response of Populus tomentosa Carr. *Forests* 10: 383.

Robert HS, Quint A, Brand D, Vivian-Smith A, Offringa R (2009) BTB and TAZ domain scaffold proteins perform a crucial function in Arabidopsis development. *The Plant Journal* 58(1): 109–121.

Rodriguez RE, Ercoli MF, Debernardi JM, Breakfield NW, Mecchia MA, Sabatini M, Cools T, De Veylder L, Benfey PN, Palatnik JF (2015) MicroRNA miR396 regulates the switch between stem cells and transit-amplifying cells in Arabidopsis roots. *The Plant Cell* 27: 3354–3366.

Roy S, Liu W, Nandety RS, Crook A, Mysore KS, Pislariu CI, Frugoli J, Dickstein R, Udvardi MK (2020) Celebrating 20 years of genetic discoveries in legume nodulation and symbiotic nitrogen fixation. *The Plant Cell* 32: 15–41.

Ruan Q, Bai X, Wang Y, Zhang X, Wang B, Zhao Y, Zhu X, Wei X (2024) Regulation of endogenous hormone and miRNA in leaves of alfalfa (Medicago sativa L.) seedlings under drought stress by endogenous nitric oxide. *BMC Genomics* 25: 229.

Sahito ZA, Wang L, Sun Z, Yan Q, Zhang X, Jiang Q, Ullah I, Tong Y, Li X (2017) The miR172c-NNC1 module modulates root plastic development in response to salt in soybean. *BMC Plant Biology* 17: 1–12.

Sehgal A, Sita K, Kumar J, Kumar S, Singh S, Siddique KHM, Nayyar H (2017) Effects of drought, heat and their interaction on the growth, yield and photosynthetic function of lentil (*Lens culinaris* Medik.) genotypes varying in heat and drought sensitivity. *Frontiers in Plant Science* 8: 1776.

Shi R, Jiao W, Yun L, Zhang Z, Zhang X, Wang Q, Li Y, Mi F (2021) Utilization of transcriptome, small RNA, and degradome sequencing to provide insights into drought stress and rewatering treatment in Medicago ruthenica. *Frontiers in Plant Science* 12: 675903.

Shivaprasad P, Akbergenov R, Trinks D, Rajeswaran R, Veluthambi K, Hohn T, Pooggin MM (2005) Promoters, transcripts, and regulatory proteins of Mungbean yellow mosaic geminivirus. *Journal of Virology* 79: 8149–8163.

Shivaprasad PV, Chen H-M, Patel K, Bond DM, Santos BA, Baulcombe DC (2012) A microRNA superfamily regulates nucleotide binding site–leucine-rich repeats and other mRNAs. *The Plant Cell* 24: 859–874.

Shu Y, Liu Y, Li W, Song L, Zhang J, Guo C (2016) Genome-wide investigation of microRNAs and their targets in response to freezing stress in Medicago sativa L, based on high-throughput sequencing. *G3: Genes, Genomes, Genetics* 6: 755–765.

Si-Ammour A, Windels D, Arn-Bouldoires E, Kutter C, Ailhas J, Meins Jr F, Vazquez F (2011) miR393 and secondary siRNAs regulate expression of the TIR1/AFB2 auxin receptor clade and auxin-related development of Arabidopsis leaves. *Plant Physiology* 157: 683–691.

Simón-Mateo C, García JA (2006) MicroRNA-guided processing impairs Plum pox virus replication, but the virus readily evolves to escape this silencing mechanism. *Journal of Virology* 80: 2429–2436.

Singh L, Kohli D, Gaikwad K, Kansal R, Dahuja A, Paul V, Bharadwaj C, Jain PK (2021) Effect of drought stress on morphological, biochemical, physiological traits and expression analysis of microRNAs in drought-tolerant and sensitive genotypes of chickpea. *Indian Journal of Genetics and Plant Breeding* 81: 266–276.

Singh L, Mehta BK, Kumar S, Tiwari S, Kushwaha AK (2017) Role of miRNAs in legume crops in response to biotic and abiotic stress. *Journal of Pure & Applied Microbiology* 11(1): 605 –614.

Singh RJ, Chung GH, Nelson RL, (2007) Landmark research in legumes. *Genome*, 50(6): 525–537.

Song Q-X, Liu Y-F, Hu X-Y, Zhang W-K, Ma B, Chen S-Y, Zhang J-S (2011) Identification of miRNAs and their target genes in developing soybean seeds by deep sequencing. *BMC Plant Biology* 11: 1–16.

Song X, Li Y, Cao X, Qi Y (2019) MicroRNAs and their regulatory roles in plant–environment interactions. *Annual Review of Plant Biology* 70: 489–525.

Sorin C, Declerck M, Christ A, Blein T, Ma L, Lelandais-Brière C Hartmann C (2014) A mi R 169 isoform regulates specific NF-YA targets and root architecture in A rabidopsis. *New Phytologist* 202(4): 1197–1211.

Sós-Hegedűs A, Domonkos Á, Tóth T, Gyula P, Kaló P, Szittya G (2020) Suppression of NB-LRR genes by miRNAs promotes nitrogen-fixing nodule development in Medicago truncatula. *Plant, Cell & Environment* 43: 1117–1129.

Sosa-Valencia G, Palomar M, Covarrubias AA, Reyes JL (2017) The legume miR1514a modulates a NAC transcription factor transcript to trigger phasiRNA formation in response to drought. *Journal of Experimental Botany* 68: 2013–2026.

Subramanian S (2012) MicroRNA regulation of symbiotic nodule development in legumes. In *MicroRNAs in Plant Development and Stress Responses*. Signaling and Communication in Plants, vol 15. Springer, Berlin, Heidelberg . https://doi.org/10.1007/978-3-642-27384-1_9

Subramanian S, Fu Y, Sunkar R, Barbazuk WB, Zhu J-K, Yu O (2008) Novel and nodulation-regulated microRNAs in soybean roots. *BMC Genomics* 9: 1–14.

Sun Z, Wang Y, Mou F, Tian Y, Chen L, Zhang S, Jiang Q, Li X (2016) Genome-wide small RNA analysis of soybean reveals auxin-responsive microRNAs that are differentially expressed in response to salt stress in root apex. *Frontiers in Plant Science* 6: 1273.

Sunkar R, Zhu J-K (2004) Novel and stress-regulated microRNAs and other small RNAs from Arabidopsis. *Plant Cell* 16: 2001–2019.

Suresh Kumar C, Nagesh Babu R, Shafia Hoor F (2023) Differential expression of miRNAs under Boron toxicity affects circadian cycle and vegetative development in French Bean. *Journal of Stress Physiology & Biochemistry* 19: 120–131.

Swiderski MR, Birker D, Jones JD (2009). The TIR domain of TIR-NB-LRR resistance proteins is a signaling domain involved in cell death induction. *Molecular Plant-Microbe Interactions* 22(2): 157–165.

Taiz L, Zeiger E (2010) *Plant Physiology*. 5th edn. Sunderland: Sinauer Associates Inc.

Tang P, Zhang Y, Sun X, Tian D, Yang S, Ding J (2010) Disease resistance signature of the leucine-rich repeat receptor-like kinase genes in four plant species. *Plant Science*, 179(4), 399–406.

Tian B, Wang S, Todd TC, Johnson CD, Tang G, Trick HN (2017) Genome-wide identification of soybean microRNA responsive to soybean cyst nematodes infection by deep sequencing. *BMC Genomics* 18: 1–13.

Tiwari M, Bhatia S (2020) Expression profiling of miRNAs indicates crosstalk between phytohormonal response and rhizobial infection in chickpea. *Journal of Plant Biochemistry and Biotechnology* 29: 380–394.

Tiwari M, Pandey V, Singh B, Bhatia S (2021) Dynamics of miRNA mediated regulation of legume symbiosis. *Plant, Cell & Environment* 44: 1279–1291.

Tiwari M, Singh B, Yadav M, Pandey V, Bhatia S (2021) High throughput identification of miRNAs reveal novel interacting targets regulating chickpea-rhizobia symbiosis. *Environmental and Experimental Botany* 186: 104469.

Trindade I, Capitão C, Dalmay T, Fevereiro MP, Santos DMd (2010) miR398 and miR408 are up-regulated in response to water deficit in Medicago truncatula. *Planta* 231: 705–716.

Tsikou D, Yan Z, Holt DB, Abel NB, Reid DE, Madsen LH, Bhasin H, Sexauer M, Stougaard J, Markmann K (2018) Systemic control of legume susceptibility to rhizobial infection by a mobile microRNA. *Science* 362: 233–236.

Tsuji H, Aya K, Ueguchi-Tanaka M, Shimada Y, Nakazono M, Watanabe R, Matsuoka M (2006) GAMYB controls different sets of genes and is differentially regulated by microRNA in aleurone cells and anthers. *The Plant Journal*, 47(3), 427–444.

Turner M, Nizampatnam NR, Baron M, Coppin S, Damodaran S, Adhikari S, Arunachalam SP, Yu O, Subramanian S (2013) Ectopic expression of miR160 results in auxin hypersensitivity, cytokinin hyposensitivity, and inhibition of symbiotic nodule development in soybean. *Plant Physiology* 162: 2042–2055.

Turner M, Yu O, Subramanian S (2012) Genome organization and characteristics of soybean microRNAs. *BMC Genomics* 13: 1–16.

Ucar S, Yaprak E, Yigider E, Kasapoglu AG, Oner BM, Ilhan E, Ciltas A, Yildirim E, Aydin M (2024) Genome-wide analysis of miR172-mediated response to heavy metal stress in chickpea (Cicer arietinum L.): Physiological, biochemical, and molecular insights. *BMC Plant Biology* 24: 1–25.

Um T, Choi J, Park T, Chung PJ, Jung SE, Shim JS, Kim YS, Choi IY, Park SC, Oh SJ (2022) Rice microRNA171f/SCL6 module enhances drought tolerance by regulation of flavonoid biosynthesis genes. *Plant Direct* 6: e374.

Valdés-López O, Arenas-Huertero C, Ramirez M, Girard L, Sanchez F, Vance CP, Luis Reyes J, Hernandez G (2008) Essential role of MYB transcription factor: PvPHR1 and microRNA: PvmiR399 in phosphorus-deficiency signalling in common bean roots. *Plant, Cell & Environment* 31: 1834–1843.

Valdés-López O, Yang SS, Aparicio-Fabre R, Graham PH, Reyes JL, Vance CP, Hernández G (2010) MicroRNA expression profile in common bean (Phaseolus vulgaris) under nutrient deficiency stresses and manganese toxicity. *New Phytologist* 187: 805–818.

Valiollahi E, Farsi M, Kakhki AM (2014) Sly-miR166 and Sly-miR319 are components of the cold stress response in Solanum lycopersicum. *Plant Biotechnology Reports* 8: 349–356.

Van Ooijen G, Mayr G, Kasiem MM, Albrecht M, Cornelissen BJ, Takken FL (2008) Structure–function analysis of the NB-ARC domain of plant disease resistance proteins. *Journal of Experimental Botany* 59: 1383–1397.

Vaucheret H, Béclin C, Fagard M (2001) Post-transcriptional gene silencing in plants. *Journal of Cell Science* 114: 3083–3091.

Várallyay É, Válóczi A, Ágyi Á, Burgyán J, Havelda Z (2010) Plant virus-mediated induction of miR168 is associated with repression of ARGONAUTE1 accumulation. *The EMBO Journal* 29: 3507–3519.

Verma V, Ravindran P, Kumar PP (2016) Plant hormone-mediated regulation of stress responses. *BMC Plant Biology* 16: 86.

Vyse, K, Faivre, L, Romich, M, Pagter, M, Schubert, D, Hincha, D. K, et al. (2020). Transcriptional and post-transcriptional regulation and transcriptional memory of chromatin regulators in response to low temperature. *Frontiers in Plant Science* 11: 39. doi: 10.3389/fpls.2020.00039

Wang J, Mao L, Li Y, Lu K, Qu C, Tang Z, Li J, Liu L (2024) Natural variation in BnaA9. NF-YA7 contributes to drought tolerance in Brassica napus L. *Nature Communications* 15: 2082.

Wang L, Sun Z, Su C, Wang Y, Yan Q, Chen J, Ott T, Li X (2019) A GmNINa-miR172c-NNC1 regulatory network coordinates the nodulation and autoregulation of nodulation pathways in soybean. *Molecular Plant* 12: 1211–1226.

Wang M, Wang Q, Zhang B (2013). Response of miRNAs and their targets to salt and drought stresses in cotton (Gossypium hirsutum L.). Gene. 2013 Nov 1; 530(1): 26–32. doi: 10.1016/j.gene.2013.08.009. Epub 2013 Aug 13. PMID: 23948080.

Wang M, Wang Q, Zhang B (2013) Response of miRNAs and their targets to salt and drought stresses in cotton (Gossypium hirsutum L.). *Gene* 530: 26–32.

Wang S, Kurepa J, Hashimoto T, Smalle JA (2011) Salt stress-induced disassembly of Arabidopsis cortical microtubule arrays involves 26S proteasome-dependent degradation of SPIRAL1. *Plant Cell* 23: 3412–3427. doi: 10.1105/tpc.111.089920

Wang T, Chen L, Zhao M, Tian Q, Zhang W-H (2011) Identification of drought-responsive microRNAs in Medicago truncatula by genome-wide high-throughput sequencing. *BMC Genomics* 12: 1–11.

Wang X, Wang Z, Zheng Z, Dong J, Song L, Sui L, Nussaume L, Desnos T, Liu D (2019) Genetic dissection of Fe-dependent signaling in root developmental responses to phosphate deficiency. *Plant Physiology* 179(1): 300–316.

Wang Y, Li P, Cao X, Wang X, Zhang A, Li X (2009) Identification and expression analysis of miRNAs from nitrogen-fixing soybean nodules. *Biochemical and Biophysical Research Communications* 378: 799–803.

Wang Y, Wang L, Zou Y, Chen L, Cai Z, Zhang S, Zhao F, Tian Y, Jiang Q, Ferguson BJ (2014) Soybean miR172c targets the repressive AP2 transcription factor NNC1 to activate ENOD40 expression and regulate nodule initiation. *The Plant Cell* 26: 4782–4801.

Wang Y, Wang Z, Amyot L, Tian L, Xu Z, Gruber MY, Hannoufa A (2015) Ectopic expression of miR156 represses nodulation and causes morphological and developmental changes in Lotus japonicus. *Molecular Genetics and Genomics* 290: 471–484.

Wang Y, Zhang C, Hao Q, Sha A, Zhou R, Zhou X, Yuan L (2013) Elucidation of miRNAs-mediated responses to low nitrogen stress by deep sequencing of two soybean genotypes. *PLoS One* 8: e67423.

Wei XS, Liu S, Sun C, Xie GS, Wang LQ (2021) Convergence and divergence: Signal perception and transduction mechanisms of cold stress in Arabidopsis and rice. *Plants* 10(9): 1864.

Wen FL, Yue Y, He TF, Gao XM, Zhou ZS, Long XH (2020) Identification of miR390-*TAS3-ARF* pathway in response to salt stress in *Helianthus tuberosus L. Gene* 738, 144460.

Wong J, Gao L, Yang Y, Zhai J, Arikit S, Yu Y, Duan S, Chan V, Xiong Q, Yan J (2014) Roles of small RNA s in soybean defense against P hytophthora sojae infection. *The Plant Journal* 79: 928–940.

Wu G, Park MY, Conway SR, Wang J-W, Weigel D, Poethig RS (2009) The sequential action of miR156 and miR172 regulates developmental timing in Arabidopsis. *Cell* 138: 750–759.

Wu J, Wang L, Wang S (2017) MicroRNAs associated with drought response in the pulse crop common bean (Phaseolus vulgaris L.). *Gene* 628: 78–86.

Xia R, Meyers BC, Liu Z, Beers EP, Ye S, Liu Z (2013) MicroRNA superfamilies descended from miR390 and their roles in secondary small interfering RNA biogenesis in eudicots. *Plant Cell* 25: 1555–1572.

Xia R, Xu J, Meyers BC (2017) The emergence, evolution, and diversification of the miR390-*TAS3-ARF* pathway in land plants. *Plant Cell* 29: 1232–1247.

Xing X, Cao C, Xu Z, Qi Y, Fei T, Jiang H, Wang X (2023) Reduced soybean water stress tolerance by MiR393a-mediated repression of GmTIR1 and abscisic acid accumulation. *Journal of Plant Growth Regulation* 42: 1067–1083.

Xu F, Liu Q, Chen L, Kuang J, Walk T, Wang J, Liao H (2013) Genome-wide identification of soybean microRNAs and their targets reveals their organ-specificity and responses to phosphate starvation. *BMC Genomics* 14: 1–30.

Xu H, Li Y, Zhang K, Li M, Fu S, Tian Y, Qin T, Li X, Zhong Y, Liao H (2021) miR169c-NFYA-C-ENOD40 modulates nitrogen inhibitory effects in soybean nodulation. *New Phytologist* 229: 3377–3392.

Xu M, Hu T, Zhao J, Park M-Y, Earley KW, Wu G, Yang L, Poethig RS (2016) Developmental functions of miR156-regulated SQUAMOSA PROMOTER BINDING PROTEIN-LIKE (SPL) genes in Arabidopsis thaliana. *PLoS Genetics* 12: e1006263.

Xu Y, Zhang X, Hou R, Zhang X, Li S, Yue F, Zhang X, Zhu X (2022) Multiple MicroRNAs are involved in regulating peanut (Arachis hypogaea L.) resistance to Sclerotium rolfsii at the early stage. *Tropical Plant Biology* 15: 276–287.

Yadav A, Kumar S, Verma R, Rai SP, Lata C, Sanyal I (2023) Target cleavage mapping and tissue-specific expression analysis of PGPR responsive miR166 under abiotic stress in chickpea (Cicer arietinum L.). *Plant Cell, Tissue and Organ Culture (PCTOC)* 154: 415–432.

Yan K, Liu P, Wu C-A, Yang G-D, Xu R, Guo Q-H, Huang J-G, Zheng C-C (2012) Stress-induced alternative splicing provides a mechanism for the regulation of microRNA processing in Arabidopsis thaliana. *Molecular Cell* 48: 521–531.

Yan Z, Hossain MS, Arikit S, Valdés-López O, Zhai J, Wang J, Libault M, Ji T, Qiu L, Meyers BC (2015) Identification of micro RNA s and their mRNA targets during soybean nodule development: Functional analysis of the role of miR393j-3p in soybean nodulation. *New Phytologist* 207: 748–759.

Yan Z, Hossain MS, Wang J, Valdés-López O, Liang Y, Libault M, Qiu L, Stacey G (2013) miR172 regulates soybean nodulation. *Molecular Plant-Microbe Interactions* 26: 1371–1377.

Yu Y, Ni Z, Wang Y, Wan H, Hu Z, Jiang Q, Sun X, Zhang H (2019) Overexpression of soybean miR169c confers increased drought stress sensitivity in transgenic Arabidopsis thaliana. *Plant Science* 285: 68–78.

Yu Y, Zhang S, Yu Y, Cui N, Yu G, Zhao H, Meng X, Fan H (2023) The pivotal role of MYB transcription factors in plant disease resistance. *Planta* 258: 16.

Yuan S, Zhao J, Li Z, Hu Q, Yuan N, Zhou M, Xia X, Noorai R, Saski C, Li S (2019) MicroRNA396-mediated alteration in plant development and salinity stress response in creeping bentgrass. *Horticulture Research,* 6, 48. https://doi.org/10.1038/s41438-019-0130-x

Zaheer U, Munir F, Salum YM, He W (2024) Function and regulation of plant ARGONAUTE proteins in response to environmental challenges: A review. *PeerJ* 12: e17115.

Zampieri M, Ceglar A, Dentener F, Toreti A (2017) Wheat yield loss attributable to heat waves, drought and water excess at the global, national and subnational scales. *Environmental Research Letters* 12: 064008.

Zandalinas SI, Mittler R, Balfagón D, Arbona V, Gómez-Cadenas A (2018) Plant adaptations to the combination of drought and high temperatures. *Physiologia Plantarum* 162: 2–12.

Zeng HQ, Zhu YY, Huang SQ, Yang ZM (2010) Analysis of phosphorus-deficient responsive miRNAs and cis-elements from soybean (Glycine max L.). *Journal of Plant Physiology* 167: 1289–1297.

Zeng Q-Y, Yang C-Y, Ma Q-B, Li X-P, Dong W-W, Nian H (2012) Identification of wild soybean miRNAs and their target genes responsive to aluminum stress. *BMC Plant Biology* 12: 1–16.

Zeng W, Sun Z, Lai Z, Yang S, Chen H, Yang X, Tao J, Tang X (2019) Determination of the miRNAs related to bean pyralid larvae resistance in soybean using small RNA and transcriptome sequencing. *International Journal of Molecular Sciences* 20: 2966.

Zhai J, Jeong D-H, De Paoli E, Park S, Rosen BD, Li Y, González AJ, Yan Z, Kitto SL, Grusak MA (2011) MicroRNAs as master regulators of the plant NB-LRR defense gene family via the production of phased, trans-acting siRNAs. *Genes & Development* 25: 2540–2553.

Zhang F, Yang J, Zhang N, Wu J, Si H (2022) Roles of microRNAs in abiotic stress response and characteristics regulation of plant. *Frontiers in Plant Science* 13: 919243.

Zhang H, Liu S, Ren T, Niu M, Liu X, Liu C, Wang H, Yin W, Xia X (2023) Crucial abiotic stress regulatory network of NF-Y transcription factor in plants. *International Journal of Molecular Sciences* 24: 4426.

Zhang H, Li Y, Zhu JK (2017) Developing naturally stress resistant crops for a sustainable agriculture. *Nature Plants* 3: 17018.

Zhang J, Xu H, Yang Y, Zhang X, Huang Z, Zhang D (2021) Genome-wide analysis of long non-coding RNAs (lncRNAs) in two contrasting soybean genotypes subjected to phosphate starvation. *BMC genomics* 22: 433.

Zhang Q, Zhang J, Wei H, Fu X, Ma L, Lu J, Yu S (2020) Genome-wide identification of NF-YA gene family in cotton and the positive role of GhNF-YA10 and GhNF-YA23 in salt tolerance. *International Journal of Biological Macromolecules* 165: 2103–2115.

Zhang X, Cai J, Wollenweber B, Liu F, Dai T, Cao W, Jiang D (2013) Multiple heat and drought events affect grain yield and accumulations of high molecular weight glutenin subunits and glutenin macropolymers in wheat. *Journal of Cereal Science* 57: 134–140.

Zhang X, Ren C, Xue Y, Tian Y, Zhang H, Li N, Sheng C, Jiang H, Bai D (2022) Small RNA and Degradome deep sequencing reveals the roles of microRNAs in peanut (Arachis hypogaea L.) cold response. *Frontiers in Plant Science* 13: 920195.

Zhang Y, Waseem M, Zeng Z, Xu J, Chen C, Liu Y, Zhai J, Xia R (2022) MicroRNA482/2118, a miRNA superfamily essential for both disease resistance and plant development. *New Phytologist* 233(5): 2047–2057.

Zhao J, Lu Z, Wang L, Jin B (2020) Plant responses to heat stress: Physiology, transcription, noncoding RNAs, and epigenetics. *International Journal of Molecular Sciences* 22: 117.

Zhao W, Li Z, Fan J, Hu C, Yang R, Qi X, Chen H, Zhao F, Wang S (2015) Identification of jasmonic acid-associated microRNAs and characterization of the regulatory roles of the miR319/TCP4 module under root-knot nematode stress in tomato. *Journal of Experimental Botany* 66(15): 4653–4667.
Zhao Y, Ma W, Wei X, Long Y, Zhao Y, Su M, Luo Q (2019) Identification of exogenous nitric oxide-responsive miRNAs from alfalfa (Medicago sativa L.) under drought stress by high-throughput sequencing. *Genes* 11: 30.
Zhou Y, Liu W, Li X, Sun D, Xu K, Feng C, Kue Foka IC, Ketehouli T, Gao H, Wang N (2020) Integration of sRNA, degradome, transcriptome analysis and functional investigation reveals gma-miR398c negatively regulates drought tolerance via GmCSDs and GmCCS in transgenic Arabidopsis and soybean. *BMC plant biology* 20: 1–19.
Zhou ZS, Huang SQ, Yang ZM (2008) Bioinformatic identification and expression analysis of new microRNAs from Medicago truncatula. *Biochemical and Biophysical Research Communications* 374: 538–542.
Zhou ZS, Zeng HQ, Liu ZP, Yang ZM (2012) Genome-wide identification of Medicago truncatula microRNAs and their targets reveals their differential regulation by heavy metal. *Plant, Cell & Environment* 35: 86–99.
Zhu Q-H, Fan L, Liu Y, Xu H, Llewellyn D, Wilson I (2013) miR482 regulation of NBS-LRR defense genes during fungal pathogen infection in cotton. *PloS One* 8: e84390.
Zhu YY, Zeng HQ, Dong CX, Yin XM, Shen QR, Yang ZM (2010) microRNA expression profiles associated with phosphorus deficiency in white lupin (Lupinus albus L.). *Plant Science* 178: 23–29w.

Section III

Future Scope

12 The Nodule Nexus
Present Status, Strategies, and Applications of Legume Engineering

Biplob Dey and Romel Ahmed

12.1 INTRODUCTION

Legume grain, a rich protein source, has a significant contribution to global food and feed security. Crops like chickpeas, lentils, and beans are rich in high-quality protein and lysine, while cereals like rice complement them with sulfur-rich amino acids, creating a balanced diet widely consumed in regions like South Asia and Africa (Kuyah et al., 2023). Rich in proteins, fibers, vitamins, and minerals, legumes are vital, especially for smallholder farmers in developing countries. Moreover, the environmental impact of animal-based protein production has increased interest in plant-based alternatives (van Loon et al., 2023). The production of meat, a major source of protein, notoriously contributes to CO_2 emissions (Poore & Nemecek, 2018), water use, and land degradation, making legumes a more sustainable option with a lower environmental footprint. In Europe, the demand for domestically grown legumes, such as faba beans, peas, and soybeans, is rising to reduce reliance on imported animal feed and to support sustainable agriculture (van Loon et al., 2023). Despite their potential as food and feed, legume engineering attempts are limited. Recent progress in molecular tools for genome editing and regulation of expression offers an opportunity to escalate legume breeding efforts.

Legume engineering holds immense potential due to the symbiotic relationship legumes form with nitrogen-fixing rhizobia bacteria, which benefits both the plants and the environment. Legumes can fix atmospheric nitrogen through this symbiosis, reducing the need for synthetic nitrogen fertilizers. This biological nitrogen fixation mitigates environmental pollution caused by excess fertilizer use (Peix et al., 2015). Annually, ~50–70 million tons of nitrogen are fixed biologically in agricultural systems, with legumes like soybeans contributing significantly to global nitrogen cycling (Herridge et al., 2008). The ability of legumes to improve soil fertility through nitrogen fixation makes them essential in sustainable agricultural practices, especially in the face of climate change and rising food demand (Dey et al., 2024, 2022). However, despite their potential, legume production lags behind that of cereals due to several factors, including relatively low yield improvements over recent decades, susceptibility to biotic and abiotic stresses, and limited breeding progress (Araújo et al., 2015). While soybeans have seen substantial advances, other legumes suffer from gaps in breeding programs due to limited genetic resources, and technical limitations, such as the lack of efficient doubled haploid protocols.

The ongoing challenges of climate change further underscore the importance of improving legume crops. Increased temperatures (Krishnamurthy et al., 2011), drought (Khandal et al., 2017), and soil salinity threaten legume yields (Chen et al., 2019), necessitating genetic engineering solutions to enhance resilience to these abiotic stresses (Das, 2018; Mishra et al., 2014; Saha et al., 2020). By enhancing genetic traits related to drought resistance, pest resistance, and nutrient uptake, legume

DOI: 10.1201/9781003477235-15

engineering can address these constraints and ensure the stability of legume-based cropping systems (Chakraborty et al., 2016; Krishna et al., 2011).

Traditional breeding methods, while historically important, are often slow, resource-intensive, and limited by the genetic variation present within breeding populations. Moreover, conventional methods require many years to achieve the desired goal, which may not be sufficient to meet the growing demand for food due to population increases. In contrast, genetic engineering offers a faster route by introducing external genes to improve specific traits. However, the introduction of transgenes has raised concerns regarding biosafety and public acceptance, particularly in regard to genetically modified organisms. Despite these challenges, recent advances in molecular techniques are reshaping crop improvement approaches.

Among the most promising technologies is CRISPR/Cas9, a revolutionary gene-editing tool that offers precision and efficiency in modifying plant genomes (Rasheed et al., 2022). CRISPR, which originally functioned as a bacterial defense mechanism, has become the preferred method for targeted gene editing due to its simplicity, accuracy, and flexibility (Watson et al., 2021). Unlike classical breeding, CRISPR does not rely on genetic variation and can directly introduce specific, novel genetic traits. The system consists of two main components: a single guide RNA (sgRNA), which directs the Cas9 protein to the target gene, and the Cas9 endonuclease, which makes a precise double-stranded break at the desired location (Rasheed et al., 2022; Watson et al., 2021). This capability allows for the creation of genetically modified legumes with enhanced traits such as pest resistance, drought tolerance, and improved nutritional content, offering a powerful tool to meet future food security challenges.

12.2 NODULE SYMBIOSIS: UNDERSTANDING THE KEY FEATURES

Nodule organogenesis is a complex symbiotic process resulting from bacterial signaling molecules, primarily Nod factors, and plant-derived signals, including hormonal regulation and intracellular pathways. This initiation begins with the release of lipochitooligosaccharides (Nod factors) by rhizobia. These molecules are synthesized through bacterial nodulation genes and serve as key signals for host recognition. In the root epidermis, Nod factors are perceived by plant LysM kinase complexes (NFP/LYK) depending on the plant species, such as nfr1, symrk, and nfr5 in *Lotus japonicus* and nfp in *Medicago truncatula* (Amor et al., 2003; Madsen et al., 2003; Ye et al., 2024). Upon recognition, these receptors activate downstream signaling pathways that include calcium spiking, actin cytoskeleton remodeling, and the induction of specific transcription factors (e.g., NSP1, NSP2, NIN) (Crespi & Frugier, 2009).

The Nod factors also trigger responses in the inner cortical and pericycle cells of the root, distinct from the epidermal infection process. These responses are mediated by the activation of genes such as *dmi1*, *dmi2*, and *dmi3 for M. truncatula*, which encode critical components of the symbiotic signaling pathway (Ané et al., 2004; Imaizumi-Anraku et al., 2004). The *dmi1* protein functions as a cation channel that facilitates calcium spiking (Ané et al., 2004), while *dmi2*, a leucine-rich repeat receptor kinase, links the perception of Nod factors to intracellular signaling (Crespi & Frugier, 2009; Stracke et al., 2002). A calcium- and calmodulin-dependent protein kinase, *dmi3*, integrates these signals, activating downstream transcriptional regulators such as NIN (Lévy et al., 2004), which are essential for initiating cortical cell division and nodule primordium formation.

The nodule primordium forms from dedifferentiated cortical cells undergoing mitotic divisions, with species-specific inner or outer cortical layer recruitment (Fournier et al., 2008). Determinate nodules, typical of tropical legumes, develop a transient meristem where cells differentiate into nitrogen-fixing cells. Indeterminate nodules in temperate legumes maintain a persistent meristem, generating new cells throughout their lifespan and creating distinct developmental zones for nitrogen fixation and senescence (Crespi & Frugier, 2009). Moreover, the activation of the cytokinin pathway alone is sufficient to trigger nodule organogenesis, as evidenced by the formation of spontaneous nodules in mutants with constitutively active cytokinin receptors. However, the interplay between Nod factor signaling and cytokinins extends beyond cortical activation (Tirichine et al., 2007). Cross-talk between these pathways likely coordinates the spatial and temporal progression of nodule development (Crespi & Frugier, 2009), integrating bacterial signals with plant hormonal regulation. Conversely, reducing cytokinin levels, for example, through the overexpression of cytokinin-degrading enzymes, significantly reduces nodulation, emphasizing the hormone's indispensability. Ethylene negatively regulates nodulation by counterbalancing cytokinin signaling (Enebe & Erasmus, 2023), restricting nodule formation and localization to ensure balanced resource allocation between plant growth and symbiosis. For example, the *sickle* mutant of *M. truncatula*, defective in ethylene signaling, exhibits increased nodulation and hyperinfection, demonstrating ethylene's inhibitory role (Crespi & Frugier, 2009; Schnabel et al., 2023).

The regulation of nodule numbers is tightly controlled by the host plant through both local and systemic mechanisms. Mutations in plant genes, such as those encoding the NARK (nodule autoregulation receptor kinase), result in a supernodulation phenotype, with an increased number of nodules (Searle et al., 2003; Singh & Verma, 2023). NARK, predominantly expressed in leaf tissues, is thought to perceive an active signal transported from roots to shoots, coordinating long-distance autoregulation of nodulation (Singh & Verma, 2023). The similarity of NARK to the CLAVATA1 receptor kinase, which regulates shoot apical meristem size in *Arabidopsis*, suggests the involvement of signaling peptides in nodule regulation (Mirzaei et al., 2017; Searle et al., 2003). Local regulation of auxin transport within roots also contributes to nodule formation. Flavonoids, secondary metabolites produced by the host plant, modulate auxin transport and influence nodulation (Van Noorden et al., 2006).

Plasmodesmata facilitate short-distance cell-to-cell communication, integrating signals from vascular tissues to nodule primordia (Bayer & Benitez-Alfonso, 2024). The ability of single-gene mutations to induce spontaneous nodulation highlights the plasticity of the regulatory network governing nodule organogenesis (Crespi & Frugier, 2009). Furthermore, overlaps between nodulation pathways and other plant processes, such as mycorrhizal symbiosis and lateral root formation, demonstrate the integration of environmental and developmental cues in shaping root architecture.

The ability of single mutations, such as those in cytokinin receptors or calcium-dependent protein kinases, to induce spontaneous nodules emphasizes the substantial sensitivity of the regulatory network governing nodule formation. Also, the overlap between pathways involved in nodulation and other plant responses, such as mycorrhizal symbiosis and lateral root formation, highlights the integration of environmental cues into root architecture modification (Tanwir et al., 2023). For instance, the DMI genes, initially identified in nodulation studies, are also essential for arbuscular mycorrhizal interactions (Miyata & Umehara, 2024; Parihar & Rakshit, 2024; Tanwir et al., 2023), showing the shared signaling components in symbiotic associations.

12.3 OBJECTIVES AND GOALS OF LEGUME ENGINEERING

Nitrogen fixation efficiency varies among legume species and is often suboptimal under stressed or non-ideal soil conditions. Also, in many parts of the world, particularly in arid and semi-arid regions, water scarcity poses a significant challenge to crop production (Dey et al., 2024). Legume crops are often sensitive to water stress and prone to attack by various insect pests, fungal pathogens, and viruses, resulting in reduced yields. Through genetic engineering, legumes can be modified to improve drought and heat tolerance by enhancing their ability to maintain turgor pressure, regulate stomatal closure, and increase root depth for better water absorption (Table 12.1). Legume engineering can enhance pest and disease resistance by introducing genes for plant defense mechanisms (Table 12.1), such as those involved in producing secondary metabolites like flavonoids, alkaloids, and phytoalexins, or by using gene-editing techniques to activate natural immune responses. Additionally, improvements in the nutritional quality of legume seeds, such as increasing protein

TABLE 12.1
MiRNA in Cellular Reprogramming of Legumes under Biotic and Abiotic Stresses to Enhance Tolerance

Stress	Species	Functions	References
Abiotic stress (nitrogen stress)	Soybean	Identified 362 known miRNA variants (158 families) and 90 novel miRNAs (55 families), with 150 known and two novel miRNAs showing differential expression linked to adaptive responses to nitrogen limitation in soybean genotype	Wang et al. (2013)
Abiotic stress (heat)	Chickpea	Heat-tolerant cultivated germplasm with increased pollen germination under stress conditions	Krishnamurthy et al. (2011)
Abiotic stress (salinity)	Soybean	Under continuous salt stress, 75% of activated or upregulated long intergenic non-coding RNAs (lncRNAs) primarily targeted proteins with binding and catalytic activities, including 3030 lincRNAs and 275 lncNATs in the Williams 82 genotype	Chen et al. (2019)
Abiotic stress (drought and salinity)	Chickpea	MicroRNA profiling in root apices under salinity and water shortage showed 284 unique miRNAs, including 60 novel ones, with 259 differentially expressed in stress. Relationships validation showed the roles in root hair re-patterning, lateral root formation, and K+ uptake under stress	Khandal et al. (2017)
Biotic stress (*Fusarium* wilt infection)	Chickpea	miR530 is substantially upregulated in response to fungal infection, targeting genes involved in zinc knuckle- and microtubule-associated proteins	Kohli et al. (2014)
Abiotic stress (salinity)	Soybean	Sixty-six salt-responsive miRNAs, including 22 novel ones, were identified in root meristems, with auxin-responsive *cis*-elements in their promoters suggesting auxin's role in regulating miRNA plasticity and root development	Sun et al. (2016)
Abiotic stress (drought)	Cowpea	Variations in gene expression levels correlate with drought sensitivity in cowpea, with miRNAs negatively regulating these genes. Additionally, miRNA responses to drought stress differ across tissues	Shui et al. (2013)
Abiotic stress (Ca deficiency)	Groundnut	A total of 132 potential miRNAs in 12 peanut-specific families, with 87 showing differential expression during early embryo development and 52 target genes linked to seed growth, cell division, hormone signaling	Chen et al. (2019)
Biotic stress (root knot nematode)	Soybean	Seventy putative miRNAs were identified, with 60 miRNAs from 25 families potentially linked to cultivar responses to soybean cyst nematode	Tian et al. (2017)
Biotic stress (pod rot)	Groundnut	Identified 334 peanut miRNAs, with 97 downregulated and 27 upregulated, and predicted 1998 target genes	Wang et al. (2020)

content or enhancing the bioavailability of essential amino acids, could increase global food supply by ~49% (Singh et al., 2025). Modifying the oil content, fatty acid profile, or micronutrient concentrations can make legumes even more valuable as a source of plant-based nutrition for both humans and livestock (Fanzo et al., 2021).

While legume engineering offers numerous advantages in agricultural contexts, its applications extend to forest restoration efforts. Natural forest regeneration is often hampered by severe nutrient limitations, most notably by nitrogen (Siddique et al., 2008; Song et al., 2010). Nitrogen-fixing leguminous trees have evolved as a cornerstone of nutrient cycling in these ecosystems, enhancing soil fertility and adapting secondary forest succession. However, their effectiveness is frequently constrained by deficiencies in phosphorus (a key nutrient for N_2 fixation), which creates a bottleneck for ecosystem productivity and nitrogen accumulation (Güsewell, 2004; Siddique et al., 2008). Engineering legumes to overcome these limitations has the potential for forest restoration strategies. Currently, leguminous forest tree species (e.g., *Albizia lebbeck*, *Desmodium rensonii*, *Sesbania grandiflora*) show efficiencies in nutrient acquisition and utilization. Some species thrive (e.g., *Albizia lebbeck, Acacia* spp.) with low tissue nutrient concentrations, demonstrating high nutrient use efficiency and adaptability to nutrient-poor soils (Aryal et al., 2001), while others require higher concentrations to support rapid growth and high photosynthetic capacity (Nizamani et al., 2024). This dichotomy suggests that legume engineering could be tailored to optimize traits for specific restoration objectives, such as balancing rapid biomass production with sustainable nutrient cycling. However, the extent to which functionally diverse versus homogeneous legume plantings benefit restoration remains unresolved (Hiremath & Ewel, 2001).

12.4 PROGRESS IN LEGUME ENGINEERING

Over the years, several techniques have been developed to overcome the barriers of genetic complexity in the transformation of desired genes. *Agrobacterium*-mediated transformation has been successfully optimized for many legumes, achieving higher efficiency through the use of refined vectors, better explants, and more optimized culture conditions (Table 12.2). For example, *Agrobacterium*-mediated transformation in pigeon peas has achieved up to 83% efficiency (Karmakar et al., 2019), which was a substantial improvement compared to previous outcomes. Also, there has been considerable progress in improving the nutritional content of legume seeds. For example, RNA interference (RNAi) technology was used in soybean to silence the raffinose synthase 2 gene, reducing raffinose content and enhancing nutritional quality for both livestock feed and human consumption (Valentine et al., 2017). Pest and disease resistance is another major focus, with the incorporation of Cry1Ac and Cry2Aa genes from *Bacillus thuringiensis* into various legumes, such as chickpea, pigeonpea, and cowpea, providing resistance to pests like *Helicoverpa armigera* and storage pests like *Callosobruchus*. These genetic improvements reduce pesticide dependence. Genome-editing technologies, particularly CRISPR/Cas9, have revolutionized legume genetic engineering for precise modifications to the genome, enabling the targeted insertion, deletion, or alteration of specific genes with high efficiency and accuracy.

12.5 TOOLS IN LEGUME ENGINEERING

Legume engineering involves modifying the genome of the intended species or introducing a foreign gene to the target legume species to improve the desired traits. It offers tremendous benefits in improving the yield and quality of the legume grains. The process of legume engineering comprises (i) the development of de novo genetic variation, (ii) the selection of the elite plant based on the defined target merit, and (iii) field trial, validation, mass propagation, and release of the improved variety. In the following section, we discuss the tools used in legume engineering, including conventional as well as the recent advanced tools with their applications.

12.5.1 Conventional and Recent Advanced Molecular Breeding

Legumes' contribution to ensuring global food security and ecosystem resilience is undeniable. Their exceptional capacity of fixing nitrogen for ameliorating soil health and serving as a supplement of animal protein are well documented. However, compared with cereal crops, legumes have gained less attention in their genetic improvement. In the recent past, conventional breeding has played a significant role in improving the yield, nutritional quality, stress tolerance (heat, drought, nutrients), pest resistance, and nitrogen fixation efficacy of hundreds of legume crops (Varshney et al., 2018). Nevertheless, conventional breeding has some limitations, such as crossing, and many undesirable

TABLE 12.2
General Overview of Advances in Genetic Engineering and Transformation for Enhancing Legume Traits

Crops	Progress/developed traits/ aims	References
Soybean	Development of resistance to multiple potyvirus strains, including soybean mosaic virus, using RNAi technology	Yang et al. (2018)
	Stable *Agrobacterium*-mediated transformation of split-seed explants (imbibed seeds) achieved ~19% efficiency using the binary plasmid vector pDAB9381 with the PAT marker gene	Pareddy et al. (2020)
	RNAi-mediated silencing of the soybean RS2 gene significantly reduced raffinose content, improving nutritional quality and metabolizable energy	Valentine et al. (2017)
	GsWRKY20 gene shows reduced stomatal density and conductance, minimizing drought-induced water loss	Ning et al. (2017)
	Overexpression of the *Perilla frutescens* γ-tocopherol methyltransferase (Pf-γ-TMT) gene enhanced α-tocopherol content which improves the nutritional quality	Arun et al. (2014)
	Optimized protocol for *Agrobacterium*-mediated transformation using the binary plasmid pTF102	Li et al. (2017)
	Sodium nitroprusside enables efficient (35%) *Agrobacterium*-mediated transformation using half-seed explants	Karthik et al. (2020)
	Efficient siRNA-mediated (small interfering RNAs) gene silencing of the FAD3 gene family significantly reduces α-linolenic acid	Flores et al. (2008)
Pea	Transgenic pea plants with AtNHX1 gene showed stable morphology, genetic integrity, and enhanced salt and frost tolerance over six generations	Ali et al. (2018)
	Transgenic plants with herbicide resistance and Mendelian inheritance	Schroeder et al. (1993)
	Identified AlphaAI-1 gene provides complete protection against pea weevil under field conditions, while alphaAI-2 gene offers partial protection by delaying larval maturation	Morton et al. (2000)
Chickpea	Iron biofortification using CaNAS2 and GmFER genes	Tan et al. (2018)
	Enhanced pest resistance, reducing *H. armigera* survival significantly	Ganguly et al.(2014; Indurker et al. (2007); Sanyal et al. (2005)
	P5CS-transformed chickpea lines showed increased proline accumulation and improved salt stress tolerance without hampering yield under salinity	Kumar Ghanti et al. (2011)
	Chickpea transformation using immature cotyledons achieved 2.08% efficiency with optimized culture conditions and successful transgene integration	Tripathi et al. (2013)
	Cry1Ac overexpression in chickpea using Ubiquitin and rbcS promoters effectively controls *Helicoverpa armigera* infestation, especially with chloroplast targeting	Chakraborty et al. (2016)
	Overexpressing miR408 enhances drought tolerance by regulating plantacyanin, DREB (dehydration-responsive element binding protein), and drought-responsive genes	Hajyzadeh et al. (2015)
	Agrobacterium-mediated plumular meristem transformation method in chickpea achieved 60% transgenic event frequency and stable T1 progeny	Ganguly et al. (2020)
	Agrobacterium-mediated transformation protocol from apical meristem achieved a 1.2% transformation efficiency with successful gene integration	Srivastava et al. (2017)
Cowpea	*Agrobacterium*-mediated transformation using a curling virus constitutive promoter and PMI selection marker achieves a 3.6% efficiency	Bakshi et al. (2012)
	Transgenic cowpea expressing VrNHX1 from mungbean showed improved salt tolerance, enhanced K^+/Na^+ ratio, and better physiological traits	Mishra et al. (2014)
	Agrobacterium-mediated stable transformation using pBSF16 vector resulted in fertile transgenic plants, with insecticidal protein expression	Bett et al. (2019)

TABLE 12.2 (Continued)
General Overview of Advances in Genetic Engineering and Transformation for Enhancing Legume Traits

Crops	Progress/developed traits/ aims	References
	A refined *Agrobacterium*-mediated transformation protocol for cowpea achieved 1.64% transgenic plant recovery, enhancing T-DNA delivery and selection	S.K. Solleti et al. (2008)
	Transgenic cowpea lines expressing RNAi constructs showed resistance to cowpea severe mosaic virus and cowpea aphid-borne mosaic virus, with homozygous plants exhibiting enhanced resistance	Cruz and Aragão (2014)
	Transgenic cowpeas expressing αAI-1 gene showed enhanced resistance to storage pests (e.g., *Callosobruchus chinensis, C. maculatus*) and α-amylase inhibition activity	Siva Kumar Solleti et al. (2008)
	Fertile transgenic cowpea lines expressing BtCry2Aa toxin show increased resistance to Maruca pod borer	Kumar et al. (2021)
	Transgenic cowpeas with Mendelian transgene inheritance were developed using pBSF16 with the *bar* gene, enabling insect resistance	Popelka et al. (2006)
Pigeonpea	A rapid *Agrobacterium*-mediated transformation achieved 83% efficiency using pCAMBIA1301 vector (selection marker *hptII)* in 35–40 days	Karmakar et al. (2019)
	Successfully transformed with cry1Ac (pod borer resistance) and nptII (selectable marker) genes, achieving 44.61% efficiency, and developed insect-resistant, fertile transgenic plants	Krishna et al. (2011)
	Transgenic pigeonpea with Cry2Aa gene developed; showed stable inheritance and 100% *Helicoverpa armigera* (Lepidopteran insect) mortality	Ghosh et al. (2017); Kaur et al. (2016)
	Overexpressing feedback-resistant P5CSF129A gene enhances proline accumulation, salt tolerance, growth, and cellular integrity under abiotic stress	Surekha et al. (2014)
	Agrobacterium-mediated plumular meristem transformation method with 72% efficiency facilitating rapid transgenic plant establishment	Ganguly et al. (2018)
	Overexpressing the dhdps-r1 gene increased free lysine content (amino acid), with no regulation of lysine catabolism	Thu et al. (2007)
Blackgram	Overexpression of the AtALMT1 (*Arabidopsis* malate transporter) enhances Al^{3+} tolerance without altering stress response mechanisms	Saha et al. (2020)
	ALDRXV4 gene showed improved stress tolerance, higher aldose reductase activity, and reduced oxidative damage	Singh et al. (2016)
	Transgenic blackgram plants with bacterial chitinase gene (CaMV 35S promoter) showed enhanced fungal resistance and no phenotypic changes	Das (2018)
	Herbicide-tolerant developed using pME 524 vector (marker bar, nptll) with stable transgene inheritance	Muruganantham et al. (2007)
	Enhanced salt stress tolerance via glyoxalase I overexpression	Bhomkar et al. (2008)
	Transgenic blackgram plants expressing barley chitinase and ribosome-inactivating protein show increased resistance to *Corynespora cassiicola* 27–47% disease acuteness reduced (vector pGJ-42, selection marker nptll)	Chopra and Saini (2014)
Mungbean	Transgenic mungbean expressing BjNPR1 (pathogenesis-related) gene showed increased resistance to dry rot disease (*Rhizoctonia solani*)	Vijayan and Kirti (2012)
	Overexpression of the ICE1 gene enhances cold tolerance	Rout et al. (2020)
	Ca^{2+}-dependent signal regulating annexin gene integrated (efficiency 4.2%) using cotyledonary node explants	Yadav et al. (2012)
	The atNHX1 gene enhances salt tolerance through improved ion compartmentalization, antioxidant activity, and growth under stress	Sahoo et al. (2016)

TABLE 12.3
A comparative Picture of Conventional versus Advanced Molecular Tools in Legume Engineering

Features	Conventional breeding tool	Advanced molecular breeding tool
Method	Phenotype-based selection of the best candidate from the progenies of outcrossed	DNA sequence-based that is linked to the traits of interest
Time required	It takes a longer time as several crossings and subsequent homozygous progeny selections are required	It takes less time as early selection of the traits is possible without waiting for the expression of the trait
Feasibility	Not feasible for all legumes, as heritability and expression are challenging for some traits	For low heritability and complex genomes, molecular-based selection is feasible
Introgression of traits	Challenging to introgress traits from another germplasm	It is not difficult to introgress genes/traits even from diverse germplasm
Elimination of undesirable linked DNA	Undesirable linked DNA is difficult to eliminate	It is not difficult to remove such undesirable linked DNA

traits that may come with the desirable ones. Besides, it is only compatible with plants belonging to the same species or, at best, the same family. Due to the slower rate of genetic gains, it can no longer sustain global demand. Comparatively better tools are marker-assisted selection (MAS) and genome-wide association (GWAS) studies for improved breeding. A comparative picture of conventional versus advanced molecular breeding is shown in Table 12.3.

Marker-assisted selection saves the time and cost of selecting traits of interest based on genetic constituents, which otherwise could be done by phenotypic observation (Hasan et al., 2021). MAS aids genetic engineering to a larger extent in fixing the strategies to improve the crop by identifying the associated marker sequences closely associated with traits. In legumes, MAS is being used in identifying traits such as disease resistance, drought tolerance, nitrogen fixation efficiency, and yield potential (Khatun et al., 2021). Some examples of MAS uses in legumes are identification of resistance to rust in bean (Wafula et al., 2023), drought tolerance of chickpeas (Istanbuli et al., 2024), and fortification of iron and zinc in lentil (Dhaliwal et al., 2021). Despite the significant progress in MAS-based trait selection in legumes, markers are lacking for many agronomically important traits, such as nutrition content, grain size, early flowering, etc., for a number of legumes.

Like MAS, GWAS is very useful in identifying the potential loci associated with important legume traits from a diverse population. An example of using GWAS in legume crops is the identification of significant single-nucleotide polymorphisms (SNPs) for 11 root system architecture (RSA) traits in soybean (*Glycine max*) from seven of its chromosomes (Mandozai et al., 2021). RSA is important for nutrient uptake, drought tolerance, and increased yield. Another study on alfalfa (*Medicago sativa*) reported 28 markers associated with forage traits of which 26 are linked with quality and two with digestibility traits (Wang et al., 2020).

12.5.2 Sequence-Based Legume Engineering

Next-generation sequencing (NGS) technologies have revolutionized crop engineering research. Incorporation of genome information harnesses the improvement of the crop. In addition, the development of the bioinformatics field facilitates the task significantly. NGS has made the reference genomes and re-sequencing of the germplasm and breeding lines available for a number of legumes, including chickpeas, pigeon peas, groundnuts, etc. (Bertioli et al., 2016; Chen et al., 2016). This advanced technology has made a paradigm shift from marker-assisted breeding to sequencing-based breeding. There are several reliable and resourceful legume databases publicly available for facilitating research and improvement of global commercial legumes. The Legume Information System (LIS; https://legumeinfo.org) is such a database providing genomic data of more than two dozen legumes of taxonomic interest (Berendzen et al., 2021). The platform also integrates various online tools supporting rigorous analyses. Among others, LegumeIP, SoyBase, LegumeBase, MedicagoGenome, Chickpea Transcriptome database, KnowPulse, and Lotus Base are worth mentioning. The information of the NGS database serves as the key resource to explore the functions, genome mapping, QTLs identifications, finding noble genes, and genome editing aiming toward trait improvement (Khandelwal & Jain, 2018).

12.5.3 RNAi

RNA interference is an excellent tool offering incredible specificity and efficacy in controlling the regulation of a gene or multiple genes. The technique offers flexibility in choosing the time and degree of silencing of the target gene at a specific period of growth phases or in a specific organ such as roots, flowers, or leaves. An example of RNAi-induced gene silencing has been exhibited in *M. truncatula* using the protocol by Floss et al. (2013). It can also be exemplified by silencing the p34 protein, a major soybean allergen (Newell-McGloughlin, 2008). RNAi-based knockdown of GmNORK in soybean provided an insightful result in understanding the symbiotic signaling and nodule development (Wang & Deng, 2016). In soybean, introduction of the GmNORK RNAi plasmid through *Agrobacterium rhizogenes*-mediated hairy-root transformation caused reduced nodulation in the transgenic roots, which indicates that interruption of fatty acid biosynthesis interferes with the signaling pathway (Wang & Deng, 2016). The role of microRNAs during nodulation studied in soybean by Yan et al. (2015) showed RNAi-induced silencing of miRNA ENOD93 in soybean reduced mRNA abundance. Consequently, modulation was significantly reduced, indicating the regulation of ENOD93 as a major factor in the nodule development of this crop. In *Phaseolus vulgaris* L, infection of *Rhizobium* and initiation of nodule meristem are greatly influenced by receptor for activated C kinase (RACK1), mRNA of which accumulates remarkably at 12–15 days post inoculation (Islas-Flores et al., 2011). When mRNA expression of RACK1 was downregulated by a specific RNAi construct, nodulation was greatly impaired (Islas-Flores et al., 2011). RNAi-based silencing of trehalase (*PvTRE1*) in common bean showed trehalose plays a key role in enhancing nodulation, nodule biomass, and nitrogenase activity. The knockdown plant showed increased accumulation of trehalose (78%) and has a strong correlation with bacteroid number. The role of phosphatidylinositol 3-kinase (PI3K) in symbiosis is revealed in a study by Estrada-Navarrete et al. (2016). Downregulation of Pv-*PI3K* declined the symbiosis process, impaired root hair growth, and obliterated the colonization by *Arbuscular mycorrhiza* fungi in *Phaseolus vulgaris* L. (Estrada-Navarrete et al., 2016). Similar to PI3K, a RANi-based study revealed the crucial role of the target of rapamycin (TOR) protein kinase in symbiotic interaction. Downregulation of TOR resulted in a severe reduction of nodule number in *Phaseolus vulgaris* (Nanjareddy et al., 2016).

12.5.4 Genome Editing in Engineering Legumes

Genome-editing tools have opened up a new window of legume engineering with the aim of increasing yield and nutritional properties. In editing, any part of the DNA sequence can be deleted, inserted, or changed to acquire the genetic gain. It could be a corrective measure or efficiency enhancement measure to achieve the target. While genome editing has been practiced for several decades, recent advancements of zinc-finger nucleases (ZFNs), transcription activator like effector nucleases (TALENs), and clustered regularly interspaced short palindromic repeats (CRISPR) with nuclease Cas9 make the process rapid, precise, and cost-effective (Table 12.4). A brief description of these tools in legume engineering follows.

12.5.4.1 Zinc-Finger Nucleases

ZFNs are a pioneering, precise, and efficient genome-editing technique developed in early 1990. The structure, design, and mechanism of editing by ZFNs are described elsewhere (Urnov et al., 2010). In legumes, the techniques were used for a number of crops. Two duplicate (paralogous)

TABLE 12.4
Comparative Features of Three Genome-Editing Tools used in Legume Breeding

	ZFNs	TALENs	CRISR-CAS
Construct	It has two parts namely a DNA-binding domain consisting of 4–6 zinc-finger proteins (ZFPs) and (b) a cleavage domain containing a nuclease named *Fok*1	Like ZNF, the nuclease Fok1 is fused with transcription activator-like effector (TALE) proteins, a pair of TALE is linked together in a threadlike structure for fusing with *Fok*1	It consists of a guide RNA (sgRNA) and a nuclear-localized Cas protein, Cas9, which is the popular Cas protein
Recognition of binding site	The mechanism of recognition is DNA-protein interaction. Zinc finger single domains recognize three base-pair sequences on DNA; a series of linked zinc finger domains (3–4 bp) can be designed to target 12–18 nucleotides	The mechanism of recognition is DNA–protein interaction. A single TALE motif recognizes a single nucleotide; therefore, to target a longer sequence (30–40 bp), and an array of TALEs can associate with a longer design	The mechanism of recognition is DNA–RNA interaction. The guide RNA, in association with the PAM sequence, can recognize the target DNA sequence of 22 bases by pairing.
Feasibility of designing the binding domain	A customized zinc finger motif is required to design for each triplet nucleotide to bind, so for binding a target DNA sequence, multiple zinc finger motifs need to be designed that work in an array. However, such an array influences the specificity of neighboring zinc fingers. This makes the task challenging	It is comparatively easy to design. A series of TALE domains is required to design to make a chain, as each domain can recognize one nucleotide. Unlike ZNF, it does not affect the binding specificity of neighboring TALEs	Easy as it is a RNA–DNA binding tool, only need to design a guide RNA complementary to the target DNA sequence
Target DNA recognition efficiency	Moderate	Moderate	High
Mode of binding the target sequence	Dimerization, as *Fok*I, works as a dimer, meaning cleavage occurs only when two ZNFs bind the DNA from two opposite directions	Like ZFNs, TALE is also associated with *Fok*I endonuclease, hence, dimerization is required for the cleavage to occur	It is a monomeric, single-work mode for binding each target DNA sequence
Off-target effect	As two engineered ZNFs are required to bind both strands of the target DNA sequence, off-target is naturally minimized. Hence off-target effect is minimal with high-quality ZFNs	Like ZNFs, TALENs are also required to bind the sequence from both directions (double strand), so off-target effects are reduced	The short 22-bp gRNA/PAM sequence causes a significant off-target effect in CRISPR-Cas9. However, several mechanisms have already been developed to minimize the effect

TABLE 12.4 (Continued)
Comparative Features of Three Genome-Editing Tools used in Legume Breeding

	ZFNs	TALENs	CRISR-CAS
Target DNA break	Induce double-stranded break of the target DNA sequence	Induce double-stranded break of the target DNA sequence	Can lead to a double-strand break in the case of a wild-type enzyme, or to a single-strand break when using mutant Cas9 variants called nickases
Complexity in polyploid plant genomes	Possible to edit paralogous genes	Possible to edit paralogous genes	Use of CRISPR/Cas is challenging in complex polyploid plant genomes containing paralogous genes
Multiplexing (inducing more than a single mutation in a single genome in a single round of mutagenesis)	Challenging	Challenging	It is possible to induce multiple editions by a single Cas9 fused with multiple sgRNAs. This has brought a radical change in genome editing task
Cost	High and time-consuming	Moderate cost but requires time to construct	Highly affordable
Delivery	Easy	Difficult	moderate

genes *DCL4a* and *DCL4b b* were successfully muted by ZFN through a hairy-root transformation system in soybean (*Glycine max*) and generated seven mutant lines (Curtin et al., 2011). The authors (Curtin et al., 2011) also developed a web-based tool to identify sites suitable for ZFN-based engineering of the soybean genome. Bonawitz et al. (2019) carried out ZFN-mediated integration of several transgenes in the soybean. These examples demonstrate the potential use of ZFN in legume crop improvement. To elucidate the role of starch debranching enzymes (isoamylase and a pullulanase) in the synthesis and degradation of starch, null mutants of this enzyme were created in peas by homologous recombination of DNA-mediated ZFN. Besides, using ZFNs has also been done in elucidating the function of miRNA genes regulating the expression of other associated genes via an RNAi silencing mechanism (Catalanotto et al., 2016).

12.5.4.2 TALEN

Besides ZFNs, an alternative genome editing tool named TALEN is being used for a wide range of crop improvements, including legumes. Like ZFN, TALEN uses a similar *Fok*1 domain as a DNA nuclease. *Fok*l is fused with a TALE protein, a DNA-binding protein that enters the nucleus and binds the target DNA sequence (Chen & Gao, 2013). The TALEN-mediated modified soybean was developed by the US company Calyxt (Calyxt Inc, 2019) and has been cultivated in the USA since 2016 for high oleic soybean oil as a high-quality food ingredient (Menz et al., 2020). High accumulation of oleic acid in peanut (*Arachis hypogaea* L.) resulted from mutation of fatty acid desaturase 2 (FAD2) by TALEN (Wen et al., 2018). Another successful application of TALEN was in *Glycine max* and *Medicago truncatula* for the modification of Dicer-like2 gene associated with small RNA processing (Curtin et al., 2018).

12.5.4.3 CRISPR/Cas9

CRISPR has brought a dramatic revolutionary achievement in the genome-editing process. The details of CRISPR technology have been described elaborately in many literatures. CRISPR is a simple and cost-effective efficient technique that uses cleaving enzymes and guide RNA. The most popular engineered nuclease is Cas9, which cuts the target DNA guided by CRISPR RNA. There are two versions of guide RNA, one is crRNA designed in such a way to make it complementary to the target DNA, a 17–20 nucleotide sequence, and the other one is called *trans*-activating CRISPR RNA (tracrRNA), acting as a binding platform for the Cas9 nuclease (Alok et al., 2021). In the context of legumes, there is tremendous potential to use CRISPR technology to improve the traits that have already been proven in different studies (Baloglu et al., 2022; Miller et al., 2022; Singh et al., 2023). The use of CRISPR as a promising technology in the field of legume studies is shown in Table 12.5. The breakthrough of CRISPR-Cas9-mediated genome editing in enhancing important agronomic traits and quality is comprehensively reviewed in Elsharawy and Refat (2023) and Singh et al. (2023). A particular example is wheat grain improvement by muting TabHLH489 (Lyu et al., 2024), abiotic and biotic stress tolerance of wheat (Elsharawy & Refat, 2023), enhanced amylose in wheat, and disease resistance (stripe rust) in wheat by knockdown of TaCIPK14 (He et al., 2023). Some other examples are the enhanced yield in soybean by editing nuclear localization signals of E1 and E1Lb (Gao et al., 2024), and enhanced nodulation and consequently boosting nutrient acquisition and yield by inducing mutations (cle1a/2a (ric1a/2a) in soybean (Zhong et al., 2024). The

TABLE 12.5
CRISPR/CAS9-Mediated Genome Editing of Legume Crops

Legume	Method used	Gene name	Phenotype/result observed	References
Chickpea	CRISPR/Cas9	4-coumarate ligase (4CL) and Reveille 7 (RVE7) genes	Understand the complex drought stress mechanism pathways	(Badhan et al., 2021)
Soybean	CRISPR/Cas9-mediated mutation	GmF3H2 (flavanone-3-hydroxylase 2) and GmFNSII-1 (flavone synthase II)	Soybean isoflavone content and resistance to soybean mosaic virus	(Zhang et al., 2020)
Pea (*Pisum sativum* L.)	Agrobacterium-mediated CRISPR/Cas9 system	Phytoene desaturase (PsPDS) gene	Development of stable transgenic lines	(Li et al., 2023)
Phaseolus vulgaris	RISPR/Cas9-mediated mutations by GoldenGate-cloning using the MoClo system	Xanthosine monophosphate phosphatase (XMPP), guanosine deaminase (GSDA), the nucleoside hydrolases 1 and 2 (NSH1, NSH2) or xanthine dehydrogenase (XDH)	Identify essential metabolites for ureide biosynthesis by analysis of CRISPR mutant nodules	(Voß et al., 2022)
Medicago sativa	CRISPR/Cas9 *Agrobacterium*-mediated transformation	PHOSPHATE2 (PHO2)	The pho2 mutant showed hyperaccumulation of phosphate	(Miller et al., 2022)
Peanuts	CRISPR/Cas9-based hairy root transformation system	AhFAD2A and AhFAD2B	Development of efficient protocol	(Neelakandan et al., 2022)
Soybean (*Glycine max* L.)	CRISPR/Cas9	GmIPK1	Muted lines showed reduced phytic acid content in soybean seeds	(Song et al., 2022)

development of a protocol for CRIPR in editing legumes is in progress; recently, Li et al. (2023) developed an *Agrobacterium*-Mediated CRISPR/Cas9 system in pea and showed successful editing of a pea phytoene desaturase (PsPDS) gene.

12.6 AI IN LEGUME ENGINEERING

Artificial intelligence (AI) applications in legume engineering have emerged as a potential tool to address longstanding challenges in plant breeding and tissue culture. Legumes, such as kabuli chickpea and desi chickpea, are often classified as recalcitrant plants due to their low regeneration potential. Issues such as low shoot length, hyperhydricity, and poor rooting make the development of efficient in vitro regeneration protocols difficult (Aasim et al., 2023; Kirtis et al., 2022). Recent studies, however, have demonstrated how AI methodologies can enhance these processes by providing data-driven solutions to optimize protocols and improve success rates. For example, Kirtis et al. (2022) applied machine learning to develop an in vitro regeneration protocol for chickpea and identified random forest as the most effective model, achieving the highest R^2 values in predicting outcomes. This highlights the advantage of machine learning in optimizing the many parameters that influence tissue culture success. Similarly, Aasim et al. (2022) used plumular apical explants to establish a regeneration protocol for common beans. They employed response surface methodology to predict outcomes and validated their results using advanced machine learning models, including XGBoost, support vector machines, and random forest. These studies emphasize the room of machine learning models to analyze and optimize complex, multifactorial systems in plant tissue culture.

Beyond regeneration, AI is used for developing stress-tolerant legume varieties. Aasim et al. (2023) used artificial neural networks to screen chickpea varieties for salt tolerance under in vitro conditions. This approach allowed for the rapid evaluation of multiple genotypes, offering insights into their stress response mechanisms. Deep learning has further broadened the horizons of legume engineering by enabling the analysis of large and complex datasets. Techniques such as convolutional neural networks (CNNs) and dense neural networks have been effectively applied to tasks like single-nucleotide polymorphism interaction identification and phenotype prediction (Montesinos-López et al., 2018). CNNs excel in capturing spatial information from genomic sequences and resolving local epistasis effects through their convolutional mechanisms, which consider interactions among neighboring SNPs. One significant advantage of deep learning over conventional approaches is its ability to handle missing data without requiring imputation. This results in more accurate predictions for quantitative phenotypes (Liu et al., 2019). Moreover, interpretability tools such as attention mechanisms and saliency maps have been integrated into deep learning models, allowing an understanding of the association of individual input features to specific traits. With the continuing refinement of AI models and the incorporation of explainable AI techniques, the future of legume engineering appears increasingly promising.

12.7 CHALLENGES OF LEGUME ENGINEERING

There is a plethora of challenges in engineering the legumes in comparison with other cereals. For example, if the target is the nutritional improvement of the seeds of legumes, it is complicated as carbohydrates, proteins, and lipids are evenly distributed in the seeds, biosynthesis of which is regulated by multiple pathways while for other cereals, a key part of the seed is carbohydrate that is easy for manipulation by altering the carbohydrate biosynthesis gene (Ahn et al., 2023; Smith et al., 2018). Moreover, indeterminate growth habits with overlapping flowering phases and fruiting patterns in pods of all nodes are quite different from the other crops that bear grains in the terminal structure of the plant and have a distinct transition zone of vegetative and reproductive growth phases. Manipulation of a single gene may not be effective for improving the harvest index or grain production (Ahn et al., 2023). Legume engineering by applying advanced molecular technology is challenging in many respects, broadly speaking, the major challenges are genetic, institutional, environmental, and socioeconomic (Kholová et al., 2021). The first and foremost constraint of legume engineering reported widely is the genetic diversity and complexity (Jha et al., 2022). Poor access to advanced molecular tools like CRISPR/Cas9 or RNAi in many developing countries is also a big hurdle in legume engineering. These countries rather use conventional breeding, however, conventional breeding of legumes is also complicated due to limitations in hybridization with the wild species; examples of peanut, chickpea, and pigeon peas are described by Ojiewo et al. (2019). Lack of genetic and genome resources, especially true for diverse unconventional legume crops, is considered a constraint as such information is the key resource for trait mapping, allele mining, and genome editing. Self-pollination in legumes is predominant, making the gene pool database narrower, less diverse, and less polymorphic, which restricts the full use of genome engineering potentiality. Many important attributes of legumes, such as drought tolerance, nitrogen fixation, and disease and pest resistance, are polygenic, making genome engineering challenging. For many legume species, efficient and reliable genetic transformation has been a bottleneck due to their recalcitrance to *Agrobacterium*-mediated transformation and difficulties in regenerating transgenic plants. For instance, while stable transformation was successful in soybeans, others like chickpeas, pigeonpeas, and cowpeas were initially difficult to transform. Besides, rapid climate change made legume engineering challenging due to the narrower adaption possibility in a newly changed environment because of a narrower genetic base (poor gene pool).

Institutional issues include there being no uniform set of rules and regulations regarding the acceptance of genome engineering crops or even naming the crop as either GMO or non-GMO. For example, the USA, Australia, and Brazil have defined engineered crops as conventional bred (GMO) if the foreign gene is not introduced, while any modification in the genome, irrespective of foreign or its own, is regarded as GMO in the European Union. Such skewed rules fuel the negative perception of genome engineering of crops, including legumes. Discriminatory regulations and certification processes in different parts of the world slow the release and adoption of new varieties developed through genome manipulation technology. Last but not least, the socioeconomic issue hinders legume engineering, like other crop improvements. Research in legumes mainly focuses on a few cereal crops; it is yet to characterize many legumes, and the genomic traits are not well known. Poor knowledge of the genetic constituents of several legumes, despite their tremendous economic value, slows down the efforts of legume engineering, and they remain underutilized.

12.8 CONCLUSION AND FUTURE DIRECTION

Land reclamation is the priority task to maximize agricultural returns; in such aspects, legumes are the best environmentally friendly options instead of chemical fertilizers that are unaffordable for more than 475 million farm smallholders (≤2 ha), the majority of whom live in Asia and Africa. Several strategies and successful stories exist for restoring and reclaiming degraded land by planting legumes. The incorporation of legumes in crop rotation in agriculture and plantation in severely degraded forest land ameliorates the ruined ecosystem. The provision of legumes in food and feed security, as well as in environmental amendments, can be enhanced manyfold by exploiting genetic resources. Keeping in view the crucial role of legumes and their genomic resources, this chapter comprehensively reviews the current knowledge in the molecular engineering of legumes, the applicability of tools in breeding, different genomic resources, and databases of legume crops. We have highlighted important genetic tools, such as marker-assisted selection, TALEN, ZFN, CRISPR/Cas9, RNAi, and micro-RNA, with cited examples of how these tools have improved legume quality and quantitative traits. The cited examples in this review underscore the importance of adopting advanced molecular tools to improve legume crops, particularly grains, to meet the global food and feed demand. New evolving technology offers ample opportunity to develop new varieties tolerant to drought and pests and better suited to climate change. Despite the significant progress in legume crop improvement through genetic engineering, particularly for a few agronomically important crops, more attention needs to be given to exploring more genetic resources from diverse germplasms, such as landraces and wild relatives of legumes. Still, the full potential of advanced technology and tools in legume engineering is underutilized and underexplored, the future direction should be driven to the possible use of genome editing, beneficial alleles/variant selection, and application of phenomics in rapid phenotyping of useful traits. All the possible avenues of molecular, genetic, and biotechnological tools need to be explored to overcome the complications of legume breeding. The priority aims in legume engineering should be the enhanced nutritional value of grains, productivity, and better climate resilience.

REFERENCES

Aasim, M., Akin, F., Ali, S.A., Taskin, M.B., Colak, M.S., Khawar, K.M., 2023. Artificial neural network modeling for deciphering the in vitro induced salt stress tolerance in chickpea (Cicer arietinum L). *Physiology and Molecular Biology of Plants* 29, 289. https://doi.org/10.1007/S12298-023-01282-Z

Aasim, M., Katirci, R., Baloch, F.S., Mustafa, Z., Bakhsh, A., Nadeem, M.A., Ali, S.A., Hatipoğlu, R., Çiftçi, V., Habyarimana, E., Karaköy, T., Chung, Y.S., 2022. Innovation in the breeding of common bean through a combined approach of in vitro regeneration and machine learning algorithms. *Frontiers in Genetics* 13, 897696. https://doi.org/10.3389/FGENE.2022.897696

Ahn, E., Botkin, J., Curtin, S.J., Zsögön, A., 2023. Ideotype breeding and genome engineering for legume crop improvement. *Current Opinion in Biotechnology* 82, 102961. https://doi.org/10.1016/J.COPBIO.2023.102961

Ali, Z., Saeed, W., Naseem, S., Ahmad, F., Akrem, A., Yasmeen, N., Jacobsen, H.J., 2018. Phenotypic evaluation of transgenic peas (Pisum sativum L.) harboring AtNHX1 demonstrates stable gene expression and conserved morphology in subsequent generations. *Turkish Journal of Botany* 42, 150–158. https://doi.org/10.3906/bot-1705-23

Alok, A., Chauhan, H., Upadhyay, S.K., Pandey, A., Kumar, J., Singh, K., Alok, A., Chauhan, H., Upadhyay, S.K., Pandey, A., Kumar, J., 2021. Compendium of plant-specific CRISPR vectors and their technical advantages. *Life*, 11, 1021. https://doi.org/10.3390/LIFE11101021

Amor, B.B., Shaw, S.L., Oldroyd, G.E.D., Maillet, F., Penmetsa, R.V., Cook, D., Long, S.R., Dénarié, J., Gough, C., 2003. The NFP locus of Medicago truncatula controls an early step of Nod factor signal transduction upstream of a rapid calcium flux and root hair deformation. *The Plant Journal* 34, 495–506. https://doi.org/10.1046/J.1365-313X.2003.01743.X

Ané, J.M., Kiss, G.B., Riely, B.K., Penmetsa, R.V., Oldroyd, G.E.D., Ayax, C., Lévy, J., Debellé, F., Baek, J.M., Kalo, P., Rosenberg, C., Roe, B.A., Long, S.R., Dénarié, J., Cook, D.R., 2004. Medicago truncatula DMI1 required for bacterial and fungal symbioses in legumes. *Science* 303, 1364–1367. https://doi.org/10.1126/SCIENCE.1092986

Araújo, S.S., Beebe, S., Crespi, M., Delbreil, B., González, E.M., Gruber, V., Lejeune-Henaut, I., Link, W., Monteros, M.J., Prats, E., Rao, I., Vadez, V., Patto, M.C.V., 2015. Abiotic stress responses in legumes: Strategies used to cope with environmental challenges. *Critical Reviews in Plant Sciences* 34, 237–280. https://doi.org/10.1080/07352689.2014.898450

Arun, M., Subramanyam, K., Theboral, J., Sivanandhan, G., Rajesh, M., Kapil Dev, G., Jaganath, B., Manickavasagam, M., Girija, S., Ganapathi, A., 2014. Transfer and targeted overexpression of γ-tocopherol Methyltransferase (γ-TMT) gene using seed-specific promoter improves tocopherol composition in Indian soybean cultivars. *Applied Biochemistry and Biotechnology* 172, 1763–1776. https://doi.org/10.1007/S12010-013-0645-9

Aryal, U.K., Hossain, M.K., Mridha, M.A.U., Xu, H.L., 2001. Nodulation status and nitrogenase activity of some legume tree species in Bangladesh. *Journal of Crop Production* 3, 325–335. https://doi.org/10.1300/J144V03N01_27

Badhan, S., Ball, A.S., Mantri, N., 2021. First report of CRISPR/Cas9 mediated DNA-free editing of 4CL and RVE7 genes in chickpea protoplasts. *International Journal of Molecular Sciences* 22, 1–15. https://doi.org/10.3390/ijms22010396

Bakshi, S., Saha, B., Roy, N.K., Mishra, S., Panda, S.K., Sahoo, L., 2012. Successful recovery of transgenic cowpea (Vigna unguiculata) using the 6-phosphomannose isomerase gene as the selectable marker. *Plant Cell Reports* 31, 1093–1103. https://doi.org/10.1007/S00299-012-1230-3

Baloglu, M.C., Celik Altunoglu, Y., Baloglu, P., Yildiz, A.B., Türkölmez, N., Özden Çiftçi, Y., 2022. Gene-editing technologies and applications in legumes: Progress, evolution, and future prospects. *Frontiers in Genetics* 13, 859437. https://doi.org/10.3389/FGENE.2022.859437/PDF

Bayer, E.M., Benitez-Alfonso, Y., 2024. Plasmodesmata: Channels under pressure. *Annual Review of Plant Biology* 75, 291–317. https://doi.org/10.1146/ANNUREV-ARPLANT-070623-093110

Berendzen, J., Brown, A.V., Cameron, C.T., Campbell, J.D., Cleary, A.M., Dash, S., Hokin, S., Huang, W., Kalberer, S.R., Nelson, R.T., Redsun, S., Weeks, N.T., Wilkey, A., Farmer, A.D., Cannon, S.B., 2021. The legume information system and associated online genomic resources. *Legume Science* 3, e74. https://doi.org/10.1002/LEG3.74

Bertioli, D.J., Cannon, S.B., Froenicke, L., Huang, G., Farmer, A.D., Cannon, E.K.S., Liu, X., Gao, D., Clevenger, J., Dash, S., Ren, L., Moretzsohn, M.C., Shirasawa, K., Huang, W., Vidigal, B., Abernathy, B., Chu, Y., Niederhuth, C.E., Umale, P., Arajo, A.C.G., Kozik, A., Do Kim, K., Burow, M.D., Varshney, R.K., Wang, X., Zhang, X., Barkley, N., Guimares, P.M., Isobe, S., Guo, B., Liao, B., Stalker, H.T., Schmitz, R.J., Scheffler, B.E., Leal-Bertioli, S.C.M., Xun, X., Jackson, S.A., Michelmore, R., Ozias-Akins, P., 2016. The genome sequences of Arachis duranensis and Arachis ipaensis, the diploid ancestors of cultivated peanut. *Nature Genetics* 48(4), 438–446. https://doi.org/10.1038/ng.3517

Bett, B., Gollasch, S., Moore, A., Harding, R., Higgins, T.J.V., 2019. An improved transformation system for cowpea (Vigna unguiculata l. Walp) via sonication and a kanamycin-geneticin selection regime. *Frontiers in Plant Science* 10, 418100. https://doi.org/10.3389/FPLS.2019.00219

Bhomkar, P., Upadhyay, C.P., Saxena, M., Muthusamy, A., Shiva Prakash, N., Pooggin, M., Hohn, T., Sarin, N.B., 2008. Salt stress alleviation in transgenic Vigna mungo L. Hepper (blackgram) by overexpression of the glyoxalase I gene using a novel Cestrum yellow leaf curling virus (CmYLCV) promoter. *Molecular Breeding* 22, 169–181. https://doi.org/10.1007/S11032-008-9164-8

Bonawitz, N.D., Ainley, W.M., Itaya, A., Chennareddy, S.R., Cicak, T., Effinger, K., Jiang, K., Mall, T.K., Marri, P.R., Samuel, J.P., Sardesai, N., Simpson, M., Folkerts, O., Sarria, R., Webb, S.R., Gonzalez, D.O., Simmonds, D.H., Pareddy, D.R., 2019. Zinc finger nuclease-mediated targeting of multiple transgenes to an endogenous soybean genomic locus via non-homologous end joining. *Plant Biotechnology Journal* 17, 750–761. https://doi.org/10.1111/PBI.13012

Calyxt Inc, 2019. *First Commercial Sale of Calyxt High Oleic Soybean Oil on the U.S. Market | Business Wire.*

Catalanotto, C., Cogoni, C., Zardo, G., 2016. MicroRNA in control of gene expression: An overview of nuclear functions. *International Journal of Molecular Sciences* 17, 1712. https://doi.org/10.3390/IJMS17101712

Chakraborty, J., Sen, S., Ghosh, P., Sengupta, A., Basu, D., Das, S., 2016. Homologous promoter derived constitutive and chloroplast targeted expression of synthetic cry1Ac in transgenic chickpea confers resistance against Helicoverpa armigera. *Plant Cell, Tissue and Organ Culture* 125, 521–535. https://doi.org/10.1007/S11240-016-0968-7

Chen, H., Yang, Q., Chen, K., Zhao, S., Zhang, C., Pan, R., Cai, T., Deng, Y., Wang, X., Chen, Y., Chu, W., Xie, W., Zhuang, W., 2019. Integrated microRNA and transcriptome profiling reveals a miRNA-mediated regulatory network of embryo abortion under calcium deficiency in peanut (Arachis hypogaea L.). *BMC Genomics* 20(1): 1–17. https://doi.org/10.1186/S12864-019-5770-6

Chen, K., Gao, C., 2013. TALENs: Customizable molecular DNA scissors for genome engineering of plants. *Journal of Genetics and Genomics* 40, 271–279. https://doi.org/10.1016/J.JGG.2013.03.009

Chen, R., Li, M., Zhang, Huiyuan, Duan, L., Sun, X., Jiang, Q., Zhang, Hui, Hu, Z., 2019. Continuous salt stress-induced long non-coding RNAs and DNA methylation patterns in soybean roots. *BMC Genomics* 20, 1–12. https://doi.org/10.1186/S12864-019-6101-7

Chen, X., Li, Hongjie, Pandey, M.K., Yang, Q., Wang, X., Garg, V., Li, Haifen, Chi, X., Doddamani, D., Hong, Y., Upadhyaya, H., Guo, H., Khan, A.W., Zhu, F., Zhang, Xiaoyan, Pan, L., Pierce, G.J., Zhou, G., Krishnamohan, K.A.V.S., Chen, M., Zhong, N., Agarwal, G., Li, Shuanzhu, Chitikineni, A., Zhang, G.Q., Sharma, S., Chen, N., Liu, H., Janila, P., Li, Shaoxiong, Wang, Min, Wang, T., Sun, J., Li, X., Li, C., Wang, Mian, Yu, L., Wen, S., Singh, S., Yang, Z., Zhao, J., Zhang, C., Yu, Y., Bi, J., Zhang, Xiaojun, Liu, Z.J., Paterson, A.H., Wang, S., Liang, X., Varshney, R.K., Yu, S., 2016. Draft genome of the peanut A-genome progenitor (Arachis duranensis) provides insights into geocarpy, oil biosynthesis, and allergens. *Proceedings of the National Academy of Sciences of the United States of America* 113, 6785–6790. https://doi.org/10.1073/PNAS.1600899113/SUPPL_FILE/PNAS.1600899113.SD17.XLSX

Chopra, R., Saini, R., 2014. Transformation of blackgram (Vigna mungo (L.) Hepper) by Barley chitinase and ribosome-inactivating protein genes towards improving resistance to corynespora leaf spot fungal disease. *Applied Biochemistry and Biotechnology* 174, 2791–2800. https://doi.org/10.1007/S12010-014-1226-2

Crespi, M., Frugier, F., 2009. De novo organ formation from differentiated cells: Root nodule organogenesis. *Science Signaling* 2, 1–8. https://doi.org/10.1126/scisignal.253er1

Cruz, A.R.R., Aragão, F.J.L., 2014. RNAi-based enhanced resistance to Cowpea severe mosaic virus and Cowpea aphid-borne mosaic virus in transgenic cowpea. *Plant Pathology* 63, 831–837. https://doi.org/10.1111/PPA.12178

Curtin, S.J., Xiong, Y., Michno, J.M., Campbell, B.W., Stec, A.O., Čermák, T., Starker, C., Voytas, D.F., Eamens, A.L., Stupar, R.M., 2018. CRISPR/Cas9 and TALENs generate heritable mutations for genes involved in small RNA processing of Glycine max and Medicago truncatula. *Plant Biotechnology Journal* 16, 1125–1137. https://doi.org/10.1111/PBI.12857

Curtin, S.J., Zhang, F., Sander, J.D., Haun, W.J., Starker, C., Baltes, N.J., Reyon, D., Dahlborg, E.J., Goodwin, M.J., Coffman, A.P., Dobbs, D., Joung, J.K., Voytas, D.F., Stupar, R.M., 2011. Targeted mutagenesis of duplicated genes in Soybean with zinc-finger nucleases. *Plant Physiology* 156, 466–473. https://doi.org/10.1104/PP.111.172981

Das, D.K., 2018. Expression of a Bacterial Chitinase (ChiB) gene enhances resistance against erysiphae polygoni induced powdery mildew disease in the transgenic black gram (Vigna mungo L.) (cv. T9). *American Journal of Plant Sciences* 9, 1759–1770. https://doi.org/10.4236/AJPS.2018.98128

Dey, B., Ferdous, J., Ahmed, R., 2024. Machine learning based recommendation of agricultural and horticultural crop farming in India under the regime of NPK, soil pH and three climatic variables. *Heliyon* 10, e25112. https://doi.org/10.1016/j.heliyon.2024.e25112

Dey, B., Masum Ul Haque, M., Khatun, R., Ahmed, R., 2022. Comparative performance of four CNN-based deep learning variants in detecting Hispa pest, two fungal diseases, and NPK deficiency symptoms of rice (Oryza sativa). *Computers and Electronics in Agriculture* 202, 107340. https://doi.org/10.1016/j.compag.2022.107340

Dhaliwal, S.S., Sharma, V., Shukla, A.K., Kaur, J., Verma, V., Singh, P., Singh, H., Abdel-Hafez, S.H., Sayed, S., Gaber, A., Ali, R., Hossain, A., 2021. Enrichment of zinc and iron micronutrients in lentil (Lens culinaris Medik.) through biofortification. *Molecules* 26, 7671. https://doi.org/10.3390/MOLECULES26247671

Elsharawy, H., Refat, M., 2023. CRISPR/Cas9 genome editing in wheat: Enhancing quality and productivity for global food security—A review. *Functional and Integrative Genomics* 23, 1–14. https://doi.org/10.1007/S10142-023-01190-1/FIGURES/5

Enebe, M.C., Erasmus, M., 2023. Susceptibility and plant immune control—A case of mycorrhizal strategy for plant colonization, symbiosis, and plant immune suppression. *Frontiers in Microbiology* 14, 1178258. https://doi.org/10.3389/FMICB.2023.1178258

Estrada-Navarrete, G., Cruz-Mireles, N., Lascano, R., Alvarado-Affantranger, X., Hernández-Barrera, A., Barraza, A., Olivares, J.E., Arthikala, M.K., Cárdenas, L., Quinto, C., Sanchez, F., 2016. An autophagy-related kinase is essential for the symbiotic relationship between phaseolus vulgaris and both rhizobia and arbuscular mycorrhizal fungi. *The Plant Cell* 28, 2326–2341. https://doi.org/10.1105/TPC.15.01012

Fanzo, J., Bellows, A.L., Spiker, M.L., Thorne-Lyman, A.L., Bloem, M.W., 2021. The importance of food systems and the environment for nutrition. *The American Journal of Clinical Nutrition* 113, 7–16. https://doi.org/10.1093/AJCN/NQAA313

Flores, T., Karpova, O., Su, X., Zeng, P., Bilyeu, K., Sleper, D.A., Nguyen, H.T., Zhang, Z.J., 2008. Silencing of GmFAD3 gene by siRNA leads to low α-linolenic acids (18:3) of fad3-mutant phenotype in soybean [Glycine max (Merr.)]. *Transgenic Research* 17, 839–850. https://doi.org/10.1007/S11248-008-9167-6

Floss, D.S., Schmitz, A.M., Starker, C.G., Stephen Gantt, J., Harrison, M.J., 2013. Gene Silencing in Medicago truncatula roots using RNAi. *Methods in Molecular Biology* 1069, 163–177. https://doi.org/10.1007/978-1-62703-613-9_12

Fournier, J., Timmers, A.C.J., Sieberer, B.J., Jauneau, A., Chabaud, M., Barker, D.G., 2008. Mechanism of infection thread elongation in root hairs of medicago truncatula and dynamic interplay with associated rhizobial colonization. *Plant Physiology* 148, 1985–1995. https://doi.org/10.1104/PP.108.125674

Ganguly, M., Molla, K.A., Karmakar, S., Datta, K., Datta, S.K., 2014. Development of pod borer-resistant transgenic chickpea using a pod-specific and a constitutive promoter-driven fused cry1Ab/Ac gene. *Theoretical and Applied Genetics* 127, 2555–2565. https://doi.org/10.1007/S00122-014-2397-5

Ganguly, S., Ghosh, G., Ghosh, S., Purohit, A., Chaudhuri, R.K., Das, S., Chakraborti, D., 2020. Plumular meristem transformation system for chickpea: An efficient method to overcome recalcitrant tissue culture responses. *Plant Cell, Tissue and Organ Culture* 142, 493–504. https://doi.org/10.1007/S11240-020-01873-8

Ganguly, S., Ghosh, G., Purohit, A., Kundu Chaudhuri, R., Chakraborti, D., 2018. Development of transgenic pigeonpea using high throughput plumular meristem transformation method. *Plant Cell, Tissue and Organ Culture* 135, 73–83. https://doi.org/10.1007/S11240-018-1444-3

Gao, Y., Zhang, Y., Ma, C., Chen, Y., Liu, C., Wang, Y., Wang, S., Chen, X., 2024. Editing the nuclear localization signals of E1 and E1Lb enables the production of tropical soybean in temperate growing regions. *Plant Biotechnology Journal* 22, 2145–2156. https://doi.org/10.1111/PBI.14335

Ghosh, G., Ganguly, S., Purohit, A., Chaudhuri, R.K., Das, S., Chakraborti, D., 2017. Transgenic pigeonpea events expressing Cry1Ac and Cry2Aa exhibit resistance to Helicoverpa armigera. *Plant Cell Reports* 36, 1037–1051. https://doi.org/10.1007/S00299-017-2133-0

Güsewell, S., 2004. N: P ratios in terrestrial plants: Variation and functional significance. *New Phytologist* 164, 243–266. https://doi.org/10.1111/J.1469-8137.2004.01192.X

Hajyzadeh, M., Turktas, M., Khawar, K.M., Unver, T., 2015. miR408 overexpression causes increased drought tolerance in chickpea. *Gene* 555, 186–193. https://doi.org/10.1016/J.GENE.2014.11.002

Hasan, N., Choudhary, S., Naaz, N., Sharma, N., Laskar, R.A., 2021. Recent advancements in molecular marker-assisted selection and applications in plant breeding programmes. *Journal of Genetic Engineering and Biotechnology* 19, 128. https://doi.org/10.1186/S43141-021-00231-1

He, F., Wang, C., Sun, H., Tian, S., Zhao, G., Liu, C., Wan, C., Guo, Jia, Huang, X., Zhan, G., Yu, X., Kang, Z., Guo, Jun, 2023. Simultaneous editing of three homoeologues of TaCIPK14 confers broad-spectrum resistance to stripe rust in wheat. *Plant Biotechnology Journal* 21, 354–368. https://doi.org/10.1111/PBI.13956

Herridge, D.F., Peoples, M.B., Boddey, R.M., 2008. Global inputs of biological nitrogen fixation in agricultural systems. *Plant and Soil* 311, 1–18. https://doi.org/10.1007/S11104-008-9668-3

Hiremath, A.J., Ewel, J.J., 2001. Ecosystem nutrient use efficiency, productivity, and nutrient accrual in model tropical communities. *Ecosystems* 4, 669–682. https://doi.org/10.1007/S10021-001-0036-X

Imaizumi-Anraku, H., Takeda, N., Charpentier, M., Perry, J., Miwa, H., Umehara, Y., Kouchi, H., Murakami, Y., Mulder, L., Vickers, K., Pike, J., Downie, J.A., Wang, T., Sato, S., Asamizu, E., Tabata, S., Yoshikawa, M., Murooka, Y., Wu, G.J., Kawaguchi, M., Kawasaki, S., Parniske, M., Hayashi, M., 2004. Plastid proteins crucial for symbiotic fungal and bacterial entry into plant roots. *Nature* 433(7025), 527–531. https://doi.org/10.1038/nature03237

Indurker, S., Misra, H.S., Eapen, S., 2007. Genetic transformation of chickpea (Cicer arietinum L.) with insecticidal crystal protein gene using particle gun bombardment. *Plant Cell Reports* 26, 755–763. https://doi.org/10.1007/S00299-006-0283-6

Islas-Flores, T., Guillén, G., Alvarado-Affantranger, X., Lara-Flores, M., Sánchez, F., Villanueva, M.A., 2011. PvRACK1 loss-of-function impairs cell expansion and morphogenesis in phaseolus vulgaris L. *Root Nodules* 24, 819–826. https://doi.org/10.1094/MPMI-11-10-0261

Istanbuli, T., Nassar, A.E., Abd El-Maksoud, M.M., Tawkaz, S., Alsamman, A.M., Hamwieh, A., 2024. Genome-wide association study reveals SNP markers controlling drought tolerance and related agronomic traits in chickpea across multiple environments. *Frontiers in Plant Science* 15, 1260690. https://doi.org/10.3389/FPLS.2024.1260690

Jha, U.C., Nayyar, H., Parida, S.K., Bakır, M., von Wettberg, E.J.B., Siddique, K.H.M., 2022. Progress of genomics-driven approaches for sustaining underutilized legume crops in the post-genomic era. *Frontiers in Genetics* 13, 831656. https://doi.org/10.3389/FGENE.2022.831656/PDF

Karmakar, S., Molla, K.A., Gayen, D., Karmakar, A., Das, K., Sarkar, S.N., Datta, K., Datta, S.K., 2019. Development of a rapid and highly efficient Agrobacterium-mediated transformation system for pigeon pea [Cajanus cajan (L.) Millsp]. *GM Crops & Food* 10, 115–138. https://doi.org/10.1080/21645698.2019.1625653

Karthik, S., Pavan, G., Manickavasagam, M., 2020. Nitric oxide donor regulates Agrobacterium-mediated genetic transformation efficiency in soybean [Glycine max (L.) Merrill]. *Plant Cell, Tissue and Organ Culture* 141, 655–660. https://doi.org/10.1007/S11240-020-01808-3

Kaur, A., Sharma, M., Sharma, C., Kaur, H., Kaur, N., Sharma, S., Arora, R., Singh, I., Sandhu, J.S., 2016. Pod borer resistant transgenic pigeon pea (Cajanus cajan L.) expressing cry1Ac transgene generated through simplified Agrobacterium transformation of pricked embryo axes. *Plant Cell, Tissue and Organ Culture* 127, 717–727. https://doi.org/10.1007/S11240-016-1055-9

Khandal, H., Parween, S., Roy, R., Meena, M.K., Chattopadhyay, D., 2017. MicroRNA profiling provides insights into post-transcriptional regulation of gene expression in chickpea root apex under salinity and water deficiency. *Scientific Reports* 7(1–7), 1–14. https://doi.org/10.1038/s41598-017-04906-z

Khandelwal, R., Jain, M., 2018. Genome engineering tools for functional genomics and crop improvement in legumes. In: Wani, S., Jain, M. (eds.). *Pulse Improvement*. Springer, Cham. https://doi.org/10.1007/978-3-030-01743-9_10. https://doi.org/10.1007/978-3-030-01743-9_10

Khatun, M., Sarkar, S., Era, F.M., Islam, A.K.M.M., Anwar, M.P., Fahad, S., Datta, R., Islam, A.K.M.A., 2021. Drought stress in grain legumes: Effects, tolerance mechanisms and management. *Agronomy* 11, 2374. https://doi.org/10.3390/AGRONOMY11122374

Kholová, J., Urban, M.O., Cock, J., Arcos, J., Arnaud, E., Aytekin, D., Azevedo, V., Barnes, A.P., Ceccarelli, S., Chavarriaga, P., Cobb, J.N., Connor, D., Cooper, M., Craufurd, P., Debouck, D., Fungo, R., Grando, S., Hammer, G.L., Jara, C.E., Messina, C., Mosquera, G., Nchanji, E., Ng, E.H., Prager, S., Sankaran, S., Selvaraj, M., Tardieu, F., Thornton, P., Valdes-Gutierrez, S.P., Van Etten, J., Wenzl, P., Xu, Y., 2021. In pursuit of a better world: Crop improvement and the CGIAR. *Journal of Experimental Botany* 72, 5158–5179. https://doi.org/10.1093/JXB/ERAB226

Kirtis, A., Aasim, M., Katırcı, R., 2022. Application of artificial neural network and machine learning algorithms for modeling the in vitro regeneration of chickpea (Cicer arietinum L.). *Plant Cell, Tissue and Organ Culture* 150, 141–152. https://doi.org/10.1007/S11240-022-02255-Y

Kohli, D., Joshi, G., Deokar, A.A., Bhardwaj, A.R., Agarwal, M., Katiyar-Agarwal, S., Srinivasan, R., Jain, P.K., 2014. Identification and characterization of wilt and salt stress-responsive MicroRNAs in Chickpea through high-throughput sequencing. *PLOS ONE* 9, e108851. https://doi.org/10.1371/JOURNAL.PONE.0108851

Krishna, G., Reddy, P.S., Ramteke, P.W., Rambabu, P., Tawar, K.B., Bhattacharya, P., 2011. Agrobacterium-mediated genetic transformation of pigeon pea [Cajanus cajan (L.) Millsp.] for resistance to legume pod borer Helicoverpa armigera. *Journal of Crop Science and Biotechnology* 14(3), 197–204. https://doi.org/10.1007/S12892-010-0063-2

Krishnamurthy, L., Gaur, P.M., Basu, P.S., Chaturvedi, S.K., Tripathi, S., Vadez, V., Rathore, A., Varshney, R.K., Gowda, C.L.L., 2011. Large genetic variation for heat tolerance in the reference collection of chickpea (Cicer arietinum L.) germplasm. *Plant Genetic Resources* 9, 59–69. https://doi.org/10.1017/S1479262110000407

Kumar, A., Jaiwal, R., Sreevathsa, R., Chaudhary, D., Jaiwal, P.K., 2021. Transgenic cowpea plants expressing Bacillus thuringiensis Cry2Aa insecticidal protein imparts resistance to Maruca vitrata legume pod borer. *Plant Cell Reports* 40, 583–594. https://doi.org/10.1007/S00299-020-02657-2

Kumar Ghanti, S.K., Sujata, K.G., Kumar, B.M.V., Karba, N.N., Reddy, K.J., Rao, M.S., Kishor, P.B.K., 2011. Heterologous expression of P5CS gene in chickpea enhances salt tolerance without affecting yield. *Biologia Plantarum* 55, 634–640. https://doi.org/10.1007/S10535-011-0161-0

Kuyah, S., Muoni, T., Bayala, J., Chopin, P., Dahlin, A.S., Hughes, K., Jonsson, M., Kumar, S., Sileshi, G.W., Dimobe, K., Öborn, I., 2023. Grain legumes and dryland cereals contribute to carbon sequestration in the drylands of Africa and South Asia. *Agriculture, Ecosystems & Environment* 355, 108583. https://doi.org/10.1016/J.AGEE.2023.108583

Lévy, J., Bres, C., Geurts, R., Chalhoub, B., Kulikova, O., Duc, G., Journet, E.P., Ané, J.M., Lauber, E., Bisseling, T., Dénarié, J., Rosenberg, C., Debellé, F., 2004. A putative Ca2+ and calmodulin-dependent protein kinase required for bacterial and fungal symbioses. *Science* 303, 1361–1364. https://doi.org/10.1126/SCIENCE.1093038

Li, G., Liu, R., Xu, R., Varshney, R.K., Ding, H., Li, M., Yan, X., Huang, S., Li, J., Wang, D., Ji, Y., Wang, C., He, J., Luo, Y., Gao, S., Wei, P., Zong, X., Yang, T., 2023. Development of an Agrobacterium-mediated

CRISPR/Cas9 system in pea (Pisum sativum L.). *The Crop Journal* 11, 132–139. https://doi.org/10.1016/J.CJ.2022.04.011

Li, S., Cong, Y., Liu, Y., Wang, T., Shuai, Q., Chen, N., Gai, J., Li, Y., 2017. Optimization of agrobacterium-mediated transformation in soybean. *Frontiers in Plant Science* 8, 247214. https://doi.org/10.3389/FPLS.2017.00246

Liu, Y., Wang, D., He, F., Wang, J., Joshi, T., Xu, D., 2019. Phenotype prediction and genome-wide association study using deep convolutional neural network of soybean. *Frontiers in Genetics* 10, 486384. https://doi.org/10.3389/FGENE.2019.01091

Lyu, J., Wang, D., Sun, N., Yang, F., Li, X., Mu, J., Zhou, R., Zheng, G., Yang, X., Zhang, C., Han, C., Xia, G.M., Li, G., Fan, M., Xiao, J., Bai, M.Y., 2024. The TaSnRK1-TabHLH489 module integrates brassinosteroid and sugar signalling to regulate the grain length in bread wheat. *Plant Biotechnology Journal* 22, 1989–2006. https://doi.org/10.1111/PBI.14319

Madsen, E.B., Madsen, L.H., Radutoiu, S., Olbryt, M., Rakwalska, M., Szczyglowski, K., Sato, S., Kaneko, T., Tabata, S., Sandal, N., Stougaard, J., 2003. A receptor kinase gene of the LysM type is involved in legumeperception of rhizobial signals. *Nature* 425(6958), 637–640. https://doi.org/10.1038/nature02045

Mandozai, A., Moussa, A.A., Zhang, Q., Qu, J., Du, Y., Anwari, G., Al Amin, N., Wang, P., 2021. Genome-wide association study of root and shoot related traits in spring Soybean (Glycine max L.) at seedling stages using SLAF-Seq. *Frontiers in Plant Science* 12, 568995. https://doi.org/10.3389/FPLS.2021.568995

Menz, J., Modrzejewski, D., Hartung, F., Wilhelm, R., Sprink, T., 2020. Genome edited crops touch the market: A view on the global development and regulatory environment. *Frontiers in Plant Science* 11, 586027. https://doi.org/10.3389/FPLS.2020.586027/BIBTEX

Miller, S.S., Dornbusch, M.R., Farmer, A.D., Huertas, R., Gutierrez-Gonzalez, J.J., Young, N.D., Samac, D.A., Curtin, S.J., 2022. Alfalfa (Medicago sativa L.) pho2 mutant plants hyperaccumulate phosphate. *G3 Genes|Genomes|Genetics* 12(6), jkac096. https://doi.org/10.1093/G3JOURNAL/JKAC096

Mirzaei, S., Batley, J., El-Mellouki, T., Liu, S., Meksem, K., Ferguson, B.J., Gresshoff, P.M., 2017. Neodiversification of homeologous CLAVATA1-like receptor kinase genes in soybean leads to distinct developmental outcomes. *Scientific Reports* 7(1–7), 1–15. https://doi.org/10.1038/s41598-017-08252-y

Mishra, S., Behura, R., Awasthi, J.P., Dey, M., Sahoo, D., Das Bhowmik, S.S., Panda, S.K., Sahoo, L., 2014. Ectopic overexpression of a mungbean vacuolar Na+/H+ antiporter gene (VrNHX1) leads to increased salinity stress tolerance in transgenic Vigna unguiculata L. Walp. *Molecular Breeding* 34, 1345–1359. https://doi.org/10.1007/S11032-014-0120-5

Miyata, K., Umehara, M., 2024. Roles of arbuscular mycorrhizal fungi for essential nutrient acquisition under nutrient deficiency in plants. In Ahammed, G.J., Hajiboland, R. (eds.). *Arbuscular Mycorrhizal Fungi and Higher Plants.* Springer, Singapore. https://doi.org/10.1007/978-981-99-8220-2_6

Montesinos-López, A., Montesinos-López, O.A., Gianola, D., Crossa, J., Hernández-Suárez, C.M., 2018. Multi-environment genomic prediction of plant traits using deep learners with dense architecture. *G3 Genes|Genomes|Genetics* 8, 3813–3828. https://doi.org/10.1534/G3.118.200740

Morton, R.L., Schroeder, H.E., Bateman, K.S., Chrispeels, M.J., Armstrong, E., Higgins, T.J.V., 2000. Bean alpha-amylase inhibitor 1 in transgenic peas (Pisum sativum) provides complete protection from pea weevil (Bruchus pisorum) under field conditions. *Proceedings of the National Academy of Sciences of the United States of America* 97, 3820–3825. https://doi.org/10.1073/PNAS.070054597

Muruganantham, M., Amutha, S., Selvaraj, N., Vengadesan, G., Ganapathi, A., 2007. Efficient agrobacterium-mediated transformation of Vigna mungo using immature cotyledonary-node explants and phosphinothricin as the selection agent. *In Vitro Cellular and Developmental Biology – Plant* 43, 550–557. https://doi.org/10.1007/S11627-007-9060-7

Nanjareddy, K., Blanco, L., Arthikala, M.K., Alvarado-Affantranger, X., Quinto, C., Sánchez, F., Lara, M., 2016. A legume TOR protein kinase regulates rhizobium symbiosis and is essential for infection and nodule development. *Plant Physiology* 172, 2002–2020. https://doi.org/10.1104/PP.16.00844

Neelakandan, A.K., Subedi, B., Traore, S.M., Binagwa, P., Wright, D.A., He, G., 2022. Base editing in peanut using CRISPR/nCas9. *Frontiers in Genome Editing* 4, 901444. https://doi.org/10.3389/FGEED.2022.901444/BIBTEX

Newell-McGloughlin, M., 2008. Nutritionally improved agricultural crops. *Plant Physiology* 147, 939–953. https://doi.org/10.1104/PP.108.121947

Ning, W., Zhai, H., Yu, J., Liang, S., Yang, X., Xing, X., Huo, J., Pang, T., Yang, Y., Bai, X., 2017. Overexpression of Glycine soja WRKY20 enhances drought tolerance and improves plant yields under drought stress in transgenic soybean. *Molecular Breeding* 37, 1–10. https://doi.org/10.1007/S11032-016-0614-4/FIGURES/5

Nizamani, M.M., Hughes, A.C., Qureshi, S., Zhang, Q., Tarafder, E., Das, D., Acharya, K., Wang, Y., Zhang, Z.G., 2024. Microbial biodiversity and plant functional trait interactions in multifunctional ecosystems. *Applied Soil Ecology* 201, 105515. https://doi.org/10.1016/J.APSOIL.2024.105515

Ojiewo, C., Monyo, E., Desmae, H., Boukar, O., Mukankusi-Mugisha, C., Thudi, M., Pandey, M.K., Saxena, R.K., Gaur, P.M., Chaturvedi, S.K., Fikre, A., Ganga Rao, N.P.V.R., SameerKumar, C.V., Okori, P., Janila, P., Rubyogo, J.C., Godfree, C., Akpo, E., Omoigui, L., Nkalubo, S., Fenta, B., Binagwa, P., Kilango, M., Williams, M., Mponda, O., Okello, D., Chichaybelu, M., Miningou, A., Bationo, J., Sako, D., Diallo, S., Echekwu, C., Umar, M.L., Oteng-Frimpong, R., Mohammed, H., Varshney, R.K., 2019. Genomics, genetics and breeding of tropical legumes for better livelihoods of smallholder farmers. *Plant Breeding* 138, 487–499. https://doi.org/10.1111/PBR.12554

Pareddy, D., Chennareddy, S., Anthony, G., Sardesai, N., Mall, T., Minnicks, T., Karpova, O., Clark, L., Griffin, D., Bishop, B., Shumway, N., Samuel, P., Smith, K., Sarria, R., 2020. Improved soybean transformation for efficient and high throughput transgenic production. *Transgenic Research* 29, 267–281. https://doi.org/10.1007/S11248-020-00198-8

Parihar, M., Rakshit, A., 2024. In vitro production of arbuscular mycorrhizal fungi: An overview. In: Parihar, M., Rakshit, A., Adholeya, A., Chen, Y. (eds.). *Arbuscular Mycorrhizal Fungi in Sustainable Agriculture: Inoculum Production and Application*. Springer, Singapore. https://doi.org/10.1007/978-981-97-0296-1_6

Peix, A., Ramírez-Bahena, M.H., Velázquez, E., Bedmar, E.J., 2015. Bacterial associations with legumes. *Critical Reviews in Plant Sciences* 34, 17–42. https://doi.org/10.1080/07352689.2014.897899

Poore, J., Nemecek, T., 2018. Reducing food's environmental impacts through producers and consumers. *Science* 360, 987–992. https://doi.org/10.1126/SCIENCE.AAQ0216

Popelka, J.C., Gollasch, S., Moore, A., Molvig, L., Higgins, T.J.V., 2006. Genetic transformation of cowpea (Vigna unguiculata L.) and stable transmission of the transgenes to progeny. *Plant Cell Reports* 25, 304–312. https://doi.org/10.1007/S00299-005-0053-X

Rasheed, A., Barqawi, A.A., Mahmood, A., Nawaz, M., Shah, A.N., Bay, D.H., Alahdal, M.A., Hassan, M.U., Qari, S.H., 2022. CRISPR/Cas9 is a powerful tool for precise genome editing of legume crops: A review. *Molecular Biology Reports* 49(6), 5595–5609. https://doi.org/10.1007/S11033-022-07529-4

Rout, G.R., Bansal, A., Swain, D., Jadhao, K.R., Shelke, R.G., Panda, S.K., 2020. Overexpression of ICE1 gene in mungbean (Vigna radiata L.) for cold tolerance. *Plant Cell, Tissue and Organ Culture* 143, 593–608. https://doi.org/10.1007/S11240-020-01944-W

Saha, B., Swain, D., Borgohain, P., Rout, G.R., Koyama, H., Panda, S.K., 2020. Enhanced exudation of malate in the rhizosphere due to AtALMT1 overexpression in blackgram (Vigna mungo L.) confers increased aluminium tolerance. *Plant Biology* 22, 701–708. https://doi.org/10.1111/PLB.13112

Sahoo, D.P., Kumar, S., Mishra, S., Kobayashi, Y., Panda, S.K., Sahoo, L., 2016. Enhanced salinity tolerance in transgenic mungbean overexpressing Arabidopsis antiporter (NHX1) gene. *Molecular Breeding* 36, 1–15. https://doi.org/10.1007/S11032-016-0564-X

Sanyal, I., Singh, A.K., Kaushik, M., Amla, D.V., 2005. Agrobacterium-mediated transformation of chickpea (Cicer arietinum L.) with Bacillus thuringiensis cry1Ac gene for resistance against pod borer insect Helicoverpa armigera. *Plant Science* 168, 1135–1146. https://doi.org/10.1016/J.PLANTSCI.2004.12.015

Schnabel, E., Thomas, J., El-Hawaz, R., Gao, Y., Poehlman, W., Chavan, S., Pasha, A., Esteban, E., Provart, N., Feltus, F.A., Frugoli, J., 2023. A laser capture microdissection transcriptome of M. truncatula roots responding to rhizobia reveals spatiotemporal tissue expression patterns of genes involved in nodule signaling and organogenesis. *bioRxiv* 2023.03.14.532650. https://doi.org/10.1101/2023.03.14.532650

Schroeder, H.E., Schotz, A.H., Wardley-Richardson, T., Spencer, D., Higgins, T.J.V., 1993. Transformation and regeneration of two cultivars of pea (Pisum sativum L.). *Plant Physiology* 101, 751–757. https://doi.org/10.1104/PP.101.3.751

Searle, I.R., Men, A.E., Laniya, T.S., Buzas, D.M., Iturbe-Ormaetxe, I., Carroll, B.J., Gresshoff, P.M., 2003. Long-distance signaling in nodulation directed by a CLAVATA1-like receptor kinase. *Science* 299, 109–112. https://doi.org/10.1126/SCIENCE.1077937

Shui, X.R., Chen, Z.W., Li, J.X., 2013. MicroRNA prediction and its function in regulating drought-related genes in cowpea. *Plant Science* 210, 25–35. https://doi.org/10.1016/J.PLANTSCI.2013.05.002

Siddique, I., Engel, V.L., Parrotta, J.A., Lamb, D., Nardoto, G.B., Ometto, J.P.H.B., Martinelli, L.A., Schmidt, S., 2008. Dominance of legume trees alters nutrient relations in mixed species forest restoration plantings within seven years. *Biogeochemistry* 88, 89–101. https://doi.org/10.1007/S10533-008-9196-5

Singh, A.K., Elango, D., Raigne, J., Van der Laan, L., Rairdin, A., Soregaon, C., Singh, A., 2025. Plant-based protein crops and their improvement: Current status and future perspectives. *Crop Science* 65, e21389. https://doi.org/10.1002/CSC2.21389

Singh, C., Kumar, Ramesh, Sehgal, H., Bhati, S., Singhal, T., Gayacharan, Nimmy, M.S., Yadav, R., Gupta, S.K., Abdallah, N.A., Hamwieh, A., Kumar, Rajendra, 2023. Unclasping potentials of genomics and gene editing in chickpea to fight climate change and global hunger threat. *Frontiers in Genetics* 14, 1085024. https://doi.org/10.3389/FGENE.2023.1085024/PDF

Singh, J., Verma, P.K., 2023. Plant transcription factors and nodule development. *Plant Transcription Factors: Contribution in Development, Metabolism, and Environmental Stress* 175–196. https://doi.org/10.1016/B978-0-323-90613-5.00020-0

Singh, P., Kumar, D., Sarin, N.B., 2016. Multiple abiotic stress tolerance in Vigna mungo is altered by overexpression of ALDRXV4 gene via reactive carbonyl detoxification. *Plant Molecular Biology* 91, 257–273. https://doi.org/10.1007/S11103-016-0464-9

Smith, M.R., Rao, I.M., Merchant, A., 2018. Source-sink relationships in crop plants and their influence on yield development and nutritional quality. *Frontiers in Plant Science* 871, 420465. https://doi.org/10.3389/FPLS.2018.01889

Solleti, Siva Kumar, Bakshi, S., Purkayastha, J., Panda, S.K., Sahoo, L., 2008. Transgenic cowpea (Vigna unguiculata) seeds expressing a bean α-amylase inhibitor 1 confer resistance to storage pests, bruchid beetles. *Plant Cell Reports* 27, 1841–1850. https://doi.org/10.1007/S00299-008-0606-X

Solleti, S.K., Bakshi, S., Sahoo, L., 2008. Additional virulence genes in conjunction with efficient selection scheme, and compatible culture regime enhance recovery of stable transgenic plants in cowpea via Agrobacterium tumefaciens-mediated transformation. *Journal of Biotechnology* 135, 97–104. https://doi.org/10.1016/J.JBIOTEC.2008.02.008

Song, C.J., Ma, K.M., Fu, B.J., Qu, L.Y., Xu, X.L., Liu, Y., Zhong, J.F., 2010. Distribution patterns of shrubby N-fixers and non-N fixers in an arid valley in Southwest China: Implications for ecological restoration. *Ecological Research* 25, 553–564. https://doi.org/10.1007/S11284-009-0685-3

Song, J.H., Shin, G., Kim, H.J., Lee, S.B., Moon, J.Y., Jeong, J.C., Choi, H.K., Kim, I.A., Song, H.J., Kim, C.Y., Chung, Y.S., 2022. Mutation of GmIPK1 Gene Using CRISPR/Cas9 Reduced Phytic Acid Content in Soybean Seeds. *International Journal of Molecular Sciences* 23, 10583. https://doi.org/10.3390/IJMS231810583/S1

Srivastava, J., Datta, S., Mishra, S.P., 2017. Development of an efficient Agrobacterium mediated transformation system for chickpea (Cicer arietinum). *Biologia (Poland)* 72, 153–160. https://doi.org/10.1515/BIOLOG-2017-0015

Stracke, S., Kistner, C., Yoshida, S., Mulder, L., Sato, S., Kaneko, T., Tabata, S., Sandal, N., Stougaard, J., Szczyglowski, K., Parniske, M., 2002. A plant receptor-like kinase required for both bacterial and fungal symbiosis. *Nature* 417(6892), 959–962. https://doi.org/10.1038/nature00841

Sun, Z., Wang, Y., Mou, F., Tian, Y., Chen, L., Zhang, S., Jiang, Q., Li, X., 2016. Genome-wide small RNA analysis of Soybean reveals auxin-responsive microRNAs that are differentially expressed in response to salt stress in root apex. *Frontiers in Plant Science* 6, 1273–1273. https://doi.org/10.3389/FPLS.2015.01273

Surekha, C., Kumari, K.N., Aruna, L.V., Suneetha, G., Arundhati, A., Kavi Kishor, P.B., 2014. Expression of the Vigna aconitifolia P5CSF129A gene in transgenic pigeonpea enhances proline accumulation and salt tolerance. *Plant Cell, Tissue and Organ Culture* 116, 27–36. https://doi.org/10.1007/S11240-013-0378-Z

Tan, G.Z.H., Das Bhowmik, S.S., Hoang, T.M.L., Karbaschi, M.R., Long, H., Cheng, A., Bonneau, J.P., Beasley, J.T., Johnson, A.A.T., Williams, B., Mundree, S.G., 2018. Investigation of baseline iron levels in australian chickpea and evaluation of a transgenic biofortification approach. *Frontiers in Plant Science* 9, 363578. https://doi.org/10.3389/FPLS.2018.00788

Tanwir, K., Abbas, S., Shahid, M., Chaudhary, H.J., Javed, M.T., 2023. Mycorrhizal symbiosis and nutrients uptake in plants. In: V.P. Singh and M.H. Siddiqui (eds.). *Plant Ionomics: Sensing, Signaling and Regulation,* 73–95. https://doi.org/10.1002/9781119803041.CH5

Thu, T.T., Dewaele, E., Trung, L.Q., Claeys, M., Jacobs, M., Angenon, G., 2007. Increasing lysine levels in pigeonpea (Cajanus cajan (L.) Millsp) seeds through genetic engineering. *Plant Cell, Tissue and Organ Culture* 91, 135–143. https://doi.org/10.1007/S11240-007-9227-2

Tian, B., Wang, S., Todd, T.C., Johnson, C.D., Tang, G., Trick, H.N., 2017. Genome-wide identification of soybean microRNA responsive to soybean cyst nematodes infection by deep sequencing. *BMC Genomics* 18, 1–13. https://doi.org/10.1186/S12864-017-3963-4

Tirichine, L., Sandal, N., Madsen, L.H., Radutoiu, S., Albrektsen, A.S., Sato, S., Asamizu, E., Tabata, S., Stougaard, J., 2007. A gain-of-function mutation in a cytokinin receptor triggers spontaneous root nodule organogenesis. *Science* 315, 104–107. https://doi.org/10.1126/SCIENCE.1132397

Tripathi, L., Singh, A.K., Singh, S., Singh, R., Chaudhary, S., Sanyal, I., Amla, D.V., 2013. Optimization of regeneration and Agrobacterium-mediated transformation of immature cotyledons of chickpea (Cicer arietinum L.). *Plant Cell, Tissue and Organ Culture* 113, 513–527. https://doi.org/10.1007/S11240-013-0293-3

Urnov, F.D., Rebar, E.J., Holmes, M.C., Zhang, H.S., Gregory, P.D., 2010. Genome editing with engineered zinc finger nucleases. *Nature Reviews Genetics* 11(9), 636–646. https://doi.org/10.1038/nrg2842

Valentine, M.F., De Tar, J.R., Mookkan, M., Firman, J.D., Zhang, Z.J., 2017. Silencing of soybean raffinose synthase gene reduced raffinose family oligosaccharides and increased true metabolizable energy of poultry feed. *Frontiers in Plant Science* 8, 258449. https://doi.org/10.3389/FPLS.2017.00692

van Loon, M.P., Alimagham, S., Pronk, A., Fodor, N., Ion, V., Kryvoshein, O., Kryvobok, O., Marrou, H., Mihail, R., Mínguez, M.I., Pulina, A., Reckling, M., Rittler, L., Roggero, P.P., Stoddard, F.L., Topp, C.F.E., van der Wel, J., Watson, C., van Ittersum, M.K., 2023. Grain legume production in Europe for food, feed and meat-substitution. *Global Food Security* 39, 100723. https://doi.org/10.1016/J.GFS.2023.100723

Van Noorden, G.E., Ross, J.J., Reid, J.B., Rolfe, B.G., Mathesius, U., 2006. Defective long-distance auxin transport regulation in the medicago truncatula super numeric nodules Mutant. *Plant Physiology* 140, 1494–1506. https://doi.org/10.1104/PP.105.075879

Varshney, R.K., Thudi, M., Pandey, M.K., Tardieu, F., Ojiewo, C., Vadez, V., Whitbread, A.M., Siddique, K.H.M., Nguyen, H.T., Carberry, P.S., Bergvinson, D., 2018. Accelerating genetic gains in legumes for the development of prosperous smallholder agriculture: Integrating genomics, phenotyping, systems modelling and agronomy. *Journal of Experimental Botany* 69, 3293–3312. https://doi.org/10.1093/JXB/ERY088

Vijayan, S., Kirti, P.B., 2012. Mungbean plants expressing BjNPR1 exhibit enhanced resistance against the seedling rot pathogen, Rhizoctonia solani. *Transgenic Research* 21, 193–200. https://doi.org/10.1007/S11248-011-9521-Y

Voß, L., Heinemann, K.J., Herde, M., Medina-Escobar, N., Witte, C.P., 2022. Enzymes and cellular interplay required for flux of fixed nitrogen to ureides in bean nodules. *Nature Communications* 13(1), 1–13. https://doi.org/10.1038/s41467-022-33005-5

Wafula, B.W., Arunga, E.E., Rotich, F., 2023. Prevalence and host resistance to common bean rust disease in Western and Central Kenya. *International Journal of Agronomy* 2023, 6064130. https://doi.org/10.1155/2023/6064130

Wang, L., Deng, L., 2016. GmACP expression is decreased in GmNORK knockdown transgenic soybean roots. *The Crop Journal* 4, 509–516. https://doi.org/10.1016/J.CJ.2016.05.005

Wang, M., Zhang, C. xin, Pan, L. juan, Wang, T., Chen, N., Chen, M. na, Yang, Z., Guo, X. gang, Yu, S. lin, Chi, X. yuan, 2020. Small RNA profiling reveal regulation of microRNAs in field peanut pod rot pathogen infection. *Biologia* 75, 1779–1788. https://doi.org/10.2478/S11756-020-00485-Z

Wang, W., Zhou, B., He, J., Zhao, J., Liu, C., Chen, X., Xing, G., Chen, S., Xing, H., Gai, J., 2020. Comprehensive identification of drought tolerance QTL-allele and candidate gene systems in Chinese cultivated Soybean population. *International Journal of Molecular Sciences* 21, 1–20. https://doi.org/10.3390/IJMS21144830

Wang, Y., Zhang, C., Hao, Q., Sha, A., Zhou, R., Zhou, X., Yuan, L., 2013. Elucidation of miRNAs-mediated responses to low nitrogen stress by deep sequencing of two Soybean genotypes. *PLOS ONE* 8, e67423. https://doi.org/10.1371/JOURNAL.PONE.0067423

Watson, B.N.J., Steens, J.A., Staals, R.H.J., Westra, E.R., van Houte, S., 2021. Coevolution between bacterial CRISPR-Cas systems and their bacteriophages. *Cell Host & Microbe* 29, 715–725. https://doi.org/10.1016/J.CHOM.2021.03.018

Wen, S., Liu, H., Li, X., Chen, X., Hong, Y., Li, H., Lu, Q., Liang, X., 2018. TALEN-mediated targeted mutagenesis of fatty acid desaturase 2 (FAD2) in peanut (Arachis hypogaea L.) promotes the accumulation of oleic acid. *Plant Molecular Biology* 97, 177–185. https://doi.org/10.1007/S11103-018-0731-Z/FIGURES/5

Yadav, S.K., Katikala, S., Yellisetty, V., Kannepalle, A., Narayana, J.L., Maddi, V., Mandapaka, M., Shanker, A.K., Bandi, V., Bharadwaja, K.P., 2012. Optimization of Agrobacterium mediated genetic transformation of cotyledonary node explants of Vigna radiata. *SpringerPlus* 1, 1–8. https://doi.org/10.1186/2193-1801-1-59

Yan, Z., Hossain, M.S., Arikit, S., Valdés-López, O., Zhai, J., Wang, J., Libault, M., Ji, T., Qiu, L., Meyers, B.C., Stacey, G., 2015. Identification of microRNAs and their mRNA targets during soybean nodule development: Functional analysis of the role of miR393j-3p in soybean nodulation. *New Phytologist* 207, 748–759. https://doi.org/10.1111/NPH.13365

Yang, X., Niu, L., Zhang, W., Yang, J., Xing, G., He, H., Guo, D., Du, Q., Qian, X., Yao, Y., Li, Q., Dong, Y., 2018. RNAi-mediated SMV P3 cistron silencing confers significantly enhanced resistance to multiple Potyvirus strains and isolates in transgenic soybean. *Plant Cell Reports* 37, 103–114. https://doi.org/10.1007/S00299-017-2186-0

Ye, K., Bu, F., Zhong, L., Dong, Z., Ma, Z., Tang, Z., Zhang, Y., Yang, X., Xu, X., Wang, E., Lucas, W.J., Huang, S., Liu, H., Zheng, J., 2024. Mapping the molecular landscape of Lotus japonicus nodule organogenesis through spatiotemporal transcriptomics. *Nature Communications* 15(1), 1–13. https://doi.org/10.1038/s41467-024-50737-8

Zhang, P., Du, H., Wang, J., Pu, Y., Yang, C., Yan, R., Yang, H., Cheng, H., Yu, D., 2020. Multiplex CRISPR/Cas9-mediated metabolic engineering increases soya bean isoflavone content and resistance to soya bean mosaic virus. *Plant Biotechnology Journal* 18, 1384–1395. https://doi.org/10.1111/PBI.13302

Zhong, X., Wang, J., Shi, X., Bai, M., Yuan, C., Cai, C., Wang, N., Zhu, X., Kuang, H., Wang, X., Su, J., He, X., Liu, X., Yang, W., Yang, C., Kong, F., Wang, E., Guan, Y., 2024. Genetically optimizing soybean nodulation improves yield and protein content. *Nature Plants* 10(5), 736–742. https://doi.org/10.1038/s41477-024-01696-x

13 From Fields to Pharmacies

Legume Engineering Impact on Food Security and Hunger Relief in the Pharmaceutical Realm

Ekta Khare

13.1 INTRODUCTION

Food security, which encompasses food availability, access, and utilization, is a multifaceted issue that is directly linked to hunger. It acts as the new point of entry for addressing the world food crisis. The stresses of widespread hunger and nutrient deficits have been relieved in recent years by the application of a variety of agricultural practices (Siddique et al., 2012). In this regard, it is essential to cultivate crops that are both extremely prominent and nutritious. However, it takes a lot of time and effort to generate more valuable kinds of crops using standard plants (Tilman et al., 2002).

Finding socially and ecologically acceptable ways to boost crop yields will contribute to the solution of this issue. Recent research has looked into using nanotechnology to increase agricultural quality and yield.

In plant science and agricultural research, engineered nanomaterial (NM)-based technologies have been used to improve plant nutrient delivery efficiency, boost crop survivability in stressful situations, and produce higher yields with better quality. In order to modify or enhance the overall functioning of plant cells and organelles, nanobionics, Vithanage et al. (2023) introduced particularly designed nanoparticle (NP) breeding procedures. As a result, nanobionics has developed with a range of plant-based applications, including self-illuminating plants, effective drug-delivery vehicles for plant hormones and nutrients, and plant-based nanosensors. The idea behind nanobionics (Vithanage et al., 2023) holds great promise for achieving the 2030 United Nations Sustainable Development Goals, including "zero hunger," by spurring innovation in clean energy generation, crop productivity, and the creation of sustainable cities and communities that promote health and well-being. Numerous results pertaining to food and crop growth were obtained from a Web of Science database keyword search for "nanomaterials, agriculture."

Feeding everyone on our planet is one of the biggest issues facing the modern world. This issue is exacerbated by a lack of micro- and macronutrients, deficiencies in food production leading to imbalances in supply and demand, and conflicts that produce insecurity in various parts of the world. However, some of these causes of hunger are treatable. Due to decreased food production rates and the added burden of feeding a population that will surpass 9 billion, the number of people who experience hunger and malnutrition, which decreased slightly from approximately one billion in the years 1990–1992 to 850 million in the years 2010–2012, may experience a sharp increase. This could result in roughly 2–3 billion people experiencing nutritional insecurity by year 2050 (Pingali, 2012).

A sustainable and varied approach to agriculture is necessary to solve these issues, with the goal of enhancing both food production (Figure 13.1) and nutritional value. Natural antioxidants may shield us against oxidative stress, which is strongly linked to cancer and cardiovascular disease, according

DOI: 10.1201/9781003477235-16

FIGURE 13.1 Rooftop farming.

FIGURE 13.2 Species of legumes possessing potential nutraceutical properties: (a) black soybean (*Glycine max L.*); (b) pigeon pea (*Cajanus cajan* L. Millspaugh); (c) mungbean (*Vigna radiata*); (d) lentil (*Lens culinaris*); (e) chickpea (*Cicer arietinum*).

to the findings of clinical and epidemiological research. For humans, legume seeds (Figure 13.2) are a great source of natural antioxidants like tocopherols, vitamin C, and phenolic compounds.

In addition to being essential for nutrition because of their combination of amino acids, legume proteins are a significant source of bioactive peptides. Leguminous proteins and peptides (lunasin) have been reported in many researches to have biological activity, with a notable ability to suppress inflammatory indicators. The majority of bioactive peptides fall into one of two categories: albumin or globulin fractions. Lunasin which contains a 43-amino acid residue is a soybean-derived polypeptide (Khare, 2023) that has the unique ability to prevent the division of malignant cells but not of healthy cells. By optimizing the expression of lunasin in genetically modified soybeans, many

TABLE 13.1
Potential Beneficial Effects of Legume Consumption on Human Health

S. no.	Benefits
1.	Enhanced control of the immune system, metabolism, and hormone regulation
2.	Antiinflammatory and anticarcinogenic (prostate, colorectal, endometrial, and breast cancers) properties
3.	Decreased risk of metabolic syndrome, obesity-related disorders, and cardiovascular disease
4.	Lower levels of cholesterol
5.	Decreased chance of developing type II diabetes
6.	Decreased chance of depression and osteoporosis

researchers are carrying out research to create a crop that not only addresses nutritional deficiencies but also contributes to disease resistance in relation to supporting food security and hunger relief. In the context of food security, lunasin-enriched legumes represent a promising and innovative strategy for tackling malnutrition, particularly in regions where resources are scarce and food security is a pressing concern. These specially enhanced legumes not only improve the bioavailability of essential nutrients such as proteins, vitamins, and minerals but also introduce a range of additional health-promoting compounds. This dual benefit is particularly crucial in combating protein-energy malnutrition, a pervasive issue affecting millions in developing countries, where access to diverse and nutritious food sources is often limited.

The incorporation of lunasin into legumes has the potential to significantly elevate their nutritional profile, making them a more effective food source for vulnerable populations. By enhancing the digestibility and absorption of vital nutrients, lunasin-enriched legumes can help address deficiencies that contribute to malnutrition and its associated health complications. This is especially important in areas where traditional diets may lack sufficient protein and micronutrients. In addition to their nutritional advantages, lunasin also plays a critical role in improving the stress tolerance of plants. This characteristic is particularly valuable in the face of climate change, which poses significant challenges to agricultural productivity. By facilitating the cultivation of more resilient crops, lunasin-enriched legumes can thrive under a variety of climatic conditions, including drought, salinity, and extreme temperatures. This resilience not only supports consistent food production but also helps farmers adapt to the changing environment, ensuring a more stable food supply.

Furthermore, the integration of lunasin into legume breeding programs can lead to the development of crop varieties that are not only nutritionally superior but also environmentally sustainable. By promoting the growth of legumes that require fewer resources and can withstand adverse conditions, we can enhance agricultural sustainability and reduce the reliance on chemical fertilizers and pesticides.

Ultimately, the strategic incorporation of lunasin into legume breeding initiatives holds great promise for significantly improving food availability and quality. This approach not only addresses immediate nutritional needs but also contributes to long-term solutions for global hunger and malnutrition. By fostering the development of nutrient-rich, resilient crops, we can create a more secure and sustainable food system that benefits communities around the world, particularly in regions most affected by food insecurity. In this way, lunasin-enriched legumes could play a pivotal role in the global effort to eradicate malnutrition and promote health and well-being for all.

A set of standards and guidelines were used to select the literature sources for the review. (i) The Web of Science, Springer, Science Direct, PubMed, and Google Scholar databases were utilized in this review to locate English-language scientific publications published without regard to the year in peer-reviewed journals. (ii) The literature search was conducted using keyword combinations that are typical of the triangle that is at the core of food safety and security: "sustainability," "agriculture," and "biotechnology." Following this literature search, the more than 400 articles that were found were carefully inspected, categorized based on whether they contained an approach or an application, and grouped based on the queries that were found in relation to the keywords.

Legume seeds, commonly known as pulses or grain legumes, play a crucial role in global nutrition, ranking just behind cereals as a vital source of sustenance for both humans and animals. These seeds are not only versatile in their culinary applications but also rich in essential nutrients. When consumed alongside cereals, legumes provide a complete protein profile, making them an excellent dietary choice for individuals seeking to meet their protein needs, especially in vegetarian and vegan diets. In terms of nutritional value, legume seeds are remarkably dense in protein, containing two to three times the protein content found in traditional cereal grains. This high protein content makes them an invaluable food source, particularly in regions where meat consumption may be limited due to economic or cultural factors. Additionally, certain legumes, such as soybeans and peanuts, are noteworthy for their high oil content, which contributes to their energy density and nutritional profile.

Common culinary products derived from legumes further showcase their versatility and appeal. Tofu, made from soybeans, serves as a popular meat substitute in many vegetarian and vegan dishes, while peanut butter, derived from peanuts, is a beloved spread enjoyed worldwide. Soymilk, another product made from soybeans, offers a dairy-free alternative for those with lactose intolerance or those seeking plant-based options.

Overall, the nutritional and culinary significance of legume seeds cannot be overstated. They provide a sustainable source of protein, essential nutrients, and culinary diversity, making them an integral part of diets across the globe. As the world continues to seek sustainable food sources, the importance of legumes is likely to grow, highlighting their role in promoting health, nutrition, and food security.

a) **Soybeans, black (*Glycine max* L.)**
   For many years, this black-seed coated soybean type has been widely utilized in Oriental medicine and as a tonic food. According to the principles of traditional Chinese medicine, the active peptide chemicals in black soybeans can help treat the blood circulation, diabetes, hypertension, anti-aging, cosmetics, and other conditions.
b) **Pigeon pea (*Cajanus cajan*) L. Millspaugh**
   This is a legume food that is high in protein, carbohydrates, calcium, manganese, fat, crude fiber, trace minerals, and starch. The contents of pigeon pea seeds include a range of nutritional components, 14% seed coat, 85% cotyledons, and approximately 1% embryo. The majority of the seed proteins are found in the embryo, while the majority of the carbohydrates are found in the cotyledons. It possesses both nutritional and therapeutic qualities. When added to coffee, scorched seeds can aid with vertigo and headaches, while fresh seeds can help with male incontinence. Also, kidney diseases are treated using immature seeds in hyperglycemic people.
c) **Mungbean (*Vigna radiata*)**
   Vitamins, carbohydrates, proteins, and amino acids are all abundant in mungbeans. They have several bioactive components that reduce the chance of developing certain illnesses.
d) **Lentil (*Lens culinaris*)**
   Lentil is a highly nutritious food that is high in dietary fiber, folic acid, vitamins, minerals, proteins, and carbohydrates, primarily resistant starches. Additionally, it includes a variety of bioactive substances, including phenolic compounds, phytates, lectins, enzyme inhibitors, and oligosaccharides. The main ingredient of lentil seed is carbohydrate. Lentil seeds are useful for treating and preventing a number of illnesses. Patients with diabetes and cardiovascular disease should consider lentil seeds because of their low glycemic response and high dietary fiber content.

e) **Chickpea (*Cicer arietinum*)**
Chickpeas are in high demand because of their highly nutritious content. Chickpeas play a significant role in the diets of those unable to purchase animal proteins in the semi-arid tropics. In addition to being low in cholesterol, chickpeas are an excellent source of protein, carbohydrates, dietary fiber (DF), and vitamins and minerals.

13.2 THE IMPACT OF BIOACTIVE SUBSTANCES PRESENT IN LEGUME SEEDS ON HUMAN HEALTH

Bioactive compounds found in legume seeds provide a variety of health advantages, encompassing antioxidant, antiinflammatory, antidiabetic, and anticancer effects, in addition to supporting cardiovascular and cognitive well-being.

Essential components such as isoflavones, bioactive peptides, saponins, phenolics, and fiber collaborate to enhance metabolic health, bone density, and gut functionality.

Nevertheless, legumes also possess antinutritional elements, including lectins, which may hinder nutrient absorption if not adequately processed.

Numerous bioactive substances could have a variety of positive effects on human health, which could explain their nutraceutical qualities.(Russo et al., 2025; Barman et al., 2019).

13.2.1 Heart-Related Conditions

Legumes have a major role in the process of reducing cholesterol. Consuming legumes is linked to a 11–22% decreased risk of cardiovascular disease and coronary heart disease. Consuming legume seeds has been shown to reduce low-density lipoprotein (LDL) cholesterol by partially blocking the enterohepatic circulation of bile acids and raising the hepatic output of cholesterol, which in turn raises cholesterol saturation.

Numerous bioactive substances, including fibers, oligosaccharides, ACE, vitamins, and minerals, are present in legume seeds and contribute to the prevention of cardiovascular disease. Research has shown that including 30% blackgram neutral detergent fiber in the diet will lower cholesterol levels more effectively than cellulose when it binds to bile acids.

13.2.2 Diabetes

One important component of diabetic treatment is legume seeds. Their high glycemic index and high fiber and oligosaccharide contents contribute to blood glucose regulation. Reduced insulin, cholesterol levels, and plasma epididymal adipocyte volume are all associated with low glycemic index. Thus, using legume seeds as an antidiabetic medication is possible (Vilakazi et al., 2025).

13.2.3 Hepatotoxicity

Legume seeds, such as lentils, chickpeas, and various types of beans, are increasingly recognized as a beneficial dietary option for promoting liver health. These seeds are not only a great source of plant-based protein, fiber, and essential nutrients, but also contain high levels of γ-aminobutyric acid (GABA). GABA is a naturally occurring neurotransmitter that plays a crucial role in various physiological processes, including the regulation of stress and anxiety.

One of the most notable benefits of GABA is its significant hepatoprotective qualities. Research has shown that GABA can help protect liver cells from damage caused by toxins, oxidative stress, and inflammation. This protective effect is particularly important in the context of liver diseases, such as fatty liver disease, hepatitis, and cirrhosis, where the liver is under constant threat from various harmful agents.

In addition to its hepatoprotective properties, GABA may also contribute to improved liver function by enhancing metabolic processes and promoting the regeneration of liver cells. By incorporating legume seeds into the diet, individuals can not only enjoy a nutritious food source but also support their liver health through the beneficial effects of GABA.

Furthermore, the high fiber content in legume seeds aids in digestion and helps maintain healthy cholesterol levels, which can further reduce the risk of liver-related issues. The combination of GABA and other nutrients found in legumes makes them an excellent choice for those looking to improve their overall health and specifically support liver function.

In summary, legume seeds are a valuable addition to a balanced diet, offering a range of health benefits, particularly for the liver. Their rich content of GABA, along with other essential nutrients, underscores their role as a functional food that can contribute to better liver health and overall well-being.

13.2.4 Osteoporosis

Legume seeds play a significant role in reducing the risk of osteoporosis, primarily because they are rich in calcium and protein. These two nutrients are essential for the formation and maintenance of strong, healthy bones. Furthermore, legume seeds are a source of isoflavones, including compounds like daidzein and genistein. These isoflavones are crucial as they contribute to the prevention of bone tissue degradation, thereby supporting overall bone health.

Legume seeds lower the incidence of osteoporosis since they are an excellent source of calcium and protein that contribute to the development of strong bones. Daidzein and genistein, two isoflavones, stop the deterioration of bones.

13.2.5 Obesity and Weight Control

Because they have a strong satiety impact, dry beans are said to help maintain a regular body weight by reducing the amount of food consumed daily. This impact has been attributed to a variety of seed components. Studies involving healthy participants have indicated that consuming more fiber from peas or soybeans increased the weight of feces.

13.2.6 Antiinflammatory Agent

Because of their beneficial effects on health when in a denatured form, BBI proteases from legume crops were approved by the FDA as experimental novel drugs. Since then, a great deal of research has been done on the possible health advantages of protease inhibitors, including the possibility of using them as an antiinflammatory.

13.2.7 High Blood Pressure

By inactivating the vasodilator bradykinin and converting the physiologically inactive angiotensin I to the powerful vasoconstrictor angiotensin II, ACE raises blood pressure. Isolated from a variety of legumes, angiotensin I-converting enzyme (ACE) inhibitor peptides have demonstrated efficacy in the management and prophylaxis of hypertension.

13.2.8 Sources of Natural Antioxidants in Legumes

Numerous studies have documented the antioxidant properties of various legumes, including common beans, lentils, chickpeas, and yellow and green peas. Procyanidins were discovered to be present in red peanut skins at a 7–9% concentration, with 50% of these being low-molecular-weight phenolic oligomers (Xu et al., 2007).

TABLE 13.2
Content of Natural Antioxidants in Leguminous Seeds and Their Extracts

S. no.	Legume	Capacity
1.	Beans extract	28.59–51.5
2.	Germinated pea	6.7–14.5
3.	Lentil	7.2–10.9
4.	Yellow pea	1.5
5.	Green pea	1.50

The bulk of the polyphenolics found in walnut seeds are the monomers gallic acid, methyl gallate, and ellagic acid. These polymers are known as hydrolyzable tannins when they are present as polymers and coupled to sugars. Tocopherols are the main lipid-soluble antioxidants present in peanuts and other oilseeds (Turkmen, 2005).

These substances have bioactive properties and have a significant role in determining the color, flavor, and taste of food (Amarowicz, 1995). They have the ability to interact with proteins and scavenge free radicals. Grain legumes are good options for developing new functional meals because of their bioactive phenolic chemicals, which are present as reactive metabolites and have a corresponding antioxidant effect (Xu et al., 2007).

13.3 PHENOLIC COMPOUNDS OF LEGUME SEEDS

Legumes include a bioactive phytochemical that has several health benefits. These substances include antioxidants, oligosaccharides, phenolics, peptides, low-molecular-weight proteins, and carotenoids (Amarowicz, 1995). Legumes have a wide range of phenolic contents, which have a significant impact on the color and pattern of the seed coat as well as the kind of bean cultivar. Both the cotyledon and the seed coat contain phenolic acids, but the amount in the latter is noticeably greater. Darker colors are generally linked to higher anthocyanin concentrations. Fifteen different types of dry beans had an average phenolic acid level of 31.2 mg/100 g; the total phenolic acid content varied from 19.1 to 48.3 mg/100 g. The phenolic acids found in legumes that are most frequently found are caffeic, p-coumaric, sinapic, ferulic, gallic, and p-hydroxybenzoic (Turkmen et al., 2005).

Legumes include a bioactive phytochemical that has several health benefits (Amarowicz, 1995). These substances include antioxidants, oligosaccharides, phenolics, peptides, low-molecular-weight proteins, and carotenoids. Legumes' phenolic content varies greatly and has a significant impact on the seedcoat's color and pattern. Ferulic acid was shown to be the most prevalent phenolic acid in black beans (Xu, 2007). Xu found that black beans had a total phenolic concentration of 63.8 mg/g. Among the primary classes of polyphenols, flavonoids which include anthocyanins and anthoxanthins, are thought to have the greatest potential for antioxidant activity.

13.3.1 Sources of Vitamins in Legume Seeds

When comparing the total number of nutrients found in vegetables including phenols or flavonoids, vitamins are a minor component. Often referred to as vitamin E, tocopherol has a significant antioxidant impact on human metabolism as an exogenous substance (Siitonen et al., 2023). Researchers in the field have found that vitamin E comes in multiple isoforms, including α-, β-, γ-, and δ-tocopherol and α-, β-, γ-, and δ-tocotrienol. In addition to the antioxidant properties of α- and β-tocopherol, γ-tocopherol plays a significant role in human health due to its antiinflammatory and anticancer

TABLE 13.3
Content of Total Phenolics in Leguminous Seeds and Their Extracts

S. no.	Legume	Total phenolics
1.	Bean	355
2.	Pea	183
3.	Lentil	3.7–4.1
4.	Yellow pea	1.13–1.6
5.	Green pea	1.07–1.53

properties; δ-tocopherol helps prevent fat accumulation and several cancer types, including breast, colon, and lung cancer (Zeng et al., 2020).

While overall folate concentrations differed significantly, the amounts of thiamin, riboflavin, and niacin in the flours from various legume species were comparatively similar. The levels of thiamin, riboflavin, and niacin in faba bean, lupin, and pea flour varied from 4.9 μg/g to 8.3 μg/g, 2.5 μg/g, and 18.7 μg/g to 22.9 μg/g. Many researchers measured the folate content of faba bean flour at 1221 ng/g and lupin flour at 1428 ng/g. On the other hand, pea flour had a noticeably reduced folate concentration (142 ng/g). Therefore, compared to pea flour, faba bean and lupin flour contained nine to ten times as much folate.

The ranges of thiamin contents of flours of faba bean, lupin, and pea have been measured. It was found that the variety of legume, the year of harvest, and the location of harvest all affected the amount of thiamin in the seeds. According to previous research, the riboflavin concentration of faba bean seeds ranges from 2.1 μg/g to 3.8 μg/g, which is consistent with these findings. Similarly, it has been noted that the riboflavin concentration in peas varies from 1 μg/g to 2.3 μg/g, and in blue lupin seeds, it varies from 1.9 μg/g to 3.1 μg/g. Furthermore, white lupin seeds had a riboflavin content of 2.3 μg/g, according to Erbaş et al. (2005).

13.4 LEGUME CROPS: A GOOD CANDIDATE FOR BIOFORTIFICATION

Legume crops are ideal candidates for biofortification, a method that improves the nutritional value of food crops. Legumes are well-known for their high protein and fiber contents, but they can also be improved to contain more critical elements including iron, zinc, and vitamin A (Siitonen et al., 2023).

Biofortification of legume crops can help combat malnutrition and micronutrient deficiencies, particularly in areas where legumes are staple foods. For example, iron-biofortified beans can aid in the treatment of iron-deficiency anemia, a major health problem in many regions of the world. Legumes are also ideal for biofortification since they can fix nitrogen in the soil, reducing the demand for synthetic fertilizers and encouraging sustainable agricultural practices.

The last few decades have seen major advancements in crop biofortification. Quality protein maize was the first biofortification success, resulting in a twofold increase in the levels of the vital amino acids tryptophan and lysine. Subsequently, staple crops like maize, wheat, rice, beans, and cassava were biofortified with minerals like zinc, iron, and provitamin A (Kumari et al., 2021). Millions of households have access to these crops, particularly in the world's poorer nations. It usually takes eight to ten years for these biofortified crops to be developed and released through conventional breeding (Li et al., 2022). Breeding attempts have included genomic selection, genome-wide association studies, and molecular markers to expedite the process.

By focusing on improving the nutritional content of legume crops through biofortification, we can improve the health and well-being of populations who rely on them as dietary staples. It is

TABLE 13.4
Legume Crops: Good Candidates for Biofortification

S. no.	Legume	Contribution to biofortification
1.	Soybean	Increase N, P, K, Ca, and Mg uptake
2.	Pea	Increase Fe, Zn, Cu, Mn, and Ca uptake
3.	Lentil	Increase Fe, N, and P uptake
4.	Mungbean	Increase N, P, and K, Fe uptake

an attractive area of study and development with significant potential for solving global nutrition concerns (Qaim et al., 2007).

Combating hunger and providing food for the world's growing population is one of the most important issues facing humanity today. Many factors contribute to this issue, including a deficiency in vital nutrients, a decline in food production that causes an imbalance between supply and demand, and conflicts that destabilize different parts of the world.

Since it appears that many nations will regulate genome-editing products less severely than transgenic crops, this could be addressed using CRISPR–Cas-mediated genome editing (Khare & Chauhan, 2020; Sun et al., 2017). Sustainable production of biofortified crops requires the implementation of national or regional policies that are effective (Ibrahim, 2022). Local seed systems should guarantee the availability of reasonably priced, high-quality biofortified seed and other inputs, as this will boost the yield and uptake of biofortified crops (Li et al., 2022). There's room to grow crops and increase the variety of nutrients available for biofortification. Future multi-nutrient crop development should employ a combination of conventional breeding methods and genetic engineering.

13.5 FACTORS AFFECTING BIOFORTIFIED CROPS TO IMPROVE MICRONUTRIENT STATUS

Iron bioavailability is further limited by condensed tannins, which are plentiful in particular bean kinds; iron is noticeably more bioavailable from white and yellow beans than from purple or black ones. Provitamin A bioavailability varies according to the crop and food preparation process. Fruit contains higher levels of provitamin A/β-carotene than roots, tubers, and leafy vegetables. It is fat-soluble, like other carotenoids, and its absorption is promoted by fat consumption while being blocked by fiber consumption. Even in short-term experiments, these variables have significant effects on the potential of biofortified crops to increase micronutrient status (Van Der Straeten et al., 2020).

Breeding zinc-rich staple crops has taken longer, because the amount of zinc in grains is significantly influenced by soil conditions, which can restrict the amount of zinc that reaches the grain. This limitation is increased by polishing, which can remove up to 60% of the zinc found in rice bran. Given these complicating circumstances, short-term agronomic biofortification using zinc-containing fertilizers has been proposed to improve the effectiveness of zinc-rich cereals. Once eaten, the bioavailability of zinc from a mixed diet is 26–34%, but only 18–26% is absorbed from an unprocessed cereal-based diet with significant quantities of phytate (Glahn et al., 2020).

13.6 CORRELATION BETWEEN LEGUME ENGINEERING AND FOOD APPLICATIONS

Legume engineering plays a crucial role in addressing food security and hunger. By utilizing advanced genetic techniques, scientists can enhance the nutritional content, yield, and resilience of legume crops. This engineering approach enables the development of legume varieties that are more resistant to pests, diseases, and environmental stresses, ensuring a stable and abundant food supply. Additionally, legumes have the unique ability to fix atmospheric nitrogen, enriching the soil and reducing the need for synthetic fertilizers. This sustainable farming practice not only improves soil health but also reduces production costs for farmers. Furthermore, legumes are a rich source of protein, essential minerals, and dietary fiber, making them an important component of a balanced diet. By engineering legumes to be more nutritious and productive, we can contribute to alleviating hunger and improving food security worldwide (Lockyer et al., 2018).

Increased Yield: By genetically altering legumes, farmers may produce more food on the same area of land. This can assist in meeting the rising food demand as the world's population grows.

Enhanced Nutritional Content: Making beans more nutritious by increasing their amounts of vital vitamins, minerals, and proteins can help prevent malnutrition and nutrient shortages. This is especially relevant in areas where beans are a primary source of dietary protein. (Lockyer et al., 2018).

Pest and Disease Resistance: Creating legume varieties that are resistant to pests and diseases can help safeguard crops and reduce yield loss. This can increase food security by providing a more consistent food source.

Climate Resilience: Engineering legumes to be more resistant to climate change, such as drought or excessive heat, can help crops survive in changing climatic conditions. This can help to protect food production and security in the face of climate variability.

Legume engineering can help to promote sustainable agriculture by minimizing the demand for fertilizers and pesticides. This can reduce farmers' production expenses while also promoting ecologically friendly farming methods. Engineering legumes for enhanced qualities has the potential to significantly increase food security by boosting productivity, improving nutritional quality, and supporting sustainable agricultural methods. It is an important instrument in the quest to provide a consistent and sufficient food supply for a growing global population.

13.7 MILLET-BASED FARMING SYSTEM FOR FOOD SECURITY

When specific millets are combined with legumes in a 3:1 proportion, they significantly enhanced the nutritional value of food by providing a balanced amino acid content with good protein digestibility, and high levels of iron (7.58 mg) and zinc (4.96 mg) with 100 g of pearl millet and calcium (400.57 mg) with 100 g of finger millet.

There are other names for millets, including "Mota Annaj," "nutri-cereals," "coarse cereals," and "cereals of the poor." "Shree Ann—the mother of all grains" is the most recent addition to this list. It was coined by India's Finance Minister, Mrs. Nirmala Sithraman, on February 1, 2023, during her budget speech. She also announced that the Indian Institute of Millets Research (IIMR) in Hyderabad would be transformed into the Centre of Excellence. The nutrient-dense, drought-tolerant Indian millets are primarily found in the country's arid and semiarid regions. These tiny seeded grasses, which are a part of the "Poaceae" botanical family, are a vital source of food and fodder for millions of farmers who lack resources. They also play a significant part in the economic security of India (Bourassa et al., 2023).

In India, millets are a class of basic crops that are vital to food security, particularly for impoverished and rural communities. Because of their great nutritional value, millets can help people live in locations where other crops are difficult to cultivate because of unfavorable weather conditions. They are a dependable food source even in harsh environmental conditions and perfect for farmers in such areas because they are tough, drought-resistant, and can be cultivated in

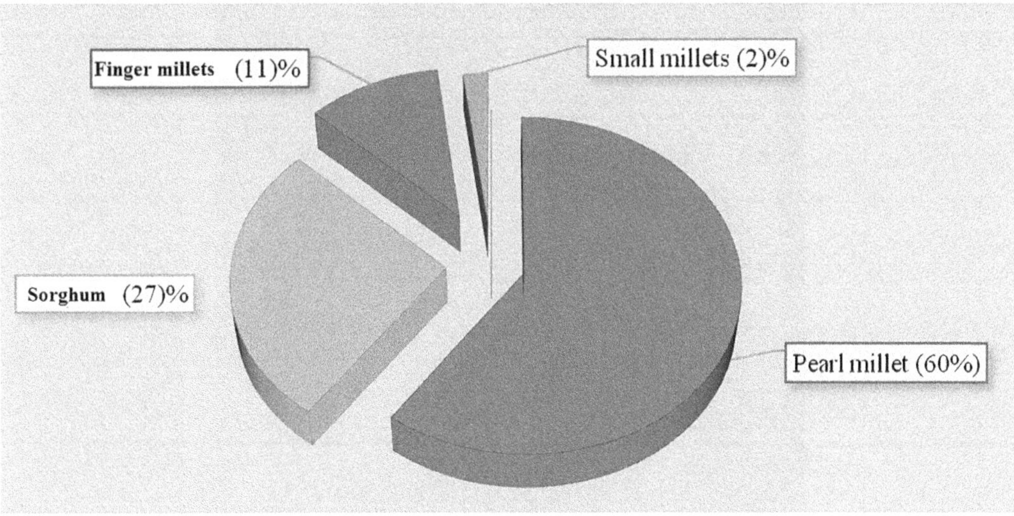

FIGURE 13.3 Millet production in India.

poor soils and areas with little rainfall. Long-term storage of millets is another benefit, since they contributes to food security during times of scarcity.

Millets appear in a range of small-seeded varieties, including finger millet (*Eleusine coracana*), proso millet (*Penicum miliaceum*), little millet (*Panicum sumatrense*), foxtail millet (*Setaria italica*), and pearl millet. Furthermore, coarse cereals such as sorghum (*Sorghum bicolor*), maize (*Zea mays*), barley (*Hordeum vulgare*), and oats (*Avena sativa)* play a significant role in addressing the global nutritional content concerns. Millets, a type of small-seeded cereal grain, are becoming increasingly popular as a potential answer to food and nutritional security. Furthermore, millets have a significant nutritious value. They are high in protein, fiber, vital minerals, and vitamins. Integrating millet-based farming systems provides a unique chance to tackle hidden hunger while also promoting sustainable agricultural practices. A 1% increase in millet output could reduce poverty by 0.65%. Rain-fed areas are 30% less productive than irrigated ones, hence improving output is especially important in these areas. Millets appear to be the key to combating poverty, malnutrition, and climate change (Jain et al., 2024).

As a result, the millet-based integrated farming system method, which includes poultry, dairy, goat farming, piggery, and apiculture in each home, would assist farmers in augmenting their income and also empower women.

Productivity of transgenic crops expressing functional foreign genes must be expanded to millets in order to produce transgenic millet varieties expressing foreign genes of agronomic value, which will be highly useful in enhancing millet productivity by conferring resistance to both biotic and abiotic stresses.

In the future, it will be possible to create a super grain by merging many agronomically relevant traits into the genome of a single millet genotype. Thus, combining current advancements in molecular breeding and genetic engineering with modern omics technologies will surely contribute to the advancement of millet research (Bouis et al., 2011).

Supporting millets is essential for enhancing health outcomes, supporting sustainable agriculture, and enhancing food security. To guarantee that millets are widely accessible and that their use is encouraged, the government and pertinent parties must collaborate. As a result, millets will become more significant in the Indian food chain and improve both human and environmental health and wellbeing. The Indian economy can gain from the promotion of millets in a number of ways,

FIGURE 13.4 Promotion of Magical Millets: pivotal to food security.

including higher agricultural output, better food security, more trade, decreased reliance on imports, and a boost to the rural economy.

Understanding biological matter's structure, organization, and functions is largely based on chemistry. The structural chemistry of living things, their metabolism or chemical reactions, and the molecular genetics of inheritance make up biochemistry in particular. The building blocks of food are made up of chemical and biological structures such as amino acids, sugars, lipids, nucleotides, vitamins, minerals, and hormones. Chemistry and the principles of thermodynamics enable plants to obtain energy from sunlight, while animals and humans obtain energy from food (Jain et al., 2024).

Understanding the chemical components is essential to meeting our current dietary and nutritional needs. Many of the chemical elements, such as nitrogen from ammonia, had been defined by the late 18th century.

The UN Millennium Declaration was adopted by the largest-ever assembly of world leaders at the Millennium Summit in September 2000, pledging their nations to a daring international alliance aimed at eradicating extreme poverty and addressing a number of deadline-bound development and health goals. These objectives, along with the Millennium Development Goals (MDGs) and accompanying indicators, could be used to track national progress and provide benchmarks.

One Millennium Development Goal aimed to halve hunger rates from 1990 to 2015. By 2010, many countries were still far from this target, highlighting global food security as a critical challenge.

The worldwide food price and economic crises, which combined to add 105 million people to the ranks of the hungry, have lately undermined the declines in hunger that were observed in developing countries during the 1990s.

It is obvious that there is still a long way to go in tackling undernutrition and hunger. A total of 200 million children under five in the developing world are malnourished. The bulk of these children reside in just 36 nations. More than 500,000 children die each year from zinc and vitamin A deficiencies alone; both deficiencies are treatable with straightforward, affordable, and efficient treatments.

The percentage of the population that consumes less than a certain amount of dietary energy and the incidence of underweight children under five are the two particular metrics used to monitor the achievement of the MDG1 hunger objective. It is difficult to determine how close the MDG1 hunger objective and "food security" are to being met, and these two metrics have problems.

13.8 CONCLUSION

Food legumes are produced using a variety of environmentally and economically viable agricultural production strategies, making them nutrient- and energy-dense food sources. Food legumes are rich in nutrients, with high quantities of protein and fiber, low levels of fat and sodium, and comparatively lower levels of carbohydrates than cereal crops. Legumes' proven nutritional value and health advantages match very nicely with the shifting consumer demand for wholesome foods. The most popular kinds of food components made from legumes have historically been protein isolates and concentrates based on soy and peas.

There has been a steady increase in the demand for protein from various legumes, including mungbeans, fava beans, lentils, and chickpeas. Furthermore, because they have unique functional qualities needed in a variety of culinary applications, legume starch and fiber fractions have also demonstrated increased commercial value.

It is commonly known that legumes are an underutilized food supply and that adding more legumes to the diet can improve its nutritional value significantly for most diets. Legumes can be used for purposes other than those associated with traditional diets and uses when they are added to various food products. Future research must concentrate on boosting the quality, increasing consumer acceptance, optimizing process/production challenges, addressing regulatory issues, and guaranteeing cost-effectiveness of legume-based meat analogs in order to successfully commercialize and sustain demand.

Millets grow better in areas with limited water resources since their cultivation uses less water than other crops. In order to battle hunger in the face of a fast changing global environment and to build agri-food systems that are resilient to climate change, millets are the world's solution for a sustainable food source.

CONFLICT OF INTEREST

The author declares no conflict of interest.

ACKNOWLEDGMENT

I would like to thank the Director of GCRG College, Lucknow, India, for providing their valuable support and encouragement throughout the work.

FUNDING

This research received no external funding.

BIBLIOGRAPHY

Amarowicz, R., Piskuła, M., Honke, J., Rudnicka, B., Troszyn ska, A. & Kozłowska, H. Extraction of phenolic compounds from lentil seeds (Lens culinaris) with various solvents. *Pol J Food Nutr Sci.* 45, 53–62 (1995).

Andrews, M. & Hodge, S. in *Climate Change and Management of Cool Season Grain Legume Crops* (eds Yadav, S. S., McNeil, D. L., Redden, R. & Patil, S. A.) Ch. 1 (2010), 1–9.

Barman, A., Marak, C. M., Barman, R. M. & Sangma, C. S. Nutraceutical Properties of Legume Seeds and Their Impact on Human Health [Internet]. In *Legume Seed Nutraceutical Research* (IntechOpen, 2019). http://dx.doi.org/10.5772/intechopen.78799

Barton, L., Thamo, T., Engelbrecht, D. & Biswas, W. K. Does growing grain legumes or applying lime cost effectively lower greenhouse gas emissions from wheat production in a semi-arid climate? *J. Clean Prod.* 83, 194–203 (2014).

Beebe, S. et al. in *Crop Adaptation to Climate Change* (eds Yadav, S. S., Redden, R. J., Hatfield, J. L., Lotze-Campen, H. & Hall, A. E.) Ch. 16 (Wiley-Blackwell, 2011).

Beyer, P. Golden Rice and 'Golden' crops for human nutrition. *New Biotechnol.* 27, 478–481 (2010).

Bhardwaj, A. K. et al. Agronomic biofortification of food crops: an emerging opportunity for global food and nutritional security. *Front. Plant Sci.* 13, 1055278 (2022).

Bouis, H. E. & Saltzman, A. Improving nutrition through biofortification: a review of evidence from HarvestPlus, 2003 through 2016. *Glob. Food Sec.* 12, 49–58 (2017).

Bouis, H. E., Hotz, C., McClafferty, B., Meenakshi, J. V. & Pfeiffer, W. H. Biofortification: a new tool to reduce micronutrient malnutrition. *Food Nutr. Bull.* 32, S31–S40 (2011).

Bourassa, M. W., Atkin, R., Gorstein, J. & Osendarp, S. Aligning the epidemiology of malnutrition with food fortification: grasp versus reach. *Nutrients* 15, 20–21 (2023).

Campion, B. et al. Isolation and characterisation of an lpa (low phytic acid) mutant in common bean (Phaseolus vulgaris L.). *Theor. Appl. Genet.* 118, 1211–1221 (2009).

Campion, B. et al. Genetic reduction of antinutrients in common bean (Phaseolus vulgaris L.) seed, increases nutrients and in vitro iron bioavailability without depressing main agronomic traits. *Field Crops Res.* 141, 27–37 (2013).

Erbaş M., Certel M., and Uslu M. K., Some chemical properties of white lupin seeds (Lupinus albus L.), *Food Chemistry*. 89(3), 341–345 (2005). https://doi.org/10.1016/j.foodchem.2004.02.040, 2-s2.0-4444314674.

Fortenberry, M., Rucker, H. & Gaines, K. Pediatric scurvy: how an old disease is becoming a new problem. *J. Pediatric Pharmacol. Ther.* 25, 735–741 (2020).

Glahn, R. P., Wiesinger, J. A. & Lung'aho, M. G. Iron concentrations in biofortified beans and nonbiofortified marketplace varieties in East Africa are similar. *J. Nutr.* 150, 3013–3023 (2020).

Gupta, R. K., Gangoliya, S. S. & Singh, N. K. Reduction of phytic acid and enhancement of bioavailable micronutrients in food grains. *J. Food Sci. Technol.* 52, 676–684 (2015).

Harjes, C. E. et al. Natural genetic variation in lycopene epsilon cyclase tapped for maize biofortification. *Science* 319, 330–333 (2008).

Huang, L.-F. et al. Plant–soil feedbacks and soil sickness: from mechanisms to application in agriculture. *J. Chem. Ecol.* 39, 232–242 (2013).

Hurrell, R. & Egli, I. Iron bioavailability and dietary reference values. *Am. J. Clin. Nutr.* 91, 1461S–1467S (2010).

Ibrahim, S. et al. CRISPR/Cas9 mediated disruption of inositol pentakisphosphate 2-kinase 1 (TaIPK1) reduces phytic acid and improves iron and zinc accumulation in wheat grains. *J. Adv. Res.* 37, 33–41 (2022).

Jain, V., Agarwal, N. & Bhatia, V. Promoting millets: charting a journey from food security to health. *Indian J. Commun. Med.* 49(1), 5–10 (2024). DOI: 10.4103/ijcm.ijcm_107_23

Japelt, R. B. & Jakobsen, J. Vitamin D in plants: a review of occurrence, analysis, and biosynthesis. *Front. Plant Sci.* 4, 136 (2013).

Jiang, L., Strobbe, S., Van Der Straeten, D. & Zhang, C. Regulation of plant vitamin metabolism: backbone of biofortification for the alleviation of hidden hunger. *Mol. Plant* 14, 40–60 (2021).

Johansen, C. et al. Integrated crop management of chickpea in environments of Bangladesh prone to botrytis grey mould. *Field Crops Res.* 108, 238–249 (2008).

Kanis J.A., Cooper C., Rizzoli R., Abrahamsen B., Al-Daghri N.M., Brandi M.L., Cannata-Andia J., Cortet B., Dimai H.P., Ferrari S., et al. Identification and management of patients at increased risk of osteoporotic fracture: Outcomes of an ESCEO expert consensus meeting. *Osteoporos. Int.* 28, 2023–2034 (2017). doi: 10.1007/s00198-017-4009-0.

Khare, E. Effect of highly efficient nano-immobilizing agents on the level of DTPA extractable – Cd & Pb. *Agric. Food e-News Lett.* 5(9), 462–465 (2023).

Khare, E. & Chauhan, P. S. Transformation of Agricultural Breeding Techniques Using Biotechnology as a Tool. In *Sustainable Agriculture Reviews 45. Sustainable Agriculture Reviews* (eds Guleria, P., Kumar, V. & Lichtfouse, E.) (vol. 45. Springer, 2020). https://doi.org/10.1007/978-3-030-53017-4_9

Kumari, M. et al. Vitamin B12 biofortification of soymilk through optimized fermentation with extracellular B12 producing Lactobacillus isolates of human fecal origin. *Curr. Res. Food Sci.* 4, 646–654 (2021).

Lassaletta, L., Billen, G., Grizzetti, B., Anglade, J. & Garnier, J. 50 year trends in nitrogen use efficiency of world cropping systems: the relationship between yield and nitrogen input to cropland. *Environ. Res. Lett.* 9, 105011 (2014).

Levis S. & Lagari V.S. The role of diet in osteoporosis prevention and management. Curr. *Osteoporos. Rep.* 10, 296–302 (2012). doi: 10.1007/s11914-012-0119-y.

Li, J. et al. Biofortified tomatoes provide a new route to vitamin D sufficiency. *Nat. Plants* 8, 611–616 (2022).

Lockyer, S., White, A. & Buttriss, J. Biofortified crops for tackling micronutrient deficiencies—what impact are these having in developing countries and could they be of relevance within Europe? *Nutr. Bull.* 43, 319–357 (2018).

Lockyer, S., White, A., Walton, J. & Buttriss, J. Proceedings of the 'Working together to consider the role of biofortification in the global food chain' workshop. *Nutr. Bull.* 43, 416–427 (2018).

Macknight, R. C. et al. Increasing ascorbate levels in crops to enhance human nutrition and plant abiotic stress tolerance. *Curr. Opin. Biotechnol.* 44, 153–160 (2017).

Malik, A. I. et al. Relay sowing of lentil (Lens culinaris subsp. culinaris) to intensify rice-based cropping. *J. Agr. Sci.* 154, 850–857 (2015).

Martin, C. & Li, J. Medicine is not health care, food is health care: plant metabolic engineering, diet and human health. *New Phytol.* 216, 699–719 (2017).

Molla, K. A., Sretenovic, S., Bansal, K. C. & Qi, Y. Precise plant genome editing using base editors and prime editors. *Nat. Plants* 7, 1166–1187 (2021).

Mourtzinis, S. et al. Climate-induced reduction in US-wide soybean yields underpinned by region- and in-season-specific responses. *Nat. Plants* 1, 14026 (2015).

Muñoz-Garach A, García-Fontana B, Muñoz-Torres M. Nutrients & Dietary Patterns Related to Osteoporosis. *Nutrients.* Jul 3, 12(7):1986 (2020). doi: 10.3390/nu12071986. PMID: 32635394; PMCID: PMC7400143.

Myers, S. S. et al. Increasing CO2 threatens human nutrition. *Nature* 510, 139–142 (2014).

Nagamine, A., Takayama, M. & Ezura, H. Genetic improvement of tomato using gene editing technologies. *J. Hortic. Sci. Biotechnol.* 98, 1–9 (2023).

Nakayasu, M. et al. Generation of α-solanine-free hairy roots of potato by CRISPR/Cas9 mediated genome editing of the St16DOX gene. *Plant Physiol. Biochem.* 131, 70–77 (2018).

Nayyar, A., Hamel, C., Lafond, G., Gossen, B. D. & Hanson, K. Soil microbial quality associated with yield reduction in continuous-pea. *Appl. Soil Ecol.* 43, 115–121 (2009).

Peltonen-Sainio, P., Jauhiainen, L., Hakala, K. & Ojanen, H. Climate change and prolongation of growing season: changes in regional potential for field crop production in Finland. *Agr. Food Sci.* 18, 171–190 (2009).

Petry, N., Egli, I., Campion, B., Nielsen, E. & Hurrell, R. Genetic reduction of phytate in common bean (Phaseolus vulgaris L.) seeds increases iron absorption in young women. *J. Nutr.* 143, 1219–1224 (2013).

Petry, N. et al. The proportion of anemia associated with iron deficiency in low, medium, and high human development index countries: a systematic analysis of national surveys. *Nutrients* 8, 693 (2016).

Pfeiffer, W. H. & McClafferty, B. HarvestPlus: breeding crops for better nutrition. *Crop Sci.* 47, S-88–S-105 (2007).

Pingali, P. L. Green revolution: impacts, limits, and the path ahead. *Proc. Natl Acad. Sci. USA* 109, 12302–12308 (2012).

Pollard, C. M. & Booth, S. Food insecurity and hunger in rich countries—it is time for action against inequality. *Int. J. Environ. Res. Public Health* 16, 1804 (2019).

Qaim, M., Stein, A. J. & Meenakshi, J. Economics of biofortification. *Agric. Econ.* 37, 119–133 (2007).

Ramirez-Cabral, N. Y. Z., Kumar, L. & Taylor, S. Crop niche modeling projects major shifts in common bean growing areas. *Agr. Forest Meteorol.* 218–219 102–113 (2016).

Reeves, T. G., Thomas, G. & Ramsay, G. *Save and Grow in Practice: Maize, Rice, Wheat. A Guide to Sustainable Cereal Production* (FAO UN, 2016).

Russo, R., Colantuono, A., Di Gaetano, S., Capasso, D., Tito, A., Pedone, E. & Pirone, L. Phaseolus coccineus seed: a valued resource for bioactive compounds targeting health and tumor cells. *Int. J. Mol. Sci.* 26, 2189 (2025). https://doi.org/10.3390/ijms26052189

Sashidhar, N., Harloff, H. J., Potgieter, L. & Jung, C. Gene editing of three BnITPK genes in tetraploid oilseed rape leads to significant reduction of phytic acid in seeds. *Plant Biotechnol. J.* 18, 2241–2250 (2020).

Sharma, D. C. & Mathur, R. Correction of anemia and iron deficiency in vegetarians by administration of ascorbic acid. *Indian J. Physiol. Pharmacol.* 39, 403–406 (1995).

Shea, Z., Ogando do Granja, M., Fletcher, E. B., Zheng, Y., Bewick, P., Wang, Z., Singer, W. M. & Zhang, B. A review of bioactive compound effects from primary legume protein sources in human and animal health. *Curr. Issues Mol. Biol.* 46, 4203–4233 (2024). https://doi.org/ 10.3390/cimb46050257.

Siddique, K. H. M., Johansen, C. & Turner, N. C. Innovations in agronomy for food legumes. *Agron. Sustain. Dev.* 32, 45–64 (2012).

Siddique, K. H. M., Li, X. & Gruber, K. Rediscovering Asia's forgotten crops to fight chronic and hidden hunger. *Nat. Plants* 7, 116–122 (2021).

Siitonen, A., Nieminen, F., Kallio, V., Tuccillo, F., Kantanen, K., Martin, J., Jouppila, K., Piironen, V., Kariluoto, S. & Edelmann, M. B vitamins in legume ingredients and their retention in high moisture extrusion. *Foods* 13(5), 637 (2023). https://doi.org/10.3390/foods13050637

Simkin, A. J. Genetic engineering for global food security: photosynthesis and biofortification. *Plants* 8, 586 (2019).

Singh, J. P., Singh, B. & Kaur, A. Bioactive compounds of legume seeds. In *Bioactive Compounds in Underutilized Vegetables and Legumes*. Reference Series in Phytochemistry (eds Murthy, H. N. & Paek, K. Y.) (Springer, 2020). https://doi.org/10.1007/978-3-030-44578-2_33-1

Stevens, G. A. et al. Trends and mortality effects of vitamin A deficiency in children in 138 low-income and middle-income countries between 1991 and 2013: a pooled analysis of population-based surveys. *Lancet Glob. Health* 3, e528–e536 (2015).

Strobbe, S. & Van Der Straeten, D. Toward eradication of B-vitamin deficiencies: considerations for crop biofortification. *Front. Plant Sci.* 9, 443 (2018).

Sun, Y. et al. Generation of high-amylose rice through CRISPR/Cas9-mediated targeted mutagenesis of starch branching enzymes. *Fronti. Plant Sci.* 8, 298 (2017).

Tilman, D., Cassman, K. G., Matson, P. A., Naylor, R. & Polasky, S. Agricultural sustainability and intensive production practices. *Nature* 418, 671–677 (2002).

Titcomb, T. J. & Tanumihardjo, S. A. Global concerns with B vitamin statuses: biofortification, fortification, hidden hunger, interactions, and toxicity. *Compr. Rev. Food Sci. Food Saf.* 18, 1968–1984 (2019).

Turkmen, N., Sari, F. & Velioglu, S. The effect of cooking methods on total phenolics and antioxidant activity of selected green vegetables. *Food Chem.* 93, 713–718 (2005).

Van Der Straeten, D. et al. Multiplying the efficiency and impact of biofortification through metabolic engineering. *Nat. Commun.* 11, 5203 (2020).

van Schoor, N. M. & Lips, P. Worldwide vitamin D status. *Best Pract. Res. Clin. Endocrinol. Metab.* 25, 671–680 (2011).

Vilakazi, B., Mafongoya, P. L., Odindo, A. O. & Phophi, M. M. The role of neglected grain legumes in food and nutrition security and human health. *Sustainability* 17, 350 (2025). https://doi.org/10.3390/su17010350

Vithanage, M., Zhang, X., Gunarathne, V., Zhu, Y., Herath, L., Peiris, K., Solaiman, Z. M., Bolan, N. & Siddique, K. H., Plant nanobionics: fortifying food security via engineered plant productivity. *Environ. Res.* 229, 115934 (2023).

Waltz, E. GABA-enriched tomato is first CRISPR-edited food to enter market. *Nat. Biotechnol.* 40, 9–11 (2022).

Wiesinger, J. A., Osorno, J. M., McClean, P. E., Hart, J. J. & Glahn, R. P. Faster cooking times and improved iron bioavailability are associated with the down regulation of procyanidin synthesis in slow-darkening pinto beans (Phaseolus vulgaris L.). *J. Funct. Foods* 82, 104444 (2021).

Xu, B. J. & Chang, S. K. C. Comparative studies on phenolic profiles and antioxidant activities of legumes as affected by extraction solvents. *J Food Sci.* 72, S159–S166 (2007).

Zeng, Z. et al. Functional dissection of HGGT and HPT in barley vitamin E biosynthesis via CRISPR/Cas9-enabled genome editing. *Ann. Bot.* 126, 929–942 (2020).

Zhang, H. et al. Genome editing of upstream open reading frames enables translational control in plants. *Nat. Biotechnol.* 36, 894–898 (2018).

Zhao, T. et al. Global burden of vitamin A deficiency in 204 countries and territories from 1990–2019. *Nutrients* 14, 950 (2022).

Zheng, X., Giuliano, G. & Al-Babili, S. Carotenoid biofortification in crop plants: citius, altius, fortius. *Biochim. Biophys. Acta Mol. Cell Biol. Lipids* 1865, 158664 (2020).

Zheng, X., Kuijer, H. N. J. & Al-Babili, S. Carotenoid biofortification of crops in the CRISPR era. *Trends Biotechnol.* 39, 857–860 (2021).

Zheng, Z. et al. Editing sterol side chain reductase 2 gene (StSSR2) via CRISPR/Cas9 reduces the total steroidal glycoalkaloids in potato. *All Life* 14, 401–413 (2021).

14 Biotechnological Innovations in Legumes

Progress, Challenges, and Sustainable Solutions for Agriculture

Aditi Dwivedi, Komal G. Lakhani*, Rasmeih Hamid*#, Bahman Panahi, Kirankumar P. Suthar#, Smrutirekha Sahu, Rajkumar B.K., and Preeti R. Parmar*

14.1 INTRODUCTION

Climate change has increasingly threatened human life by adversely affecting global crop productivity and diminishing the nutritional quality of food in recent times [1]. Since the onset of the Green Revolution, the widespread use of synthetic pesticides and fertilizers has exacerbated environmental pollution. The excessive application of these inputs disrupts ecosystems by degrading water quality and soil health [2,3]. This has led to declines in beneficial soil microorganisms and vital pollinators such as honeybees, which are essential for enhancing crop yields. In addition to these human-induced factors, environmental stressors including rising global temperatures, erratic monsoon patterns, and heightened occurrences of pests and diseases continues to diminish crop productivity. The rapid increase in the global human population, expected to reach 9.5 billion by 2050, further exacerbates these challenges [4]. Consequently, innovative and sustainable agricultural practices are urgently needed to ensure food security, conserve the environment, and utilize natural resources efficiently.

Legumes, members of the Fabaceae family, are the second-largest plant family on Earth and play a crucial role in agricultural systems due to their ability to fix atmospheric nitrogen in the soil. With over 19,400 species across 730 genera, legumes are integral to both human nutrition and sustainable agriculture angiosperm [5]. Within the Fabaceae family, three subfamilies have historically been recognized: Mimosoideae, Caesalpinieae, and Papilionoideae, the latter encompassing the majority of grain legumes [6]. Based on their growing season, legumes are classified into warm-season crops – such as faba bean, common bean, soybean, and mungbean – and cool-season crops including broad beans, lentils, chickpeas, and green peas [7].

Legumes are consumed globally in fresh (vegetable legumes) and dried (grain legumes) forms. They are a significant source of essential nutrients, including phenolic acids, dietary fiber, vitamin C, magnesium, zinc, iron, copper, phosphorus, and manganese [8]. They are also cholesterol-free, rich in polyunsaturated fats, and have a low glycemic index. These attributes make legumes beneficial in preventing conditions such as obesity, type II diabetes, and high blood pressure [9]. Known as the "poor man's meat," legumes are a rich source of protein (13–36 g/100 g), particularly beneficial for vegans to meet daily protein requirements. Legumes provide balanced amino acid profiles, such

DOI: 10.1201/9781003477235-17

as lysine, which is typically deficient in cereal crops. Furthermore, their indigestible carbohydrates act as prebiotics, enhancing gut health by reaching the colon intact. In many countries, legumes are consumed after cooking, but they are also widely consumed as sprouts, particularly mungbeans, valued for their health benefits.

Beyond their role in human nutrition, legumes support diverse markets with value-added products such as tortillas, bread, chips, extruded snacks, spreads, and doughnuts [10]. Additionally, legume-based drinks, such as soy milk, tofu, and yogurt serve as alternatives to milk-based products for individuals intolerant to lactose or following vegan diets [11]. Legumes are also utilized in industrial applications, including the production of biodegradable plastics, dyes, and gums. For instance, Sesbania and Cyamopsis are sources of galactomannan gums, employed as thickening agents in the pill and paper industries [12].

The ecological significance of legumes extends beyond food and industrial applications. By fixing atmospheric nitrogen through rhizomal nodules, legumes enhance soil fertility and support the proliferation of beneficial soil bacteria [13]. They reduce the need for synthetic fertilizers, increase humus and organic carbon content, and mitigate greenhouse gas emissions, including N_2O and CO_2 [14]. Certain legume species with deep root systems improve nutrient solubilization and water infiltration into soil layers, enhancing both soil and plant health. Furthermore, the secretion of phosphorus-modulating compounds like malate and citrate increases soil phosphorus availability, further contributing to soil fertility and plant nutrition [14].

Since the advent of agriculture, legumes have been integral to human diets and agricultural systems, particularly in regions such as the Mediterranean, Asia, and the Americas. In 2020, India emerged as the world's largest producer and consumer of grain legumes, contributing 28.12% to global production across 29 million hectares of cultivated land, yielding approximately 23.37 million tonnes [15]. Canada ranks as the second-largest producer, followed by Myanmar, Nigeria, and the Russian Federation (www.fao.org/faostat/en). India also remains one of the largest cultivators of legume crops, with chickpea leading at 9.89 million hectares followed by black gram (4.81 mha), pigeon pea (4.72 mha), green gram (4.61 mha), lentil (1.43 mha), and field pea (0.75 mha). In 2022, India became the world's largest importer of legumes, with imports of 2,496,000 tonnes, which surged by 90% to 45,000,000 tonnes in 2023. The United States is the leading importer of legumes, accounting for trade valued at $297 million, followed by Hong Kong, Spain, the Netherlands, and France (www.ec.world/en/profile/hs/legumes). This growing demand underscores the nutritional value of legumes, their role in sustainable agriculture, and their importance in addressing global protein shortages in plant-based diets.

Despite their remarkable health-promoting properties and ecological benefits, legume crops face significant challenges in terms of production and consumer demand. Unfavorable environmental conditions coupled with increasing pest and disease pressures contribute to their limited productivity. Rising global temperatures and prolonged drought conditions further exacerbate these issues. Consequently, agricultural researchers are focusing on developing new legume varieties capable of withstanding environmental stresses and resisting pest and disease infections.

14.2 TRADITIONAL BREEDING VS. GENETIC ENGINEERING

Traditional breeding has been the foundation of crop improvement, relying on selecting plants with superior phenotypic traits. This process typically utilizes a genetically related gene pool, transferring desirable traits through techniques such as cross-breeding, selection, back-crossing, and hybridization. However, these methods are highly time-intensive, requiring 6–12 years to develop and release new cultivars [16]. Despite its historical significance, traditional breeding has limitations in addressing the increasingly complex demands of modern agriculture.

The primary challenge lies in the time required to achieve significant progress. Limited genetic variation, especially when identical parental lines are used, reduces genetic diversity and often leads to inbreeding depression, where offspring display reduced vigor and yield [17]. Moreover, unintended mutations and genotype–environment (G × E) interactions can influence phenotypic traits, complicating breeding efforts. The manual handling and reliance on observational skills further increase the likelihood of human error resulting in inconsistent outcomes [18]. These constraints highlight the need for advanced, efficient, and precise approaches to crop improvement.

Modern breeding techniques, including marker-assisted selection (MAS), gene sequencing, and genetic engineering, have emerged to overcome these challenges (Figure 14.1) [19]. Among these, genetic engineering and gene-editing tools such as CRISPR/Cas9 offer unparalleled precision and efficiency, significantly reducing the complexity of breeding procedures. These tools enable faster, more targeted improvements to meet the demands of modern agriculture [20,21].

Genetically modified (GM) crops first gained prominence in the 1990s, revolutionizing crop improvement by enabling the insertion of novel genes from unrelated species or the removal of specific genes within target organisms. For instance, GM soybeans have been engineered for herbicide tolerance and insect resistance [22]. This technology has facilitated advancements in yield, nutritional content, and pest resistance, which are particularly important for legume crops facing challenges such as low productivity and susceptibility to environmental stresses (see Figure 14.1) [23].

GM technology encompasses two primary approaches: cisgenesis and transgenesis [24]. Cisgenesis involves introducing genes from the same species, as seen in cisgenic soybean varieties developed for enhanced drought tolerance [25,26]. Transgenesis, on the other hand, introduces genes from unrelated organisms, exemplified by Bt cowpea engineered for resistance to *Helicoverpa armigera* and Chickpea Chlorotic Dwarf Virus [27,28]. While cisgenesis is often more publicly accepted due to the absence of foreign DNA, transgenesis addresses specific challenges enabling traits like herbicide tolerance, pest resistance, and novel metabolite production that are otherwise difficult to achieve through conventional breeding [29,30].

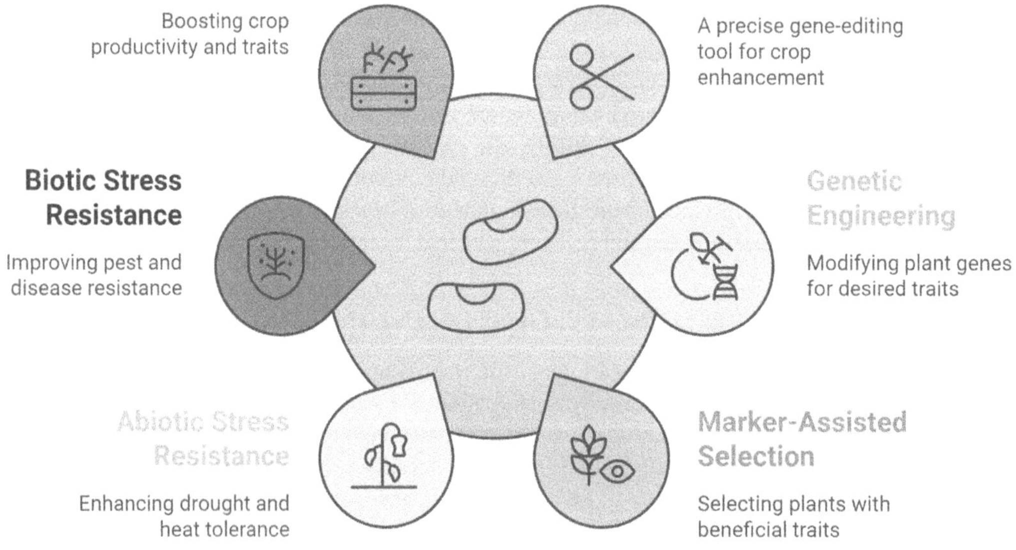

FIGURE 14.1 Biotechnology's impact on leguminous crops.

The precision of genetic engineering enables immediate evaluation of specific traits and the identification of stable transgenics with minimal unintended insertions. For instance, GM soybean plants with modified oil composition have been developed to improve nutritional value – a goal that would take considerably longer using traditional breeding methods [31]. However, concerns remain regarding environmental risks, such as gene flow to wild relatives in crops like chickpea.

Gene-editing technologies such as ZFN, TALEN, and CRISPR/Cas9 represent a significant evolution in genetic engineering. CRISPR/Cas9 in particular has proven highly effective for legumes, enabling precise mutagenesis of genes associated with drought tolerance and pest resistance [32]. This technology reduces the dependency on extensive genetic crosses and genotyping, saving time and resources. For example, CRISPR/Cas9 has been successfully used to improve stress resilience in soybean and lentils, showcasing its potential to address the limitations of traditional breeding [33].

14.3 BIOTECHNOLOGICAL ADVANCES IN LEGUMES

14.3.1 Abiotic Stress Resistance

Abiotic stresses such as drought, salinity, heat, cold, and heavy metal toxicity significantly reduce the productivity of leguminous crops with yield losses reaching up to 50% globally [34]. These stresses impair plant growth, development, and seed quality, posing a critical challenge to food security. For example, drought adversely affects biomass allocation to reproductive organs, leading to reduced seed weight and quality [35].

To counter water deprivation, legumes like chickpea and soybean exhibit survival mechanisms, including reduced stomatal opening, decreased photosynthetic activity, and increased synthesis of osmoprotectants such as soluble sugars (e.g., sucrose), sugar alcohols (e.g., pinitol), organic acids, and amino acids (e.g., proline). These compounds aid in osmotic adjustment and reactive oxygen species (ROS) scavenging, protecting plants under stress [35]. Advances in high-throughput sequencing have identified stress-responsive genes enabling the development of transgenic plants with enhanced drought tolerance. Transcription factor families such as DREB, MYB, NAC, and WRKY have been targeted in crops like soybean and lentils [36,37]. Non-invasive phenotyping tools such as spectral and thermal imaging have further distinguished drought-tolerant phenotypes based on water-use efficiency and canopy temperature [38].

Salinity stress, often associated with dwindling water levels, also limits legume productivity by disrupting ionic balance and osmoprotectant accumulation. Genetic engineering of transporter genes like *NHX* and *HKT1* has improved salt tolerance in legumes such as chickpea and lentils [39]. Heat stress, driven by global warming, necessitates the development of heat-tolerant legumes, while cold tolerance remains underexplored. For instance, overexpression of superoxide dismutase (*SOD*) genes in alfalfa has enhanced recovery from winter stress, while transgenic mungbeans with ICE1 overexpression showed improved cold tolerance by increasing chlorophyll, proline, and lipid content [40–42]. Similarly, heavy metal stress has been mitigated through overexpression of genes like *NRAMP* and *MT1*, which enable legumes to sequester toxic metals such as cadmium, as demonstrated in transgenic alfalfa [43].

14.3.2 Biotic Stress Resistance

Biotic stresses, including pests, pathogens, and parasitic weeds, cause over 30% of yield losses in leguminous crops. Key pests such as pod borers (*Helicoverpa armigera*), aphids (*Aphis craccivora*), leaf miners, thrips, and bruchids reduce seed quality and overall yield. Additionally, fungal pathogens like *Phytophthora*, *Ascochyta*, and *Cercospora*, along with viruses and parasitic weeds like *Striga* and *Orobanche* pose severe challenges [44].

Transgenic approaches have been employed to address these threats. For example, Bt soybean expressing *Bacillus thuringiensis* (Bt) delta-endotoxin Cry14Ab effectively controls soybean cyst nematode (SCN) [45]. Cry proteins, proteinase inhibitors, trypsin inhibitors, and cystatins have also been integrated into crops like chickpea and cowpea, providing resistance to pod borers and aphids [20]. Genes such as chitinase, glucanase, and osmotin have been used to enhance fungal and bacterial disease resistance. In common bean, *PvPOX1* gene transfer via *Agrobacterium rhizogenes* improved resistance to *Fusarium* wilt [46]. Viral resistance has been achieved through coat protein genes and RNA interference (RNAi) [3]. CRISPR/Cas9 gene editing has further enhanced resistance to soybean mosaic virus (SMV) by targeting genes like *GmF3H1*, *GmF3H2*, and *GmFNSII-1* [47]. Similarly, CRISPR/Cas9 has been used to edit conserved regions of the Chickpea Chlorotic Dwarf Virus genome, offering promising resistance strategies [28].

14.3.3 Enhancing Yield and Agronomic Traits

Plant and root architecture, flowering time, and nodulation significantly influence the yield potential of leguminous crops. Genetic engineering has focused on optimizing the source-to-sink ratio to improve nutrient allocation and biomass production.

Root architecture impacts water and nutrient uptake, while plant architecture determines light interception and resource distribution. For instance, CRISPR/Cas9-mediated deletion of *FLOWERING LOCUS T* homologs (*GmFT2a* and *GmFT5a*) in soybean delayed flowering, enhancing crop adaptability and productivity [48]. Similarly, targeting *GmmiR156b* using CRISPR/Cas9 increased node numbers and pod-bearing capacity, improving yield [49].

Nodulation, a critical trait for nitrogen fixation, has also been targeted. Genes such as *Symbiotic Nitrogen Fixation* (SNF) influence nodule size and number, enhancing nitrogen assimilation. Genetic modifications in SNF pathways have shown potential to improve nodulation efficiency in crops like chickpea and lentil.

14.3.4 Nutritional Value Improvements

The rising demand for plant-based protein has highlighted the nutritional significance of legumes. Legumes like soybean, chickpea, and lentils are excellent sources of high-quality protein and essential amino acids. Soybean, containing 40% protein and 20% oil on a dry weight basis, is particularly valued for its role in food security [50]. High-protein soymeal, enriched with digestible amino acids, addresses hidden hunger in low-income populations [51].

Fatty acids play vital roles in seed development and resilience. Oils derived from crops like soybean and canola, rich in oleic acid, have nutritional and industrial applications [52]. Genetic engineering has further enhanced the nutritional profiles of legumes. For example, transgenic alfalfa expressing GsPPCK3 and SCMRP showed increased methionine content, improving feed quality [53]. Similarly, the overexpression of *Corynebacterium* DHDPS and lysine-insensitive *E. coli* Aspartokinase in soybean significantly increased seed lysine content [54]. Advanced tools like CRISPR/Cas9 and TALENs have also improved the nutritional profiles of *Glycine max* and *Medicago truncatula*, creating mutant lines with enhanced traits [55,56].

14.4 Environmental and Health Impacts

Genetic engineering of leguminous crops offers significant potential in addressing global food security challenges. However, it also raises concerns regarding environmental and health impacts that need careful consideration.

TABLE 14.1
Improved Traits using Various GM Techniques

Crop	Methodologies	Traits developed/outcomes	Genes involved	References
Peanut	CRISPR/Cas9	Enhanced oleic acid content	AhFAD2	[57]
	TALEN	Nodule formation	AhNFR1 and AhNFR5	[58]
		Seed content improvement	AhFAD2	[59]
Soybean	*Agrobacterium-* mediated	Double loci mutants showing dome-shaped trifoliate leaves,	GmPPD1	[60]
	transformation		GmPPD2	[49]
	(site-directed	extremely twisted pods, and	GmSPL9	[61]
	mutagenesis) using	produced few seeds	GmF3H1, GmF3H2, GmFNSII-1	[62]
	CRISPR/Cas9	Altered node and branched number of mutants		[63]
			GmKasI	[64]
	CRISPR/Cas9	Doubled isoflavone content	GmPDS	[65]
	CRISPR/Cas9	Novel seed composition	GmLox1, GmLox2, GmLox3	[66]
	(multiplexed editing)	Albino and dwarf mutant	GmPRR37	
	CRISPR/Cas9	Reduction in beany flavor due to loss of lipoxygenase activity	FAD2-1A/B	
	CRISPR/Cas9, TALENs			
		Changes in flowering time		
	CRISPR/Cas9	Increased oleic acid and reduced linolenic acid content		
	CRISPR/Cas9			
	TALEN			
Cowpea	CRISPR/Cas9	Mutants have reduced nodule formation	SNF	[67]
	(Hairy root transformation)		VuSPO11-1	[68]
		Meiosis mutation	SPO11-1, REC8 and OSD1	[69]
	CRISPR/Cas9	Male and Female Sterilities		
	CRISPR/Cas9			
Chickpea	CRISPR/Cas9	Drought tolerance	4CL and RVE7	[70]
	CRISPR/Cas9 multiplex	Resistance to Chickpea Chlorotic Dwarf Virus (CpCDV)	LIR, Rep and RepA	[28]
Lotus japonica	CRISPR/Cas9	Non production of triterpenoids	CYP716A51	[71]
	CRISPR/Cas9	Reduced nodule formation	LjCZF1 and LjCZF2	[72]
Medicago truncatula	CRISPR/Cas9	Regulation of floral development	MtSUP	[73]
	CRISPR/Cas9	Mutant nodules are ineffective and smaller	NPD genes	[74]
	CRISPR/Cas9		MtPDS	[75]
		Albino plants		
Beans	CRISPR/Cas9	Ureide biosynthesis	XMPP, GSDDA, XDH, NSH1, NSH2	[76]

14.4.1 Genetic Contamination and Biodiversity Loss

One of the primary environmental concerns associated with genetically modified (GM) legumes is the potential for genetic contamination through pollen transfer [77]. This gene flow between GM crops and wild relatives or non-GM crops could lead to unintended ecological consequences. For example, the transfer of herbicide tolerance from transgenic soybeans to wild relatives may introduce "superweeds," resistant to herbicides, thereby complicating weed management efforts and reducing biodiversity. The dominance of GM crops such as Bt chickpeas engineered for pest resistance may also impact the genetic diversity of wild populations, which are valuable for future breeding programs.

The genetic uniformity of GM legumes could leave agricultural systems more vulnerable to unforeseen threats such as emerging fungal pathogens. Reduced biodiversity within legume crops might impair the ecosystem's ability to maintain stability, particularly in legume-based farming systems that rely on nitrogen fixation for soil health.

14.4.2 Allergenicity

Health concerns regarding GM legumes are largely centered around allergenicity, particularly for individuals who are prone to food allergies. Genetic modifications can introduce new proteins into crops, which might trigger allergic reactions in sensitive individuals [78]. For example, transgenic soybeans incorporating a methionine-rich protein from Brazil nuts caused allergic reactions in some individuals, highlighting the need for comprehensive allergenicity testing before GM legumes are introduced to the market [79,80]. The unpredictable nature of allergenicity in GM crops necessitates further investigation to ensure that these crops do not pose additional risks to human health [81].

14.4.3 Antibiotic Resistance

The use of antibiotic resistance markers in genetic engineering has raised concerns about the potential transfer of antibiotic-resistant genes to bacteria in the human gut. While most modern GM foods no longer contain such markers, the widespread use of antibiotics in animal feed remains a significant issue, contributing to the development of antibiotic-resistant bacteria [82]. This can exacerbate the growing problem of antibiotic resistance in human populations. Further studies are needed to assess the risks of cross-contamination between GM crops and non-GM crops, particularly in relation to the transfer of antibiotic resistance markers [83].

14.5 ENVIRONMENTAL RISKS OF GM LEGUMES

The environmental impact of GM legumes also extends to ecosystem dynamics. The widespread use of GM plants engineered for pest resistance or herbicide tolerance can lead to the development of resistance in pests and weeds over time, necessitating the development of new genetic traits. This cycle could undermine the long-term sustainability of GM crops, increase the cost of agricultural inputs, and strain global agricultural systems [84]. Additionally, the release of GM crops into the environment without adequate containment measures could lead to unintended ecological changes affecting the balance of ecosystems and biodiversity [85].

In conclusion, while genetic engineering presents valuable tools for improving leguminous crops, it is essential to balance these advances with environmental stewardship and rigorous health and safety assessments. Ongoing research is necessary to mitigate potential risks associated with GM crops, ensuring that their adoption contributes positively to global food security without compromising human health or the environment.

14.6 CHALLENGES IN COMMERCIALIZATION

The commercialization of genetically engineered (GE) leguminous crops faces several challenges, including regulatory hurdles, socioeconomic issues, and public perception concerns.

14.6.1 Regulatory Challenges

The regulatory framework for GE crops is complex and varies widely between countries. While some regions, such as the United States, have more relaxed regulations, others, like the European Union, require extensive biosafety assessments, which can delay the approval process for years

[86]. Regulatory approval processes are often time-consuming, economically burdensome, and inconsistent between trading partners, which can limit the international trade of GM legumes. Furthermore, genome-edited crops, especially those that do not contain foreign DNA, may face fewer restrictions in some countries, but still face challenges in others such as those that enforce stricter biosafety standards. These discrepancies create trade barriers and complicate market access for GE crops across borders.

14.6.2 Socioeconomic Issues

The commercialization of GE leguminous crops also raises socioeconomic concerns, particularly in developing countries. While GE crops can offer benefits such as higher yields and pest resistance, they often come with high initial costs for patented seeds and chemical inputs [87,88]. Small-scale farmers in developing regions may struggle to afford these inputs, exacerbating income inequality and increasing dependence on multinational corporations for seeds and agricultural products. This dependency can lead to financial burdens and limit the autonomy of local farmers, further contributing to the cycle of poverty in agricultural communities [89].

14.6.3 Public Perception

Public perception remains one of the most significant barriers to the commercialization of GM crops, including legumes. Despite scientific evidence supporting the safety of GM crops, many consumers, particularly in Europe and other developed regions, are skeptical about their safety and ethical implications [90]. Concerns about the "unnaturalness" of genetic modification and the dominance of biotech corporations in the seed market fuel public opposition [91]. Additionally, the requirement for labeling GM products in many countries, particularly in the European Union, further hampers consumer acceptance by creating a stigma around GE food products [92].These public concerns can lead to reduced market demand, making it difficult for companies to successfully commercialize GM legumes.

In conclusion, while the potential of genetically engineered legumes to address global food security issues is vast, their commercialization is hindered by a range of regulatory, socioeconomic, and public perception challenges. Effective communication, transparent regulatory processes, and public engagement are essential to overcoming these barriers and ensuring the successful integration of GM crops into global agricultural systems.

14.7 FUTURE PERSPECTIVES AND CONCLUSION

The role of biotechnology in agriculture, particularly in leguminous crops, is pivotal to addressing global challenges related to food security, sustainability, and climate change. As the global population continues to grow, and environmental stresses such as droughts, salinity, and heat intensify, biotechnological innovations are increasingly recognized as crucial tools for developing climate-resilient and high-yielding crops.

Biotechnology, especially through genome-editing technologies like CRISPR/Cas9, holds immense promise in transforming the agricultural landscape. These tools allow for precise modifications to legume genomes, improving traits such as drought tolerance, disease resistance, and nutrient content. The ability to enhance key traits in legumes like soybean, chickpea, and lentils not only improves crop resilience but also contributes to nutritional enhancement, addressing deficiencies in critical amino acids and proteins. Such advancements align with sustainable agriculture practices by reducing dependency on chemical fertilizers and pesticides, thus minimizing environmental impact technology continues to evolve, the integration of genetic engineering with traditional breeding methods will likely play a central role in developing leguminous crops that can thrive in changing climates and adverse conditions. The precision of modern tools such as base and prime editing promises to overcome current limitations offering more sustainable agricultural systems. By enhancing nitrogen-fixing abilities and reducing the need for synthetic fertilizers, GM legumes could significantly contribute to the sustainability of farming practices.

However, significant challenges persist that hinder the widespread adoption and commercialization of genetically modified legume. Regulatory frameworks vary across regions, and the approval processes are often slow and costly, creating barriers for farmers and developers. Socioeconomic issues including the high cost of GM seeds and inputs may disproportionately affect smallholder farmers, particularly in developing countries. Public perception and ethical concerns about the safety of GM crops also pose significant obstacles to their acceptance.

Looking ahead, agricultural biotechnology lies in striking a balance between innovation, regulation, and societal acceptance. International collaboration, transparent communication of risks and benefits, and a commitment to ethical responsibility will be crucial in ensuring the successful integration of biotechnological innovations into global food systems. With continued research and responsible deployment, biotechnology can be a cornerstone in the achievement of the United Nations Sustainable Development Goals (SDGs), particularly in eradicating hunger, improving food security, and promoting environmental sustainability.

In conclusion, biotechnology present unities for leguminous crops, enhancing their role in sustainable agriculture. By overcoming regulatory, socioeconomic, and public perception challenges, biotechnology can play a key role in securing a resilient and sustainable food future for all.

AUTHOR CONTRIBUTION

Aditi Dwivedi and Komal G. Lakhani contributed equally to this work and share first authorship. AD, KGL and SS: Conceptualization, Writing – Original Draft; RH and BP: Conceptualization, Writing – Review and Editing; KPS, RBK, and PP: Corrected and edited the manuscript. All authors read and approved the final manuscript.

ACKNOWLEDGMENT

We sincerely acknowledge the SERB, DST, and CII for providing the prime minister's fellowship for Doctoral Research to Ms. Komal G. Lakhani.

CONFLICT OF INTEREST

The authors declare no conflict of interest.

NOTES

* Authors contributed equally to this work.

\# Corresponding authors of the manuscript.

REFERENCES

1. Lakhani KG, Hamid R, Motamedi E, Marviya GV. A review on plant metabolite-mediated nanoparticle synthesis: sustainable applications in horticultural crops. *Frontiers in Nanotechnology*. 2025;7:1545413. https://doi.org/10.3389/fnano.2025.1545413
2. Hossain ME, Shahrukh S, Hossain SA. Chemical fertilizers and pesticides: impacts on soil degradation, groundwater, and human health in Bangladesh. *Environmental Degradation: Challenges and Strategies for Mitigation*: Springer; 2022. p. 63–92. https://doi.org/10.1007/978-3-030-95542-7_4

3. Jha UC, Nayyar H, Parida SK, Bakır M, von Wettberg EJ, Siddique KH. Progress of genomics-driven approaches for sustaining underutilized legume crops in the post-genomic era. *Frontiers in Genetics*. 2022;13:831656. https://doi.org/10.3389/fgene.2022.831656
4. Choudhury A, Rajam MV. Genetic transformation of legumes: an update. *Plant Cell Reports*. 2021;40:1813–30. https://doi.org/10.1007/s00299-021-02749-7
5. Saikia P, Nag A, Anurag S, Chatterjee S, Khan ML. Tropical legumes: status, distribution, biology and importance. *The Plant Family Fabaceae: Biology and Physiological Responses to Environmental Stresses*. 2020:27–41. https://doi.org/10.1007/978-981-15-4752-2_2
6. Mousavi-Derazmahalleh M, Bayer PE, Hane JK, Valliyodan B, Nguyen HT, Nelson MN, et al. Adapting legume crops to climate change using genomic approaches. *Plant, Cell & Environment*. 2019;42(1):6–19. https://doi.org/10.1111/pce.13203
7. Sbai H, Hajib A, Msairi S, Amalich S, Bouyahya A, Lee L-H, et al. Fungal infections of legume crops: challenges and management approaches. *Journal of Agriculture and Food Research*. 2024:101447. https://doi.org/10.1016/j.jafr.2024.101447
8. Siddiqui SA, Ucak İl, Jain S, Elsheikh W, Ali Redha A, Kurt A, et al. Impact of drying on techno-functional and nutritional properties of food proteins and carbohydrates–a comprehensive review. *Drying Technology*. 2024;42(4):592–611. https://doi.org/10.1080/07373937.2024.2303580.
9. Lisciani S, Marconi S, Le Donne C, Camilli E, Aguzzi A, Gabrielli P, et al. Legumes and common beans in sustainable diets: nutritional quality, environmental benefits, spread and use in food preparations. *Frontiers in Nutrition*. 2024;11:1385232. https://doi.org/10.3389/fnut.2024.1385232.
10. Graham PH, Vance CP. Legumes: importance and constraints to greater use. *Plant Physiology*. 2003;131(3):872–7. https://doi.org/10.1104/pp.017004
11. Mefleh M, Pasqualone A, Caponio F, Faccia M. Legumes as basic ingredients in the production of dairy-free cheese alternatives: a review. *Journal of the Science of Food and Agriculture*. 2022;102(1):8–18. https://doi.org/10.1002/jsfa.11502.
12. Yadav H, Gupta L, Dubey S, Maiti S, Yadav N. Economical importance, structural diversity, and properties of natural gums. *Natural Gums*: Elsevier; 2023. pp. 23–53. https://doi.org/10.1016/B978-0-323-99468-2.00002-4.
13. Jena J, Maitra S, Hossain A, Pramanick B, Gitari HI, Praharaj S, et al. Role of legumes in cropping system for soil ecosystem improvement. *Ecosystem Services: Types, Management and Benefits:* Nova Science Publishers; 2022. p. 415. ISBN: 978-1-68507-614-6 ©.
14. Stagnari F, Maggio A, Galieni A, Pisante M. Multiple benefits of legumes for agriculture sustainability: an overview. *Chemical and Biological Technologies in Agriculture*. 2017;4:1–13. https://doi.org/10.1186/s40538-016-0085-1.
15. Gayacharan, Parida SK, Mondal N, Yadav R, Vishwakarma H, Rana JC. Mining legume germplasm for genetic gains: an Indian perspective. *Frontiers in Genetics*. 2023;14:996828. https://doi.org/10.3389/fgene.2023.996828.
16. Ahmar S, Gill RA, Jung K-H, Faheem A, Qasim MU, Mubeen M, et al. Conventional and molecular techniques from simple breeding to speed breeding in crop plants: recent advances and future outlook. *International Journal of Molecular Sciences*. 2020;21(7):2590. https://doi.org/10.3390/ijms21072590.
17. Hamdan YA. Genetic diversity analysis of Palestinian safflower (Carthamus tinctorius L.) using inter simple sequence repeat (ISSR) markers. *Genetic Resources and Crop Evolution*. 2024:1–13. https://doi.org/10.1007/s10722-024-01962-z.
18. Teressa T, Semahegn Z, Bejiga T. Multi environments and genetic-environmental interaction (GxE) in plant breeding and its challenges: a review article. *International Journal of Research Studies in Agricultural Sciences*. 2021;7(4):11–8. https://doi.org/10.20431/2454-6224.0704002.
19. Lamichhane S, Thapa S. Advances from conventional to modern plant breeding methodologies. *Plant Breeding and Biotechnology*. 2022;10(1):1–14. https://doi.org/10.9787/PBB.2022.10.1.1.
20. Aziz MA, Masmoudi K. Molecular breakthroughs in modern plant breeding techniques. *Horticultural Plant Journal*. 2024; 11(1): 15-41. https://doi.org/10.1016/j.hpj.2024.01.004.
21. Kumar R, Das SP, Choudhury BU, Kumar A, Prakash NR, Verma R, et al. Advances in genomic tools for plant breeding: harnessing DNA molecular markers, genomic selection, and genome editing. *Biological Research*. 2024;57(1):80.

22. Rahman SU, McCoy E, Raza G, Ali Z, Mansoor S, Amin I. Improvement of soybean: a way forward transition from genetic engineering to new plant breeding technologies. *Molecular Biotechnology*. 2023;65(2):162–80. https://doi.org/10.1007/s12033-022-00456-6.
23. Baloglu MC, Celik Altunoglu Y, Baloglu P, Yildiz AB, Türkölmez N, Özden Çiftçi Y. Gene-editing technologies and applications in legumes: progress, evolution, and future prospects. *Frontiers in Genetics*. 2022;13:859437. https://doi.org/10.3389/fgene.2022.859437.
24. Koul B. *Cisgenics and Transgenics: Strategies for Sustainable Crop Development and Food Security*. Springer, 2022. https://doi.org/10.1007/978-981-19-2119-3.
25. Ha CV, Le DT, Nishiyama R, Watanabe Y, Tran UT, Dong NV, et al. Characterization of the newly developed soybean cultivar DT2008 in relation to the model variety W82 reveals a new genetic resource for comparative and functional genomics for improved drought tolerance. *BioMed Research International*. 2013;2013(1):759657. https://doi.org/10.1155/2013/759657.
26. Chen Z, Yang X, Tang M, Wang Y, Zhang Q, Li H, et al. Molecular characterization and drought resistance of GmNAC3 transcription factor in Glycine max (L.) Merr. *International Journal of Molecular Sciences*. 2022;23(20):12378. https://doi.org/10.3390/ijms232012378.
27. Surender Khatodia SK, Kirti Bhatotia KB, Behl R. Helicoverpa resistant chickpea plants: from Bt toxins to plant-mediated RNAi. 2017, 3(1):52 - 60.
28. Malik MAM, Haider MS, Zhai Y, Khan MAU, Pappu HR. Towards developing resistance to chickpea chlorotic dwarf virus through CRISPR/Cas9-mediated gene editing using multiplexed gRNAs. *Journal of Plant Diseases and Protection*. 2023;130(1):23–33. https://doi.org/10.1007/s41348-022-00677-6.
29. S Telem R, Wani H, B Singh N, Nandini R, Sadhukhan R, Bhattacharya S, et al. Cisgenics–a sustainable approach for crop improvement. *Current Genomics*. 2013;14(7):468–76. https://doi.org/10.2174/13892029113149990006.
30. Vasudevan S, Pooja S, Raju TJ, Damini C. Cisgenics and intragenics: boon or bane for crop improvement. *Frontiers in Plant Science*. 2023;14:1275145. https://doi.org/10.3389/fpls.2023.1275145.
31. Clemente TE, Cahoon EB. Soybean oil: genetic approaches for modification of functionality and total content. *Plant Physiology*. 2009;151(3):1030–40. https://doi.org/10.1104/pp.109.146282.
32. Mushtaq M, Ahmad Dar A, Skalicky M, Tyagi A, Bhagat N, Basu U, et al. CRISPR-based genome editing tools: insights into technological breakthroughs and future challenges. *Genes*. 2021;12(6):797. https://doi.org/10.3390/genes12060797.
33. Yao D, Zhou J, Zhang A, Wang J, Liu Y, Wang L, et al. Advances in CRISPR/Cas9-based research related to soybean [Glycine max (Linn.) Merr] molecular breeding. *Frontiers in Plant Science*. 2023;14:1247707. https://doi.org/10.3389/fpls.2023.1247707.
34. Détain A, Bhowmik P, Leborgne-Castel N, Ochatt S. Latest biotechnology tools and targets for improving abiotic stress tolerance in protein legumes. *Environmental and Experimental Botany*. 2022;197:104824. https://doi.org/10.1016/j.envexpbot.2022.104824.
35. Arya H, Singh MB, Bhalla PL. Towards developing drought-smart soybeans. *Frontiers in Plant Science*. 2021;12:750664. https://doi.org/10.3389/fpls.2021.750664.
36. You D, Ren Y, Huo S, Dong J, Ren S, Shi T. A kinetic analysis of oxidation of the antioxidant N-acetyl-l-cysteine (NAC) by Pt (IV) complexes. *Transition Metal Chemistry*. 2016;41:295–304. https://doi.org/10.1007/s11243-016-0021-z.
37. Sosa-Valencia G, Palomar M, Covarrubias AA, Reyes JL. The legume miR1514a modulates a NAC transcription factor transcript to trigger phasiRNA formation in response to drought. *Journal of Experimental Botany*. 2017;68(8):2013–26. https://doi.org/10.1093/jxb/erw380.
38. Crusiol LGT, Nanni MR, Furlanetto RH, Sibaldelli RNR, Cezar E, Sun L, et al. Classification of soybean genotypes assessed under different water availability and at different phenological stages using leaf-based hyperspectral reflectance. *Remote Sensing*. 2021;13(2):172. https://doi.org/10.3390/rs13020172.
39. Parveen K, Saddique MAB, Rehman SU, Ali Z, Aziz I, Shamsi IH, et al. Identification and characterization of salt stress-responsive NHX gene family in chickpea. *Plant Stress*. 2023;10:100266. https://doi.org/10.1016/j.stress.2023.100266.
40. McKersie BD, Bowley SR, Jones KS. Winter survival of transgenic alfalfa overexpressing superoxide dismutase. *Plant Physiology*. 1999;119(3):839–48.

41. McKersie BD, Murnaghan J, Jones KS, Bowley SR. Iron-superoxide dismutase expression in transgenic alfalfa increases winter survival without a detectable increase in photosynthetic oxidative stress tolerance. *Plant Physiology*. 2000;122(4):1427–38. https://doi.org/10.1104/pp.122.4.1427.
42. Rout GR, Bansal A, Swain D, Jadhao KR, Shelke RG, Panda SK. Overexpression of ICE1 gene in mungbean (Vigna radiata L.) for cold tolerance. *Plant Cell, Tissue and Organ Culture (PCTOC)*. 2020;143:593–608. https://doi.org/10.1007/s11240-020-01944-w.
43. Rai KK, Pandey N, Meena RP, Rai SP. Biotechnological strategies for enhancing heavy metal tolerance in neglected and underutilized legume crops: a comprehensive review. *Ecotoxicology and Environmental Safety*. 2021;208:111750. https://doi.org/10.1016/j.ecoenv.2020.111750.
44. Acharjee S, Higgins TJ. Genetic engineering of grain legumes: their potential for sustainable agriculture and food and nutritional security. *Genetic Enhancement in Major Food Legumes: Advances in Major Food Legumes*. 2021:91–121. https://doi.org/10.1007/978-3-030-64500-7_4.
45. Kahn TW, Duck NB, McCarville MT, Schouten LC, Schweri K, Zaitseva J, et al. A Bacillus thuringiensis Cry protein controls soybean cyst nematode in transgenic soybean plants. *Nature Communications*. 2021;12(1):3380. https://doi.org/10.1038/s41467-021-23743-3.
46. Xue R, Wu X, Wang Y, Zhuang Y, Chen J, Wu J, et al. Hairy root transgene expression analysis of a secretory peroxidase (PvPOX1) from common bean infected by Fusarium wilt. *Plant Science*. 2017;260:1–7. https://doi.org/10.1016/j.plantsci.2017.03.011.
47. Zhang P, Du H, Wang J, Pu Y, Yang C, Yan R, et al. Multiplex CRISPR/Cas9-mediated metabolic engineering increases soya bean isoflavone content and resistance to soya bean mosaic virus. *Plant Biotechnology Journal*. 2020;18(6):1384–95. https://doi.org/10.1111/pbi.13302.
48. Cai Y, Chen L, Liu X, Guo C, Sun S, Wu C, et al. CRISPR/Cas9-mediated targeted mutagenesis of GmFT2a delays flowering time in soya bean. *Plant Biotechnology Journal*. 2018;16(1):176–85. https://doi.org/10.1371/journal.pone.0136064.
49. Bao A, Chen H, Chen L, Chen S, Hao Q, Guo W, et al. CRISPR/Cas9-mediated targeted mutagenesis of GmSPL9 genes alters plant architecture in soybean. *BMC Plant Biology*. 2019;19:1–12. https://doi.org/10.1186/s12870-019-1746-6.
50. Singer WM, Lee YC, Shea Z, Vieira CC, Lee D, Li X, et al. Soybean genetics, genomics, and breeding for improving nutritional value and reducing antinutritional traits in food and feed. *The Plant Genome*. 2023;16(4):e20415. https://doi.org/10.1002/tpg2.20415.
51. Olaniyan OT, Adetunji CO, Adetunji JB, Oloke JK, Olaleye OO, Hefft DI, et al. Soy-based food products consumed in Africa: a panacea to mitigate food insecurity and health challenges. *Fermentation and Algal Biotechnologies for the Food, Beverage and Other Bioproduct Industries*: CRC Press; 2022. p. 87–103. https://doi.org/10.1201/9781003178378-5.
52. Msanne J, Kim H, Cahoon EB. Biotechnology tools and applications for development of oilseed crops with healthy vegetable oils. *Biochimie*. 2020;178:4–14. https://doi.org/10.1016/j.biochi.2020.09.020.
53. Sun M, Sun X, Zhao Y, Zhao C, DuanMu H, Yu Y, et al. Ectopic expression of GsPPCK3 and SCMRP in Medicago sativa enhances plant alkaline stress tolerance and methionine content. *PLoS One*. 2014;9(2):e89578. https://doi.org/10.1371/journal.pone.0089578.
54. Falco S, Guida T, Locke M, Mauvais J, Sanders C, Ward R, et al. Transgenic canola and soybean seeds with increased lysine. *Bio/technology*. 1995;13(6):577–82. https://doi.org/10.1038/nbt0695-577.
55. Ullah Z, Iqbal J, Abbasi BA, Ijaz S, Sharifi-Rad J, Yaseen T, et al. How genome editing can be helpful in the biofortification of legumes. In: Faiz, S., Ashraf, U., Attia, K.A., Amir, R.M. (eds) Climate Smart Agriculture for Future Food Security. Springer, Singapore. Springer; 2023. p. 207–32. https://doi.org/10.1007/978-981-96-4499-5_7.
56. Bhowmik P, Konkin D, Polowick P, Hodgins CL, Subedi M, Xiang D, et al. CRISPR/Cas9 gene editing in legume crops: opportunities and challenges. *Legume Science*. 2021;3(3):e96. https://doi.org/10.1002/leg3.96.
57. Yuan M, Zhu J, Gong L, He L, Lee C, Han S, et al. Mutagenesis of FAD2 genes in peanut with CRISPR/Cas9 based gene editing. *BMC Biotechnology*. 2019;19:1–7. https://doi.org/10.1186/s12896-019-0516-8.
58. Shu H, Luo Z, Peng Z, Wang J. The application of CRISPR/Cas9 in hairy roots to explore the functions of AhNFR1 and AhNFR5 genes during peanut nodulation. *BMC Plant Biology*. 2020;20:1–15. https://doi.org/10.1186/s12870-020-02614-x.

59. Wen S, Liu H, Li X, Chen X, Hong Y, Li H, et al. TALEN-mediated targeted mutagenesis of fatty acid desaturase 2 (FAD2) in peanut (Arachis hypogaea L.) promotes the accumulation of oleic acid. *Plant Molecular Biology*. 2018;97:177–85. https://doi.org/10.1007/s11103-018-0731-z.
60. Kanazashi Y, Hirose A, Takahashi I, Mikami M, Endo M, Hirose S, et al. Simultaneous site-directed mutagenesis of duplicated loci in soybean using a single guide RNA. *Plant Cell Reports*. 2018;37:553–63. https://doi.org/10.1007/s00299-018-2251-3
61. Zhang M, Chen C, Hu W, Chen L, Zhan J. Influence of source credibility on consumer acceptance of genetically modified foods in China. *Sustainability*. 2016;8(9):899. https://doi.org/10.3390/su8090899
62. Virdi KS, Spencer M, Stec AO, Xiong Y, Merry R, Muehlbauer GJ, et al. Similar seed composition phenotypes are observed from CRISPR-generated in-frame and knockout alleles of a soybean KASI ortholog. *Frontiers in Plant Science*. 2020;11:1005. https://doi.org/10.3389/fpls.2020.01005
63. Du H, Zeng X, Zhao M, Cui X, Wang Q, Yang H, et al. Efficient targeted mutagenesis in soybean by TALENs and CRISPR/Cas9. *Journal of Biotechnology*. 2016;217:90–7. https://doi.org/10.1016/j.jbiotec.2015.11.005
64. Wang J, Kuang H, Zhang Z, Yang Y, Yan L, Zhang M, et al. Generation of seed lipoxygenase-free soybean using CRISPR-Cas9. *The Crop Journal*. 2020;8(3):432–9. https://doi.org/10.1016/j.cj.2019.08.008.
65. Wang L, Sun S, Wu T, Liu L, Sun X, Cai Y, et al. Natural variation and CRISPR/Cas9-mediated mutation in GmPRR37 affect photoperiodic flowering and contribute to regional adaptation of soybean. *Plant Biotechnology Journal*. 2020;18(9):1869–81. https://doi.org/10.1111/pbi.13346.
66. Haun W, Coffman A, Clasen BM, Demorest ZL, Lowy A, Ray E, et al. Improved soybean oil quality by targeted mutagenesis of the fatty acid desaturase 2 gene family. *Plant Biotechnology Journal*. 2014;12(7):934–40. https://doi.org/10.1111/pbi.12201.
67. Ji J, Zhang C, Sun Z, Wang L, Duanmu D, Fan Q. Genome editing in cowpea Vigna unguiculata using CRISPR-Cas9. *International Journal of Molecular Sciences*. 2019;20(10):2471. https://doi.org/ 10.3390/ijms20102471.
68. Che P, Chang S, Simon MK, Zhang Z, Shaharyar A, Ourada J, et al. Developing a rapid and highly efficient cowpea regeneration, transformation and genome editing system using embryonic axis explants. *The Plant Journal*. 2021;106(3):817–30. https://doi.org/10.1111/tpj.15202.
69. Juranić M, Nagahatenna DS, Salinas-Gamboa R, Hand ML, Sánchez-León N, Leong WH, et al. A detached leaf assay for testing transient gene expression and gene editing in cowpea (Vigna unguiculata [L.] Walp.). *Plant Methods*. 2020;16:1–17. https://doi.org/ 10.1186/s13007-020-00630-4.
70. Badhan S, Ball AS, Mantri N. First report of CRISPR/Cas9 mediated DNA-free editing of 4CL and RVE7 genes in chickpea protoplasts. *International Journal of Molecular Sciences*. 2021;22(1):396. https://doi.org/10.3390/ijms22010396.
71. Suzuki H, Fukushima EO, Shimizu Y, Seki H, Fujisawa Y, Ishimoto M, et al. Lotus japonicus triterpenoid profile and characterization of the CYP716A51 and LjCYP93E1 genes involved in their biosynthesis in planta. *Plant and Cell Physiology*. 2019;60(11):2496–509. https://doi.org/10.1093/pcp/pcz145.
72. Cai K, Yin J, Chao H, Ren Y, Jin L, Cao Y, et al. A C3HC4-type RING finger protein regulates rhizobial infection and nodule organogenesis in Lotus japonicus. *Journal of Integrative Plant Biology*. 2018;60(9):878–96. https://doi.org/10.1111/jipb.12703.
73. Rodas AL, Roque E, Hamza R, Gómez-Mena C, Minguet EG, Wen J, et al. MtSUPERMAN plays a key role in compound inflorescence and flower development in Medicago truncatula. *The Plant Journal*. 2021;105(3):816–30. https://doi.org/10.1111/tpj.15075.
74. Trujillo DI, Silverstein KA, Young ND. Nodule-specific PLAT domain proteins are expanded in the Medicago lineage and required for nodulation. *New Phytologist*. 2019;222(3):1538–50. https://doi.org/10.1111/nph.15697.
75. Meng Y, Hou Y, Wang H, Ji R, Liu B, Wen J, et al. Targeted mutagenesis by CRISPR/Cas9 system in the model legume Medicago truncatula. *Plant Cell Reports*. 2017;36:371–4. https://doi.org/10.1007/s00299-016-2069-9.
76. Alamillo JM, López CM, Rivas FJM, Torralbo F, Bulut M, Alseekh S. Clustered regularly interspaced short palindromic repeats/CRISPR-associated protein and hairy roots: a perfect match for gene

functional analysis and crop improvement. *Current Opinion in Biotechnology*. 2023;79:102876. https://doi.org/ 10.1016/j.copbio.2022.102876.
77. Oliver MJ, Li Y. *Plant Gene Containment*: John Wiley; 2012. https://doi.org/10.1002/9781118352670.
78. Abdul Aziz M, Brini F, Rouached H, Masmoudi K. Genetically engineered crops for sustainably enhanced food production systems. *Frontiers in Plant Science*. 2022;13:1027828. https://doi.org/10.3389/fpls.2022.1027828.
79. Wiederstein M, Baumgartner S, Lauter K. Soybean (Glycine max) allergens— a review on an outstanding plant food with allergenic potential. *ACS Food Science & Technology*. 2023;3(3):363–78. https://doi.org/10.1021/acsfoodscitech.2c00380.
80. Organisms EPoGM, Mullins E, Bresson JL, Dalmay T, Dewhurst IC, Epstein MM, et al. Scientific Opinion on development needs for the allergenicity and protein safety assessment of food and feed products derived from biotechnology. *EFSA Journal*. 2022;20(1):e07044. https://doi.org/10.2903/j.efsa.2022.7044.
81. Bakshi A. Potential adverse health effects of genetically modified crops. *Journal of Toxicology and Environmental Health, Part B*. 2003;6(3):211–25. https://doi.org/10.1080/10937400306469.
82. Goldstein D, Tinland B, Gilbertson L, Staub J, Bannon G, Goodman R, et al. Human safety and genetically modified plants: a review of antibiotic resistance markers and future transformation selection technologies. *Journal of Applied Microbiology*. 2005;99(1):7–23. https://doi.org/10.1111/j.1365-2672.2005.02595.x.
83. Price B, Cotter J. The GM Contamination Register: a review of recorded contamination incidents associated with genetically modified organisms (GMOs), 1997–2013. *International Journal of Food Contamination*. 2014;1:1–13. https://doi.org/10.1186/s40550-014-0005-8.
84. Azadi H, Ghanian M, Ghoochani OM, Rafiaani P, Taning CN, Hajivand RY, et al. Genetically modified crops: towards agricultural growth, agricultural development, or agricultural sustainability? *Food Reviews International*. 2015;31(3):195–221. https://doi.org/10.1080/87559129.2014.994816.
85. Tsatsakis AM, Nawaz MA, Kouretas D, Balias G, Savolainen K, Tutelyan VA, et al. Environmental impacts of genetically modified plants: a review. *Environmental Research*. 2017;156:818–33. https://doi.org/10.1016/j.envres.2017.03.011.
86. Qaim M, Kouser S. Genetically modified crops and food security. *PloS One*. 2013;8(6):e64879. https://doi.org/10.1371/journal.pone.0064879.
87. Greene C, Wechsler SJ, Adalja A, Hanson J. *Economic Issues in the Coexistence of Organic, Genetically Engineered (GE), and Non-GE Crops*; 2016. https://doi.org/10.22004/ag.econ.232929.
88. Sciences NAo, Earth Do, Studies L, Crops CoGE, Experience P, Prospects F. *Genetically Engineered Crops: Experiences and Prospects*: Washington, DC: National Academies Press; 2017. https://doi.org/10.17226/23395.
89. Brookes G, Barfoot P. *GM Crops: Global Socio-Economic and Environmental Impacts 1996-2018*. PG Econ Ltd, UK; 2020: p. 1–213. https://doi.org/10.17226/23395.
90. Kato-Nitta N, Maeda T, Inagaki Y, Tachikawa M. Expert and public perceptions of gene-edited crops: attitude changes in relation to scientific knowledge. *Palgrave Communications*. 2019;5(1):137. https://doi.org/10.1057/s41599-019-0328-4.
91. Alalwan AA, Abu-Romman S, Al-Weshah G, Dwivedi YK, Albanna H. Examining the key determinants of the jordanian customer's adoption of genetically modified food. *Heliyon*. 2023;9(6): e16920. https://doi.org/10.1016/j.heliyon.2023.e16920.
92. Grossman MR. Traceability and labeling of genetically modified crops, food, and feed in the European Union. *Journal of Food Laws & Policy*. 2005;1:43. https://doi.org/10.54119/jflp.wznq2733.

15 Enhancing Sustainability

Genetic Engineering of Legumes, Consumer Acceptance, and Environmental Impact Analysis

Manali Chakraborty, Savita Budhwar, and Suneel Kumar

15.1 INTRODUCTION

Sustainability has become an imperative global priority due to climate change and population growth. With traditional practices, agricultural sectors are contributing to increasing environmental footprints. Besides primary crops like rice and wheat, another essential part of the human diet and agricultural systems is legumes. These hold a unique place in the sustainability agenda due to their nitrogen-fixing abilities, rich nutritional value, and adaptability to diverse environments (Rubiales & Mikic, 2015; Goyal et al., 2021). However, with traditional cultivation practices, legume production is affected due to limitations in terms of resilience to climate stress, pests, and diseases. Along with this, large-scale farming also contributes to the environmental footprint.

Genetic engineering could offer promising solutions to address such challenges by enabling the development of genetically engineered or modified (GM) legume varieties. These varieties can be designed to lower emissions, resist adverse weather conditions, pest infestations, and disease conditions, as well as enriching the nutritional composition (Choudhury & Rajam, 2021). Such an approach can contribute significantly to implementing the United Nations Sustainable Developmental Goals (SDGs) agenda focusing on zero hunger, responsible consumption, and climate actions. However, achieving the success of GM legumes goes beyond scientific modifications. It relies on several other factors, such as consumer acceptance which plays a pivotal role in commercial success. In addition, knowledge of the environmental impact, food safety, and ethical concerns of the genetically modification process are also crucial factors. Hence, translucent interaction is required about the associated benefits and risks to bridge the gap between the scientific community and consumer population. Nevertheless, to design an approach, it is important to have knowledge about the current situation and what can be the targeted tactic to minimize the challenges associated with it. This chapter provides a detailed outline describing the multifaceted role of genetic engineering in contributing to agricultural sustainability, focusing on the legume cultivation system. It will also address the ecological impact of GM legume cultivation and traditional practices by comparing insights into the life cycle assessment and environmental footprint data. The chapter also attempts to shed light on the roles of consumer acceptance and regulatory frameworks to ensure the responsible use of genetic engineering technology to offer substantial potential to reduce the environmental impact, and improve food security and overall health benefits.

15.2 GENETICALLY MODIFIED LEGUMES: CURRENT SCENARIO

Genetic engineering refers to tweaking the genome through lab-based technology to acquire beneficial traits. In legumes, these traits might include better resistance to pests, increased tolerance to

 DOI: 10.1201/9781003477235-18

drought or salty soils, and enhanced nutritional value. Unlike traditional breeding, which relies on crossbreeding over several generations, genetic engineering enables precise changes at the molecular level, speeding up the process of developing new varieties.

Legume crops – such as chickpeas, soybeans, beans, and lentils – play a key role in global food systems (Raina et al., 2023). Not only are legumes rich in protein, they also improve soil quality by fixing atmospheric nitrogen and are fundamental to crop rotation systems. However, due to environmental stress, infections, and other obstructions, the legume cultivation process is affected, and the yields can be reduced significantly. The integration of genetic engineering could address these issues to promote a sustainable process. So far, globally, the concept of GM legumes is not as popular as for other staple crops like maize and rice (Ojiewo et al., 2018). However, there are exceptions. Herbicide-resistant GM soybeans (e.g., Roundup Ready® soybeans), especially in the United States, are widely used. Nevertheless, researchers are keen to expand genetic engineering in legumes, focusing on developing more resilient and nutrient-dense varieties.

GM soybeans are the most commercially successful, and account for a significant fraction of global soybean production.

Apart from tolerance to harmful herbicides, other traits are also being explored in legumes to achieve broader sustainability goals (Brookes & Barfoot, 2017; Nandula, 2019). Pest infestation resistance development in GM cowpea led to less or negligible chemical pesticide use and proved to be beneficial for the ecosystem and human health. GM legumes might offer significant solutions to enhance climate resilience (Ye et al., 2018). Varieties of resistance to drought and salinity are also being put into trial with the aim of assisting farmers in arid and semi-arid regions and coping with water scarcity without having the yield constrained. These climate-resilient varieties can safeguard food systems against the consequences of the climate impact (Acevedo et al., 2020).

Apart from the climate aspects, GM varieties can be a boon to human health and could be a potential solution to combat malnutrition, especially in countries where legumes are the major crop. Technology is being used to enhance and improve nutrient concentrations by increasing the levels of essential amino acids, vitamins, and/or minerals (iron, zinc) (Dissanayaka, 2021) and to make legumes more nutri-dense. The modification in GM legumes not only improves their quality, but also maintains the yield. Although these approaches are in trial phases or have mostly been attempted in lab-scale research experiments, GM legumes can contribute to the development of a sustainable and healthy society.

15.3 IMPACT ON SUSTAINABLE DEVELOPMENT GOALS (SDGS)

The concept and implication of GM legumes in daily life can support the achievement of several United Nations Sustainable Development Goals (Figure 15.2), particularly those associated with zero hunger, fostering good health and well-being, responsible consumption and production, and climate action (Stagnari et al., 2017; Smyth, 2022). GM legumes can influence these global targets in the following ways:

- Zero Hunger (SDG 2): GM legumes with resistance to adverse climate effects like drought, salinity, and pest infestations can address the challenges of productivity in vulnerable regions. They can increase the yields of crops with rich nutrient composition, maintaining the quality of legumes. This approach can maintain the fluidity of the food supply to overcome food scarcity, leading to zero hunger worldwide, especially in low-income regions.
- Good Health and Well-being (SDG 3): Alterations in the genetic make-up of legumes not only can enhance the nutrient quality, but also contribute to better access to essential nutrients. Iron and zinc deficiencies are found most often, especially in rural areas. GM legumes with iron and zinc biofortification can address widespread deficiencies and improve overall public health, particularly in areas with high cultivation and production of legumes. Furthermore,

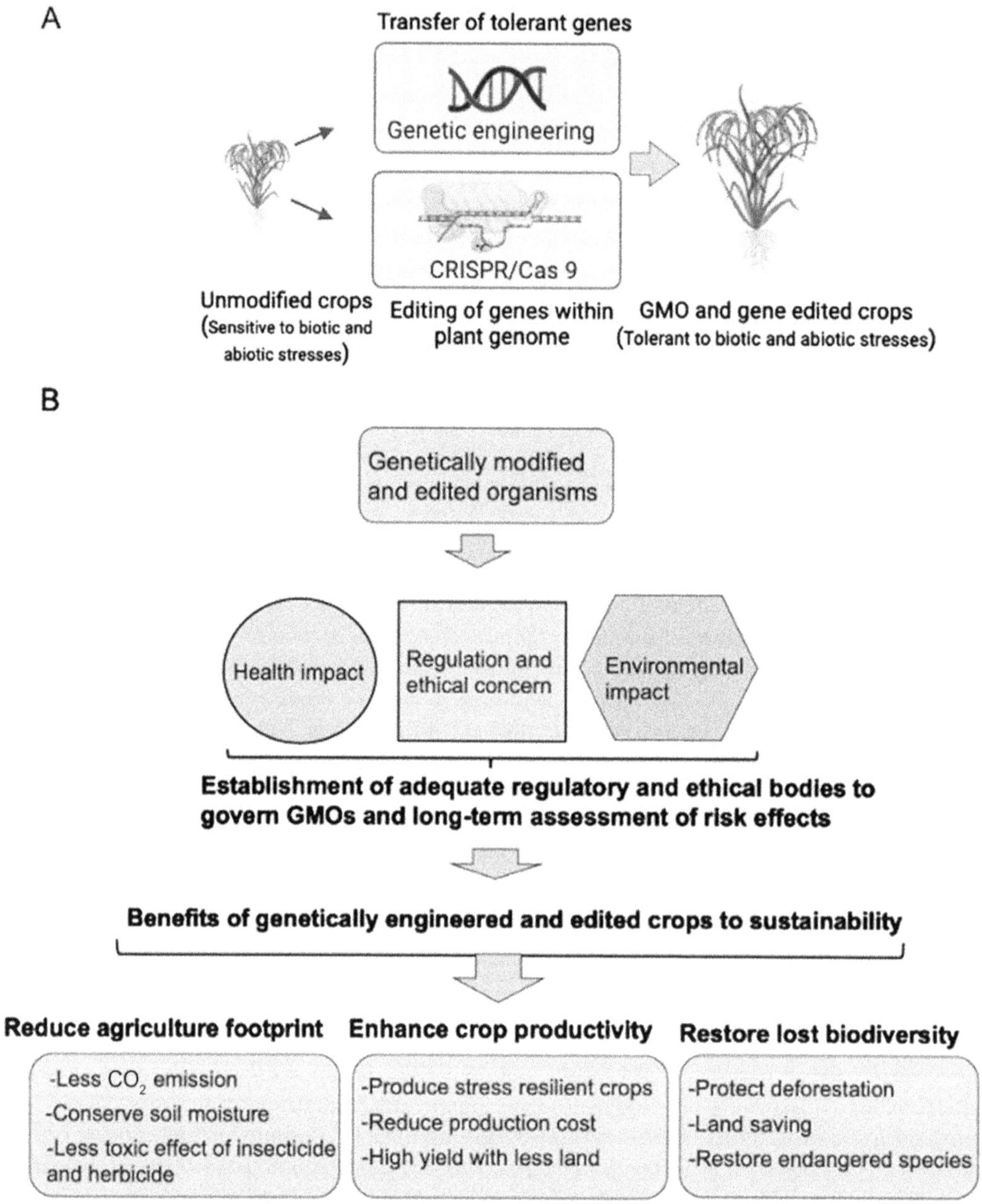

FIGURE 15.1 Benefits of genetic modification in agriculture (Mmbando & Ngongolo, 2024).

less or negligible use of chemical pesticides for the cultivation of pest infestation-resistant GM legumes will strengthen the safe food system.

- Responsible Consumption and Production (SDG 12): The advanced technologies to modify the genetic materials of legumes can contribute to a low environmental footprint for legume cultivation systems. Lower use of chemicals, water, and other resources will also promote sustainable agricultural practices.

FIGURE 15.2 Sustainable Development Goals – Zero hunger, Good health and well-being, Responsible consumption and production, Climate action – these goals can be achieved through genetically modified legumes with strategic approaches. (UN SDG.)

- Climate Action (SDG 13): Climate-resilient GM legumes can mitigate the impacts of climate change on agriculture. By adjusting crops to endure extreme weather conditions, genetic engineering can help maintaining food production in spite of environmental disruptions.

Genetic engineering casts a powerful tool for improving the sustainability of legume crops. Through targeted alterations, researchers can develop varieties that are more resilient to climate change, nutrient dense, and less reliant on redundant agricultural inputs (Bailey-Serres et al., 2019). This not only supports the long-term sustainability of food systems but also contributes to achieving critical global goals for food security, health, and environmental conservation.

15.4 CONSUMER ACCEPTANCE OF GM LEGUMES

Consumer acceptance is a big hurdle for all genetically modified foods – legumes included. While the safety and possible benefits of GM crops have been widely accepted within the scientific community, public perception of these products often seems to be based more on health, environmental, ethical, or social factors. Understanding these factors is vital for enabling and promoting successful adoption and market acceptance of GM legumes.

The obvious health risks associated with consuming genetically modified foods are an apparent factor influencing consumer views (Lusk et al., 2018). In spite of numerous studies indicating that GM crops are as safe as traditional crops, the perception of GM foods as artificial or risky persists among certain consumer populations. Misinformation regarding probable long-term health problems, allergic reactions, and gene transfer between GM foods and human cells contributes to this disbelief. As a result, consumers prioritizing health and safety may be hesitant to buy GM legume products unless they are assured of their safety through rigorous analyzing protocols and direct communication.

Many consumers also have dual thoughts about the contribution of GM crops to sustainability. Despite the established research studies, certain populations believe that GM crops will impose negative effects on the environment, especially toward biodiversity, and lead to inadvertent ecological consequences (Raman, 2017). For GM legumes to gain acceptance, consumers need clear evidence that their environmental benefits, such as reduced resource use, lower greenhouse gas (GHG) emissions, and soil protection, offset any potential risks.

Ethical concerns also play a considerable role in determining consumer views toward GM legumes. Some consumers object to genetic modification on the justification that it is an unnatural interference in the food system (Rose et al., 2020). Additionally, there is a debate regarding the agribusiness focusing on GM seeds and small-scale farmers that, by integrating GM crops, it can take control over seed supply chains and potentially harm smallholder farmers.

Consumer acceptance of GM foods varies widely across regions and cultures. In the United States, for instance, GM crops such as soybeans and corn have been largely accepted and are widely used in food industries. However, European countries like France and Germany have seen strict regulatory frameworks and public debates regarding the acceptance of GM foods (Madsen et al., 2003). Similarly, acceptance of GM legumes in developing countries may be shaped by cultural values, access to information, and the economic interests of local farmers.

One of the fundamental barriers to encouraging consumers' acceptance of GM legumes is the prevalence of misinformation (McFadden & Lusk, 2016). Understanding of modern technologies, particularly genetic engineering, is often limited, leading to myths and misconceptions among consumers. The debate surrounding GM crops is frequently framed as white or black – as either completely beneficial or entirely harmful – without acknowledging the complexities of the underlying science. Consequently, many consumers make their decisions based on fear or doubt rather than a balanced understanding of the evidence, despite the scientific consent indicating the safety and benefits of GM legumes or any other foods.

15.5 IMPORTANCE OF FOOD SAFETY COMMUNICATION

Clear, transparent, and evidence-based communication is crucial for addressing consumer concerns and promoting the acceptance of genetically modified legumes. A critical aspect of this communication involves informing consumers about the rigorous safety testing and regulatory oversight that GM crops undergo before reaching the market (Chassy, 2014). Typically, GM foods are subject to extensive safety evaluations, including assessments for allergenicity, toxicity, and nutritional content, conducted by independent regulatory agencies such as the US Food and Drug Administration, the European Food Safety Authority, and other international bodies.

Effective communication must go beyond simply presenting scientific data. Consumers need to understand not only the safety of GM legumes but also their potential benefits (Arpaia, 2012). For instance, GM legumes engineered for pest resistance can reduce the need for chemical pesticides, leading to safer working conditions for farmers and lower pesticide residue levels in food products. Similarly, GM legumes with improved drought tolerance can contribute to more sustainable agriculture in water-scarce regions, thereby reducing the environmental impact of food production. Furthermore, messaging should address ethical and social concerns, emphasizing that genetic engineering is a tool that can be used to address pressing global challenges, such as food security, climate change, and resource conservation (Roa et al., 2016). Public outreach campaigns involving scientists, farmers, policymakers, and consumer advocacy groups can help create a more balanced dialogue about GM foods. In countries where GM foods face skepticism, it may be necessary to involve trusted local voices, such as community leaders, health professionals, and educators, to help build trust and convey accurate information (Lassoued et al., 2019). These efforts should be proactive, engaging consumers before opposition takes root, and they should be tailored to the specific cultural and regional contexts in which they are implemented.

15.6 STRATEGIES FOR IMPROVING CONSUMER TRUST AND MARKET ACCEPTANCE

Providing clear and accurate labeling on GM legumes can help establish consumer trust. Many countries mandate labeling of GM products, allowing consumers to make informed choices (Matthew, 2002). However, this must be paired with public education to ensure consumers understand the meaning and safety of genetic modification. Fostering open discussions about the benefits and risks of GM crops can build trust and reduce opposition (National Academies of Sciences, Medicine, Division of Behavioral, Social Sciences, Committee on the Science of Science Communication, & A Research Agenda, 2017). Creating platforms for debate and exchange among scientists, farmers, and consumers can demystify genetic engineering and promote informed decision-making, such as through public forums, webinars, and media campaigns. Consumers are more likely to accept GM legumes when they can perceive tangible advantages, such as improved nutrition, lower costs, or environmental sustainability (Lusk, 2014). Highlighting attributes like nutrient enhancement or reduced environmental impact can resonate with health-conscious and environmentally aware consumers. Partnering with non-governmental organizations, consumer advocates, and environmental groups can lend credibility to the acceptance of GM legumes (Raman, 2017). If these trusted entities endorse such products, consumers may be more willing to accept them. Demonstrating the use of genetic engineering for social good, such as enhancing food security, can also improve acceptance. Consumer acceptance of GM legumes is a multifaceted issue, influenced by factors beyond the scientific aspects of genetic engineering, including health concerns, environmental perceptions, ethical considerations, and misinformation. Building consumer trust requires transparent communication, addressing ethical and cultural concerns, and emphasizing the tangible benefits of GM legumes. Effective public outreach can facilitate greater acceptance of these technologies, contributing to more sustainable and resilient food systems globally.

15.7 ENVIRONMENTAL IMPACT AND LIFE CYCLE ANALYSIS (LCA) OF GM LEGUMES

Life cycle assessment is a critical analytical tool that evaluates the environmental impacts of a product, process, or system from its inception to its end-of-life, commonly referred to as a "cradle-to-grave" analysis (Durairaj, 2002). LCA provides a comprehensive assessment by quantifying the energy and materials required, as well as the emissions and waste generated throughout the entire life cycle of a product. In the context of genetically modified legumes, LCA can help determine whether the claimed environmental benefits, such as reduced greenhouse gas emissions, lower pesticide use, and decreased water consumption, are substantiated when compared to conventional legume farming practices.

The relevance of LCA in assessing GM legumes lies in its ability to offer a holistic view of the ecological footprint associated with their production, processing, distribution, and consumption (Borghi et al., 2017). By comparing the life cycle impacts of GM legumes with their non-GM counterparts, stakeholders, including farmers, policymakers, and consumers, can make informed decisions about the sustainability of adopting these crops. Furthermore, LCA helps identify opportunities for improving resource efficiency and minimizing environmental damage, which is crucial for achieving long-term agricultural sustainability and meeting climate resilience goals.

Numerous life cycle assessment studies have compared the environmental impacts of GM and non-GM legume production, providing valuable insights into the sustainability implications of genetic engineering. A study on GM soybean farming in Argentina found that incorporating herbicide-tolerant varieties in the cultivation system resulted in a 28% reduction in GHG emissions and 45% decrease in fossil energy use, mainly due to less tillage and pesticide use (Brookes & Barfoot, 2017). Another comprehensive life cycle assessment of GM cowpea production in sub-Saharan Africa estimated significant decreases in various environmental impacts compared to traditional farming methods. The study found that the GM cowpea varieties could lead to a considerable reduction in pesticide use, as well as significant decreases in water consumption and GHG emissions (Pretty et al., 2005).

Legumes are a rich source of protein for humans and their livestock, yet the cultivation of most legumes has been neglected (Acharjee & Higgins, 2021). However, due to the high demand for plant-based protein, soybeans have received considerable attention in recent decades, including the development of GM soybeans.

Agricultural practices worldwide are creating increased environmental and economic pressures, which are becoming a hurdle for sustainable development (Nabavi-Pelesaraei & Damgaard, 2023).

Few studies have made an attempt to understand the current scenario by analyzing a combined approach toward life cycle assessment and life cycle cost (LC-IMPACT approach). In a study by Nabavi-Pelesaraei and Damgaard (2023), the same combination was implied to assess irrigated chickpea production and the eco-environmental scenario of the dry-farming system in the Kermanshah province of Iran. The result indicated that dry-farming chickpea production was more environment-friendly than irrigated production. Such approach-based studies can be helpful to identify the key hotspot contributing to GHG emissions and help in decision-making to enhance the sustainability of chickpea production systems.

15.8 COMPARISON BETWEEN GM AND TRADITIONAL LEGUME FARMING

A key question explored through life cycle assessment is the comparative environmental sustainability of genetically modified legume farming versus traditional practices. Studies on various GM crops, including legumes, have demonstrated that genetic engineering can yield significant environmental benefits, particularly by reducing the need for chemical inputs and enhancing resource use efficiency (Tsatsakis et al., 2017; Schulman, 2020; Raman, 2017). One major advantage of GM legumes is their built-in pest resistance, achieved through the introduction of genes that protect the plants from specific pests. This can substantially decrease the need for chemical pesticides. GM cowpea varieties engineered to resist the Maruca pod borer have been shown to cut down on pesticide application. Reduced pesticide use not only lowers input costs for farmers but also minimizes harmful chemical runoff into water bodies, preserving local ecosystems and improving soil health. In many cases, GM legumes, particularly herbicide-tolerant varieties, enable reduced or non-tillage farming. Conventional agriculture often relies on tillage to control weeds, but this practice leads to soil erosion, disrupts soil microbial communities, and releases stored carbon into the atmosphere. No-till farming, facilitated by herbicide-tolerant GM crops, helps preserve soil structure, reduces erosion, and enhances carbon sequestration in the soil. This practice also reduces fuel use for farm equipment, further decreasing greenhouse gas emissions. GM legumes tend to have higher yields compared to non-GM varieties, particularly in environments where they face abiotic stresses like drought or saline conditions. The increased productivity per unit of land means that fewer resources are needed to produce the same amount of food, leading to a reduction in the overall environmental footprint. Higher yields also contribute to better food security, especially in regions where legume crops are a staple source of nutrition.

However, it is important to recognize that the environmental benefits of GM crops can vary depending on different functional units, such as the type of legume, local agricultural practices, and the regulatory environment.

15.9 GREENHOUSE GAS (GHG) EMISSIONS AND ENERGY USE IN GM LEGUME PRODUCTION

Agricultural activities are a significant source of global greenhouse gas emissions, contributing nearly 10–12% of all anthropogenic emissions (Jensen et al., 2012). The production of genetically modified legumes can help mitigate this environmental footprint by reducing the need for energy-intensive farming practices and improving the efficiency of resource utilization. Conventional agriculture relies heavily on fossil fuels for various mechanized operations, such as tillage, planting, pesticide application, and harvesting. The adoption of GM legumes, particularly herbicide-tolerant varieties, can lead to a reduced need for tillage, resulting in lower fuel consumption. Additionally, the decreased use of pesticides in GM legume farming lowers the energy required for the manufacture, transportation, and application of these chemicals. Collectively, these factors contribute to lowered greenhouse gas emissions associated with GM legume production compared to traditional farming methods. Legumes are naturally capable of nitrogen fixation, converting atmospheric nitrogen into a form usable by plants. GM legumes can further enhance this trait, improving their efficiency in nitrogen fixation and thereby reducing the need for synthetic nitrogen fertilizers. Nitrogen fertilizers are a significant source of nitrous oxide, a potent greenhouse gas with a global warming potential nearly 300 times that of carbon dioxide. By minimizing fertilizer use, GM legumes can help mitigate the release of nitrous oxide and lower the overall carbon footprint of agriculture. No-till farming practices, facilitated by GM legumes, not only prevent soil erosion but also enhance the soil's ability to sequester carbon. Healthy, undisturbed soils act as carbon sinks, storing significant amounts of CO_2 that would otherwise contribute to atmospheric greenhouse gas concentrations. In contrast, conventional tillage disrupts the soil and accelerates the decomposition of organic matter, releasing carbon back into the atmosphere.

15.10 REDUCING ECOLOGICAL FOOTPRINTS

The environmental benefits of genetically modified legumes hold great potential for advancing sustainable agriculture (Stagnari et al., 2017). Through further refinement of genetic engineering techniques, researchers can obtain legumes that are even more efficient in resource utilization and environmentally friendly.

GM legume varieties could be modified to fix nitrogen even more efficiently, substantially reducing the need for synthetic nitrogen fertilizers. This would not only mitigate greenhouse gas emissions but also prevent water pollution from nitrate runoff. Combining multiple desirable traits, such as drought tolerance, pest resistance, and improved nitrogen fixation, in a single GM legume variety can lead to the creation of highly resilient, resource-efficient, and sustainable production options. These multi-trait GM legumes require fewer inputs, consume less water, and generate fewer emissions, all while maintaining high yields. Precision agriculture techniques, including data-driven irrigation and nutrient management, can be integrated with GM legume farming to further optimize resource utilization. By modifying water and fertilizer application to the specific needs of the crop, farmers can minimize waste and maximize efficiency, resulting in an even lower environmental footprint.

15.11 REGULATORY FRAMEWORKS AND PUBLIC DEBATE

15.11.1 Global Regulatory Approaches to GM Crops

Despite the potential environmental benefits of genetically modified legumes, their widespread adoption faces a complex regulatory landscape and ongoing public debate. Many countries have developed regulatory frameworks to assess the safety and potential risks of GM crops, including considerations for environmental impacts. The European Union, for example, has a comprehensive approval process that evaluates the agricultural, environmental, and health implications of GM crops before authorizing their cultivation and commercialization.

Divergent Standards: Different nations have varying regulations governing genetically modified organisms. For instance, the US has a more permissive approach compared to the European Union's stringent approval processes (Bennett et al., 2013). In many countries, specific organizations assess the safety and effectiveness of genetically engineered crops. International Guidelines (Paoletti et al., 2008)

- Codex Alimentarius: This international food standards body provides guidelines for food safety that influence national GMO regulations.
- Biodiversity Protocols: Agreements like the Cartagena Protocol on Biosafety govern the global trade of GMOs, ensuring they do not adversely impact biodiversity.
- Importance of Well-Informed Public Discussion on Genetic Engineering

- Current Debates and Policy Challenges
- The Role of Stakeholders: Government, Academia, Industry, and Civil Society

15.11.2 Global Regulatory Landscape for GM Crops

The regulatory frameworks overseeing genetically modified organisms, including GM legumes, vary significantly across the world. These frameworks aim to ensure the safety of GM crops for human consumption, animal feed, and the environment, while also addressing ethical and socio-economic concerns.

In the US, GM crop regulation falls under three main agencies: the USDA, the FDA, and the EPA. The USDA oversees cultivation and environmental safety, the FDA ensures food safety, and the EPA evaluates pesticides, including those produced by GM crops (McHughen & Smyth, 2007). The US has a relatively permissive stance, with several GM legume varieties, including GM soybeans, widely grown and consumed. The EU takes a more cautious approach to GM crops. GM organisms undergo rigorous risk assessments by the European Food Safety Authority before approval (Brandt, 2003). Many European countries also have national restrictions or bans on GM crop cultivation, reflecting widespread public skepticism. However, the EU allows the import of GM crops for animal feed, including GM soybeans from the US and South America. In developing nations, GM crop regulatory frameworks are often influenced by international trade and food security considerations. For example, countries like Brazil and Argentina have adopted GM soybean varieties for commercial cultivation, recognizing their economic and environmental benefits. Conversely, some African countries have hesitated to adopt GM crops due to concerns.

15.12 ETHICAL CONSIDERATIONS IN GM LEGUME ADOPTION

The adoption of genetically modified legumes raises several ethical considerations that must be carefully examined in the context of sustainable agriculture, food security, and socioeconomic impacts (Phillips, 2017; Thompson, 2012; National Academies of Sciences, 2016; Weale, 2010), as outlined below.

Equity and Access: A key ethical concern is whether smallholder farmers, particularly in developing countries, will have equitable access to GM legume technologies. The patenting of GM seeds by multinational corporations can limit farmers' ability to save and replant seeds, increasing their dependency on commercial seed suppliers. This could exacerbate disparities between large commercial farmers and smallholders, who may lack the financial means to afford the technology.

Intellectual Property Rights: The intellectual property rights associated with GM crops can lead to legal and economic challenges for farmers. Patent holders often impose restrictions on the use of GM seeds, preventing farmers from saving and reusing them. This has sparked ethical debates about the concentration of power in the hands of a few large biotech companies and the potential loss of traditional farming practices.

Environmental Ethics: While GM crops can reduce the environmental impact of agriculture, there are concerns about unintended ecological consequences. For instance, the development of herbicide-resistant weeds due to over-reliance on herbicide-tolerant crops can lead to increased herbicide use in the long term, potentially harming biodiversity. Ethical questions also arise regarding the potential impacts of GM crops on non-target species and ecosystems.

Cultural Acceptance: In some cultures, the genetic modification of plants is perceived as an unnatural intervention in nature. Ethical debates about "tampering" with natural processes can influence public attitudes toward GM crops, affecting their acceptance and adoption in certain regions. Addressing these cultural concerns is crucial for the successful deployment of GM legumes globally.

15.13 PUBLIC DEBATE AND THE IMPORTANCE OF INFORMED DIALOGUE ON GM LEGUMES

The adoption of genetically modified legumes is not merely a scientific and agricultural issue, but also a societal one, shaped by public perception, media portrayal, and political discourse. The public debate on GM crops has been polarized, with strong arguments on both sides regarding the potential benefits and risks of genetic engineering (Mathiowetz & Jones, 2015; Lassoued et al., 2018; Rose et al., 2020; Gaskell et al., 2004). For GM legumes to be successfully integrated into sustainable agriculture, it is essential to foster well-informed public discussions that are grounded in scientific evidence, transparent communication, and consideration of societal values.

Public concerns about GM crops often revolve around food safety, environmental sustainability, and ethical considerations, as discussed earlier. These concerns can be amplified by misinformation, particularly through social media and advocacy groups that may oppose GMOs on ideological grounds. As a result, promoting constructive dialogue that addresses both the scientific and ethical aspects of GM legumes is crucial for gaining public trust.

Effective communication strategies are vital for bridging the gap between scientific knowledge and public perception of GM legumes. Scientists, policymakers, and industry stakeholders must engage with the public to explain the benefits of GM legumes, such as improved nutrition, enhanced climate resilience, and reduced environmental impact.

- Transparency and Trust: Transparency is a cornerstone of effective communication. The public must be informed about the safety assessments, regulatory processes, and scientific evidence supporting the adoption of GM legumes. Providing clear, accessible information about how GM crops are tested for safety and environmental impacts can help alleviate concerns about potential risks.
- Addressing Ethical and Cultural Concerns: In addition to presenting the scientific case for GM legumes, it is important to address ethical and cultural concerns. For instance, discussions about intellectual property rights, the impact on smallholder farmers, and the preservation of traditional farming practices should be part of the public dialogue. Acknowledging and respecting cultural perspectives on genetic modification can help foster more inclusive and constructive debates.
- Inclusive Public Engagement: Public debates should not be limited to scientists and policymakers. Farmers, consumers, environmental groups, and community leaders should also be involved in the conversation. Engaging with diverse stakeholders ensures that different perspectives are heard, and it can allow a broader range of viewpoints to be considered, leading to more informed and solid decisions.

15.14 CHALLENGES AND FUTURE DIRECTIONS

a. Public Perception and Acceptance
    - Consumer Concerns: Consumers often exhibit skepticism and resistance toward genetically modified organisms, driven by concerns about safety, ethical considerations, and misinformation.
    - Labeling Requirements: In some countries, the mandatory labeling of GMO products can influence market dynamics and consumer purchasing decisions.

b. Environmental Impact Assessments
    - Biodiversity Risks: Assessing the potential impact of genetically engineered legumes on local ecosystems and biodiversity is a significant challenge, necessitating extensive research.

   - Long-Term Effects: Comprehensive long-term studies are crucial to understand the ecological ramifications of introducing genetically modified legumes into various environments.

c. Intellectual Property Issues
   - Patents and Access: The patenting of genetically engineered traits can create barriers for smallholder farmers and restrict their access to advanced agricultural technologies.
   - Licensing Costs: The costs associated with licensing genetically modified seeds can be prohibitive for small-scale farmers, leading to inequities in agricultural practices.

d. Technical and Scientific Challenges
   - Precision and Off-Target Effects: Ensuring precision in genetic modifications is crucial to avoid unintended consequences, which requires the use of advanced techniques and thorough testing.
   - Scaling Up Production: Translating laboratory successes into viable agricultural practices can be challenging, necessitating significant investment and infrastructure.
     - o Technical and Ethical Challenges in Engineering Sustainable Legumes
     - o Prospects for GM Legumes in Climate Adaptation
     - o Long-Term Considerations for Environmental and Food Security Path Forward
   - Collaborative Approaches: Engaging diverse stakeholders, including scientists, policymakers, farmers, and consumers, in discussions can help address concerns and promote understanding.
   - Adaptive Regulations: Developing flexible regulatory frameworks that can adapt to new technologies while ensuring safety can foster innovation and acceptance.
   - Education and Outreach: Public education about the benefits and safety of genetically engineered legumes can help alleviate concerns and promote informed decision-making.

15.15 CONCLUSION

Life cycle assessment offers a valuable framework for evaluating the environmental impacts of cultivating genetically modified legumes compared to traditional farming practices. The introduction of GM legumes has demonstrated promise in reducing chemical pesticide use, enhancing resource efficiency, lowering greenhouse gas emissions, and improving soil and water conservation. However, these benefits are contingent upon responsible management strategies and ongoing research to address potential challenges, such as biodiversity loss and herbicide resistance. Moving forward, continued advancements in genetic engineering techniques, coupled with sustainable farming methods, can ensure that GM legumes play a pivotal role in the global transition toward more sustainable and climate-resilient agriculture. By adopting a life cycle perspective, policymakers, farmers, and consumers can gain a better understanding of the long-term environmental benefits of GM legumes and make informed decisions that contribute to a sustainable future.

ACKNOWLEDGMENTS

The authors are thankful to the University of Copenhagen, Denmark and the Central University of Haryana, India.

CONFLICT OF INTEREST

The authors declare no conflict of interest.

REFERENCES

Acevedo, M., Pixley, K. V., Zinyengere, N., Meng, S., Tufan, H., Cichy, K. A., Bízikova, L., Isaacs, K., Ghezzi-Kopel, K., & Porciello, J. (2020). A scoping review of adoption of climate-resilient crops by small-scale producers in low- and middle-income countries [Review of A scoping review of adoption of climate-resilient crops by small-scale producers in low- and middle-income countries]. *Nature Plants*, 6(10), 1231. https://doi.org/10.1038/s41477-020-00783-z

Acharjee, S., & Higgins, T. J. (2021). Genetic engineering of grain legumes: Their potential for sustainable agriculture and food and nutritional security. In Saxena, K. B., Saxena, R. K., & Varshney, R. K. (eds.) *Genetic Enhancement in Major Food Legumes: Advances in Major Food Legumes* (pp. 91–121). Springer Cham. https://doi.org/10.1007/978-3-030-64500-7

Arpaia, S. A., Nicholas, A., Birch, E., Chesson, A., Du Jardin, P., Gathmann, A., Gropp, J., Herman, L., Hoen-Sorteberg, Hi-G. Jones, H., Kiss, J., Kleter, G., Lovik, M., Messéan, A., Naegeli, H., Kaare, M., Nielsen, J., Ovesna, J., Perry, N., Rostoks, C., Christodoulidou, A., Devos, Y., Fernández Dumont, A., Gennaro, A., Gomes, A., Mestdagh, S., Rodighiero, S.. (2012). Scientific Opinion updating the risk assessment conclusions and risk management recommendations on the genetically modified insect resistant maize MON 810. *EFSA Journal*, 10(12), 3017, 98 pp. https://onlinelibrary.wiley.com/resolve/doi?DOI=10.2903/j.efsa.2012.3017

Bailey-Serres, J., Parker, J. E., Ainsworth, E. A., Oldroyd, G., & Schroeder, J. I. (2019). Genetic strategies for improving crop yields [Review of Genetic strategies for improving crop yields]. *Nature*, 575(7781), 109. https://doi.org/10.1038/s41586-019-1679-0

Bennett, A. B., Chi-Ham, C. L., Barrows, G., Sexton, S., & Zilberman, D. (2013). Agricultural biotechnology: Economics, environment, ethics, and the future. *Annual Review of Environment and Resources,* 38(1), 249. https://doi.org/10.1146/annurev-environ-050912-124612

Borghi, A. D., Strazza, C., Magrassi, F., Taramasso, A. C., & Gallo, M. (2017). Life Cycle Assessment for eco-design of product–package systems in the food industry—The case of legumes. *Sustainable Production and Consumption,* 13, 24. https://doi.org/10.1016/j.spc.2017.11.001

Brandt, P. (2003). Overview of the current status of genetically modified plants in Europe as compared to the USA. *Journal of Plant Physiology*, 160(7), 735. https://doi.org/10.1078/0176-1617-01031

Brookes, G., & Barfoot, P. (2017). Environmental impacts of genetically modified (GM) crop use 1996–2015: Impacts on pesticide use and carbon emissions. *GM Crops & Food,* 8(2), 117. https://doi.org/10.1080/21645698.2017.1309490

Chassy, B. M. C. (2014). Food safety evaluation of crops produced through biotechnology. *Journal of the American College of Nutrition*, 21(sup3), 166S-173S www.tandfonline.com/doi/abs/10.1080/07315724.2002.10719261

Choudhury, A., & Rajam, M. V. (2021). Genetic transformation of legumes: An update. *Plant Cell Reports*, 40, 1813–1830.

National Academies of Sciences, Medicine, Division of Behavioral, Social Sciences, Committee on the Science of Science Communication, & A Research Agenda. (2017). Communicating science effectively: A research agenda (Vol. 6). The National Academies Press, Washinton DC, 152 pages. DOI: https://doi.org/10.17226/23674 http://op.niscair.res.in/index.php/JST/article/view/22808

Dissanayaka, D. M. S. B. D., Rankoth, L. M., Gunathilaka, W. M. N. D., Prasantha, B. D. R., & Marambe, B. (2021). Utilizing food legumes to achieve iron and zinc nutritional security under changing climate. *Journal of Crop Improvement*, 35(5), 700–721. www.tandfonline.com/doi/full/10.1080/15427528.2021.1872754

Durairaj, S. (2002). Evaluation of life cycle cost analysis methodologies. *Corporate Environmental Strategy*, 9(1), 30. https://doi.org/10.1016/s1066-7938(01)00141-5

Gaskell, G., Allum, N., Wagner, W., Kronberger, N., Torgersen, H., Hampel, J., & Bardes, J. (2004). GM foods and the misperception of risk perception. *Risk Analysis*, 24(1), 185. https://doi.org/10.1111/j.0272-4332.2004.00421.x

Goyal, R. K., Mattoo, A. K., & Schmidt, M. A. (2021). Rhizobial–host interactions and symbiotic nitrogen fixation in legume crops toward agriculture sustainability. *Frontiers in Microbiology*, 12, 669404.

Jensen, E. S. et al. (2023). Legumes for mitigation of climate change and the provision of feedstock for biofuels and biorefineries. A review. *Agronomy for Sustainable Development*, 32(2), 329–364. https://link.springer.com/article/10.1007/s13593-011-0056-7

Lassoued, R., Smyth, S. J ., Phillips, P. W., & Hesseln, H. (2018). Regulatory uncertainty around new breeding techniques. *Frontiers in Plant Science*, 9, 1291. www.frontiersin.org/articles/10.3389/fpls.2018.01291/pdf

Lassoued, R., Macall, D. M., Hesseln, H., Phillips, P. W. B., & Smyth, S. J. (2019). Benefits of genome-edited crops: Expert opinion. *Transgenic Research*, 28(2), 247. https://doi.org/10.1007/s11248-019-00118-5

Lusk, J. L., Mcfadden, B. R., & Rickard, B. J. (2014). Which biotech foods are most acceptable to the public? *Biotechnology Journal*, 10(1), 13–16. https://onlinelibrary.wiley.com/doi/10.1002/biot.201400561

Lusk, J. L., McFadden, B. R., & Wilson, N. (2018). Do consumers care how a genetically engineered food was created or who created it? *Food Policy*, 78, 81. https://doi.org/10.1016/j.foodpol.2018.02.007

Madsen, K. H., Lassen, J., & Sandøe, P. (2014). Genetically modified crops: A US Farmer's versus an EU Citizen's point of view. *Acta Agric Scand (B)*, 53(S1), 60–67. www.tandfonline.com/doi/abs/10.1080/1659140310015120

Mathiowetz, A., & Jones, J. M. (2015). Genetically modified grains and the consumer. In *Elsevier eBooks* (pp. 50–56). Elsevier BV. https://doi.org/10.1016/b978-0-08-100596-5.00075-5

Matthew, C. R., Matthew, C., Huffman, W. E., Shogren, J. F., & Tegene, A. (2002). Are us consumers tolerant of gm foods? 2002 Annual meeting, July 28–31, Long Beach, CA 19889, American Agricultural Economics Association (New Name 2008: Agricultural and Applied Economics Association). https://ideas.repec.org/p/ags/aaea02/19889.html

McFadden, B. R., & Lusk, J. L. (2016). What consumers don't know about genetically modified food, and how that affects beliefs. *The FASEB Journal*, 30(9), 3091. https://doi.org/10.1096/fj.201600598

McHughen, A., & Smyth, S. J. (2007). US regulatory system for genetically modified [genetically modified organism (GMO), rDNA or transgenic] crop cultivars [Review of US regulatory system for genetically modified [genetically modified organism (GMO), rDNA or transgenic] crop cultivars]. *Plant Biotechnology Journal*, 6(1), 2. https://doi.org/10.1111/j.1467-7652.2007.00300.x

Mmbando, G. S., & Ngongolo, K. (2024). The current status of the use of genetic modification and editing to improve biodiversity and ecological sustainability. *All Life*, 17(1), 2417191.

Nabavi-Pelesaraei, A., & Damgaard, A. (2023). Regionalized environmental damages and life cycle cost of chickpea production using LC-IMPACT assessment. *Environmental Impact Assessment Review*, 103, 107259.

Nandula, V. K. (2019). Herbicide resistance traits in maize and soybean: Current status and future outlook. https://mdpi-res.com/d_attachment/plants/plants-08-00337/article_deploy/plants-08-00337.pdf?version=1568008372

National Academies of Sciences, Engineering, and Medicine; Division on Earth and Life Studies; Board on Agriculture and Natural Resources; Committee on Genetically Engineered Crops: Past Experience and Future Prospects. (2016, May). Regulation of current and future genetically engineered crops: Past experience and future prospects. Washington (DC): National Academies Press (US). ISBN-13: 978-0-309-43738-7ISBN-10: 0-309-43738-5. www.ncbi.nlm.nih.gov/books/NBK424533/

Ojiewo, C. O., Monyo, E. S., Desmae, H., Boukar, O., Mukankusi, C., Thudi, M., Pandey, M. K., Saxena, R. K., Gaur, P. M., Chaturvedi, S. K., Fikre, A., rao, N. G. prsada, Sameerkumar, C. V., Okori, P., Janila, P., Rubyogo, J. C., Chigeza, G., Akpo, E., Omoigui, L. O., … Varshney, R. K. (2018). Genomics, genetics and breeding of tropical legumes for better livelihoods of smallholder farmers [Review of Genomics, genetics and breeding of tropical legumes for better livelihoods of smallholder farmers]. *Plant Breeding*, 138(4), 487. https://doi.org/10.1111/pbr.12554

Paoletti, C., Flamm, E. L., Yan, W., Meek, S., Renckens, S., Fellous, M., & Kuiper, H. A. (2008). GMO risk assessment around the world: Some examples. *Trends in Food Science & Technology*, 19. https://doi.org/10.1016/j.tifs.2008.07.007

Phillips, P. P., Contents Development and Commercialization of Genetically Modified Plants Ethics and Genetically Modified Crops Food Safety Issues Food Security Intellectual Property Rights Ownership of Plant Genetic Resources Plant Genetic Engineering. (2017). Development and commercialization of genetically modified plants. *Encyclopedia of Applied Plant Sciences* (Second Edition), vol 2, pp. 1–7. www.sciencedirect.com/science/article/pii/B9780123948076000344

Pretty, J., Noble, A., Bossio, D., Dixon, J., Hine, R., Vries, F. W. T. P. de, & Morison, J. (2005). Resource-conserving agriculture increases yields in developing countries. *Environmental Science & Technology*, 40(4), 1114. https://doi.org/10.1021/es051670d

Raina, A., Laskar, R. A., Khan, S., Tomlekova, N., Ravelombola, W., & Thudi, M. (2023). Editorial: Legume breeding in transition: Innovation and outlook. *Frontiers in Genetics*, 14. https://doi.org/10.3389/fgene.2023.1221551

Raman, R. (2017). The impact of Genetically Modified (GM) crops in modern agriculture: A review [Review of The impact of Genetically Modified (GM) crops in modern agriculture: A review]. *GM Crops & Food*, 8(4), 195. https://doi.org/10.1080/21645698.2017.1413522

Raman, R. (2023). The impact of Genetically Modified (GM) crops in modern agriculture: A review. *GM Crops & Food*, 8(4), 195–208 www.ncbi.nlm.nih.gov/pmc/articles/PMC5790416/

Roa, C., Hamilton, R. S., Wenzl, P., & Powell, W. (2016). Plant genetic resources: Needs, rights, and opportunities [Review of plant genetic resources: Needs, rights, and opportunities]. *Trends in Plant Science*, 21(8), 633. https://doi.org/10.1016/j.tplants.2016.06.002

Rose, K. M., Brossard, D., & Scheufele, D. A. (2020). Of society, nature, and health: How perceptions of specific risks and benefits of genetically engineered foods shape public rejection. *Environmental Communication,* 14(7), 1017. https://doi.org/10.1080/17524032.2019.1710227

Rubiales, D., & Mikic, A. (2015). Introduction: Legumes in sustainable agriculture. *Critical Reviews in Plant Sciences*, 34(1–3), 2–3.

Schulman, A. H. (2020). The impact of GM crops on agriculture. In *Genetically modified and irradiated food* (pp. 195–213). Academic Press. https://doi.org/10.1016/b978-0-12-817240-7.00012-7

Smyth, S. J. (2022). Contributions of genome editing technologies towards improved nutrition, environmental sustainability and poverty reduction [Review of contributions of genome editing technologies towards improved nutrition, environmental sustainability and poverty reduction]. *Frontiers in Genome Editing*, 4. https://doi.org/10.3389/fgeed.2022.863193

Stagnari, F., Maggio, A., Galieni, A., & Pisante, M. (2017). Multiple benefits of legumes for agriculture sustainability: An overview. *Chemical and Biological Technologies in Agriculture*, 4(1). https://doi.org/10.1186/s40538-016-0085-1

Thompson, P. B. T. (2012). Genetically modified food: Ethical issues. https://onlinelibrary.wiley.com/doi/10.1002/9780470015902.a0024153

Tsatsakis, A., Nawaz, M. A., Kouretas, D., Balias, G., Savolainen, K., Tutelyan, V. A., Golokhvast, K. S., Lee, J., Yang, S. H., & Chung, G. (2017). Environmental impacts of genetically modified plants: A review [Review of Environmental impacts of genetically modified plants: A review]. *Environmental Research*, 156, 818. https://doi.org/10.1016/j.envres.2017.03.011

Weale, A. (2010). Ethical arguments relevant to the use of GM crops [Review of Ethical arguments relevant to the use of GM crops]. *New Biotechnology*, 27(5), 582. https://doi.org/10.1016/j.nbt.2010.08.013

Ye, H., Roorkiwal, M., Valliyodan, B., Zhou, L., Chen, P., Varshney, R. K., & Nguyen, H. T. (2018). Genetic diversity of root system architecture in response to drought stress in grain legumes [Review of Genetic diversity of root system architecture in response to drought stress in grain legumes]. *Journal of Experimental Botany*, 69(13), 3267. https://doi.org/10.1093/jxb/ery082

Index

Note: Page numbers in *italics* indicate figures and in **bold** indicate tables on the corresponding pages.

S

T

U

V

W

Y

Z

For Product Safety Concerns and Information please contact our EU
representative GPSR@taylorandfrancis.com
Taylor & Francis Verlag GmbH, Kaufingerstraße 24, 80331 München, Germany

www.ingramcontent.com/pod-product-compliance
Ingram Content Group UK Ltd.
Pitfield, Milton Keynes, MK11 3LW, UK
UKHW052219190926
475702UK00006B/1106

* 9 7 8 1 0 3 2 7 5 7 4 4 5 *